AF871369

PRINCIPLES OF BIOLOGY

VOLUME 1:
FOUNDATIONS OF LIFE AND CELLULAR SYSTEMS

Dr. Snigdha Singh and
Dr. Shubham Sewariya

ALLIED
HEALTH
PRESS

Principles of Biology
Volume 1: Foundations of Life and Cellular Systems

Authors: **Dr. Snigdha Singh and Dr. Shubham Sewariya**

Allied Health Press
4164 Lakeshore Road
Burlington ON L7L 1A4
Canada
www.alliedhealthpress.com
Email: orders@arclereducation.com

ISBN: 978-1-77956-850-2

ABOUT THE AUTHORS

Dr. Snigdha Singh is an accomplished researcher and educator in medicinal chemistry and nanoscience, currently serving as Assistant Professor in the Department of Chemistry at the University of Delhi and Deputy Director of the Institute of Nano Medical Sciences, University of Delhi. With a distinguished academic trajectory and extensive international research exposure, she has established herself as a leading young scientist in the field of chemical and biomedical research.

She earned her Ph.D. in Medicinal Chemistry from the University of Delhi (2021), under the supervision of Prof. Ramesh Chandra, where her thesis focused on the synthesis and development of novel heterocycles and their biological evaluation. Prior to that, she completed her M.Tech. in Chemical Synthesis and Process Technologies (CSPT) at the University of Delhi with first division (2016) and her B.Sc. (Hons) in Chemistry from Hindu College, University of Delhi (2013), graduating with top merit distinctions. Her academic excellence was recognized early with a merit fellowship and multiple university-level awards.

Dr. Singh's professional journey reflects a unique blend of teaching, research, and leadership roles. Beyond her faculty appointment, she has served as Resident Tutor at Ambedkar-Ganguly Students House for Women, and has held prestigious international positions, including Postdoctoral Associate at Rensselaer Polytechnic Institute (USA), Guest Scientist at Humboldt University (Germany), and Visiting Scientist at Lobachevsky University (Russia). She was also selected as an Indo-UK Newton Bhabha Scholar at the University of Oxford and an Indo-Italy Exchange Scholar at the University of Siena, further expanding her global research collaborations. Her research contributions span drug design, nanostructures for biomedical applications, and innovative drug delivery systems. She has published extensively in high-impact journals such as ACS Sustainable Chemistry & Engineering, RSC Advances, Bioorganic & Medicinal Chemistry, and Inorganic Chemistry Communications, alongside several book chapters and authored textbooks. She is also a co-inventor on an international patent related to hydroxyethylamine-based piperazine compounds (2022). In addition, she has mentored Ph.D. candidates and supervised more than 27 graduate and postgraduate students.

Dr. Singh's work has been recognized through numerous honors, including the Indo Asian Young Women Researcher Award (2024), Young Researcher in Medicinal Chemistry Award (2024), and the Teacher Excellence Award (2023). She has received multiple international travel fellowships and was ranked 19th All-India in the UGC-CSIR NET Examination (2017).

An active member of leading scientific societies such as the Royal Society of Chemistry, American Chemical Society, and Indian Science Congress Association, Dr. Singh also serves as a reviewer for several international journals. Her career reflects a deep commitment to advancing chemical sciences while fostering global academic exchange and mentoring the next generation of researchers.

Dr. Shubham Sewariya is a Research Scientist at the University of Delhi, India, with expertise in medicinal chemistry and drug discovery. His research focuses on developing novel anticancer drug leads by combining synthetic chemistry, biological insights, and computational approaches. He holds an Integrated BS–MS degree in Chemistry from the Indian Institute of Science Education and Research (IISER), Thiruvananthapuram, and has completed his Ph.D. jointly at University of Delhi and the University of Lancashire, UK. His doctoral work centers on the discovery and development of secondary metabolites, including noscapinoids and coumarins, as potential anticancer agents. Dr. Sewariya has received several prestigious fellowships and awards, including the Commonwealth Split-site PhD Scholarship (2023–24) funded by the UK Foreign, Commonwealth & Development Office, the CSIR Senior Research Fellowship, and the INSPIRE SHE Fellowship from Department of Science and Technology, Govt of India. He has also presented his work at international platforms such as the American Chemical Society (ACS) conferences in San Francisco and Chicago. With more than a dozen peer-reviewed publications, his contributions span synthetic chemistry, nanomedicine, and biophysical studies, particularly in designing intelligent nanocarriers and molecular probes for cancer therapy. Beyond the laboratory, Dr. Sewariya is deeply committed to advancing translational research that bridges science and medicine for the betterment of human health.

Brief Table of Contents

Getting Started with This Book

Chapter Openers

A list of Chapter Outline is presented providing a brief Introduction to what will be covered in the chapter.

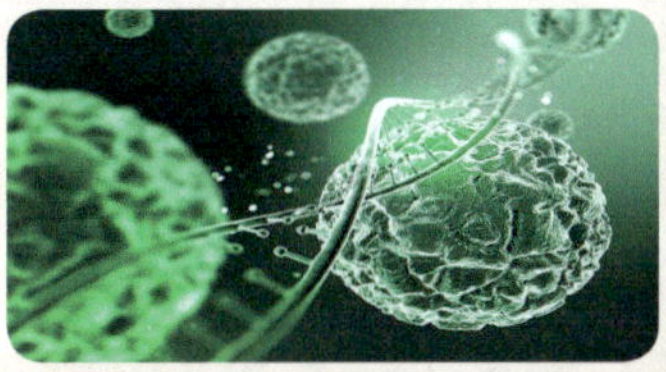

LEARNING OBJECTIVES

After studying this chapter, you will be able to:

1. Discuss the scientific method in biology;
2. Explain the cell theory;
3. Discuss the genetics and heredity;
4. Evolution and natural selection;
5. Focus on ecology and ecosystems;
6. Learn about human biology;
7. Evaluate the plant biology;
8. Explain the microbiology;
9. Focus on current trends in biology.

Introduction

Often referred to as the science of life, biology is a broad field of study aimed at comprehending the several kinds of life on Earth, from tiny creatures to sophisticated multicellular entities. Comprising several subfields, including molecular biology, ecology, and evolutionary biology, each of which advances our knowledge of living systems, this field clarifies the mechanisms controlling life events, including development, reproduction, and interaction with the surroundings, which is the central goal of biology. Moreover, biology fills in the void between the natural sciences and medicine by offering a necessary understanding of health, illness, and the interactions between organisms and their environments. Dealing with urgent global issues such as climate change, biodiversity loss, and new diseases depends on an awareness of these complex biological interactions, therefore highlighting the critical part biology plays in modern civilization. Fundamentally, biology explores and tests ideas about the living world using the scientific method. Through methodical observation and experimentation,

1.1. The Scientific Method in Biology

Learning Outcome

1. Describe the steps of the scientific method;
2. Focus on hypothesis formation;
3. Learn about experimental design.

Learning Outcomes

The beginning of each section of every chapter contains a set of Learning Outcomes that inform students of concepts they should understand. These outcomes help students understand what they should be able to do if they have mastered the material in that section.

Test Your Knowledge

Throughout each chapter, Test Your Knowledge cues you to note which topics have been covered and to self-test your understanding of the material presented.

Test Your Knowledge

1. Explain why correlation does not imply causation in scientific experiments.
2. How does the scientific method help reduce the influence of personal opinions in research?

Chapter Summary

The Summary Section appear at the end of each chapter and summarize main concepts covered throughout the chapter. These points provide key concepts in an engaging and meaningful way.

Summary

The investigation of biological principles highlights the importance of a comprehensive view of life by reflecting the intricate interactions that occur within living systems. The complexity of biology goes beyond individual species or systems, as this essay highlights; it also includes the interactions that mold ecosystems and affect biodiversity. Researchers can conduct in-depth analyses of metabolomic data using sophisticated technologies like MetaboAnalyst, which offer insights into the molecular mechanisms underlying these interactions. As demonstrated by the developments in MetaboAnalyst, for example, the integration of diverse omics data promotes a thorough understanding of metabolic pathways and helps identify important biological markers necessary for research on development and disease (Chong et al., 2018). Additionally, the phenomenon of ecological invasions is a key component of biological research and a vital field of study. The introduction of non-native species into unfamiliar habitats has significant ramifications and calls for multidisciplinary management strategies. Examining the origins and effects of these biological incursions, invasion science has become a crucial field that highlights the need to comprehend ecological dynamics and community interactions. As interest in biological invasions increases, new findings are emerging that put conventional ecological theories and practices to the test. The importance of incorporating various scientific viewpoints is shown by this developing discipline, which also highlights how invasive species shape ecosystems and advances our knowledge of ecological resilience (David et al., 2013). When taken as a whole, these discoveries from invasion science and metabolomics highlight the value of multidisciplinary biology research. Fostering cross-disciplinary collaboration and utilizing technological breakthroughs to improve our comprehension of intricate biological interactions are crucial as we look to the future of biological research. In addition to deepening our understanding of the complexities of life, this method lays the groundwork for tackling urgent environmental issues. Future studies can ensure the continued relevance of biology in a world that is constantly changing by acknowledging the interconnection of biological systems and opening the door to creative solutions that support sustainability and biodiversity conservation. We might imagine a thorough comprehension of life that serves both academia and society at large through ongoing investigation and integration of scientific knowledge.

Knowledge Assessment

The chapter includes material designed to review and reinforce each content. It includes Critical Thinking Questions to help test of your understanding of information presented in the chapter and Short answer Questions to review your understanding of the key concepts presented in the chapter.

Critical Thinking Questions

1. **Who is known as the "Father of Modern Biology"?**
   a. Aristotle
   b. Charles Darwin
   c. Gregor Mendel
   d. Louis Pasteur
2. **What is the basic unit of life?**
   a. Atom
   b. Molecule
   c. Cell
   d. Organism
3. **Which branch of biology studies the interaction of organisms with their environment?**
   a. Genetics
   b. Ecology
   c. Physiology
   d. Botany
4. **Which macromolecule is the main source of energy for living organisms?**
   a. Proteins
   b. Lipids
   c. Carbohydrates
   d. Nucleic acids
5. **DNA is primarily responsible for which function in living organisms?**
   a. Energy production
   b. Protein synthesis
   c. Storing genetic information
   d. Regulating body temperature

Short Answer Questions

1. What is biology, and why is it important to study biology?
2. What are the three main domains of life, and how are they different from one another?
3. Define homeostasis and give an example of how it works in living organisms.
4. What is the cell theory, and why is it fundamental to biology?
5. How do autotrophs and heterotrophs differ in the way they obtain energy?
6. What is the scientific method, and how is it applied in biological research?
7. Discuss how biology helps address global challenges such as climate change, disease, and food security.
8. Explain the levels of biological organization from the smallest unit to the largest. Provide examples at each level.
9. Describe the relationship between structure and function in biology, giving specific examples from cells or organs.
10. How has the discovery of genetics shaped our understanding of inheritance and variation among organisms?

Supplementary Materials

Key Terms: A list of key terms is given for quick reference

KEY TERMS FROM THIS CHAPTER

Annotation enrichment analysis	Biological structure	Cell
Cell theory	Cellular structure	Ecology
Ecosystems	Genetics	Human biology
Microbiology	Natural selection	Organism
Plant biology		

Further Resources: Recommendations for additional readings, websites, and multimedia resources.

References

1. Argelaguet, R., Velten, B., Arnol, D., Dietrich, S., Zenz, T., Marioni, J. C., Buettner, F., et al. (2018). Multi-omics factor analysis—A framework for unsupervised integration of multi-omics data sets. *Molecular Systems Biology, 14.* https://doi.org/10.15252/msb.20178124.
2. Bacher, S., Blackburn, T. M., Essl, F., & Genovesi, P. (2017). *Socio-Economic Impact Classification of Alien Taxa (SEICAT).* https://core.ac.uk/download/286567949.pdf (accessed on 8th March 2025).
3. Barnett, R., Brace, S., Cucchi, T., & Dobney, K. (2013). *Divergent Evolutionary Processes Associated with Colonization of Offshore Islands.* https://core.ac.uk/download/20664364.pdf (accessed on 8th March 2025).

Detailed Table of Contents

CHAPTER 5

Membranes and Transport 157

CHAPTER 6

Energy, Enzymes, and Metabolism 197

LIST OF FIGURES

PREFACE

Life sciences constitute the foundation of our understanding of life's fundamental mechanisms, from the chemical bonds that drive a single cell to the extensive networks that govern our world's ecosystems. Bridging knowledge from chemistry, physics, and mathematics, this field illuminates how organisms grow, evolve, and respond to their environment. Through studies of cell function laws, heredity, energy exchange, and ecological equilibrium, students and scientists acquire the information to solve problems in medicine, agriculture, conservation, and biotech innovation. Here, excitement over unveiling nature's processes generates innovation, and careful experimental design and attention to detail ensure that findings remain defensible under critical scrutiny.

The study of living organisms reveals patterns and exceptions that shape our view of the natural world. This discipline examines how molecules assemble into macromolecules, how these macromolecules give rise to structures and organelles, and how cells coordinate their functions to produce tissues, organs, and entire organisms. Understanding these relationships enables us to develop new therapeutics, engineer crops resistant to disease, and conserve biodiversity in the face of climate change pressures. By exploring both universal principles and organism-specific solutions, students develop an appreciation for the diversity and beauty of life.

Principles of Biology, Volume 1, is written to guide students through the fundamental concepts that underlie biological systems. It provides clear explanations, supported by diagrammatic examples and current illustrations drawn from recent research. Each chapter builds upon the previous one, allowing major facts to accumulate logically and conveniently. Whether you're encountering these ideas for the first time or refreshing your familiarity, this book provides a guide through the highlights of modern biology.

Chapter 1 provides an overview of the scope of biological inquiry, its main subdisciplines, and the scientific method applied to living systems. Chapter 1 summarizes the history of biological thought, the evolution of principal experimental tools, and the role of hypothesis testing in expanding our knowledge of life's complexity.

Chapter 2 addresses the chemical fundamentals required for biological functioning. Atomic structure, chemistry of bonding, reaction energetics, and properties of water will be learned by readers. Emphasis is placed on how chemical principles explain the behavior of macromolecules and biochemical pathways.

Chapter 3 investigates the macromolecules that provide the cell's structural and functional framework. Carbohydrates, lipids, proteins, and nucleic acids are given complete coverage, including their monomeric units, three-dimensional conformations, and the cellular processes in which they participate.

Chapter 4 examines cell structure, including prokaryotic and eukaryotic models, and outlines the functions of organelles. Microscopy techniques are covered, and the dynamic organization of cytoskeletal components is highlighted.

Chapter 5 covers membrane structure and transport processes, including diffusion, osmosis, active transport, and endo- and exocytosis. The chapter covers how cells regulate their internal environments and communicate with their environments.

Chapter 6 covers enzymes and energy conversion, describing how catalysts lower the activation energy, how metabolic pathways are organized, and how energy carriers, such as ATP, enable work within the cell.

Chapter 7 explains the pathways for cellular respiration after the stages of glycolysis, the citric acid cycle, and oxidative phosphorylation. Electron transport, proton gradients, and ATP synthesis are explained in detail.

Chapter 8 explains photosynthesis in terms of the light-dependent reactions that convert solar energy to chemical potential and the Calvin cycle reactions that fix carbon into organic molecules. Comparative observations reveal differences among plant, algal, and bacterial systems.

Chapter 9 discusses the stages of cell division, including mitotic events, checkpoint controls, and cell cycle regulation. It emphasizes the need for precise chromosome segregation to provide genetic stability.

Chapter 10 addresses the specialized division events of meiosis and the principles of inheritance it establishes. Topics under discussion include homologous recombination, independent assortment, and how gametogenesis facilitates sexual reproduction.

Chapter 11 discusses the molecular organization of DNA and how it is accurately replicated. Chapter 11 explains DNA polymerases, replication forks, proofreading function, and the challenges of replicating linear chromosomes.

Chapter 12 examines gene expression and the new science of biotechnology. It discusses transcriptional regulation, RNA processing, and translation and addresses recombinant DNA techniques, genome editing, and therapeutic and agricultural applications.

This book is designed for undergraduate life science students, as well as those in pre-medical and pre-dental courses and anyone seeking comprehensive information about biology. It is suitable for teachers as a textbook for lecture classes and lab seminars, whereas students can use it as a reliable reference. Guiding readers through the laws that govern living systems, this book aims to equip its readers with the information and tools for analysis that will enable them to continue their studies further and apply biological principles to research, industry, and other fields.

INTRODUCTION TO BIOLOGY

1

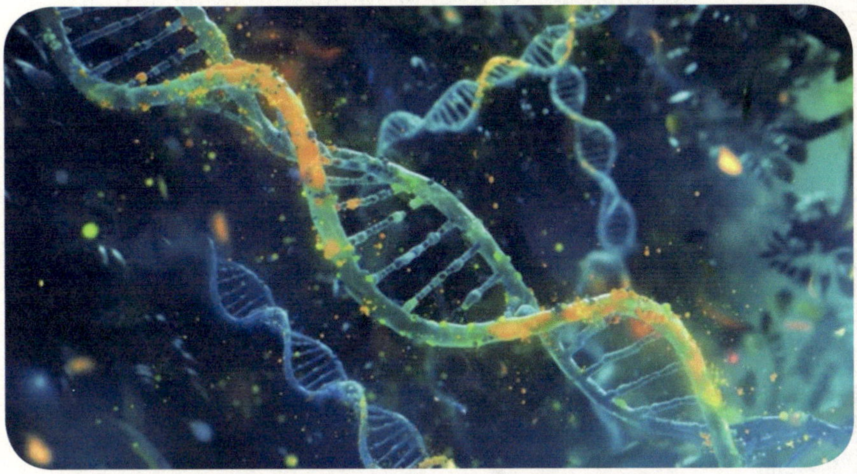

LEARNING OBJECTIVES

After studying this chapter, you will be able to:

1. Discuss the scientific method in biology;
2. Define cell theory and cell structure;
3. Explain genetics and heredity;
4. Describe evolution and natural selection;
5. Explore ecology and ecosystems;
6. Illustrate human biology and physiology.

Introduction

Biology as a fundamental science is the scientific investigation of life and living organisms, their structure, function, growth, evolution, and distribution. The general branch comprises all the sub-disciplines such as botany, zoology, microbiology, and ecology, with different types of life and interaction among species and between them and the environment. The scope of biology also encompasses molecular biology and biochemistry, where scientists study the biochemical processes that regulate cell function. Studying the complex relationship of organisms, biology attempts to explain the machinery of life and thereby facilitates the creation of a greater understanding of micro and macro levels of living systems. The applicability of biology is of fundamental significance, as not only does it add to scientific knowledge, but it also addresses everyday issues such as health care, conservation of the environment, and loss of biodiversity. There are some fundamental principles behind the science of biology in the wake of scientific inquiry. At the core of biological inquiry lies the concept of the cell as the building block of life, followed by questions regarding cellular function and cell-cell interaction. Organisms are classified according to their shared features and evolutionary history, which is examined from genetic and evolutionary biological perspectives. In addition, biology also involves the examination of ecosystems and the interdependence of living things, analyzing how organisms evolve in accordance with their environments in the long run. With the growing demand for scientific proof, biology must also face the issue of ethics, most notably in fields like genetic engineering and animal experimentation. Journals such as the BJP are already embracing guidelines to promote compliance of research with ethical standards and reproducibility, where great emphasis is on thorough experimental design reporting and animal welfare (Lilley et al., 2020).

Figure 1.1. Definition and scope of biology.

Source: By Fir0002 - Own work, GFDL 1.2, https://commons.wikimedia.org/w/index.php?curid=7613324

In addition, the scope of biology has been greatly expanded with advances in technology underpinning pioneering research and discovery. Methods like genome sequencing, CRISPR, and bioinformatics have transformed our comprehension of biological systems and disease processes. Such technologies allow scientists to edit genetic information and dissect complicated data sets, opening the door to medical, agricultural, and conservation breakthroughs. With globalization and climate change increasingly destabilizing classic biological models, multidisciplinary cooperation becomes more critical. By combining genetic, ecological, and environmental information, biologists can formulate comprehensive strategies to offset the impact of human activity on ecosystems. Lastly, the continuous evolution of biology reflects mankind's search for the secrets of life, answering both the short-term need and the long-term sustainability goals (Szklarczyk et al., 2020) (Figure 1.1).

Explanation of Biology as a Science

Biology as a science is based on the systematic study of life and living things, through the integration of several sub-disciplines, which make it a complete study. This science examines biological processes using various methods, such as observation, experiment, and analysis. Utilizing the scientific process, biologists can make hypotheses and test these ideas rigorously against experimental evidence, a cornerstone of science that sets biology apart from other sciences. Effective biology also requires an interdisciplinary model, incorporating contributions from chemistry, physics, and mathematics, to explain complex biological concepts. Moreover, the history of biological science reveals its adaptation to new paradigms and theories, reflecting its dynamism in response to questions about the world of life. Over the past few decades, the study of complicated systems of life has gained more attention, reflecting the complexity and interdependence within ecosystems and organisms. Complexity, as experienced through complicated frameworks, commands our comprehension of biological interactions beyond the linear cause-and-effect model. This perspective is mirrored in the broader understanding of intricate systems, as reflected in the interaction of the various elements in biological systems. Ecological systems, for instance, reflect how individual organisms, like plants and animals, exhibit behavior not only as a result of their intrinsic nature but also as a consequence of their interaction with the broader surroundings. Dealing with this complexity enables biologists to draw conclusions that are required to address pressing world problems such as climate change and biodiversity conservation, to stress the relevance of biological research in societies (Newman et al., 2011). The epistemology of biology is also confronted with enormous challenges in coping with the dual experiences of descriptive and perceived complexity. Since complexity itself has emerged as a concept within scientific discourse, it is imperative to define the way by which these methods characterize our understanding of biological occurrences. Descriptive complexity addresses the multiform layers and intertwinements experienced within biologic complexes, while perceived complexity addresses the way by which such complexities are subjectively perceived by scientists. This distinction is central to developing systems practice, which equips scientists with the tools to manipulate and interpret the intricate realities of living systems. As researchers rightly argue, as much as it is important to understand the forest of biological interactions, it is also as important to study individual trees. By the settlement of those epistemological issues, as a science of biology can possibly possess a more advanced and efficient approach to coping with natural systems, firming up both theoretical and practical knowledge (Ison et al., 2004).

Overview of Various Branches of Biology

Biology is a broad branch of study with various fields focusing on different aspects of living things and how they interact with their surroundings. For instance, molecular biology explores the intricate activities of biological molecules and how DNA, RNA, and proteins operate and interact at the cellular level. Molecular biology has been hugely important in recent years, with developments in biotechnology and genetic engineering. Understanding molecular mechanisms allows for amazing progress in medicine, agriculture, and ecology. By examining the very constituents of life, molecular biology provides data that tells everything else in the biological sciences something, emphasizing the interrelatedness of biological information. Ecology instead examines the relationship between organisms and their surroundings, with ecosystems as interactive webs of relationships. Ecologists study events such as population increase, community structure, and energy flow in ecological systems. Ecology is important for addressing topical global concerns such as climate change and habitat loss because it informs conservation efforts and sustainable use. Through an in-depth understanding of ecological concepts, biologists can develop methods to undo human effects on the environment, reflecting the important role that ecology plays in promoting biodiversity and preserving natural resources for future generations (Kennedy & Henderson, 2003). Lastly, evolutionary biology seeks to understand the mechanisms that render life on Earth varied. Through the exploration of concepts such as natural selection, adaptation, and speciation, evolutionary biologists explain the way living things change over time and how evolution results in the variety of life forms that exist today. This science intertwines with genetics, paleontology, and ecology to give a comprehensive picture of how life has developed. Also, computational developments have enhanced research into biological networks with quantitative information of evolutionary processes and their possible influence on biological modeling. The intricate interaction among various branches, as enlightened by analysis of biological networks (Bourdon & Eveillard, 2007), underscores the necessity for a multidisciplinary approach in order to capture the full spectrum of biological processes (Figure 1.2).

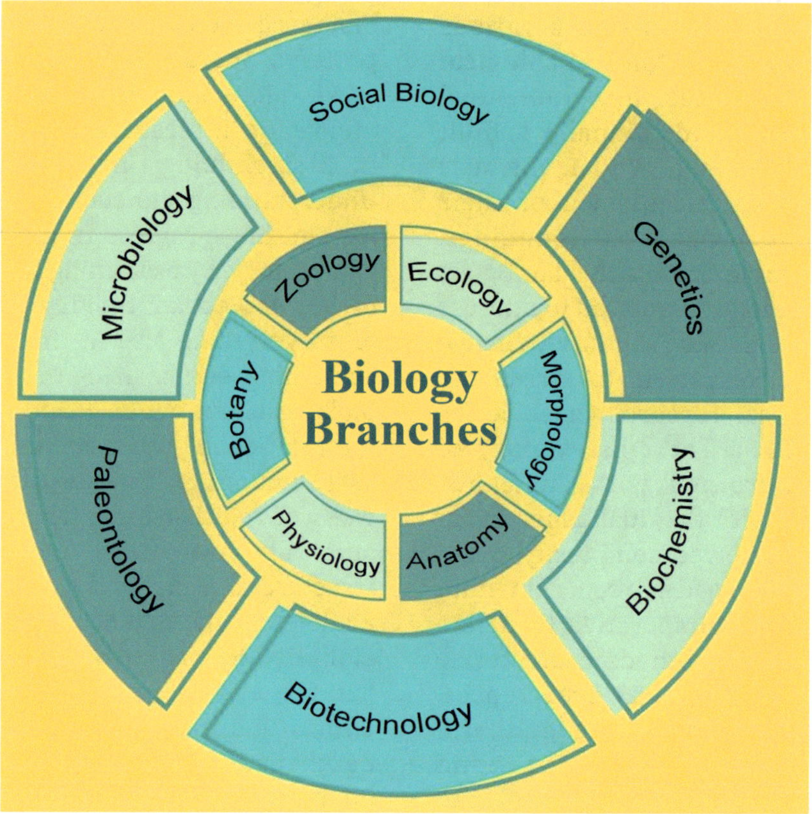

Figure 1.2. Branches of biology.

Source: https://media.licdn.com/dms/image/v2/C4D22AQEJIk9ag1zg7Q/feedshare-shrink_800/feedshare-shrink_800/0/1672592686384?e=2147483647&v=beta&t=KioNOphzAlUxCpR_jcMVq3fIWQIDxaLcRcqssEbjjog.

Importance of Biology in Understanding Life

Biology has been a foundation of complexity of life and it shows how multifaceted life interacts with its surroundings. This discipline does not only deal with who is what and what is what, it goes deeper into the complex interrelationships that rule the life processes. As an example, modern biology focuses on such concepts as phenotypic plasticity that demonstrates the way organisms respond to different environments, affecting their reproduction and survival. As described by Gare et al. (2013), understanding these dynamics provides insight into evolutionary mechanisms while outlining foundational concepts that encapsulate biological diversity. Such comprehensive exploration not only helps in explaining the current shapes of life but also helps in foreseeing the future trends of evolutionary changes, which cements the position of biology as one of the core subjects in unraveling the mysteries of life. In addition, the importance of biology can be extrapolated to its contribution to other applied sciences and this has defined disciplines like medicine, environmental science, and biotechnology. Biological research and its insights have brought revolutionary change in the field of healthcare, especially in the manner in which diseases are perceived and treated. Reductionist strategy in biological studies has, in the past, enabled greater understanding of the cellular and molecular processes. Yet, as outlined by Gare et al. (2013), this perspective is evolving into a more integrated view, recognizing the limitations of reductionism and the necessity for a holistic comprehension of biological systems. Recent studies tend to focus on interdisciplinary studies, where biology is integrated with physics, chemistry, and mathematics to promote innovations that can solve global problems, including climate change, epidemics that threaten public health, and the protection of biodiversity. Lastly, biological literacy is the foundation of important conservation and sustainability measures. With the increasing severity of the threats of climate change and habitat destruction, a solid understanding of biological principles is required to create effective management and conservation strategies. The interaction of the species, ecosystems and their environments needs to be known so as to sustain ecological processes that are important in keeping the planet healthy. The exploration of concepts like evolutionary capacitance and niche construction, as discussed by Martins et al. (2018), adds depth to our comprehension of biodiversity and ecological stability. With the ever-growing unraveling of life mysteries, the use of biology in solving environmental issues is getting more and more crucial, which makes this study even more essential in ensuring the future of our planet.

1.1. The Scientific Method in Biology

Learning Outcome

1. Steps of the scientific method;
2. Role of hypothesis and experimentation;
3. Importance of data analysis and interpretation.

The scientific method is a rigorous process that guides biological research and experimentation. It enables scientists to formulate hypotheses, i.e., testable predictions based on existing knowledge or observations. In biology, scientists would first query living organisms, for example, how genetic variations affect adaptation in a changing environment. Through controlled conditions and careful observation, biologists utilize variable isolation and cause-and-effect determinations. Not only is this required for comprehension of complex biological systems, but it also promotes the reproducibility of results, and thereby the validity of science. Current advances in biotechnology, particularly as a reaction to the global challenges of the COVID-19 pandemic, emphasize the value of the scientific process in biology. The scientific study of viral diseases is a reflection of how scientists apply the scientific method in discovering viral genomics and immunogenicity. A knowledge of viruses' genetic codes helps develop vaccines, which are thoroughly tested and verified using many experimental protocols before public distribution. As outlined in recent research, synthetic biology borrows from systems biology to necessitate multidisciplinary approaches founded on the scientific method towards maximizing vaccine preparations and improving public health against emerging challenges like COVID-19 (Moghaddasi et al., 2020). Further, the intersection of fields within biological sciences presents evidence of the integrative nature of the scientific method. Educational campaigns that highlight the methodological rigor of the scientific method are instrumental in cultivating scientific literacy. Through the refutation of pseudoscience and the promotion of critical thinking, scientific method courses aim to prepare students with the resources for unbiased analysis. This interdisciplinary course not only enhances students' knowledge of biology but also prepares students to differentiate between authentic scientific debate and misinformation. As more eloquently stated under respective project aims, the ultimate aim in the long term is not merely to relay scientific knowledge, but to foster a good appreciation of methodologies which form the basis of our understanding of the world around us (Dowse et al., 2000) (Figure 1.3).

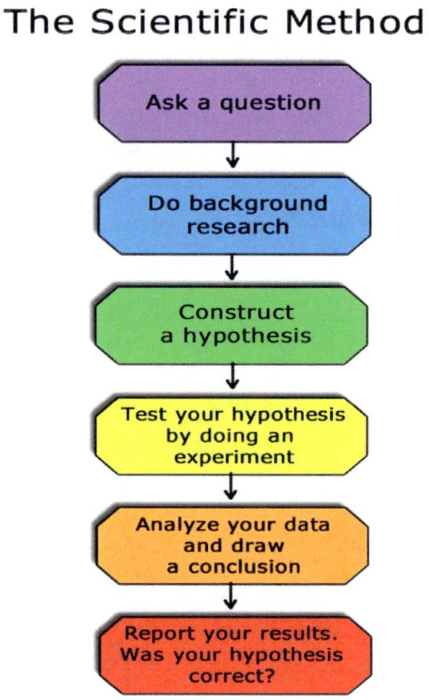

Figure 1.3. The scientific method in biology.

Source: https://www.joanwink.com/wp-content/uploads/2017/09/anchor_chart_sci_method_lg.jpg.

1.1.1. Steps of the Scientific Method

The scientific method is the basis of biological inquiry, providing a systematic means to inquire into events and acquire reliable findings. The first step in the scientific method is observation, which leads to formulating a question that solves a specific issue or problem. For instance, biologists may observe unusual patterns of plant growth and subsequently inquire into possible environmental elements that caused the phenomenon. This first critical phase is the foundation of the remainder of the steps, for it compels researchers to engage with the world of nature in a meaningful manner. Being aware of the nature of science is crucial for the establishment of a solid scientific question, which most often requires a long discussion among instructors to address the complexity and depth of scientific concepts, as discussed by Sundaryono et al. (2021). Following the formulation of a question, researchers create a hypothesis that makes an explanatory or predictive assertion regarding the phenomenon they are observing. A hypothesis must be falsifiable and testable so that empirical research can validate or refute it. A biologist, for example, might create a hypothesis that increased sunlight leads to increased growth rates in plants. To verify this hypothesis, controlled experiments are set up wherein various groups of plants are subjected to varying levels of sunlight. Systematic experimentation is employed to collect and examine data for the purpose of establishing whether the results verify the hypothesis. This circular process not only strengthens the authority of critical thinking but also clarifies the importance of refining hypotheses in relation to results in experiments and thereby instills more insight into scientific principles emphasized by Altintas et al. (2014). Once empirical data have been collected and processed, researchers make conclusions that validate or invalidate their hypotheses. This important step allows researchers to put findings into the context of their original question and assess the broader implications of what they have learned. If information validates the hypothesis, it can be sent for peer review publication, contributing to the scientific community's growing base of knowledge. When the hypothesis is disproven, it leads to further exploration, suggesting new questions and refining existing theories. This responsive and reflective approach is an excellent demonstration of the active scientific process of inquiry in which conclusions guide additional hypotheses and additional research. The scientific process is then a continuous cycle of inquiry that increases our understanding of biological mechanisms, underlining the importance of evidence-based results as well as the social dimension of science, as observed by Sundaryono et al. (2021) (Figure 1.4).

Figure 1.4. Steps of the scientific method in biology.

Source: https://biologydictionary.net/wp-content/uploads/2017/03/The-Scientific-Method.jpg.

1.1.2. Role of Hypothesis and Experimentation

In biological research, the creation of hypotheses is a starting point for scientific investigation. A hypothesis is simply an informed prediction that defines a clear path for experimentation. This predication quality is important in leading investigators to study intricate biological systems and events. For example, in a study that aimed to establish the role of heart of glass homolog 1 (HEG1) in vascular biology, researchers made informed hypotheses that HEG1 mediates atheroprotective endothelial reactions to stable blood flow in the hope of understanding its possible role towards atherosclerosis prevention (Ian et al., 2023). Through making such hypotheses, researchers are in a position to devise a framework for data collection and interpretation, allowing them to navigate seas of data with a discerning eye. Therefore, not only does the hypothesis guide the experimental design, but also the general understanding of biological phenomena. Experimentation is the process by which hypotheses are under a strict test and verification. Through the use of controlled experimental conditions, variables can be isolated, providing clarity to the measured effects. A good example of this approach is offered by the study of the HEG1 expression in endothelial cell-specific knockout mice with inducible HEG1, which allows for direct comparison with the control groups (Ian et al., 2023). This stringent experiment allowed for the investigation of the impact of HEG1 on atherosclerotic plaque formation. These experiments are not merely hypothesis testing; they also refine scientific knowledge and create new questions that can lead to other research. Experimentation is therefore important in biology since it provides the empirical evidence needed to prove or disprove a hypothesis while simultaneously expanding the scientific debate. Further, novel approaches such as quantitative systems pharmacology (QSP) illustrate the forward-thinking approach of hypothesis-based experimentation in biology. By integrating data from various disciplines, QSP affords scientists the opportunity to try out many hypotheses through complex computational models, thereby reducing their reliance on potentially costly and time-consuming experimental procedures. This new approach enables better probing of biological processes with higher efficiency, allowing scientists to make well-informed predictions and rationalize drug development approaches. To illustrate, QSP has been applied to model drug-drug interactions and predict efficacies under clinical conditions, reflecting its utility in translational biology. Ultimately, the role of hypothesis and experiment in the world of biology is dynamic and multifaceted, constantly pushing science forward and expanding our understanding of living systems (Figure 1.5).

Figure 1.5. Role of hypothesis and experimentation.

Source: https://contenthub-static.grammarly.com/blog/wp-content/uploads/2021/11/how-to-write-a-hypothesis.jpeg.

1.1.3. Importance of Data Analysis and Interpretation

Data analysis and interpretation are an important part of biology whose role can hardly be overestimated in light of the fact that scientific inquiry is turning into a data-driven endeavor. With researchers gathering huge volumes of biological information, the capacity to process and make sense of this data assumes great importance. Scientific conclusions may be inaccurate or even misleading without a rigorous data analysis, and this affects not only the particular study but also the whole field of biology. As an illustration, in research that involves an evaluation of ecological effects, inappropriate interpretation of data may result in erroneous judgments concerning the resilience of species or the health of ecosystems. Hence, there is a need to incorporate sound statistical practices so that biological studies can culminate in sound findings that can be replicated and are pivotal in making breakthroughs in the field of life sciences. This focus on data integrity is representative of the increasing need for interdisciplinary collaboration, whereby biologists need to interact meaningfully with data scientists, in order to promote the integrity of research findings. Furthermore, the intricacy of the biological data requires a profound association of the statistical tools and biological knowledge, as it has been noted in the recent literature. The intersection between life sciences and statistics points to a cooperative strategy that is becoming more critical in contemporary research. In areas such as genomics and systems biology, scientists must now create complex experiments and interrogate data using different statistical paradigms. Such significance of the data interpretation is also stressed in the environment where the innovative approaches can revitalize such analyses. This helps researchers to negotiate the complex interactions that characterize biological systems because it closes the gap between experimental design and statistical analysis. The cooperative relationship between biology and statistics stimulates scientists to develop a level of understanding with each other to facilitate the communication and adoption of the methodologies between the disciplines. With the increased use of technological advances and models in order to forecast biological results, data analysis has ceased to be a supplementary role to a foundation of biological study. With the organisms and ecosystems being subjected to high-throughput technologies, the need to have competent data analysis gives a backbone of essence to the validation of experimental results. Nevertheless, analysis of the vast amount of generated data cannot be done without a solid knowledge base of biological concepts as well as data interpretation methods. Workshops and training have emerged as effective strategies to enhance this interdisciplinary partnership, demonstrating how collaboration can aid in developing innovative solutions to obscure biological queries (Steele et al., 1998). Moreover, this strategy assists in solving the possible mismatch that can occur because of a communicative gap among scientists belonging to different backgrounds. Ultimately, the ability to analyze and interpret biological data accurately is not merely an academic exercise; it is imperative for advancing our understanding of life and addressing real-world biological challenges (Steele et al., 1998).

Test Your Knowledge

1. What are the steps of the scientific method?
2. Define the role of hypothesis and experimentation.

1.2. Cell Theory and Cell Structure

Learning Outcome

1. Historical development of cell theory;
2. Types of cells: prokaryotic vs. eukaryotic;
3. Organelles and their functions.

The foundation of modern biology is built upon the principles of cell theory, which articulates that all living organisms are composed of cells, the fundamental units of life. This theory emerged in the mid-19th century, primarily through the contributions of scientists such as Schleiden, Schwann, and Virchow. Cell theory holds that not only is the cell the basic structural and functional unit of life, but also that it is produced only from already formed cells through the process of cell division. The extent of these principles goes far beyond mere classification; they stress that all cellular complexity and diversity are products of cellular interactions, offering a paradigm that guides both biological investigation and medical advancement, for example, research methods in molecular biology that progressively expose the complexities of cell dynamics and function, thus fundamentally transforming our understanding of life itself (Mansouri & Parker, 2025). The cell structure is crucial in the functioning of living organisms, with different cell structures having specific functions. Eukaryotic cells, for instance, are characterized by a well-organized structure that has membrane-bound organelles such as the nucleus, mitochondria, and endoplasmic reticulum (ER). Each of these organelles regulates important processes, from energy production to protein production, facilitating the overall function of the cells. Conversely, prokaryotic cells, like bacteria, have a less complex structure without membrane-bound organelles but instead use the cell membrane and cytoplasm to carry out functions required by the cell. Learning about these structural variations improves our understanding of how different forms of life function and survive in their worlds. Such findings benefit research in molecular biology, where studying the processes and interactions of cells yields useful data that can be applied to genetics and biotechnology (Williams, 2020).

Figure 1.6. Cell theory and cell structure.

Source: https://img.clickviewapp.com/v2/thumbnails/W134Lr?size=medium.

Cell theory and structure learning not only educates on biological knowledge but also benefits scientific progress of methods and applications. With molecular biology, researchers are able to study cellular processes at resolutions previously impossible, with emerging technologies such as high-throughput sequencing and gene editing. Such methods allow researchers to explore gene expression, cellular response, and even the genetic origins of disease. For instance, understanding based on cellular biology has given the role of specific mutations in cancer's initiation, with a view to identifying targeted medications and new therapy avenues. As such, the unification of cell theory and molecular approaches is reshaping medical principles and enhancing therapeutic interventions, particularly in managing genetic diseases and cancers, leading to improved health care outcomes among individuals globally (Williams, 2020; Mansouri & Parker, 2025) (Figure 1.6).

1.2.1. Historical Development of Cell Theory

The 19th-century formulation of cell theory is a milestone moment in the science of biology, if only because it gave the elementary foundation for understanding life at the cellular level. The theory evolved from the combined work of several scientists in the 19th century, who all rectified earlier suppositions about the make-up of living things. Key figures such as Matthias Schleiden and Theodor Schwann were responsible for the establishment of the basic principles of cell theory, namely that all living organisms are composed of cells, and that the cell is the basic building block of life. Their research, coupled with advances in microscopy, revolutionized biological inquiry by allowing cells to be seen and providing a context in which the complexities of life were able to be investigated. This collective work illuminated the complicated interrelations between various forms of life and their cellular building blocks, providing a prime knowledge foundation that would be necessary for biological investigation to proceed. Further advances in cell theory were motivated by scientists such as Rudolf Virchow, who developed the theory that all cells grow from current cells. This idea not only cemented the fact that cells are self-reproducing units but also amplified the focus on the continuity of life and cell division in growth and repair processes. The application of these principles to the scientific community provided windows for further study of cell function and pathology. As cell theory unfolded, it was apparent that multidisciplinary research—utilizing expertise from fields such as chemistry and physics—

was required in an effort to break down the complexity of cellular structures and processes. Multidisciplinary research has since become a core feature of modern biological education, as seen in the planning of innovative curricula that integrate different fields of science to enable student understanding and engagement with life sciences (Camfield & Land, 2017). Cell theory application and modification have had profound consequences not just in the biosciences but also in a variety of other scientific disciplines. The theory has evolved from a simple notion to a sophisticated model that intersects with advances in genetics, molecular biology, and biotechnology. Since it is now more deeply understood how the molecular building blocks within cells, like DNA, RNA, and proteins, work, the role of cells within the larger context of biological systems has become better understood. This development points to pedagogical adjustments in teaching biology because current pedagogy must be adjusted to address these complex scientific developments. In particular, the integration of writing and active-learning approaches has been shown to significantly enhance student interest and comprehension, developing a new breed of biologists with the potential to fuel the changing cell theory and its applications in a variety of scientific settings (Camfield & Land, 2017) (Figure 1.7).

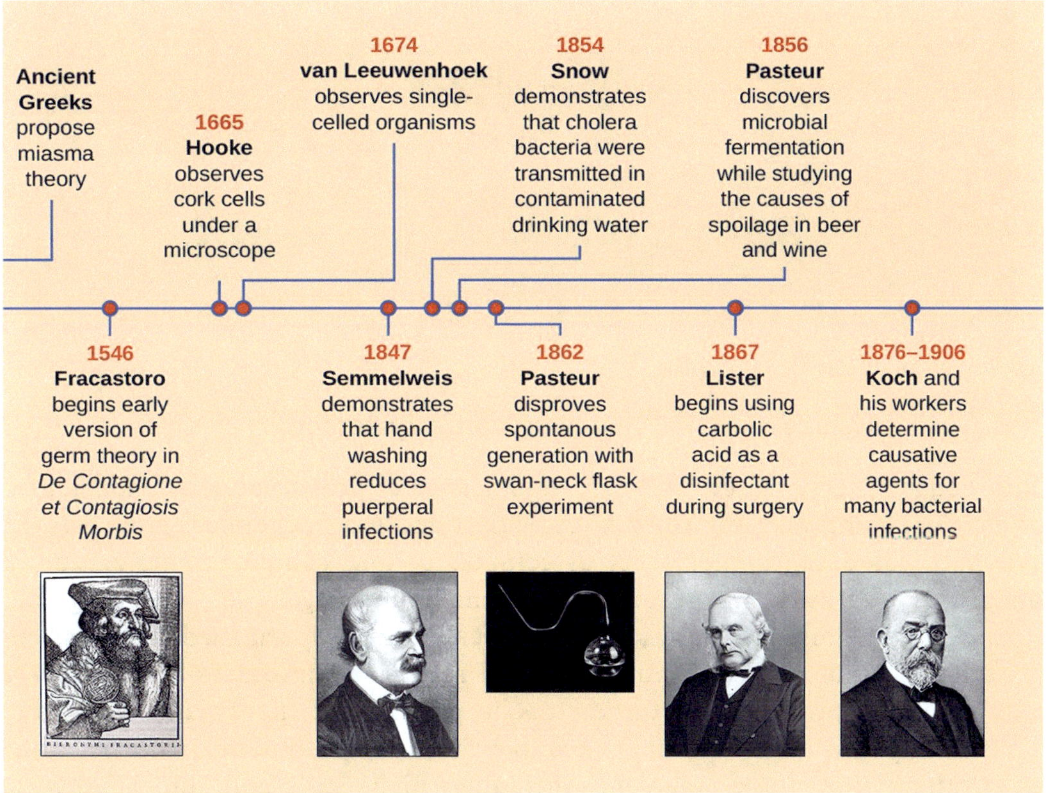

Figure 1.7. Historical development of cell theory.

Source: https://library.fiveable.me/_next/image?url=https%3A%2F%2Fstorage.googleapis.com%2Fstatic.prod.fiveable.me%2Fsearch-images%252F%2522Key_discoveries_in_cell_theory_Robert_Hooke_Leeuwenhoek_Schleiden_Schwann_historical_cell_biology_images%2522-Timeline.jpeg.jpg&w=828&q=75.

1.2.2. Types of Cells: Prokaryotic versus Eukaryotic

Cell division into prokaryotic and eukaryotic types is a fundamental field of biological study. Prokaryotic cells are typically characterized by the fact that they are uncomplicated; they are generally smaller, lack well-defined nuclei, and possess fewer organelles than their eukaryotic counterparts. The prokaryotic DNA is not encapsulated in a membrane-bound nucleus, which practically makes them different from eukaryotic cells, whose nucleus serves as a compartment to enclose and protect DNA. Prokaryotic cells were initially believed to be primitive, but nowadays, complexities that cannot be simplified have already been discovered in them, like their very advanced mechanisms of segregation of DNA and intracellular arrangement (Berlanga et al., 2010).

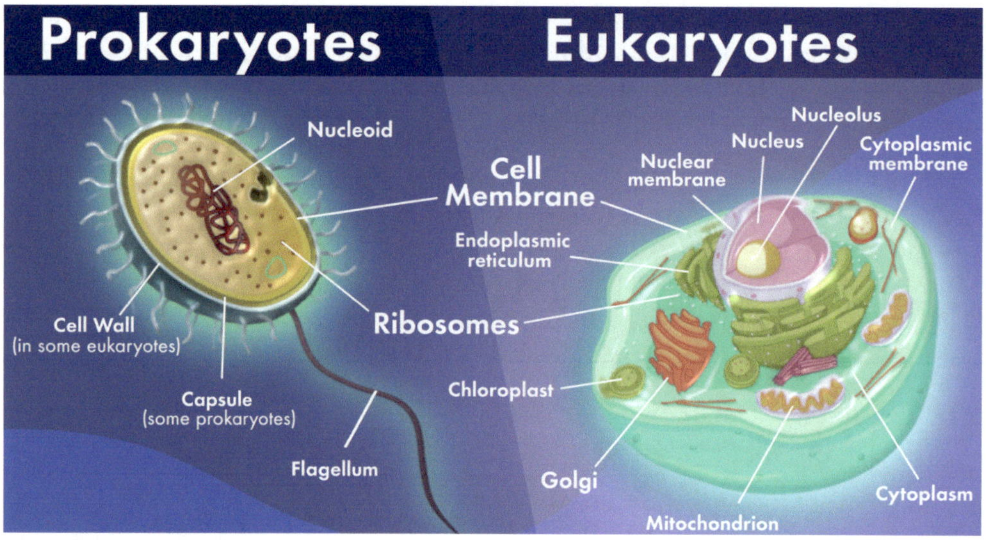

Figure 1.8. Types of cells: Prokaryotic versus eukaryotic.

Source: https://cdn.hswstatic.com/gif/Prokaryotesvseukaryotes.jpg.

This knowledge reflects the evolutionary significance of prokaryotes, a fundamental pillar in the tree of life (TOL), and play key roles in ecological and biogeochemical processes. Eukaryotic cells, meanwhile, possess a higher level of complexity and compartmentalization. They are larger than prokaryotic cells and possess membrane-bound organelles, including the nucleus, mitochondria, ER, and Golgi apparatus, which are engaged in specialized functions in the cell. This compartmentalization supports more advanced regulatory processes and energy-generating pathways to emerge, contributing to the metabolic potential of the organisms. Eukaryotic cell formation has been proposed to have happened by way of endosymbiotic processes with the early eukaryotes engulfing prokaryotic organisms, including chloroplasts and mitochondria, leading to enhanced functionality (Berlanga et al., 2010). This evolutionary transition not only paved the way for multicellularity, but it also created avenues for life forms to become more complex, ultimately resulting in more diversity of form and function throughout biological history. Prokaryotic and eukaryotic cells are not merely structurally different; they are fundamentally different in terms of life processes such as DNA replication and segregation. Prokaryotic cells possess specialized systems to ensure that their genetic material is accurately segregated at cell division, a key function in the maintenance of genetic stability. Sophisticated segregation systems in prokaryotes have been shown by current research to be vital for viability in various diverse environments. These results demonstrate that prokaryotic cells, long thought of as simple, possess sophisticated mechanisms that relate to their competency and adaptability. Understanding these cell types is the foundation for learning about the complexities of life, including evolutionary biology and metabolic pathways that sustain various forms of life, thereby highlighting the significance of both prokaryotes and eukaryotes in broader biological research (Figure 1.8).

1.2.3. Organelles and Their Functions

Organelles are cell structures that contain specialized functions that are necessary for the survival of the cell and the general well-being of the organism. Among these organelles, the nucleus plays a significant role as the command center of eukaryotic cells, which contains genetic material and controls gene expression. By contrast, prokaryotic cells without a defined nucleus have a more decentralized control of their genetic information.

Figure 1.9. Organelles and their functions.

Source: By Kelvinsong - Own work, CC0, https://commons.wikimedia.org/w/index.php?curid=22952603

The organelles each play a role in homeostasis and metabolic functioning and are therefore important in the study of cellular biology. Educators often introduce these concepts at various educational stages, emphasizing the differences between prokaryotic and eukaryotic cells, as seen in biology curricula designed for younger students (Bos et al., 2016). This factual understanding leads to advanced interaction with other more involved biological issues. Other than the nucleus, the organelles that are of great significance are mitochondria and ribosomes, which are involved in the production of energy and the synthesis of protein, respectively. Mitochondria, often referred to as the powerhouses of the cell, are responsible for converting nutrients into adenosine triphosphate (ATP) through cellular respiration, thereby providing energy for various cellular activities. Ribosomes, which may be found freely in the cytoplasm or attached to the ER, play a central role in the process of translating messenger RNA into polypeptides, which will finally fold into functional proteins. The way these organelles interact with each other represents a very well-orchestrated set of systems in action and is evidence of the complexity of cellular life. An understanding of cell organelles often begins with practical experiences, allowing students to observe these structures under microscopes, thus reinforcing theoretical knowledge with tangible evidence (Bos et al., 2016). Specialization of organelle functions is seen in specialized functions to meet specialized cell requirements, as in the case of some parasitic organisms. An example is the use of special secretory organelles that aid their movement and invasion of host cells by Toxoplasma and Plasmodium, representing a high level of adaptation to specific ecological niches. These apical secretory organelles, which are micronemes and rhoptries, are also involved in the life cycle and pathogenicity of these parasites, which can give information on how organelles can influence not only cellular functioning but also inter-organism interactions. It has also been recently demonstrated that the secretion pathways of these organelles necessitate the alterations of the classical endosomal systems, which highlights the evolutionary impact of organelles on the strategies of survival in the changing environment. Appreciation of these complexities also leads to an improved appreciation of the rest of the science of biology and the varieties of life (Figure 1.9).

Test Your Knowledge

1. What are the types of cells: prokaryotic vs. eukaryotic?
2. What are organelles and their functions?

1.3. Genetics and Heredity

Learning Outcome

1. Basics of DNA structure and function;
2. Mendelian genetics and inheritance patterns;
3. Modern genetics and biotechnology applications.

Genetics and the study of heredity form the basis of understanding biological processes that trigger evolution and the diversification of life forms. Gregor Mendel's early 19th-century research established the basis for understanding heredity today in the form of pea plant experiments. Mendel came up with rules that define the transmission of traits between parents and their offspring, using terms such as dominant and recessive alleles. While his work was mostly ignored at first, it would subsequently be at the forefront of the construction of the contemporary synthesis of evolutionary biology, merging his genetic findings with Charles Darwin's natural selection theory. The synthesis points out the significance of Mendelian genetics in explaining not just how traits are passed down from one generation to another but also how they are responsible for the adaptability and speciation of organisms across varied environments (Smýkal et al., 2016). Later advances in molecular biology have revolutionized genetics forever, enabling geneticists to understand the complexity of gene expression and variation. The post-genomic era has further emphasized the point that genetic information alone cannot completely explain the varied phenotypes observed between organisms. Instead, the interplay between genetic makeup and environmental factors has a great influence on determining characteristics. This notion is seconded in developmental biology, which subscribes to the belief that the development of an organism's properties is a dynamical process susceptible to molecular interaction in its cellular environment. Consequently, focus has been laid on genetic interaction with their environments, broadening the boundaries of traditional genetic studies to encompass genetic and developmental processes and their environments (Layer, 2011). Current techniques in genetics also illustrate the significance of genetic diversity in evolutionary and agricultural programs. This has attracted attention from leading researchers in this area, and the need for geneticists to prioritize innovation in genetics has been taken seriously.

Techniques such as genome-wide selection and marker-trait associations are based on Mendelian concepts for the detection and enhancement of desirable traits in other organisms. In addition, the application of emerging technologies such as gene editing tools is employed to enhance genetic diversity and yield in crops by transferring desirable traits from wild relatives. The continuous exploration of genetic variation, using traditional breeding and recent biotechnologies, highlights the dynamic nature of genetics and heredity. This comprehension is crucial in resolving the issues related to food security today, the conservation of biodiversity, and health, showing the delicate web spun by genes and the environments they inhabit in creating life on our planet (Smýkal et al., 2016) (Figure 1.10).

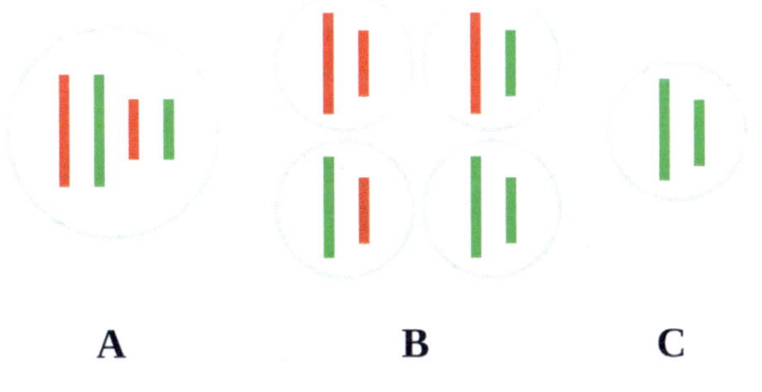

Figure 1.10. Genetics and heredity.

Source: By Original: Darekk2, SVG: Palosirkka - File:Hybridogenesis in water frogs gametes. gif, CC BY-SA 4.0, https://commons.wikimedia. org/w/index.php?curid=67142604

1.3.1. Basics of DNA Structure and Function

The structure of DNA serves as a foundation for its role in biological activities. DNA consists of two lengthy strands that are coiled together in the form of a double helix. DNA consists of four types of nucleotides represented by the bases adenine (A), thymine (T), cytosine (C), and guanine (G). Nucleotides are arranged in a sequence to contain genetic information. The specific pair rules, adenine with thymine and cytosine with guanine, are essential in replication and transcription to allow cells to copy genetic information as the cell divides. The double helix is held in place by paired bases via hydrogen bonds and by the sugar-phosphate backbone. This elegant architecture not only ensures the integrity of transmission of genetic information but also enables such transmission to occur in directing protein synthesis, thus linking the architecture of DNA to its multilateral roles in organisms. The functional roles of DNA transcend genomic storage of genetic information. The complex sequences of the DNA allow for a sophisticated gene expression system coordinating protein production needed for cellular processes. The interaction between the components, such as the promoters and enhancers, allows regulation of when and how genes get activated. Furthermore, studies on DNA revealed interesting details about evolutionary biology, particularly within the aspect of comparative genomic studies. For instance, odorant receptors were found to identify evolutionary relationships between species, which exhibit slight differences in DNA sequences that are critical to understanding divergence and adaptation. These studies prove the significance of computational biology in demystifying such gene complexities, as exemplified in studies exploring evolutionary networks between Mice, Chimpanzees, and Humans (Hassan et al., 2012). Progressive computational models and methods, such as cellular automata (CA), have also enlightened

DNA structure, evolution, and functionality. By these models, scientists can conceptually think of DNA as a multi-state one-dimensional system that can be easily analyzed in terms of its sequences. Under these conditions, the four nucleotide states can be expressed in a manner such that given rules are employed to simulate the evolution of DNA generations. Notably, studies have found that certain DNA sequences extracted from CA simulations are not found in existing genomic databases, prompting more investigations of the inherent processes of DNA transformation (Hassan et al., 2012). Observations through such studies are critical in understanding the fluidity of genetic material since, in addition to tracing the evolutionary past, they point towards the potential for new genetic patterns which have yet to be unraveled. Through such revolutionary approaches, the union of mathematical and evolutionary inquiry is still broadening our knowledge of the underlying principles that make DNA function (Figure 1.11).

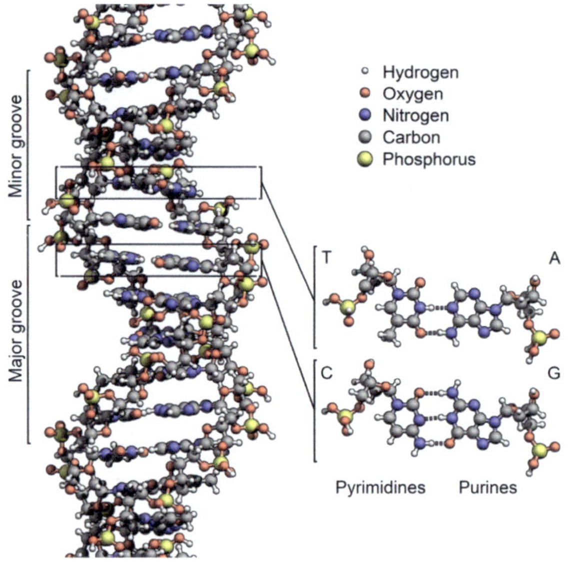

Figure 1.11. DNA structure and function.

Source: By Zephyris - Own work, CC BY-SA 3.0, https://commons.wikimedia.org/w/index.php?curid=15027555

1.3.2. Mendelian Genetics and Inheritance Patterns

Mendelian genetics is the fundamental pillar in the broader scope of biology that accounts for characters being inherited from one generation to the next. Gregor Mendel's pioneering experiments involving pea plants during the 19th century laid down the laws of inheritance, which are still applicable today. Mendel's experiments revealed dominant and recessive traits, leading to the establishment of key laws such as the Law of Segregation and the Law of Independent Assortment. The principles demonstrate orderly patterns in inheritance, allowing biologists to make predictions of phenotypic ratios in offspring. The significance of Mendelian genetics extends beyond mere academic curiosity because it is of crucial significance in such uses as agriculture, medicine, and conservation biology, where genetic inheritance needs to be comprehended for breeding programs and preservation of species variation. As important as it was, Mendelian genetics may be difficult to extend to more complex models of inheritance. Inheritance in the real world is polyallelic, with incomplete dominance and epistasis, which render the simplicity observed in Mendel's initial experiments unrealistic in the real world. Various teaching methods, including task-based learning (TBL), have been shown to enhance students' understanding of these complicated matters. Through active engagement in learning, researchers created evidence that TBL significantly improved students' performance and theoretical grasp of Mendelian genetics and inheritance patterns better than traditional lecture-based education (Borja et al., 2024). This shift towards emerging teaching styles demonstrates that educators must modify methods leading to intrinsic motivation and increased participation, bridging the gap between theoretical concepts and real-world applications in genetics. Applications of Mendelian theory extend into current genetic research and the practice of medicine. An awareness of the pattern of inheritance, for example, can provide insight into the inheritance of congenital disease, which does not exhibit simple Mendelian inheritance. More recent studies of dermatoglyphics, the study of fingerprint patterns—have drawn some firm links between genetic disease and dermatoglyphic pattern (Shivaleela et al., 2019). This research not only emphasizes the relevance of Mendelian genetics to the investigation of inherited disease but also underscores the need for an interdisciplinary approach to biology. Through the integration of concepts from genetics, anthropology, and medicine, investigators are able to build inclusive diagnostic and therapeutic instruments and strategies that reflect the ongoing impact of Mendelian genetics on our understanding of biology and human disease (Figure 1.12).

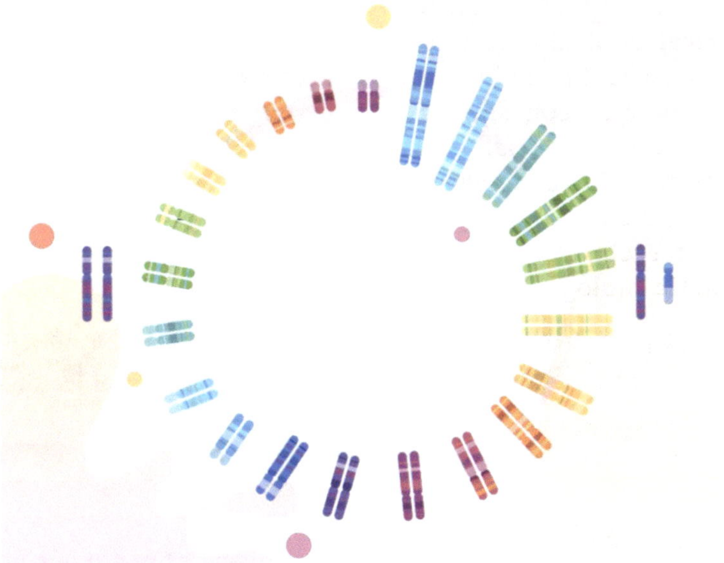

Figure 1.12. Mendelian genetics and inheritance patterns.

Source: https://geneticeducation.co.in/wp-content/uploads/2018/08/Genetic-inheritance-.005.jpeg.

1.3.3. Modern Genetics and Biotechnology Applications

Modern genetics has triggered a revolution in the area of biotechnology and has offered new solutions to many fields such as agriculture, medicine, and environmental management. By harnessing the principles of genetic engineering, scientists can now manipulate DNA to enhance desirable traits in organisms, leading to the creation of genetically modified organisms (GMOs). As an example, crops modified to be pest-resistant not only lessen the usage of chemical pesticides but also lead to sustainable farming. Moreover, recent technologies in DNA sequencing have democratized the genetic information, such that researchers and even farmers can develop more resilient crop varieties and increase food security. This coupling of genetics and biotechnology signifies a paradigm shift in how we approach challenges in food production, showcasing the potential of genetic innovations to meet the demands of a growing global population (Frietsch & Meng, 2025). Modern genetics is also important in the medical field, where it is used to develop personalized medicine, which is a form of medicine that is designed depending on the unique genetic composition of an individual. With the possibility to define genetic markers of different diseases, health providers will be able to define the risks and find more efficient interventions. Gene therapy has also advanced and this has provided opportunities to cure conditions that were historically incurable, and the methods involved allow repair or replacement of faulty genes. Such innovations demonstrate how profound genetic research has been on the general health of people, making it necessary to learn how genes work and interact with each other to support prevention and new therapies. Nevertheless, the rapid expansion of these biotechnological applications raises ethical considerations, such as the implications of gene editing and its potential consequences on human health and biodiversity (Hall et al., 2012). Modern genetics and biotechnology are also important in environmental applications, which help solve urgent ecological problems. As an example, synthetic biology is an area that combines engineering concepts and biology to design organisms capable of performing certain environmental services, including bioremediation and waste remediation. This novel solution is potentially successful in cleaning up toxic environments and reducing the effects of climate change by capturing carbon using technologies. In addition, there is also genetic engineering of microbes that have potential in the production of biofuels, which can provide an alternative to fossil fuels and are more sustainable. These uses demonstrate the flexibility of contemporary genetics in supporting ecological sustainability and conservation. However, it is crucial to monitor and regulate the deployment of these biotechnologies to ensure they align with ecological principles and safeguard biodiversity (Frietsch & Meng, 2025).

Test Your Knowledge

1. What are the basics of DNA structure and function?
2. What are Mendelian genetics and inheritance patterns?

1.4. Evolution and Natural Selection

Learning Outcome

1. Overview of evolutionary theory;
2. Mechanisms of natural selection;
3. Evidence supporting evolution.

The theory of evolution is a pillar in biology that explains the slow formation of species over time by the mechanism of variation, both under the influence of genes and environmental pressures. The essence of this idea rests in the mechanism of natural selection, which postulates that those individuals possessing helpful traits are likely to survive and reproduce. By favoring these traits, natural selection brings about the gradual enhancement of beneficial traits in a population, basically shaping the organism's adaptation to its environment. Through this process, there is achieved a compelling argument is achieved for the diverse range of life forms existing today, as every species has unique adaptations honed through countless generations. The knowledge derived from evolutionary biology has fundamental significance to grasping ecological interactions, biodiversity, and the complex relationships between organisms and the environment, which are all integral parts of the overall study in biology. Natural selection functions through a number of basic mechanisms, such as variation, competition, and heredity. Genetic variation acts as the material for evolution, with mutations and gene combinations creating new characters among organisms. As the finite resources such as food, shelter, and mates are competed for by organisms, those with beneficial adaptations have a greater chance to thrive in their respective niches. This benefit is particularly valuable in constantly changing environments, such as most ecosystems subject to ecological change through climatic events or human interventions. Notably, however, it is not simply the survival of the fittest that controls evolutionary success, but the capability of organisms to reproduce and pass favorable traits onto subsequent generations, an understanding highlighted through host-pathogen interaction research, as detailed in recent scholarship (Philip et al., 2020). The eventual evolutionary spectrum shapes not only the organisms themselves, but the ecosystems they occupy. Besides, an understanding of evolution and natural selection extends beyond theoretical importance and is of applied significance to other fields like medicine, agriculture, and conservation biology. For instance, evolutionary concepts are crucial in the formulation of strategies to combat the spread of newly emerging infectious diseases, such as SARS-CoV-2, which still pose gigantic public health concerns globally (Charles et al., 2018). Research into such pathogens emphasizes the importance of learning about the rate at which evolutionary alterations might influence the effectiveness of vaccines and treatment protocols. Selective breeding techniques in agriculture employ evolutionary concepts to enhance yield and resistance in plants. Conservation, on the other hand, focuses on maintaining genetic diversity within endangered species as a method of countering habitat loss and climate change. Thus, the comprehension of evolution and natural selection not only makes us more well-informed regarding the life on our planet but also equips us with the necessary tools to face the challenges of the present so that human societies and ecosystems can be made sustainable.

1.4.1. Overview of Evolutionary Theory

The foundation of evolutionary theory rests upon the principles of variation, inheritance, and natural selection. The concepts, originally popularized by Charles Darwin in the 19th century, have evolved into a well-established theory for the explanation of the diversification of life on Earth. Evolutionary theory proposes that populations of organisms experience genetic alteration from one generation to the next, primarily due to variations induced by mutations, gene flow, and sexual reproduction. These contrasts may possess specific advantages or disadvantages in a specific situation, influencing the survival and reproductive fitness of an individual and, through these, the genetic makeup of subsequent generations. This recurrent process, often referred to as descent with modification, emphasizes the adaptability and persistence of species in shifting ecological niches, representing a master process through which diversity arises and persists in the realm of nature. As evolutionary theory evolved over the years, it incorporated knowledge from various disciplines, making it more explanatory and applicable. In the 21st century, heritable variation and selection concepts extended beyond biology by impacting anthropology, medicine, psychology, and even economics. Scholars understand that evolutionary conceptualizations offer rich lenses for examining behavior and transformation in both natural and social worlds. For instance, evolutionary theory informs our understanding of human behavior, casting light on the delicacies of social complexity and cultural evolution. Cross-disciplinary endorsement of evolutionary theory is a product of a broad appreciation for its explanatory power in understanding phenomena in a broad array of contexts to provide an astute explanation of how organisms develop and thrive due to countless pressures. Such integration of ideas continues to spur additional research and development in the areas of science, validating the central role of evolution in the academic establishment. Evolving breakthroughs in evolutionary robotics demonstrate the dynamic nature of evolutionary theory and its application in light of modern-day issues. With the integration of biological concepts into robotic system design and the optimization of robotic systems, researchers harness the mechanisms of embodied evolution in terms of facilitating learning within robotic societies. Embodied evolution allows these robots to acquire their behavior by means of collective interactions, maximizing their efficiency and problem-solving skills. Evolutionary theory here not only further enhances our understanding of biological processes but also shapes technological innovation and application. Such cross-disciplinary undertakings provoke questions at the basic level regarding the future of robotics and artificial intelligence (AI) in relation to biological systems, a point that underscores the revolutionary potential of evolutionary principles. By filling the gap between the two domains of biology and technology, ongoing research continues to extend evolutionary theory, revealing new uses that test our assumptions about intelligence and flexibility in natural as well as artificial systems (Bredeche et al., 2018; Hanisch & Eirdosh, 2020) (Figure 1.13).

Figure 1.13. The theory of evolution and natural selection.

Source: https://i.ytimg.com/vi/ADVZVk5ZD-E/hq720.jpg?sqp=-oaymwEhCK4FEIIDSFryq4qpAxMIARUAAAAAGAElAADIQj0AgKJD&rs=AOn4CLB57no0XNSKHpiBD9bNVRGZarMrsQ.

1.4.2. Mechanisms of Natural Selection

Natural selection, one of the major mechanisms of evolution, operates through differential survival and reproduction of individuals within a population based on inherited traits. The phenomenon is inherently linked to the concept of heritable variation, where differing traits contribute to an organism's fitness in an environment. As theorized by contemporary evolutionary theory, these differences can be accounted for by a set of factors that include biological, environmental, and social forces. Natural selection is not an isolation; rather, it competes with other processes of evolution, like genetic drift and gene flow, in shaping the evolutionary trajectory of species. The ever-evolving discussion around these mechanisms reflects their significance in biological study and learning, as researchers keep venturing into the unknown to comprehend the intricacies of evolutionary processes from fields as diverse as biology to economics and beyond (Hanisch & Eirdosh, 2020). Phenotypic plasticity, a critical aspect of evolutionary biology, allows organisms to adjust to varying environmental conditions without genetic modification (GM). This capacity can enhance fitness, as those with greater plasticity will be better equipped to make use of resources or fend off environmental stresses. The relationship between phenotypic plasticity and natural selection is complex because plastic traits are either positive or negative depending on the conditions in the environment. Past research has shown that plasticity not only serves the current demands of adaptation but may also have evolutionary implications in the long term, speculating on how the convergence of various concepts in evolutionary biology is thought through maps. Such frameworks shed light on the dynamic interactions between phenotypic diversity, plasticity, and natural selection, thus allowing for an integrated understanding of how species evolve to their environments (Martins et al., 2018). The interplay between ecological processes and natural selection emphasizes the need to place evolutionary mechanisms in context. Environmental fluctuations help dictate the selective forces that organisms face, resulting in specially adapted solutions to enhance survival and reproductive success. In their interactions with their environments, the influence of natural selection on populations extends beyond the individual to influence processes at the community level and diversity levels. These encounters open the door to the intricacy of ecological interactions and adaptive mechanisms in maintaining the balance of ecology. An understanding of natural mechanisms of selection not only enlightens evolutionary theory but also provides essential information on conservation and management interventions for preserving biological diversity in a dynamic world (Martins et al., 2018; Hanisch & Eirdosh, 2020) (Figure 1.14).

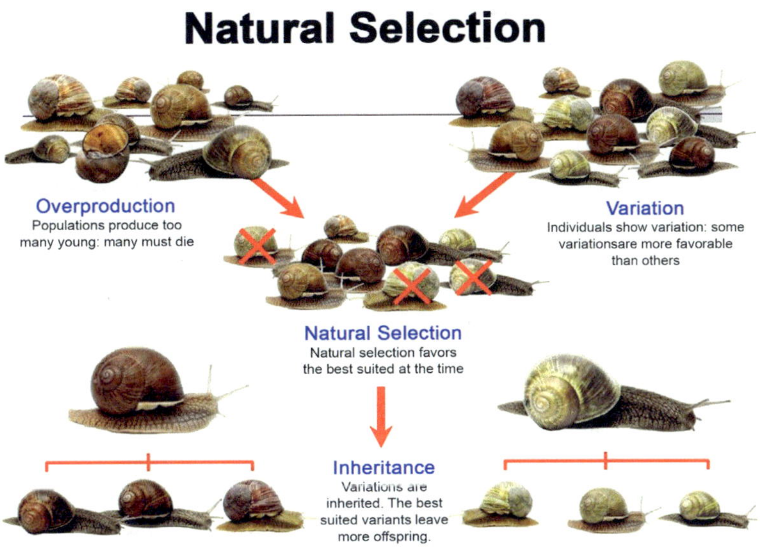

Figure 1.14. Mechanisms of natural selection.

Source: https://blogs.ubc.ca/mrpletsch/files/2019/10/Screen-Shot-2019-10-18-at-6.51.04-PM.png.

1.4.3. Evidence Supporting Evolution

The evolutionary theory is largely based on numerous pieces of evidence that have been obtained in different areas of biology which, when taken together, help to paint a picture of how life on earth has adapted. Among the strongest pieces of evidence is the fossil record, which provides evidence of the small changes that happen to species over a period of millions of years. Fossils show a distinct chronological sequence of the simple forms of life evolving to the complex forms, thus showing a chronology or record of biological change. This fact speaks in favor of the idea that species have adjusted to the environment, which resulted in the great variety of organisms that can be seen nowadays. Also, the radiometric dating methods are used to determine the ages of these fossils, which supports the time dimension of evolution and enables the scientists to determine the morphological and physiological changes that occurred in accordance with the environmental changes. Comparative anatomy and molecular biology form another major evidence for evolution. The study of homologous structures, body parts that have a similar structure, but differ in their functions in different species, indicates a shared ancestry of the species. As an example, the forelimbs of mammals, birds and reptiles have a similar structure of the bones, which means that they have common evolutionary roots. Further, the rise of molecular biology has given instruments to compare DNA sequences in a range of organisms, showing genetic similarities that highlight evolutionary relationships. Such genetic comparisons have forced the reexamination and refinement of conventional taxonomy, and have shown the profound relationships of life forms, and have supported the concept of descent with modification first proposed by Darwin. New disciplines, like social evolution and cooperation, also provide important information on how evolution works. The paradigm was always focused on competition, but recent studies highlight the significance and commonness of cooperative species interactions. Incorporation of social behaviors in evolutionary models has provided a complex dynamic that is difficult to accept in the traditional perspectives. Recent research indicates that cooperation similar to competition, is a crucial factor in species adaptation and survival. The modeling of such interactions within the framework of complex systems illustrates how biological entities not only compete but also collaborate, thereby enhancing their reproductive success and sustainability within ecosystems (DeSouza & Miramontes, 2014). Understanding these multifaceted relationships expands the traditional view of evolution and aligns with contemporary biological perspectives on the interconnectedness of life (Bacher et al., 2017).

Test Your Knowledge

1. Give an overview of evolutionary theory.
2. What are the mechanisms of natural selection?

1.5. Ecology and Ecosystems

Learning Outcome

1. Definition of ecology and its importance;
2. Components of ecosystems: biotic and abiotic factors;
3. Human impact on ecosystems and biodiversity.

It is via knowledge of ecology and ecosystems that the complex interactions that form the foundation of life on our planet are understood. Ecology, a scientific profession concerned with the interplay of organisms and their surroundings, has several levels of impact, from individual organisms to complex ecosystems. Ecosystems themselves consist of populations of living organisms interacting with physical environments to create dynamic systems with energy flow and cycling of nutrients. These interactions contribute to maintaining biodiversity, climate conditions, and the delivery of ecosystem services critical to the survival of human beings. Understanding the importance of ecological research from the science community, synthesis of findings into conservation to be utilized in order to preserve our planet's biodiversity and ecosystem's health is critical. In the past decade, the use of invasive plant control has taken off in ecological and restoration contexts. According to recent studies, despite the fact that restoration ecology mainly concerns restoring degraded ecosystems, invasive species management is often given less importance. Specifically, it has been revealed by research that just 8% of ecosystem restoration research was actually focused on managing alien invasions, while aliens normally outcompete native species and result in extensive ecological decline (Esler et al., 2012). The widespread use of mechanical control measures that has been witnessed indicates a lack of holistic restoration strategies, since the measures usually fail to include intrinsic ecological complexities.

The findings highlight the need for increased collaboration and communication between invasion ecologists and restoration practitioners to create holistic plans that look at symptoms and root causes of ecological decline. Climate change is another major driver affecting the health of ecosystems, particularly freshwater ecosystems, which are among the most vulnerable globally.

Recent studies show that climate change has significant impacts on the structure and functioning of freshwater ecosystems, making it essential for the assessment of species adaptive capacities (Allen et al., 2018). In addition, epigenetic mechanism research also has the potential to explain how species evolve to tolerate environmental change, particularly in the face of climatically induced stressors. This new science suggests that the deciphering of gene-environment dynamics on the molecular level can potentially provide useful conservation strategies for maintaining biodiversity under change.

Emphasizing the need for new research avenues, ecologists look to better understand how both epigenetic processes and traditional evolutionary responses interplay with each other to create resilience in freshwater ecosystems under climate change impacts (Figure 1.15).

Figure 1.15. Ecology and ecosystems.

Source: By Copyright (c) 2004 Richard Ling, CC BY-SA 3.0, https://commons.wikimedia.org/w/index.php?curid=78275

1.5.1. Definition of Ecology and Its Importance

Ecology is therefore the scientific study of interactions among organisms and their environment, both biotic and abiotic. Ecology investigates varying levels of biological organization, ranging from individuals and populations to groups and ecosystems, hence acting as a coherent framework for understanding the variety of life. By examining interactions between species and their environments, ecology sheds light on the forces that structure ecological communities and the complex web of life that maintains them. Beyond this, studies in ecology are also central to furthering our knowledge about biodiversity, ecosystem processes, and the likely effects of environmental change. Using such a perspective, the value of ecology becomes more apparent, whereby ecology informs conservation and sustainability to be able to sustain robust ecosystems despite anthropogenic activities and global climate change. The value of ecology goes beyond simple definitions and categorizations; it is essential for tackling key issues facing the world. One of the most well-known areas of research is biological invasions, particularly agricultural invasions, where non-native organisms can drastically alter ecosystems and push native biodiversity to extinction. The researchers emphasize reconstructing invasion routes through population genetics to manage the biological invasions effectively (Guillemaud et al., 2011). Ecological impacts of invasions in understanding yield more effective mechanisms for counteracting their impacts on agriculture, natural resources, and livelihoods. Therefore, ecology not only brings us closer to studying ecological interactions but also provides practical solutions to real-world problems, underscoring its importance in contemporary biology. Furthermore, the concepts of genetic diversity and intra-individual genetic heterogeneity (IGH) are also increasingly being discovered as key constituents in ecology. Investigating IGH gives insights into estimates of biodiversity and challenges reductive definitions of biological individuality and the intricacies of the relationship between genetic diversity and ecological functioning (Cárdenas-Flores et al., 2019). With this evolved understanding, ecologists are able to better appreciate associations between diverse biological individuals and their ecological roles. This leads to extensive knowledge of ecosystems and evolutionary processes associated with them. Lastly, the application of ecology is essential in promoting a greater appreciation for biodiversity's value, pushing for actions that ensure the integrity of ecosystems within an ever-changing world.

1.5.2. Components of Ecosystems: Biotic and Abiotic Factors

Ecosystems are complex assemblages of biotic and abiotic entities that interact to maintain life. Biotic components include all living things in an ecosystem, ranging from plants and animals to fungi and microorganisms. These organisms have several interactions that may take the form of predation, competition, or symbiosis, which play major roles in community structure and function. An example is the presence of a variety of plant species, which can add complexity to habitats and create niches for different animals and organisms. Conversely, abiotic factors—water, temperature, soil type, and sunlight—are the foundation for life, governing the physical environment upon which such biotic interactions occur. It is crucial to recognize how they intersect when studying ecological dynamics and stability of or alterations within ecosystems over time. The dynamics between biotic and abiotic elements can create unique ecological implications, particularly regarding ecosystem resilience. Resilience is a measure of the capacity of ecosystems to endure disturbances and yet not lose their fundamental functions. Experts have shown that biological diversity corresponds directly with greater system stability. For the sake of this argument, biodiversity encompasses species richness, but changes also include genetic and functional diversity. Greater diversity in species makes ecosystem processes such as nutrient cycling and energy flow more efficient and essential for sustaining life. Furthermore, biotic elements, through feedback mechanisms, can control abiotic factors. For example, plants can influence local climates through the alteration of temperature and humidity. The interaction among these factors ultimately determines the ecosystem services that underpin human societies, which underscores the importance of understanding these dynamics in environmental change, as discussed by Dyke & Weaver (2013). The importance of abiotic and biotic components extends beyond strict ecological interest; it is very relevant to human sustainability and environmental management. If urbanization and climate change threaten ecosystems, understanding interactions between organisms and physical environments becomes imperative. The biodiversity in soils, for instance, sustains a range of ecosystem services that are critical to human societies, including nutrient cycling, water filtration, and carbon sequestration. Robust soils depend on numerous organisms that contribute to their structure and fertility, so the significance of preserving biodiversity is even more reason to be undertaken as part of general environmental conservation efforts. The conclusions derived from examining these factors illuminate the way complex feedback loops of ecosystems may allow or restrict stability and functioning, aligning with the concerns articulated regarding the pivotal role embraced by biodiversity (Van der Putten et al., 2010) (Figure 1.16).

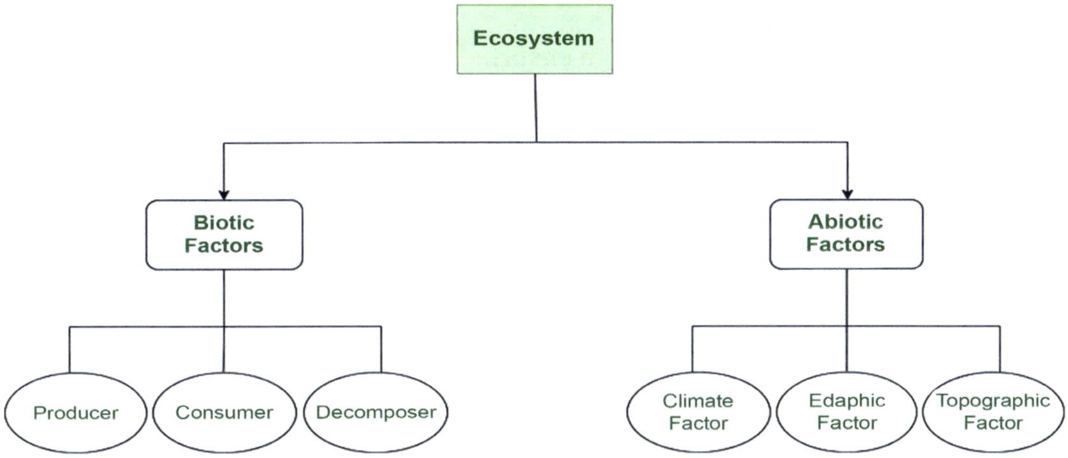

Figure 1.16. Components of ecosystems.

Source: https://media.geeksforgeeks.org/wp-content/uploads/20221220143326/Ecosystem.jpg.

1.5.3. Human Impact on Ecosystems and Biodiversity

Human activities and biodiversity are connected in a complex relationship that is characterized by a complex set of effects, including habitat destruction and climate change. The processes of urbanization, industrialization and expansion of agriculture have caused major changes in the natural ecosystems. Such alterations tend to lead to habitat fragmentation, whereby continuous habitats are divided into smaller and isolated fragments, which compromise the viability of different species. Such habitat loss not only endangers the existence of plants and animals but also decreases the genetic diversity, thus predisposing the populations to disease and alterations in the environment. By degrading ecosystems, humans compromise their resilience and the ecosystem services they provide, including clean air, water purification, and nutrient cycling, all critical for sustaining human life and activities (Van der Putten et al., 2010). Besides the direct destruction of habitat, climate change is another major human effect on biodiversity. The increased global temperatures and the changed precipitation patterns have extensive consequences to ecological communities. The change of the climate conditions may cause the redistribution of species, which means that organisms will have to adapt, to migrate, or risk extinction. The interplay between climate change and ecological dynamics was particularly notable during the Quaternary period, where fluctuations in climate influenced species evolution and extinction rates (Stewart, 2010). With species scrambling to keep up with the fast-changing environment, the general stability of the ecosystem is affected and therefore, the complex network of life on which mankind depends to sustain its existence is further threatened. Addressing climate change is, therefore, paramount for safeguarding biodiversity. Finally, biodiversity conservation is vital to the ecological well-being as well as to the maintenance of ecological systems that are viable to support human utilization and recreation. Soil biodiversity is at the basis of the many benefits that people get from ecosystems, including food, materials, and climate regulation. Soil organisms also enhance the fertility and sustainability of agricultural systems and there is a need to ensure that these invisible but very crucial ecosystem components are conserved. These vital ecosystem services can be impaired by the loss of soil biodiversity, which is why more attention and efforts should be raised towards conservation efforts. As human pressures on the environment continue to escalate, integrating sustainable practices and understanding the significance of biodiversity becomes critical to ensuring the health and viability of both ecosystems and human society (Van der Putten et al., 2010) (Figure 1.17).

Figure 1.17. Human impact on ecosystems and biodiversity.

Source: https://images.zapnito.com/users/17998/posters/aa9e5efcb5975cf74ba95450ec725e5e/deforestation_large.jpg.

Test Your Knowledge

1. What are the components of ecosystems: biotic and abiotic factors?
2. What is human impact on ecosystems and biodiversity?

1.6. Human Biology and Physiology

Learning Outcome

1. Overview of human body systems;
2. Homeostasis and its significance;
3. Common diseases and their biological basis.

Information on the intricacies of human physiology and biology is essential in knowing how the human body acts and responds to things around it. Human biology includes the study of the structure, functions, and processes of the cells that help maintain life. Physiology, on the other hand, delves into the processes through which these biological systems interact and maintain homeostasis. This complex dependency is crucial in understanding disease mechanisms, such as diabetes and assessments of health. For example, monitoring systems can contribute to the establishment of health disparities in some populations, as evident through the monitoring of students' eye health in Kazakhstan. Such initiatives highlight the importance of educational infrastructure in ensuring preventive health practices and awareness in the population (Beisembaevna et al., 2021). To this end, people's knowledge of human biology and physiology not only enhances scholastic progress but also enables individuals to become agents of healthy living and habits. Furthermore, the role of education in human biology cannot be exaggerated. University programs in biological sciences promote critical thinking as well as problem-solving abilities that future health practitioners will inevitably require. In these courses, students also participate in laboratory experiments and field studies that build their understanding of physiological principles. For instance, human physiology courses are central to preparing future teachers who will be imparting vital health-related information. These learning spaces also create spaces for extended professional growth among faculty, researchers, and practitioners of human biology. Also, there are specialized catalogs for organizing courses and programs that focus on water and water-related resources, which mostly overlap with human physiology since water plays a crucial role in body functions (Steele et al., 1998). The overlapping calls for multidisciplinary teaching in biology with the incorporation of theoretical knowledge and practice.

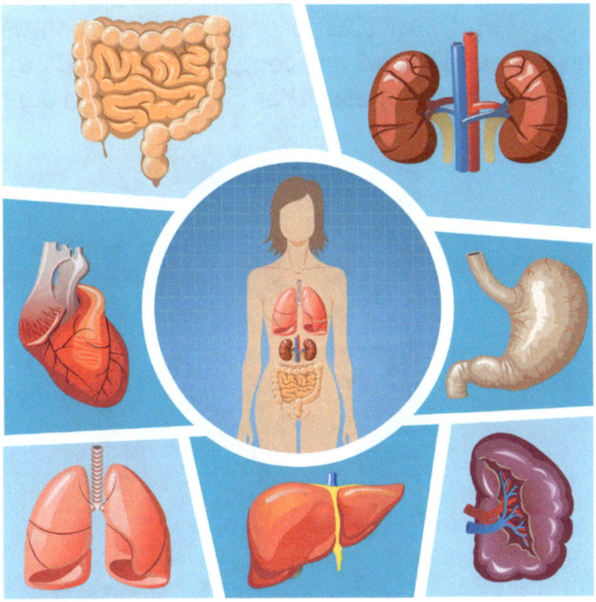

Figure 1.18. Human biology and physiology.

Source: https://cdn.entrance360.com/media/uploads/2019/07/15/shutterstock_1260845158.jpg.

As society grapples with creating solutions for newly arising health problems, human biology and physiology are of greater significance. Greater emphasis is placed on understanding the impact of the environment on human health, necessitating an education system that can adequately address the challenges. This is particularly observed within regions with health inequalities, where particular research, such as monitoring the visual domains of students, reveals critical information that supports intervention measures (Beisembaevna et al., 2021). This type of research not only indicates present health issues but also fosters a culture of continuous enhancement in healthcare delivery. Lastly, an integrative model of learning human biology and physiology can optimize the health literacy of students so that they can become well-educated advocates of their own health as well as the health of their communities and become part of an increasingly scientifically educated society (Figure 1.18).

1.6.1. Overview of Human Body Systems

The human body is an complex network of numerous systems cooperating harmoniously to maintain life and facilitate maximum functioning. Of these systems, the circulatory system is pivotal, delivering oxygen and nutrients to organs while simultaneously removing waste products. The heart, blood vessels, and blood are components of this system working together as a whole to facilitate the transportation of essential substances throughout the body.

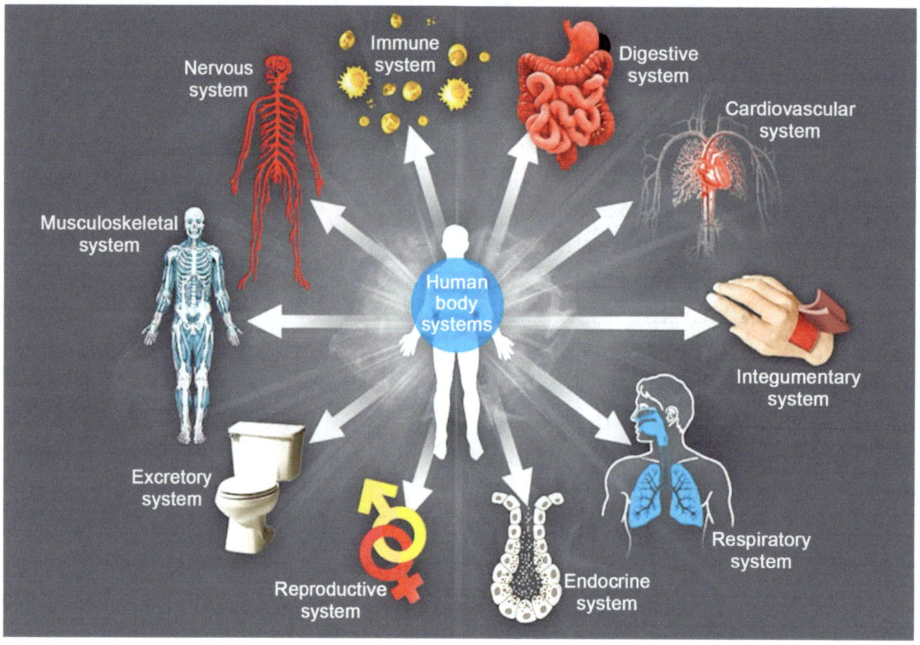

Figure 1.19. Human body systems.

Source: https://www.datocms-assets.com/117510/1722387503-the-body-s-systems20161104-16595-1aca5q9.jpg.

The operation of the circulatory system is crucial not only to support individual cells but also to enable complex interactions with other body systems. For instance, the circulatory system interacts with the respiratory system in exchange of gases, which reflects the interdependent nature of human biology and makes the need to regard these systems as an integrating concept in the study of physiology and disease amply clear (Newman et al., 2011). Alongside the circulatory system, the nervous system is the body's information highway, which makes possible rapid responses to stimuli and homeostasis. The nervous system includes the brain, spinal cord, and peripheral nerves and deciphers sensory information and controls voluntary and involuntary functions. The complexity of the nervous system allows it to influence and regulate many body functions, from heartbeats to digestion, by means of electrical impulses. In addition, developments in the understanding of this system have revealed significant insights into neural plasticity and its influence on learning and memory. Such a dynamical nature itself shows not only the body's capacity but also the important role played by the nervous system in maintaining overall well-being and responding to changes in the environment (Newman et al., 2011). Another critical component of the human body's system set is the integumentary system, which includes the skin, hair, nails, and associated glands. This system serves the body as its first line of defense against environmental threats and is crucial to protection against infection, temperature regulation, and prevention of dehydration. The capacity for healing and regeneration of the skin is important for general health, as well as acting as a barrier protecting against both physical and environmental damage while facilitating sensation by way of nerve endings. Moreover, recent advances in technology, including the creation of wireless body area networks (WBAN), represent the convergence of biology and technology in allowing ongoing observation of skin as well as other bodily processes. Such advances illustrate the shifting landscape of human biology, demonstrating how different systems collaborate to optimize health and well-being (Figure 1.19).

1.6.2. Homeostasis and Its Significance

Homeostasis is a key concept in biology that summarizes the processes by which organisms maintain an internal environment that is relatively stable despite external change. Stability is necessary for survival since it allows organisms to function optimally, independent of the circumstances. Man, for instance, maintains body temperature, blood acidity, and electrolyte levels with elaborate mechanisms by the use of hormones and feedback loops. These mechanisms demonstrate the body's ability to dynamically adjust to internal and external changes, thereby facilitating that physiological functions can occur effectively. Homeostasis not only preserves the integrity of bodily functions but also aids in higher biological systems, creating a consistent network in which health is preserved by precise regulating mechanisms, hence standing out in its significance in biological science as well as health science. The significance of homeostasis is not mere stability, it is an active, dynamic interaction among physiological processes that evolve across time. Homeostasis is not just a synchronic mechanism for preserving the status quo but also a diachronic system that dictates how organisms develop across generations. Understanding homeostasis in this intergenerational context uncovers the complex relationships forged through cellular signaling in development and repair processes. These findings emphasize the importance of a systems perspective of biology whereby homeostasis is not only an equilibrium state but as a dynamic and adaptive mechanism that evolved to deal with the dynamics of a continuously changing environment. This interdisciplinarity provides major insights into physiological processes and evolutionary modifications, further establishing its key position in biological sciences (Torday, 2015). Further, considering homeostasis in organizational closure terms also contributes to enhancing the profundity of understanding its complexity. The theory is based on interdependencies in various biological systems working in coordination to achieve homeostatic balance. Unlike what is described under traditional feedback mechanisms, organizational closure offers a more sophisticated description of how metabolic processes such as blood glucose control are achieved in an organization of structures with interdependent components. Through comprehension of such interactions as self-reinforcing constraints, scientists can build more detailed models to explain endocrine regulation and other homeostatic processes. Therefore, this alternative method not only argues against mainstream views but also opens a door for more complex mathematical modeling in biology, ultimately resulting in a deeper grasp of how organisms acquire and maintain homeostasis (Bich et al., 2020) (Figure 1.20).

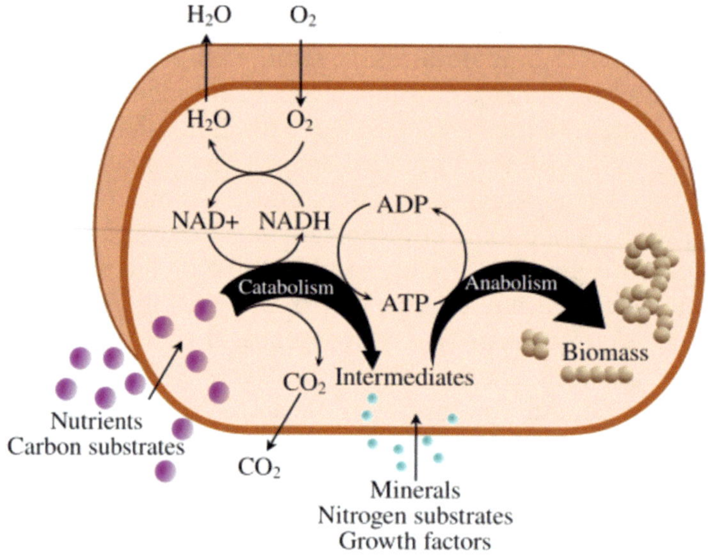

Figure 1.20. Homeostasis.

Source: By Д.Ильин: vectorization - File:Metabolism.png by Linares-Pastén, J. A., CC0, https://commons.wikimedia.org/w/index.php?curid=150135192

1.6.3. Common Diseases and Their Biological Basis

The biological basis of common diseases forms the basis of knowing the etiology of the disease, as well as the possible treatment. Complications like diabetes, high blood pressure and heart ailments have their roots in a complicated combination of hereditary, environmental and lifestyle choices. As an example, diabetes may develop due to insulin resistance or insulin deficiency, and it depends on genetic constitution and diet. In the same way, environmental stressors, including high sodium consumption and obesity, can aggravate hypertension in combination with genetic susceptibilities. It is important to acknowledge these multi-dimensional biological foundations since they influence the predisposition of individuals and require different prevention efforts and therapeutic approaches. In addition to improving the understanding of the disease process, this strategy will highlight the need to have an interdisciplinary approach between geneticists, nutritionists, and medics in coming up with informed health policies and interventions. The other essential feature of the common diseases is their infectious nature and the central role in it plays microorganisms. Examples of infectious disease agents that cause a significant threat to health include influenza, tuberculosis, and HIV/AIDS, which use biological systems to gain and sustain infection. The infectious disease ontology (IDO) serves as a comprehensive framework for understanding these complex interactions, providing well-defined categories and relationships relevant to pathogens and diseases (Babcock et al., 2021). Through disease classification and the determination of biological determinants of diseases, scientists may understand infection and host response mechanisms in a better way. Such a systematic process contributes to the creation of diagnostic tools, vaccines and antiviral agents that can limit the effects of infectious diseases on the health of the population. Thus, a biological approach to studying diseases can not only guide treatment but also allow us to improve our prevention and reaction to possible outbreaks in the future. The new diseases and the biology behind them pose further threats to the population and highlight the importance of a well-developed research and data-sharing infrastructure. A recent and emotional illustration of how quickly changing pathogens can alter societal norms is the current COVID-19 pandemic. The progress of biological research, including genomics, bioinformatics, etc., leads to better comprehension of the virological behavior, transmission, and mutation. For instance, the creation of specific ontologies such as the coronaviruses infectious disease ontology (CIDO) has facilitated more effective information-sharing among researchers and healthcare providers (Babcock et al., 2021). Such ontological structures encourage data discovery and reproducibility in infectious disease research, which is critical to developing effective population health responses to not only the present but also the future health crises. Biology, through the incorporation of data sciences in biological research, can transform to address the challenges of the emerging infectious diseases and lead to the development of innovative solutions that can increase the resilience of the health of the population.

Test Your Knowledge

1. What re the human body systems?
2. What is homeostasis and its significance?

Summary

The field of biology is a pillar that gives the basic foundation for comprehending the complexities of life. By incorporating data from various departments, such as genetics, ecology, and evolutionary biology, the students have an in-depth view of living organisms. Biology outlines how various species adapt and thrive in their environments, displaying the complex interactions that sustain ecosystems. As we have learned, biological principles extend beyond the classroom into key areas such as medicine, conservation, and biotechnology. Not only does this emphasize short-term usefulness and enhance education, but also teaches future generations their responsibility towards ecological stewardship and scientific discovery. Therefore, emphasizing the interconnectedness of living organisms requires future generations to respect and preserve biodiversity, further integrating biology into the framework of global sustainability initiatives. The educational journey through biology ultimately reveals the dynamic processes behind life on Earth. As students learn about cellular operation and evolutionary forces, they discover the underlying tendencies that govern biological systems. The rigorous scientific procedures employed in biological research delineate the importance of empirical evidence and critical thinking. Considering recent theoretical and research evidence, for instance, regarding neural structures and mental schemata, once again evidences the degree to which psychology and neuroscience overlap with biology. Such cross-disciplinary experience contributes to a broader understanding of sophisticated concepts, for instance, the nature of meaning and thought, as explored within the broad field of cognitive theory (Gelepithis et al., 2009). Therefore, students are not only taught facts but also the analytical skills necessary for more advanced scientific investigation. Besides, since the need for training across disciplines persists to be more and more recognized, the role of biology in constructing an integral educational strategy is hardly overstated. Interactions between scientific disciplines, e.g., neuroinformatics, highlight the vast potential for innovation and scientific progress (Willshaw, 2011). The integration of technological tools and techniques enhances the research into life processes, opening up new avenues for research and understanding. The construction of specialized graduate programs that involve all aspects of biology ensures that students are well equipped with the depth of modern scientific issues. Therefore, final thoughts on the biology study highlight its essential role in both personal development and the overall development of society, calling for further interaction with the life sciences in schools and in practice.

Biology, which is one of the central pillars of natural sciences, offers a broad category of concepts that are fundamental in knowledge about life and living organisms. The prominent concepts that fall in this category include evolution principles, cellular structure and function, genetics, and ecology. Evolution represents the core of biological sciences and provides insight into why diversity occurs in life on Earth. Explanation of cellular structure is also crucial as it directs to the complex framework which enables organisms to perform optimally. Cells, the building blocks of life, perform rudimentary functions from metabolism to communication, and the study of their characteristics is paramount in any biological inquiry. Moreover, the principles of genetics examine heredity and variation, detailing how characteristics are inherited across generations, giving meaning to the tale of biological evolution. It also relies on knowledge of systems and interactions within organisms and their environment. Ecology, which is the study of interactions among organisms and their environment, emphasizes ecosystems and diversity. It unfolds not only the interactions among living organisms but also adaptations to various environmental changes. Further, the concept of threshold concepts in biology can illuminate student identification of troublesome knowledge that is troublesome but revolutionary in attaining their disciplines, as examined by research into learning encounters from various disciplines like Geography and Medical Sciences (Jamieson et al., 2011). Focusing on troublesome knowledge is involved in a stronger understanding of core biological principles, illuminating the interconnectedness of knowledge in the discipline. Also, the meaning notion in biological contexts, as cognitive frameworks depict, is a core area of research. Research into communication, thinking, and understanding in biological terms points to the psychological processes underlying the basis of information conception and exchange. This connection takes a focal role in building a cohesive understanding of complex biological processes, such as the inter-level structures of primitives that structure cognitive engagement with biological substance (Gelepithis et al., 2009). Central concepts in biology thus not only define what is fundamental in terms of concepts but also prescribe the methods by which students and experts engage with the discipline. This multifaceted understanding of biology's key components operates to drive future investigation and application across several scientific disciplines and demonstrates the interactive nature of biological research and its relevance in a larger context.

KEY TERMS FROM THIS CHAPTER

Adaptation	Autotroph	Biodiversity
Biology	Cell	Chromosome
Deoxyribonucleic Acid)	Ecology	Evolution
Gene	Genetics	Heterotroph
Homeostasis	Metabolism	Mutation
Natural selection	Organelle	Organism
Reproduction	Species	

Critical Thinking Questions

1. Which of the following best defines biology?

a. The study of rocks and minerals
b. The study of celestial objects
c. The study of living organisms
d. The study of past human cultures

2. Which of the following is not a characteristic of living organisms?

a. Reproduction
b. Growth and development
c. Inability to adapt
d. Response to stimuli

3. The basic unit of life is the:

a. Atom
b. Cell
c. Molecule
d. Organ

4. Which of the following levels of biological organization is the most complex?

a. Tissue
b. Organ
c. Cell
d. Organism

5. Which branch of biology deals with the classification of organisms?

a. Ecology
b. Genetics
c. Taxonomy
d. Physiology

Short Answer Questions

1. What is biology?
2. Define biology and explain what it studies.
3. List the characteristics of living things.
4. What are the levels of biological organization from smallest to largest?
5. What are the three domains of life?

Answers to Critical Thinking Questions

1. (c); 2. (c); 3. (b); 4. (d); 5. (c)

References

1. Smýkal, P., K Varshney, R., K Singh, V., Coyne, C. J., Domoney, C., Kejnovský, E., & Warkentin, T. (2016). From Mendel's discovery on pea to today's plant genetics and breeding : Commemorating the 150th anniversary of the reading of Mendel's discovery. *TAG. Theoretical and applied genetics. Theoretische und angewandte Genetik*, *129*(12), 2267–2280. https://doi.org/10.1007/s00122-016-2803-2 (accessed on 13th June 2025).
2. Bredeche N, Haasdijk E and Prieto A (2018) Embodied Evolution in Collective Robotics: A Review. Front. Robot. AI 5:12. doi: 10.3389/frobt.2018.00012 (accessed on 13th June 2025).
3. Allen, A., Allendorf, F., Alley, R., Andrewartha, H., Anway, M., Araújo, M., Ardura, A., et al. (2018). Synthesizing the role of epigenetics in the response and adaptation of species to climate change in freshwater ecosystems. https://core.ac.uk/download/158347914.pdf (accessed on 13th June 2025).
4. Altintas, I., Chen, R., Crawl, D., Ellisman, M., et al. (2014). EPiK–a workflow for electron tomography in Kepler. https://core.ac.uk/download/323064369.pdf (accessed on 13th June 2025).
5. Thomas Guillemaud, Marc Ciosi, Eric Lombaert, Arnaud Estoup. Biological invasions in agricultural settings: insights from evolutionary biology and population genetics. Comptes Rendus. Biologies, 2011, 334 (3), pp.237-246. ⟨10.1016/j.crvi.2010.12.008⟩. ⟨hal-02646885⟩ (accessed on 13th June 2025).
6. Babcock, S., Beverley, J., Cowell, L. G., Smith, T., et al. (2021). The infectious disease ontology in the age of COVID-19. https://core.ac.uk/download/305120399.pdf (accessed on 13th June 2025).
7. Bacher, S., Blackburn, T. M., Essl, F., Genovesi, P., et al. (2017). Socio-economic impact classification of alien taxa (SEICAT). https://core.ac.uk/download/286567949.pdf (accessed on 13th June 2025).
8. Beisembaevna, A. S., Ibadullayeva, S. Z., & Karabalayeva, A. B. (2021). Monitoring of the visual system of students in the southern region of Kazakhstan and the implementation of the results in the methods of teaching biology. https://core.ac.uk/download/427151471.pdf (accessed on 13th June 2025).
9. van der Putten, W. H., Mudgal, S., Turbé, A., de Toni, A., Lavelle, P., Benito, P., & Ruiz, N. (2010). *Soil biodiversity: functions, threats and tools for policy makers.* Bio Intelligence Service. http://ec.europa.eu/environment/archives/soil/pdf/biodiversity_report.pdf (accessed on 13th June 2025).
10. Berlanga, M., & Guerrero, R. (2010). The hidden side of the prokaryotic cell: Rediscovering the microbial world. https://core.ac.uk/download/159084328.pdf (accessed on 13th June 2025).
11. Bich, L., Mossio, M., & Soto, A. M. (2020). Glycemia Regulation: From Feedback Loops to Organizational Closure. *Frontiers in physiology*, *11*, 69. https://doi.org/10.3389/fphys.2020.00069 (accessed on 13th June 2025).
12. Ullah, S., Higgins, H., Braem, B., Latre, B., Blondia, C., Moerman, I., Saleem, S., Rahman, Z., & Kwak, K. S. (2012). A comprehensive survey of Wireless Body Area Networks : on PHY, MAC, and Network layers solutions. *Journal of medical systems*, *36*(3), 1065–1094. https://doi.org/10.1007/s10916-010-9571-3 (accessed on 13th June 2025).

13. Borja, E. A., & Mutya, R. C. (2024). Enhancing student's conceptual understanding on the patterns of Mendelian genetics through task-based learning. *JPBI (Journal Pendidikan Biology Indonesia.* https://www.semanticscholar.org/paper/4669c7054d1c6e26679f742e56978751d9f6fa3d (accessed on 13th June 2025).

14. Bos, S. (2016). *Prokaryotic and Eukaryotic Cells (6th Grade).* https://core.ac.uk/download/216388565.pdf (accessed on 13th June 2025).

15. Bourdon, J., & Eveillard, D. (2007). Toll based measures for dynamical graphs. https://core.ac.uk/download/53022132.pdf (accessed on 13th June 2025).

16. Camfield, E., & Land, K. (2017). The Evolution of Student Engagement: Writing Improves Teaching in Introductory Biology. *UC Merced.* Retrieved from https://escholarship.org/uc/item/0d55d7wb (accessed on 13th June 2025).

17. Cárdenas-Flores, A., Roger, A., Roger, A., Gonzalez, A. V., Rinkevich, B., Rinkevich, B., Santelices, B., et al. (2019). Considering intra-individual genetic heterogeneity to understand biodiversity. https://core.ac.uk/download/286357369.pdf (accessed on 13th June 2025).

18. DeSouza, O., & Miramontes, O. (2014). Social evolution: New horizons. http://arxiv.org/abs/1404.6267 (accessed on 13th June 2025).

19. Dowse, H. B. (2000). A new course promoting science as a way of knowing. https://core.ac.uk/download/217120917.pdf (accessed on 13th June 2025).

20. Dyke, J. G., & Weaver, I. S. (2013). The emergence of environmental homeostasis in complex ecosystems. https://core.ac.uk/download/18935006.pdf (accessed on 13th June 2025).

21. Hanisch, S., & Elrdosh, D. (2020, March 27). Conceptual clarification of evolution as an interdisciplinary science. https://doi.org/10.35542/osf.io/vr4t5 (accessed on 13th June 2025).

22. Esler, K. J., Fisher, J. L., Gaertner, M., Sharma, G. P., et al. (2012). Insights into invasion and restoration ecology: Time to collaborate towards a holistic approach to tackle biological invasions. https://core.ac.uk/download/18325893.pdf (accessed on 13th June 2025).

23. Fomovska, A., Topolska, A. E., Lige, B., Shen, B., Kafsack, B. F., Luft, B. J., Lord, C., et al. (2013). Protein trafficking through the endosomal system prepares intracellular parasites for a home invasion. https://core.ac.uk/download/19785875.pdf (accessed on 13th June 2025).

24. Frietsch, Rainer & Meng, Yu, 2010. "**Indicator-based reporting on the Chinese innovation system 2010: Life sciences in China**," Discussion Papers "Innovation Systems and Policy Analysis" 26, Fraunhofer Institute for Systems and Innovation Research (ISI).(accessed on 13th June 2025).

25. Gare, A. (2013). Overcoming the Newtonian paradigm: The unfinished project of theoretical biology from a Schellingian perspective. https://core.ac.uk/download/131214004.pdf (accessed on 13th June 2025).

26. Gelepithis, P. A. M. (2009). Outline of a new approach to the nature of mind. https://core.ac.uk/download/pdf/87224.pdf (accessed on 13th June 2025).

27. Hassan, S. S., Choudhury, P. P., Guha, R., Chakraborty, S., & Goswami, A. (2012). DNA sequence evolution through Integral Value Transformations. *Interdisciplinary sciences, computational life sciences*, *4*(2), 128–132. https://doi.org/10.1007/s12539-012-0103-3Hall, G. M., & Howe, J. (2012). The impact of synthetic biology in chemical engineering–Educational issues. https://core.ac.uk/download/6112945.pdf (accessed on 13th June 2025).

28. Hofner, B., Lawo, J. P., Müller, T., Repsilber, D., et al. (2012). Biologists meet statisticians: A workshop for young scientists to foster interdisciplinary teamwork. http://arxiv.org/abs/1208.5597 (accessed on 13th June 2025).

29. Hürtgen, D. (2018). Towards synthetic life: Establishing a minimal Segro some for the rational design of biomimetic systems. https://core.ac.uk/download/287653452.pdf (accessed on 13th June 2025).

30. Ison, R., & Schlindwein, S. L. (2004). Human knowing and perceived complexity: Implications for systems practice. https://core.ac.uk/download/20.pdf (accessed on 13th June 2025).

31. Jamieson, S., MacAndrew, S. B. G., & Spedding, N. (2011). How was it for you? A cross-disciplinary study of 'troublesome knowledge' as identified by undergraduate students and lecturers in geography, medical science and psychology. Report for Grduates for the 21st Century, Quality Enhancement Theme, HE Academy Scotland. https://core.ac.uk/download/228176308.pdf (accessed on 13th June 2025).

32. Layer, P. G. (2011). "Life" shaped by genes that depend on their surrounds. https://core.ac.uk/download/131210435.pdf (accessed on 13th June 2025).

33. Leil, T., & Ermakov, S. (2015). Editorial: The emerging discipline of quantitative systems pharmacology. *Frontiers in Pharmacology, 6.* https://www.semanticscholar.org/paper/106b67ef704a4c75d29665fb72dbf160b949a192 (accessed on 13th June 2025).

34. Lilley, E., Stanford, S. C., Kendall, D., Alexander, S. P. H., Cirino, G., Docherty, J. R., George, C. H., et al. (2020). ARRIVE 2.0 and the *British Journal of Pharmacology*: Updated guidance for 2020. *British Journal of Pharmacology, 177*, 3611–3616. https://doi.org/10.1111/bph.15178 (accessed on 13th June 2025).

35. Mansouri, V., & Parker, N. (2025). Brief introduction of molecular biology research. Brief Introduction of Molecular Biology Research. (2025). *Life Studies*, *1*(1), 1-22. https://doi.org/10.71204/73w5en60(accessed on 13th June 2025).

36. Martins, R. P. (2018). The conceptual structure of evolutionary biology: A framework from phenotypic plasticity. https://core.ac.uk/download/286139307.pdf (accessed on 13th June 2025).

37. Moghaddasi H, Rezaei-Tavirani M, Samadi L, The Role of Systems Biology and Synthetic Biology in Appearing and Managing COVID-19.J Cell Mol Anesth.2020;6(1):e149669.https://doi.org/10.22037/jcma.v6i1.33152.(accessed on 13th June 2025).

38. Newman, M. E. J. (2011). Complex systems: A survey. http://arxiv.org/abs/1112.1440 (accessed on 13th June 2025).

39. Ravarani, C. N. J., Erkina, T. Y., De Baets, G., Dudman, D. C., Erkine, A. M., & Babu, M. M. (2018). High-throughput discovery of functional disordered regions: Investigation of transactivation domains. *Molecular Systems Biology, 14.* https://doi.org/10.15252/msb.20188190 (accessed on 13th June 2025).

40. Redish, E. F., Bauer, C., Carleton, K. L., Cooke, T. J., & Cooper, M. (2014). NEXUS/Physics: An interdisciplinary repurposing of physics for biologists. https://core.ac.uk/download/73341621.pdf (accessed on 13th June 2025).

41. Shivaleela, C., Nandyal, B. V., LakshmiPrabha, S., & Kumar, G. V. (2019). A comparative study of dermatoglyphic patterns of fingers in normal children and in children with congenital

heart diseases. *Academia Anatomica International.* vol. 5, no. 1, 2019, pp. 82–86. https://www.semanticscholar.org/paper/a5832e91fbaad61c8c2d78a7b8b2c9765ea8f075 (accessed on 13th June 2025).

42. Steele, K. F. (1998). *Directory of Water Related Courses Offered at Colleges and Universities in Arkansas as of November 1998.* https://core.ac.uk/download/145191276.pdf (accessed on 13th June 2025).

43. Stewart, J. R. (2010). Understanding evolutionary processes during past Quaternary climatic cycles: Can it be applied to the future? https://core.ac.uk/download/4898204.pdf (accessed on 13th June 2025).

44. Sundaryono, A., Karyadi, B., & Nursaadah, E. (2021). Penguatan aspek-aspek nature of science and technology dalam pembelajaran bagi MGMP IPA Kabupaten Kaur Provinsi Bengkulu. https://core.ac.uk/download/573888742.pdf (accessed on 13th June 2025).

45. Szklarczyk, D., Gable, A. L., Nastou, K., Lyon, D., Kirsch, R., Pyysalo, S., Doncheva, N. T., et al. (2020). The STRING database in 2021: Customizable protein–protein networks, and functional characterization of user-uploaded gene/measurement sets. *Nucleic Acids Research, 49*, D605–D612. https://doi.org/10.1093/nar/gkaa1074 (accessed on 13th June 2025).

46. Tamargo, I. A., Baek, K., Kang, D., Kim, Y., Tsuji, S., & Jo, H. (2023). Abstract 18133: Inducible, endothelial cell-specific knockout of *Heg1* exacerbates atherosclerosis development in mice. *Circulation.* https://www.semanticscholar.org/paper/c6f4d4b28fc97efe61361c9b086dcfa987e7eab2

47. Kennedy, T.J., & Henderson, S. (2003). The GLOBE Program: Bringing Together Students, Teachers and Scientists to Increase Scientific Understanding of the Earth through Research. *Children, Youth and Environments 13*(2), 217-227. https://dx.doi.org/10.1353/cye.2003.0039. (accessed on 13th June 2025).

48. Torday J. S. (2015). Homeostasis as the Mechanism of Evolution. *Biology*, *4*(3), 573–590. https://doi.org/10.3390/biology4030573V'kovski, P., Kratzel, A., Steiner, S., Stalder, H., & Thiel, V. (2020). Coronavirus biology and replication: Implications for SARS-CoV-2. *Nature Reviews Microbiology, 19*, 155–170. https://doi.org/10.1038/s41579-020-00468-6 (accessed on 13th June 2025).

49. Williams, N. (2020). Science and technology. In *Mark Twain in Context.* https://www.semanticscholar.org/paper/2acae5e0b100863f717016118740a51e75a6fa09 (accessed on 13th June 2025).

50. Willshaw, D. (2011). INCF workshops on needs for training in neuroinformatics: Extended and short course provision. 2nd and 3rd INCF Workshops on "Needs for Training in Neuroinformatics: Extended and Short Course Provision" . https://core.ac.uk/download/pdf/289803.pdf (accessed on 13th June 2025).

51. Younas, A. (2020). Finding a consensus between philosophy of applied and social sciences: A case of biology of human rights. https://core.ac.uk/download/323306547.pdf (accessed on 13th June 2025).

THE CHEMISTRY OF LIFE

2

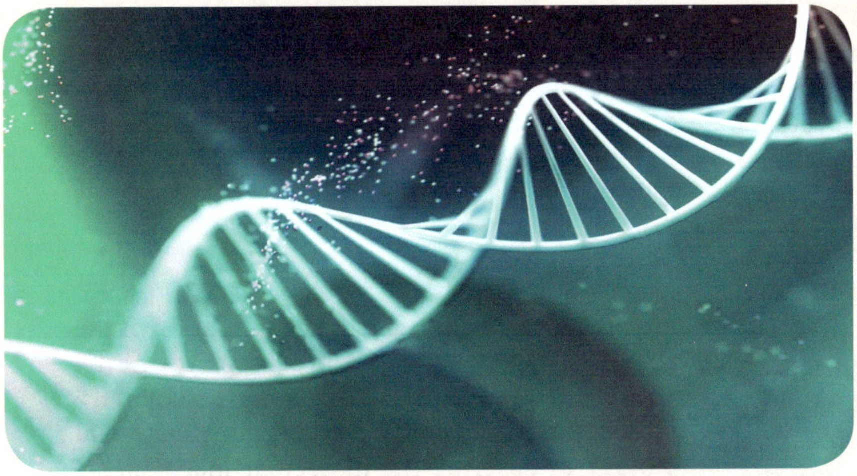

LEARNING OBJECTIVES

After studying this chapter, you will be able to:

1. Illustrate the building blocks of life;
2. Describe enzymes and metabolism;
3. Discuss about cellular respiration;
4. Explain photosynthesis;
5. Define genetic information and DNA;
6. Explore biochemical signaling.

Introduction

In studying complicated relationships within biological systems, one recognizes that life's chemistry is a building block of all living organisms. It encompasses a staggering quantity of interactions on the molecular level, connecting ideas from a multitude of sciences, including physics, chemistry, and biology. The importance of such interactions is underscored by the progress in fields such as nanotechnology, where research is rapidly shifting to reveal new data about the molecular structure upon which life exists. The journal set that provides a snapshot of nanoscience and nanotechnology on the interfaces among applied physics, chemistry, and the life sciences is growing rapidly with the introduction of new journals. The corresponding contributions of nations are expected to fluctuate with the interpretations of the corresponding interfaces between sets of journals (Leydesdorff et al., 2008). This interdisciplinarity is critical in understanding the complex net of reactions and interrelations that govern life processes. Core to life chemistry is the unusual character of elements making up biological molecules. While carbon is constantly lauded as the foundation of the chemistry of life due to its unparalleled ability to synthesize such a diverse range of stable compounds, the role of metal ions cannot be overestimated. There are more than 20 metals within the human body, and the majority of them have significant functions in many biochemical processes. These metals confer special catalytic activities, facilitate electron transfer, and stabilize structural components of proteins and enzymes. Life is dependent upon metals. While carbon, in terms of abundance and versatility, is THE element of life, the vast diversity and richness of chemistry found in living systems would be impossible to achieve without metal ions (Freisinger et al., 2019).

This collaboration of metals and carbon illustrates the biochemical intricacy of pathways and the significance of familiarity with these components in understanding more profound principles of living systems. Besides, the study of life chemistry is beyond the identification of significant elements; it points toward the significance of the interaction of biological macromolecules and their surroundings. By observing how these molecules respond to chemical stimuli, scientists can analyze mechanisms that drive metabolic processes, signal cascades, and genetic regulation. The clarification of these interactions not only increases our insight into the fundamental processes of life but also offers new avenues for therapeutic development. Science continues to unveil the dynamic richness of biological chemistry, with progress in areas such as synthetic biology and the design of medicines. As we advance toward a deeper understanding of the complex interactions of the biological realm, increasingly we need an interdisciplinary research program that coherently brings together chemistry, physics, and biology in a tale of life itself (Figure 2.1).

Figure 2.1. The chemistry of life biology.

Source: https://classful.com/wp-content/uploads/2022/06/62bb66e8e10819865490001656448754958-Biochem.jpg.

Definition of Biochemistry

Biochemistry is basically the examination of the chemical compounds and processes occurring within living organisms. This inter-disciplinary science fuses elements of both chemistry and biology to describe the molecular mechanisms of physiological functions. Through the examination of the structure and dynamics of biomolecules such as proteins, nucleic acids, carbohydrates, and lipids, biochemistry seeks to shed light on the behavior of such entities and how they interact with each other to sustain life. By enzymatic processes and metabolic pathways, biochemistry accounts for how living things convert energy and maintain homeostasis and describes the fine balance of life at the molecular level. Biochemistry also involves how biochemical patterns differ with changes in physiological states, thus relating molecular structure to molecular function in health and disease. The role of metabolism in biochemistry is fundamental since it basically controls the energetic and material exchanges that support life. Metabolism can be described as a complex network of biochemical reactions structured into modular units typified by specific chemical substrates and pathways. Such an organization enables living organisms to utilize energy from their environment to power vital processes such as growth, reproduction, and cellular maintenance. As emphasized in current research, not only is metabolic organization modular but also hierarchical and shows a profound relationship between metabolism and the broader systems of life, including cellular integration and ecological interactions (Braakman & Smith, 20132). In its comprehension of these metabolic modules and their evolutive significance, biochemistry gains a complex perspective on how life has evolved over time to fill various ecological niches. Besides, the informative character of biochemistry is crucial to understanding life's specific characteristics. Organisms are defined by complex information-handling capabilities governing their development, behavior, and response to the environment. The complexity arises from a series of causative interactions whereby higher-level biological processes influence the cell dynamics of molecular interactions (Cairns-Smith et al., 2012). The unique dynamics and architecture of biomolecules thus allow for context-dependent causality, emphasizing the interconnection between biochemistry and large-scale evolutionary histories. Through the consideration of these hierarchical architectures, biochemistry not only explains contemporary biological phenomena but also predicts new lines of inquiry, particularly concerning the origins of life and the workings of living systems. With this viewpoint, biochemistry is an indispensable pillar of our understanding of the chemistry of life, revealing the underlying principles that control the universe.

Importance of Chemistry in Biological Processes

The intricacy of biological processes is intrinsically rooted in chemistry, underscoring the pivotal role that concepts of chemistry play in life. Cellular processes are determined by the interactions of molecules since biochemical reactions are substantially founded upon chemical bonds and molecular shape. A case in point involves the solubility of organic compounds determining nutrient transport and absorption within organisms, as these properties determine how substances operate in highly complex biological systems. It is crucial to comprehend the mechanisms of these interactions because they do not just support life but also support ecological systems. Research on how solubility and polarity impact biological function can better explain the delicate balance in ecosystems and the significance of biochemical pathways that allow life to persist, making evident the profound connection between chemistry and biology (Cláudia et al., 2013). Also, the use of chemistry in biological processes is reflected in medicine and environmental science, and the significance of chemistry has its place in all fields. Biochemical reactions play an integral role in drug development, where the interaction between drugs and biological targets is an explicit demonstration of chemical principles. The molecular design of drugs typically relies heavily upon a thorough understanding of chemical interaction, solubility, and stability for efficacy and safety of therapeutic application. Environmental biochemistry also addresses the chemical underpinnings of ecological interactions and pollutant fate, with a focus on the role that these chemical processes have in sustaining health. The Water-based courses within curricula also imply the interdisciplinary nature of the studies as they cross over the boundaries between chemistry, biology, and environmental science (Steele et al., 1998). Broadly, the contribution of chemistry to biological processes implies the interconnected nature of all living things and the environment around them. The biochemical makeup of organisms supports complex processes such as metabolism, cellular respiration, and photosynthesis that are essential to generate energy and develop. The processes depict how simple chemical reactions not only drive life at the cellular level but also influence planetary biogeochemical cycles. Moreover, the uses of studying such chemical interactions are combatting global warming and preserving biodiversity, making chemistry an effective tool in addressing some of the most pressing concerns that the world is facing today. Above all, the understanding of the contribution of chemistry to biological processes increases the appreciation of life itself, and the necessity to further research and study this field is reinforced (Figure 2.2).

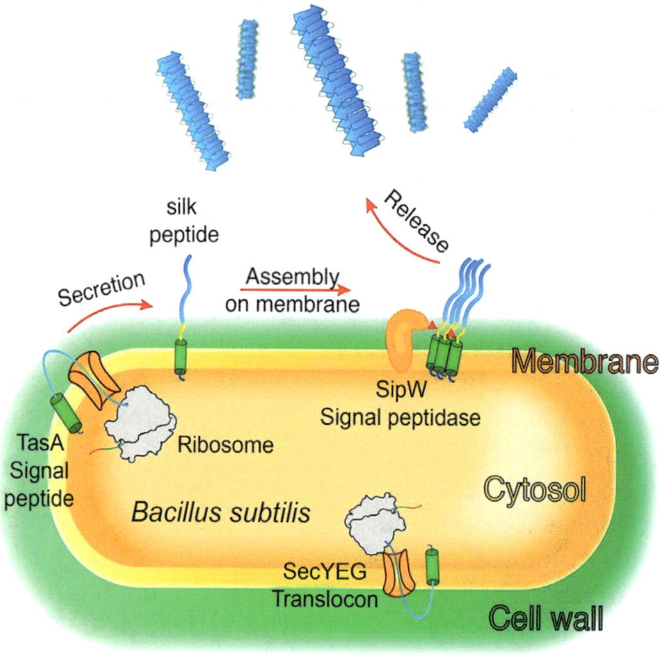

Figure 2.2. Chemistry in biological processes.

Source: https://chemistry.jhu.edu/wp-content/uploads/sites/72/2024/03/friedlab2024.jpg.

2.1. The Building Blocks of Life

Learning Outcome

1. Atoms and molecules in living organisms;
2. The role of water in biological systems;
3. Macromolecules: proteins, carbohydrates, lipids, and nucleic acids.

The biomolecules that are the fundamental building blocks of life may be divided into four fundamental classes: carbohydrates, proteins, lipids, and nucleic acids. The different classes have individual but connected roles in biological processes. An example is carbohydrates, which are mainly concerned with energy storage and provision, mainly in the form of glucose and starch. They also play structural functions in the cell walls, particularly in plants where cellulose is the core. Amino acid sequences form proteins that are essential in catalyzing biochemical reactions, supporting and maintaining structure, and the regulating of cellular processes. Lipids serve as a concentrated energy storehouse and also as major constituents of cell membranes. Lastly, nucleic acids, such as DNA and RNA, are required to store, transmit and express genetic information, and are essentially the set of instructions to construct and operate living organisms. The complexity of the interaction of these biomolecules is supported by the fact that it is through the chemical interactions that these biomolecules are able to facilitate numerous processes that sustain life. As an example, the proteins catalyze enzymatic reactions that frequently require carbohydrate substrates or lipid cofactors to proceed effectively, resulting in the synthesis or degradation of essential metabolites.

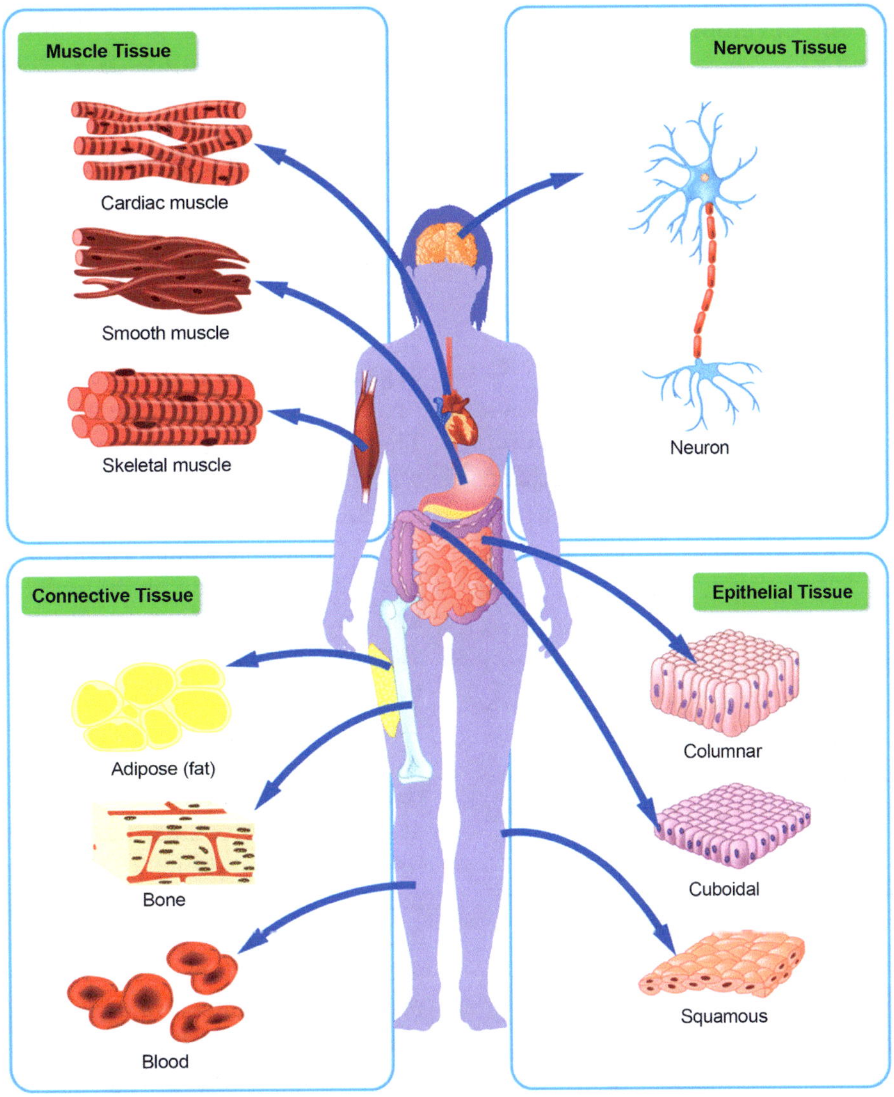

Figure 2.3. Building blocks of life.

Source: https://dr282zn36sxxg.cloudfront.net/datastreams/f-d%3Ab62670b421748527892c0221319f55238f3ec14dd6e708c91fd93ebd%2BIMAGE_TINY%2BIMAGE_TINY.1.

Further, the construction of more complicated structures, such as cell membranes, where lipids organize themselves to form a barrier through which the passage of substances in and out of the cell is controlled, which is closely related to the proteins incorporated into these membranes. With the continued research, developments in bioinformatics and cheminformatics are showing the potential of synthesizing new biomolecules and being able to comprehend their interaction on a new level. Innovative approaches, including those highlighted in the LOTUS initiative, aim to unify disparate molecular data, fundamentally reshaping how we interpret and utilize biomolecular information in life sciences, thereby enhancing our understanding of the chemistry of life (Rutz et al., 2022). Recent developments in AI and machine learning (ML) are further propelling our exploration of the building blocks of life. By offering the ability to forecast the characteristics of molecules, identify previously unknown reaction pathways and create new ones, these technologies are changing the very landscape of chemical research. Traditional molecular representations, such as Smiles, though valuable, often yield limitations in interpreting complex chemical information accurately (Krenn et al., 2022). The introduction of new molecular languages, like Self-referencing embedded string (Selfies), promises to overcome these challenges by ensuring robust and interpretable representations. The advancement not only boosts computational predictions but also facilitates the process of sharing data across different fields of science. Through teamwork and innovation, these technological discoveries will further perfect our knowledge of how biomolecules interact with each other and lead to the identification of novel therapeutic approaches that will enhance our molecular-level understanding of life (Figure 2.3).

2.1.1. Atoms and Molecules in Living Organisms

Atoms are the simple building materials of all living organisms and are the building blocks of the fundamental construction materials upon which life is dependent. The simple particles are combined to become molecules, which then participate in a range of biological processes required to sustain life. Familiar atoms, such as carbon, hydrogen, oxygen, and nitrogen, form the backbone of biomolecules, including proteins, lipids, carbohydrates, and nucleic acids. All these molecules are essential; for example, proteins function as enzymes to catalyze biochemical reactions, and nucleic acids like DNA are crucial for the storage and transmission of genetic information. The complex interaction among these atoms and molecules illustrates the complexity of biological systems, demonstrating to us an intimate network of chemical interactions that form the basis of life's processes and adaptations. At a biological level, molecular interactions also involve supramolecular structures made up of assemblies of molecules aggregating through weak chemical bonds. Supramolecular chemistry is particularly significant in explaining water behavior in living organisms, as water molecules possess unique properties that facilitate biochemical reactions. The study of molecular protonics, or the concern with how protons are transferred between these structures, further highlights the sophistication of molecular interactions in biology. Research has elaborated on the generation of structures known as polyatomic Rydberg molecules and suggested that their presence in biological organisms could affect various biochemical processes and physiological consequences. The exploration of such molecular configurations reveals the nuanced and dynamic interplay of atoms and molecules that form the basis of life itself (Avakyan et al., 2023). Advanced imaging techniques, such as electron cryomicroscopy, are revolutionizing our understanding of the molecular structures that comprise living organisms. This technology allows researchers to visualize individual molecules at unprecedented resolutions, unraveling the complexities of biomolecular architecture. As researchers delve into the details of molecular shapes, they can discover how the arrangements of atoms can affect biological function. These excursions are part of the project of structural biology, an attempt to visualize the delicate scaffolds that give rise to life. The ability to visualize these atomic structures not only creates a more profound understanding of the components of life but also holds promise to shape medical and technological advancement. By bridging the gap between molecular structure and biological function, scientists are actively advancing our collective knowledge about the chemistry of life (Segev et al., 2023) (Figure 2.4).

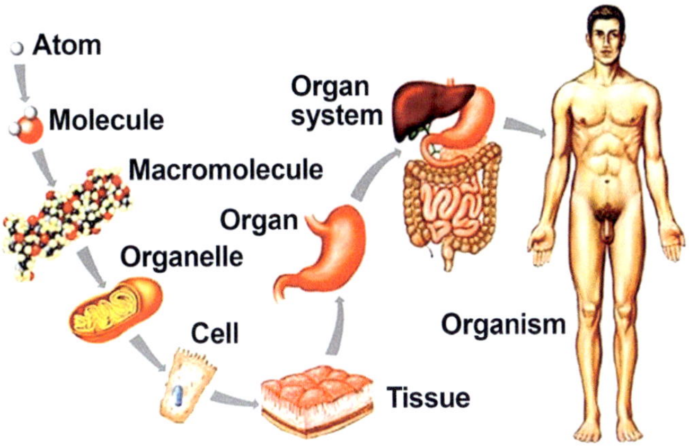

Figure 2.4. Atoms and molecules in living organisms.

Source: https://external-content.duckduckgo.com/iu/?u=https%3A%2F%2Fo.quizlet.com%2Fi%2FBro7LKSk-k-rZpTi_s2Esw.jpg&f=1&nofb=1&ipt=93ee38843c3a7ea48eaadc8811e4b5af133b1384102e392d39a46fde7d31130e.

2.1.2. The Role of Water in Biological Systems

Water is generally seen as the basis of biological processes since it possesses unique properties necessary for life. As a polar molecule, water exhibits a broad array of behaviors, such as cohesion and adhesion, which play crucial roles in cellular processes. Cohesion allows water molecules to stick to one another, making droplets and aiding nutrient and waste transport within organisms. Further, adhesion facilitates water interaction with cellular components, enhancing vital biochemical reactions. This molecular interaction capability highlights water's function as a solvent, where it dissolves a variety of substances used in metabolic reactions, thus sustaining life on our planet (Horowitz et al., 1974). Such characteristics illustrate how water's molecular properties directly affect biological functions and the general chemistry of life. In living systems, water serves as a medium in which biochemical reactions occur, and hence it is crucial in maintaining homeostasis. The ability of water to absorb and release heat without rapid temperature changes is significant in the regulation of cellular functions and metabolic activities. This property, which can also be termed high heat capacity, allows for living organisms to maintain stable internal conditions irrespective of the changes in external temperatures. Besides, the density of water plays an essential role in aquatic ecosystems, enabling organisms to float and affecting circulation in the ocean and lake (Barthel et al., 2013). By facilitating the existence of different forms of life in varied environments, water indirectly sustains the genetic diversity that drives evolution. Its function thus extends beyond mere provision to the fundamental mechanisms sustaining living systems. The significance of water in biochemical processes goes further to show itself through its interaction with ion channels in cell membranes. Ion channels are selective pores that allow for the transport of ions required for the generation of electrical signals in cells. Water molecules ensure the transport of these ions by acting as a channel whereby they can move through and thereby impact essential functions such as the conduction of nerve impulses and muscle contraction. It is this unique interaction between water and proteins and ions that renders the channels so efficient, similar to the functioning of transistors in electronic devices (Barthel et al., 2013). Thus, the dynamic interplay among biological macromolecules, ions, and water is an accurate reflection of how water's chemistry serves the complicated processes inherent in life, which underscores its essential character in the chemistry of life (Figure 2.5).

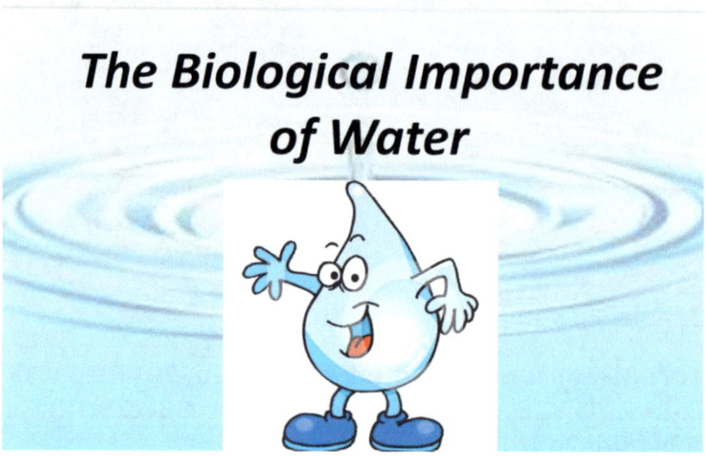

Figure 2.5. The role of water in biological systems.

Source: https://external-content.duckduckgo.com/iu/?u=https%3A%2F%2Fslideplayer.com%2F12553421%2F75%2Fimages%2Fslide_1.jpg&f=1&nofb=1&ipt=65896cc69238e9d250da9798224af40090a162a2594df6282326cf1c3532c88f.

2.1.3. Macromolecules: Proteins, Carbohydrates, Lipids, and Nucleic Acids

Macromolecules are enormous, complex molecules that play vital roles in life in biological systems, and they fall under four main categories: proteins, carbohydrates, lipids, and nucleic acids. All the macromolecules possess various functions that are vital for survival. Proteins, for instance, are composed of amino acids and are required for various biological processes, including catalysis, support, and transportation. Their functional diversification is due to different sequences and arrangements of amino acids, leading to different three-dimensional shapes. Carbohydrates primarily act as energy and structural molecules. They are built on the simple sugars, and their complexity ranges from monosaccharides to polysaccharides, demonstrating the diversity of their functions in life processes. Lipids are another critical group of macromolecules, most well-known for energy storage and membrane development. Lipids are hydrophobic molecules, and this property allows them to serve as insulation and long-term energy storage. Lipid structure, which consists of fatty acid chains and glycerol, is why phospholipid bilayers are necessary in cellular membranes that are tasked with permeability as well as cell-to-cell interaction. This characteristic is vital in maintaining cellular integrity and function. Nucleic acids, such as DNA and RNA, are responsible for the storage and transmission of genetic information. These are polymers made up of nucleotide monomers, where DNA contains hereditary information and RNA is involved in protein synthesis, and this is an indication of the inter-relatedness of macromolecules in life chemistry. Research on the origin of life has revealed some fascinating interconnections between these macromolecules and the environment, particularly concerning the catalytic function of water in biochemical processes. It has been found that water pH and temperature play critical roles in the synthesis of biological macromolecules from simple organic molecules. For instance, in alkaline mineral water, temperatures of up to 65°C to 95°C and pH between 9 and 11 are favorable for the polymerization of molecules into more complex organic structures such as proteins and polysaccharides, thereby predicting the mode of origin of early life forms in such environments (Ignatov et al., 2015). In the same vein, investigations of organisms that have adapted to survive on substrates of carbon, including heavy water, have demonstrated the pathways of metabolism that provide key biomolecules like amino acids, proteins, carbohydrates, and lipids, indicating the chemical sophistication of life (Ignatov et al., 2015). This interaction between the macromolecules and their environment points toward the complex network of interactions that characterize the chemistry of life (Figure 2.6).

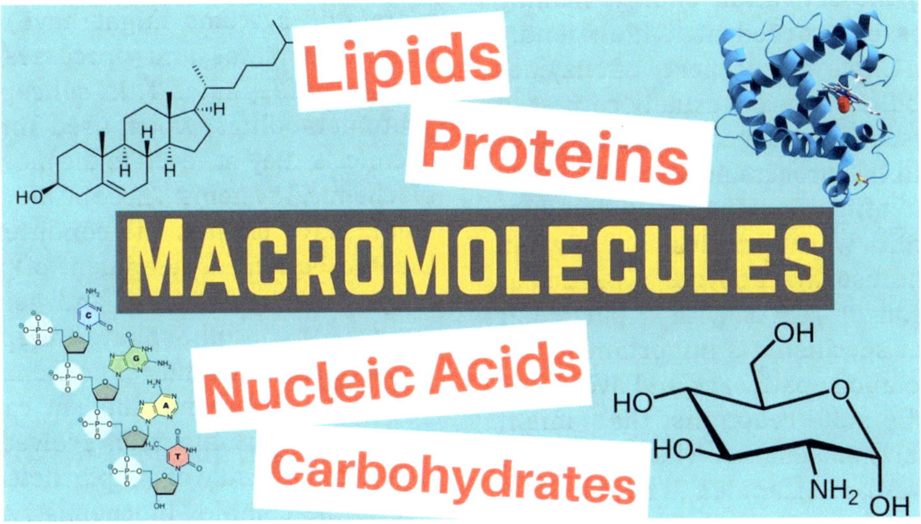

Figure 2.6. Macromolecules.

Source: https://i.ytimg.com/vi/V5hhrDFo8Vk/hq720.jpg?sqp=-oaymwEhCK4FEIIDSFryq4qpAxMIARUAAAAAGAElAADIQj0AgKJD&rs=AOn4CLD6UH34hGPAzFPeHsmS9Mau_8Ao-g.

Test Your Knowledge

1. What is the role of water in biological systems?
2. What are macromolecules?

2.2. Enzymes and Metabolism

Learning Outcome

1. Function and structure of enzymes;
2. The role of enzymes in metabolic pathways;
3. Factors affecting enzyme activity.

Enzymes are critical in the occurrence of metabolism as they are biological catalysts, which speed up the chemical reactions that support life. Enzymes increase the rate of reactions by decreasing the activation energy, making cellular processes proceed at a life-friendly rate. This is an important property of enzymes because metabolic pathways usually consist of a cascade of reactions that convert substrates to products and this process needs to be tightly regulated and efficient. Moreover, enzymes are very specific, which implies that a given type of enzyme usually catalyzes a particular type of reaction or reacts with a particular substrate. Such specificity is important to the integrity of metabolic pathways and avoidance of undesirable side reactions that might perturb cellular homeostasis. The modularity of these pathways, as examined in metabolism, highlights how enzymes can be grouped into functional units, efficiently orchestrating the elaborate dance of molecular transformations that are fundamental to life (Braakman & Smith, 2013). In addition to catalysis, enzymes are also involved in more complex regulation mechanisms, controlling metabolic pathways. Such regulatory mechanisms make sure that metabolic reactions become dynamically responsive to the varying needs of the cells and their environment. Allosteric regulation, covalent modification, and enzyme concentration can also play an important role in enzymatic activity, highlighting the significance of control in metabolism. Also, another complexity to this regulation is the existence of cofactors, which are molecules that help the enzymes in catalysis. As an example, some cofactors can determine the substrate specificity of enzymes or the reaction mechanism used and thereby regulate the metabolic dynamics in the cell. This interplay between enzymes and cofactors highlights the hierarchical organization of metabolism, further emphasizing the profound impact that complex biochemical networks have on cellular function and organismal survival (Braakman & Smith, 2013). Studies of metabolic pathways have also provided insight into the evolutionary consequences of enzyme function and adaptation. The idea of metabolic fossils, discussed in recent research, is that early metabolic systems might have evolved with other substrates and processes that do not rely on phosphate. This concept states that protometabolites, which used simple prebiotic chemicals, may have led to the more complicated biochemical systems. These primordial processes would have created the conditions leading to the subsequent development of more complex enzymatic functionality and genetic systems by creating a phosphate-independent core metabolism. The study of how enzymes evolved and their role in metabolism can give useful clues as to how life itself evolved, underlining how the chemistry of simple molecules can give rise to the complex biochemistry of life that is seen today (Figure 2.7).

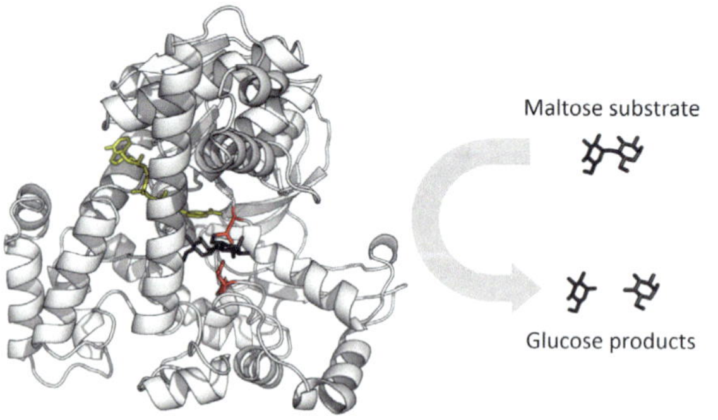

Figure 2.7. Enzymes and metabolism.

Source: By Thomas Shafee - Own work, CC BY 4.0, https://commons.wikimedia.org/w/index.php?curid=38738689

2.2.1. Function and Structure of Enzymes

Enzymes are biological catalysts, accelerating biochemical reactions by lowering the activation energy required for the reaction to occur. Shape determines their function in deciding; enzymes tend to be globular proteins that have a unique three-dimensional structure required for substrate attachment. All enzymes have an active site, a specially shaped region where substrates attach and are chemically modified. This precision also makes enzymes substrate-selective, as in the case of a lock-and-key mechanism.

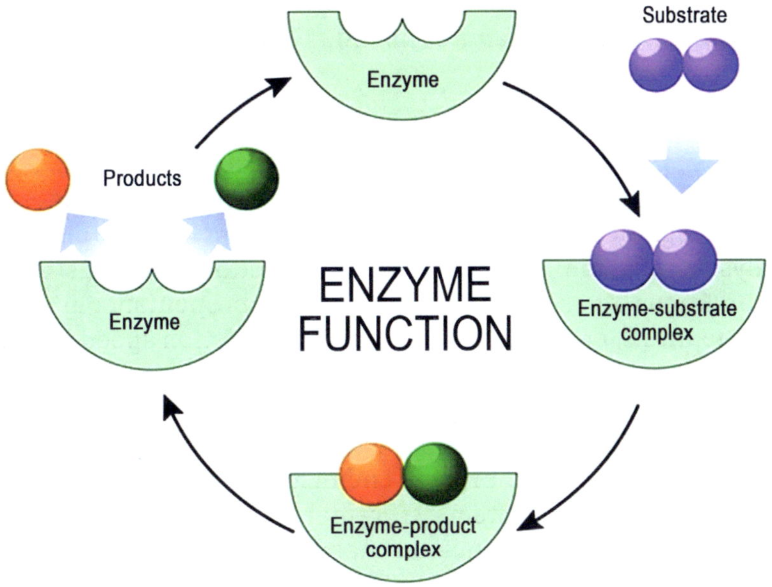

Figure 2.8. Function and structure of enzymes.

Source: https://d2jx2rerrg6sh3.cloudfront.net/image-handler/picture/2018/8/shutterstock_760624435.jpg.

As such, slight modifications in the enzyme structure significantly impact its catalytic capability. This structural precision is typical of metabolic processes, where enzymes regulate and facilitate vital reactions for life, demonstrating their ultimate function in the chemistry of life. Modularity of enzyme structure also helps to reveal their central function in metabolism. Many enzymes act in complex metabolic pathways, as modules consisting of connected reactions. Such pathways include conserved sets of substrates and enzymes that occur repeatedly with the same type of reaction mechanisms, imparting metabolic processes robustness and efficiency. Enzymes typically interact with cofactors, small, non-protein molecules necessary for their catalysis, an extra turn to enhance their operation. Metabolic pathway modular architecture is an example of how enzymes not only catalyze the reaction but also serve functions in the hierarchical organization of metabolic networks. The enzyme structure versus its functional modules interaction points towards the evolutionary significance of these biochemical entities as they expose how they evolve to facilitate the emergence of complex biochemical systems within living organisms (Braakman & Smith, 2013). New advances in biotechnology have elevated the level of enzyme function to new uses in a variety of fields. Techniques such as genetically engineered immobilization of enzymes and microencapsulation enhance stability and reusability in industrial use, particularly in agricultural and medical biotechnology. Immobilized enzymes can be used to produce functional foods that will enhance health impacts while maintaining processing efficiency (Baianu et al., 2004). Genetic engineering and molecular modeling are also at the core of designing enzymes with characteristics tailored to suit particular requirements, increasing their use in pharmaceuticals and diagnostics. Ongoing research and development of enzymes prove their active nature in the chemistry of life, indicating not only the critical biological importance but also the promise to revolutionize industries through biotechnological advances (Baianu et al., 2004) (Figure 2.8).

2.2.2. The Role of Enzymes in Metabolic Pathways

In biochemistry, enzymes are vital control elements of metabolic processes, catalyzing the complex reactions that make life possible. As catalysts in the living organism, these proteins speed up chemical reactions to a significant degree by lowering the activation energy required. Through this process of catalysis, cells can conduct essential functions, such as energy generation and biosynthesis, under conditions otherwise insufficient to sustain spontaneous reactions. By organizing complex arrays of biochemicals, enzymes not only catalyze reaction speed but also contribute to the modular and hierarchical structure of metabolism. This organization demonstrates how enzymes directly engage with substrates, illustrating the merit of structural compatibility, which in turn allows for a better flow of the metabolism. This functional specificity argues in favor of the notion that enzymes are essential components of life's chemistry, to make possible a range of metabolic processes and balancing in cellular systems extremely delicately. The evolutionary significance of enzymes further enhances their influence on metabolic pathways, parallel to their role in the adaptation of organisms and the variety of forms of life. One of the most impressive characteristics of metabolic organization is the modularity of enzyme-catalyzed reactions, and often these utilize conserved enzymes across a great variety of organisms. These modules also can regulate critical processes, such as carbon fixation, which have evolved relatively little over the ages. Moreover, as observed, modules organizing metabolism suggest an extreme evolutionary history, with ancestral metabolic types utilizing an elementary set of innovations to adapt to their environments. Such an evolutionary perspective emphasizes the part played by enzymes not only in shaping metabolic processes within contemporary organisms but also in creating the biochemical diversity that exists throughout the biosphere. In addition, in trying to trace the origins of enzymatic activities, research has broadened the possibility of phosphate-independent metabolic conditions evolving from primordial stages. As postulated, finding a straightforward prebiotic-based core metabolism suggests that early biochemical reactions could have been present before the advent of enzymatic reactions involving phosphate. Based on this hypothesis, it is conjectured that certain metabolic networks, for example, iron-sulfur cluster-based ones, could be a legacy of ancient biochemical systems prior to sophisticated enzyme activity. These discoveries serve to highlight the elasticity and resiliency of primitive metabolic function, indicating that enzymes have remained active in their development to refine and maximize the chemistry of the building blocks of life. As such, the study of these primitive mechanisms not only offers an expanded context for understanding enzyme activities but also discloses the close coupling between metabolic evolution and the chemistry of life itself (Figure 2.9).

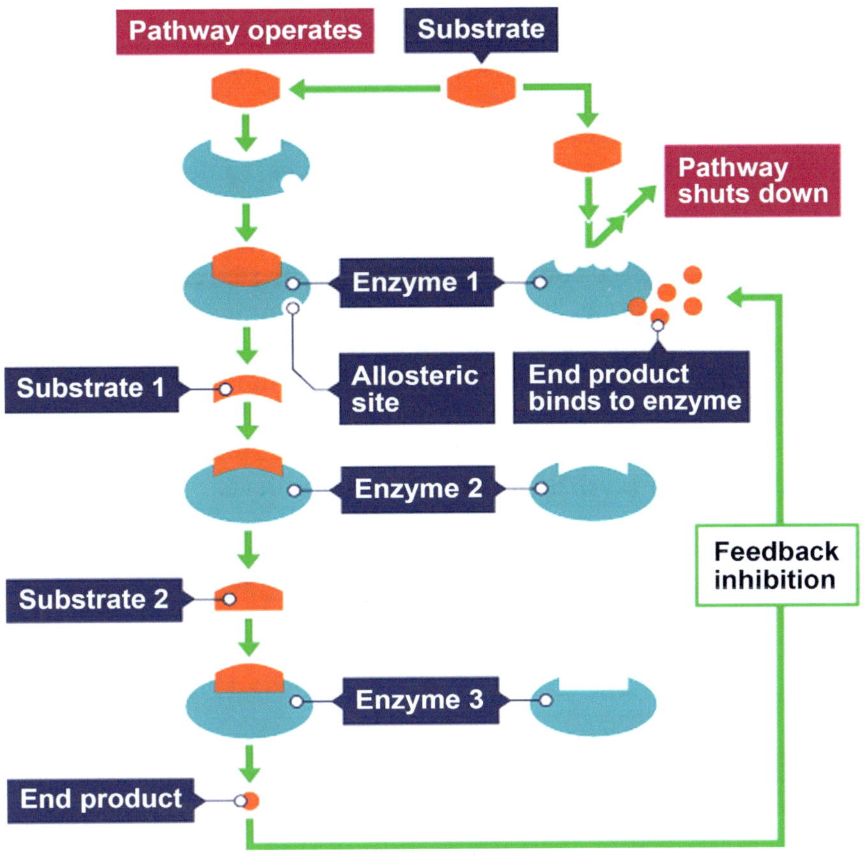

Figure 2.9. The role of enzymes in metabolic pathways.

Source: https://bam.files.bbci.co.uk/bam/live/content/zss8mp3/medium.

2.2.3. Factors Affecting Enzyme Activity

Enzymes are biological catalysts crucial to the acceleration of chemical reactions and are highly susceptible to many factors. Temperature and pH are some of the factors that significantly determine the regulation of enzyme activity. Each enzyme has an optimum temperature and pH at which its functional efficiency is highest. Deviations from such optimum levels can result in changes in the structure of the enzyme, altering its active site and, hence its catalytic efficiency. Especially, gross deviations can cause denaturation, which is the process by which enzymes lose their functional conformation and hence their activity. Such factors are of prime importance in applications in biotechnology and medicine and highlight the need to control environmental conditions so that enzyme efficiency is maintained in such processes as digestion and industrial fermentation. Substrate concentration is another vital factor that affects enzyme activity. As substrate concentration increases, the rate of reaction usually increases to a level where a maximum velocity, or Vmax, is reached. The rate of reaction is then at a constant level since all active sites of the enzyme molecules are utilized. The saturation of the active sites of the enzyme molecules is a limiting condition for the availability of the substrate in enzyme reactions. However, if the concentration continues to increase unabated, the effect may not yield a directly proportional increase in rates of reaction, and other mechanisms may need to be utilized to optimize enzyme activity. Certain competitive inhibitors also share structural similarity with substrates and act at the active site, decreasing the total activity of enzymes, emphasizing the multifactorial regulation of enzymes in biochemical reactions. Environmental components such as ionic strength and the presence of activators or inhibitors also influence the activity of enzymes. Ionic strength can affect substrate binding to the enzyme and the structural integrity of the enzyme, leading to variations in catalytic efficiency. Biophysical reactions also often include regulatory features that modulate enzyme activity through allosteric interactions. These controls can energize or repress activity, depending on their nature and concentration, showing the dynamic regulation of metabolic processes. The fine balance between enzymatic activity and regulatory components reflects the necessity of understanding enzyme performance in the overall scheme of cellular function and general organismal health. For instance, responses to oxidative stress, as discussed in the study on reactive oxygen species (ROS) (Hasanuzzaman et al., 2020) and ferroptosis (Li et al., 2020), show how enzyme activities are woven into complex survival mechanisms at the cellular level.

Test Your Knowledge

1. What are the functions and structure of enzymes?
2. What are the factors affecting enzyme activity?

2.3. Cellular Respiration

Learning Outcome

1. Overview of cellular respiration processes;
2. The role of ATP in energy transfer;
3. Anaerobic vs. aerobic respiration.

Cellular respiration is a fundamental biochemical process that enables organisms to convert energy stored in macromolecules into usable forms, primarily ATP. Depending on the availability of oxygen, this multi-step pathway can be divided into aerobic and anaerobic respiration. In eukaryotic cells, aerobic respiration takes place in the mitochondria and involves glycolysis, the citric acid cycle, and oxidative phosphorylation. By contrast, anaerobic respiration, common in some prokaryotes or in conditions limiting oxygen availability results in the accumulation of lactic acid or ethanol, depending on the organism. The amount of energy that is produced in these processes underlines the centrality of these processes in the ongoing cellular functionality, in the sense that the chemistry of life is closely intertwined with the efficiency of energy conversions in biological systems. Cellular respiration is not only important in the generation of energy; it is also central in the regulation of cellular homeostasis as well as other metabolic processes. During this process, the balance between ROS and nitric oxide (NO) is critical. Elevated levels of ROS can lead to cellular dysfunction and death, as illustrated in conditions like Leber's hereditary optic neuropathy (LHON), where the accumulation of ROS correlates with impaired cell viability (Arese et al., 2015). Moreover, a balanced existence of these reactive species can promote signaling pathways, which control apoptosis and cell survival. Therefore, the study of cellular respiration not only expounds on the dynamics of energy, but also its consequences in health and disease, hence a very important topic in the field of biochemistry and molecular biology.

Moreover, the complexities of cellular respiration involve hierarchical networks of interactions between macromolecules, cells and tissues, and the chemistry of life in general. Cellular metabolism is cooperative, such that energetically favorable reactions occur efficiently in cellular systems. This cooperation parallels findings that underscore the importance of local and global interactions among biological entities, influencing overall system robustness and adaptability (Braeckman et al., 2009). As cellular respiration proceeds to evolve further in accordance with the variations in the environment, it highlights the evolutionary importance of the cooperation between metabolic pathways and organisms. Therefore, the knowledge of cellular respiration can be used as a prism through which the viewer can perceive the connectedness of life processes, and that cooperation and energy relations are the basics of biological complexity and stability (Figure 2.10).

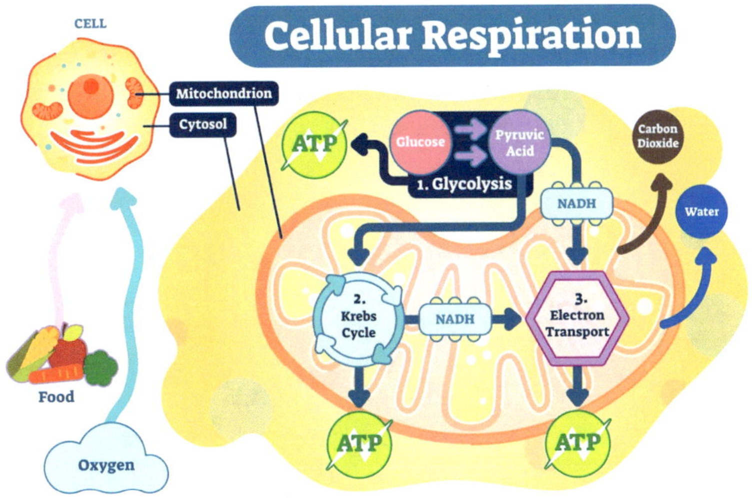

Figure 2.10. Cellular respiration.

Source: https://cdn-ildfibj.nitrocdn.com/qaWeLntkRcyxNUcEoWbLpZLQkcgLFsiJ/assets/images/optimized/www.biologyonline.com/wp-content/uploads/2020/02/Cell-Respiration-768x529.jpg.

2.3.1. Overview of Cellular Respiration Processes

Cellular respiration is a critical biochemical process in which biochemical energy from the nutrients is converted into ATP, the cellular currency, by living organisms. The complex process begins with glycolysis, in the cytoplasm, where glucose is cleaved enzymatically into pyruvate, producing a small amount of ATP and NADH. Subsequently, in oxygen conditions, pyruvate is transferred to the mitochondria for the citric acid cycle (Krebs cycle), where it is further oxidized. Additional molecules of NADH and $FADH_2$ are generated by this process, which are essential for the second stage of cellular respiration. The combined rate of ATP synthesis is maximized in oxidative phosphorylation, where the electrons of NADH and $FADH_2$ are transferred through the electron transport chain (ETC) to generate ATP by chemiosmosis. The synchronization of the various cellular respiration processes demonstrates the highly developed control and flexibility of metabolic pathways of life. Oxygen-based aerobic respiration is the most efficient method of producing ATP and can generate about 36–38 ATP molecules from each glucose molecule. Anaerobic respiration, on the other hand, occurs in the absence of oxygen and leads to fermentation reactions that yield fewer molecules of ATP and products like ethanol or lactic acid. These alternative pathways illustrate how organisms can inhabit various ecological niches and change metabolic strategy so that energy generation takes place under varying conditions. Shifts between aerobic and anaerobic respiration also emphasize the position of oxygen as a central player in energy metabolism in all forms of life. Cellular respiration is not simply an energy conversion process but also a critical element of biogeochemical cycling and ecological balance. For instance, the carbon dioxide (CO_2) released during respiration serves as an important input for photosynthesis in plants, establishing a cycle of relationship between the two processes. That cellular respiration is central to life's overall chemistry by linking ecological systems and restoring balance within the environment is what this interdependence demonstrates. Besides, knowledge about cellular respiration has motivated green synthesis studies, linking metabolic processes to green ways of material science, pinpointing the necessity of green methods in producing essential materials of manufacture (Singh et al., 2018). The complex balance of cellular respiration mirrors how closely connected energy conversion, environmental sustainability, and chemistry are that propels life itself (JingguZhang et al., 2018) (Figure 2.11).

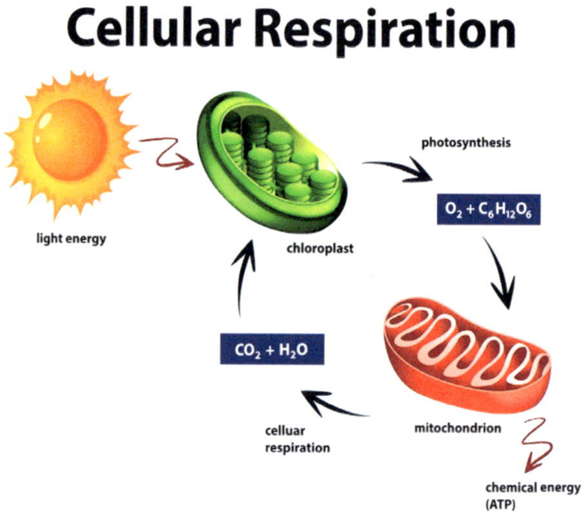

Figure 2.11. Cellular respiration processes.

Source: https://fatty15.com/cdn/shop/articles/Understanding_the_Process_of_Cellular_Respiration_5784a5aa-4b03-42dd-8399-d4a800ee0393_1200x1200.png?v=1747225503.

2.3.2. The Role of ATP in Energy Transfer

The function of ATP in energy transfer is central to the sustenance of life, as it is the fundamental energy currency in biological systems. ATP stores energy that is made available through the release from metabolic activity, specifically cellular respiration and photosynthesis. It is obtained by hydrolysis of ATP, which releases energy that can be used for a variety of cellular activities, including muscle contraction, active transport, and biosynthesis. Again, here, the function of ATP is not limited to an energy transmitter; it is an essential component of the regulation of metabolic pathways and biochemical processes coordination, emphasizing the close interrelationship between ATP metabolism and the biochemical structure of life. The modularity of metabolism, as explained, illustrates how ATP functions within higher-order metabolic frameworks, which influence cellular function and evolutionary processes that underlie life (Braakman & Smith, 2013). In addition, ATP does not function by itself. ATP is joined by cofactors and other molecules to facilitate different biochemical reactions that allow life. Such interactions are significant because ATP is facilitative of forming complex networks that tie diverse metabolic pathways together.

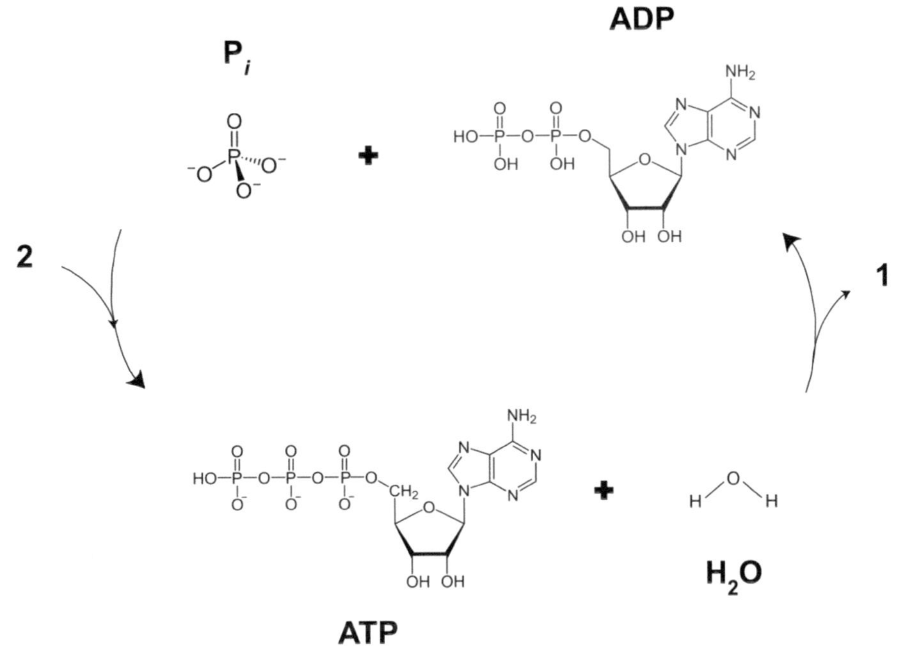

Figure 2.12. The role of ATP in energy transfer.

Source: By Butrboy - Own work, Public Domain, https://commons.wikimedia.org/w/index.php?curid=79504120

The versatility of ATP makes it utilizable as a substrate for the incorporation into many enzymatic processes, most of which are sites of regulation on metabolic pathways. Significantly, this regulatory role is emphasized by the capacity of ATP to couple exergonic reactions, which yield energy, to endergonic reactions, which require energy input. Coupling becomes vital for organisms to efficiently manage resource utilization and energy, demonstrating ATP's role in bioenergetics. This connectivity illustrates how ATP is not only an energy donor but also a master metabolic coordinator of cellular functions (Braakman & Smith, 2013). In general, an understanding of ATP's role as an energy transmitter illuminates such broad biological processes as cell growth and division. Maintaining the correct balance of ATP production and utilization is paramount to a cell's capability to sustain cellular homeostasis and respond to environmental variation. For example, varying cellular energy demands provoke ATP synthesis modifications, in many cases mediated through complex feedback. Initial metabolic pathways were supposed to operate thioester-type energy couplings before the evolution of phosphate-based systems, in that a hypothesis is given, implying the energy transfer mechanisms in evolution. Therefore, ATP is no longer merely a source of energy but a key player in the evolutionary history of life's complexity. In its unique chemistry and its critical functions, ATP draws out the chemistry of life itself, both as a metabolic staple and an evolutionary remnant (Figure 2.12).

2.3.3. Anaerobic versus Aerobic Respiration

In cellular metabolism, energy is acquired by organisms through two broad mechanisms: anaerobic respiration and aerobic respiration. Anaerobic respiration is the mechanism that occurs in the absence of oxygen and typically yields partial breakdown of glucose, leading to energy and products like lactic acid or ethanol.

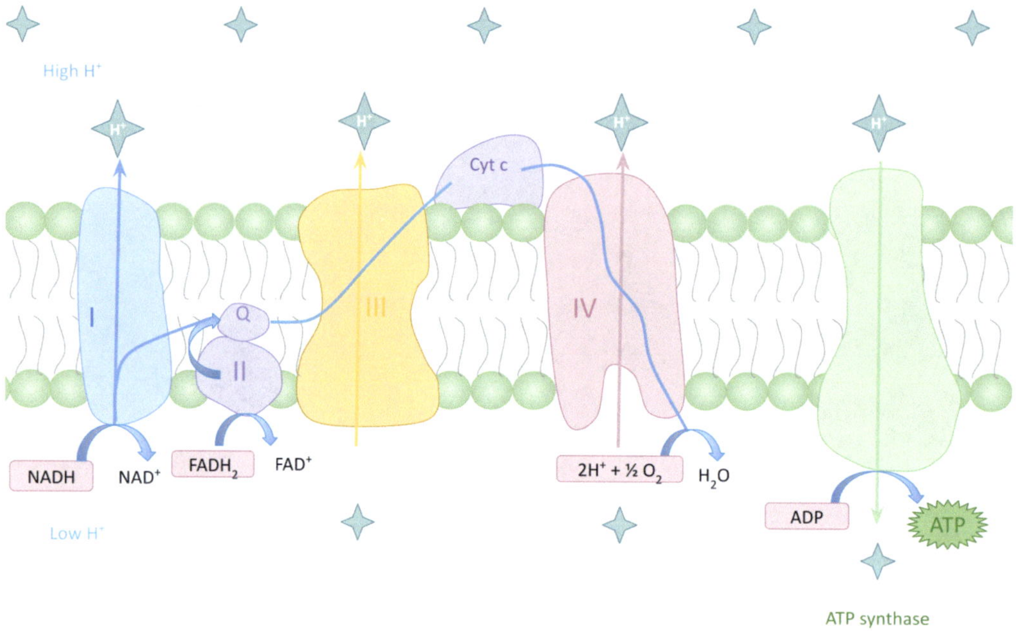

Figure 2.13. Aerobic versus anaerobic respiration.

Source: By Irinapickle - Own work, CC BY-SA 4.0, https://commons.wikimedia.org/w/index.php?curid=130463092

This mechanism is much less efficient in terms of energy, as it yields only two molecules of ATP per molecule of glucose as opposed to aerobic respiration. Aerobic respiration, employing oxygen as a terminal electron acceptor, is the most efficient in extracting energy from glucose and yields up to 36 molecules of ATP. The relative efficiencies of these two processes largely establish the ecological niches to which different organisms become adapted, exhibiting one of the elementary characteristics of the life chemistry. With the production of energy, the biochemical process of aerobic respiration and anaerobic respiration varies greatly. Aerobic respiration involves a series of steps: glycolysis, the Krebs cycle, and the ETC. All three stages are nicely arranged to facilitate stepwise oxidation of glucose, releasing stored energy in the form of ATP. Anaerobic respiration, on the other hand, bypasses the Krebs cycle and instead relies heavily on glycolysis followed by fermentation reactions, which recycle NAD^+ for the continuation of ATP production. This biochemistry comes in no less handy in low-oxygen environments, allowing such organisms as certain bacteria and yeasts to exist and contribute to their environments. Such metabolic adaptability demonstrates the richness of life's chemistry. The variations between anaerobic and aerobic respiration apply not only to individual organisms but to ecological and environmental contexts as well. The prevalence of anaerobic processes in certain ecosystems, such as marshes or deep-sea hydrothermal vents, could lead to novel biochemical cycles and community processes. Anaerobic bacteria, for example, play key roles in nutrient cycling and decomposition, ultimately controlling soil health and plant productivity. Furthermore, the release of gases during anaerobic respiration raises questions about greenhouse gas emissions and global climate change. As scientists continue to study these metabolic pathways, as with complementary research concerning the impact of environmental stresses like salinization, the better the understanding is of their ecological impact and living organisms' dependence on one another becomes, showing the complicated chemistry behind all living organisms (Ellen et al., 2015; María et al., 2021) (Figure 2.13).

Test Your Knowledge

1. What is the role of ATP in energy transfer?
2. Differentiate anaerobic vs. aerobic respiration.

2.4. Photosynthesis

Learning Outcome

1. The process of photosynthesis in plants;
2. The significance of chlorophyll and light energy;
3. The impact of photosynthesis on the ecosystem.

Photosynthesis is one of the most important biochemical processes, which forms the basis of life on Earth as it captures light energy and stores it as chemical energy in the form of glucose. In plant cells, this transformation takes place primarily in chloroplasts, where the sunlight is harvested by chlorophyll pigments. The CO_2 in the atmosphere and water in the soil are used in this process (photosynthesis), which involves a set of complicated reactions, and the end products are glucose and oxygen is given off as a byproduct. This is a process that not only supports the existence of plants but also supports the existence of other diverse life forms that rely on plants to survive in terms of the supply of oxygen and organic matter. The significance of photosynthesis may be considered as the basis of the complex life web, as a way of energy transformation and storage in ecosystems, and as an important source of the atmospheric composition of the Earth. The purpose of organic pigments in photosynthesis is not just the absorption, but they also increase the efficiency of photon dissipation, which is crucial to maintain life.

The thermodynamic dissipation theory has suggested that the necessity to enhance the dissipation rate of solar photons around the globe has resulted in the evolution of organic pigments. Over time, these pigments have evolved to maximize their utility, in terms of increasing their effective photon cross sections per size and in reducing the lifetime of their excited electronic states. Also, they have been subjected to evolutionary pressures, to quench non-radiative de-excitation channels, and to absorb a wider range of solar radiation.

This adaptation not only underscores the complex relationship between photosynthesis and thermodynamic principles but also highlights how life has evolved to exploit energy efficiently, furthering the complexity of ecological interactions on Earth (Michaelian et al., 2014). By introducing the notion of energy into the analysis of photosynthesis, one will be able to gain a better insight into the role that it plays in biological systems. Since energy transitions are central to the chemistry of life, education programs, including capstone courses in energy, can help shed light on the complex photosynthesis process.

The combination of the theoretical information and the practical examples will allow the students to understand the way the energy moves in the ecosystems, being affected by the photosynthetic activity. This learning model not only focuses on the biochemical process of photosynthesis, but it also expands the debate on the role of energy sustenance in life. Through concepts of energy, learners can appreciate the thermodynamic principles that govern life's processes, ultimately leading to enhanced comprehension of the delicate balance between energy capture and utilization in the biosphere (Arce et al., 2001) (Figure 2.14).

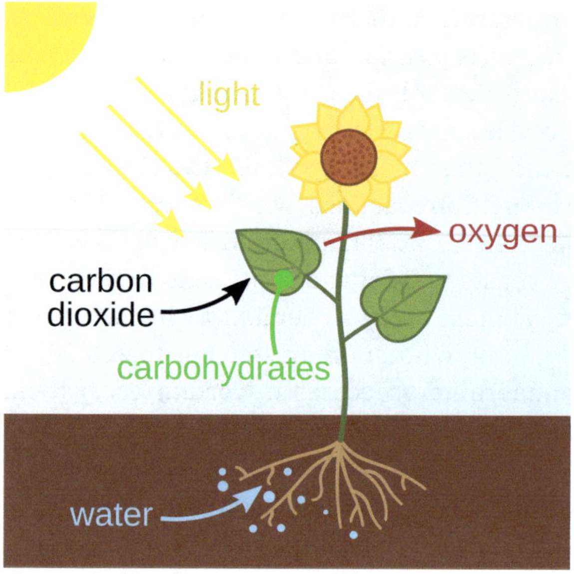

Figure 2.14. Photosynthesis.

Source: By At09kg, Wattcle, NefronusAt09kg: originalWattcle: vector graphicsNefronus: redoing the vector graphics - This file was derived from: Photosynthesis.gif:, CC BY-SA 4.0, https://commons.wikimedia.org/w/index.php?curid=49183032

2.4.1. The Process of Photosynthesis in Plants

Photosynthesis is a vital biological process whereby light energy is trapped and stored as chemical energy, sustaining life on Earth. This complex process is largely performed in chloroplasts of plant cells, and in which chlorophyll, the green pigment used by plants for the absorption of sunlight, plays a central role. When photosynthesis takes place, plants take in CO_2 from the atmosphere and water from the soil and experience a series of complex reactions. The initial phase, the light-dependent reactions, harvest photons and translate them into energy-rich compounds, ATP and nicotinamide adenine dinucleotide phosphate (NADPH). The energy carriers are then utilized by the following light-independent reactions, also known as the Calvin cycle, which form glucose through a series of enzyme-catalyzed reactions, effectively storing energy in a stable form for use at a later time. The biochemical photochemical processes of photosynthesis are regulated by non-equilibrium thermodynamic laws, and these add to the energy dynamics of the planets. As pointed out by Michaelian et al. (2014), organic pigment evolution was compelled by the requirement to increase dissipation of solar energy into the environment. Through optimized solar spectrum capture, plants optimized their photochemical activity and thereby increased biomass yield and oxygen release. Not only does it foster plant growth, but it also maintains oxygen balance in the atmosphere, vital for the existence of aerobic organisms. Evolutionary adaptations of chlorophyll and other pigments reflect an complex relationship between life and energy circulation, emphasizing photosynthesis as the prime mechanism controlling terrestrial ecosystems and the Earth's atmosphere. Besides, photosynthesis implications are not just a matter of mere energy conversion; they lie at the core of ecological health and sustainability. The glucose produced during photosynthesis serves as a precursor to other organic molecules that sustain the continuation of food chains. The oxygen produced is a waste product that allows for respiration in animals and other organisms, evidence of the dependency of life forms. Educational paradigms that emphasize principles of energy, as described by Arce et al. (2001), can further students' knowledge of this fundamental process. By illuminating the fundamental processes of photosynthesis, we can realize its key role in making life possible, and therefore, an increased awareness of environmental management and the delicate balance of ecosystems (Figure 2.15).

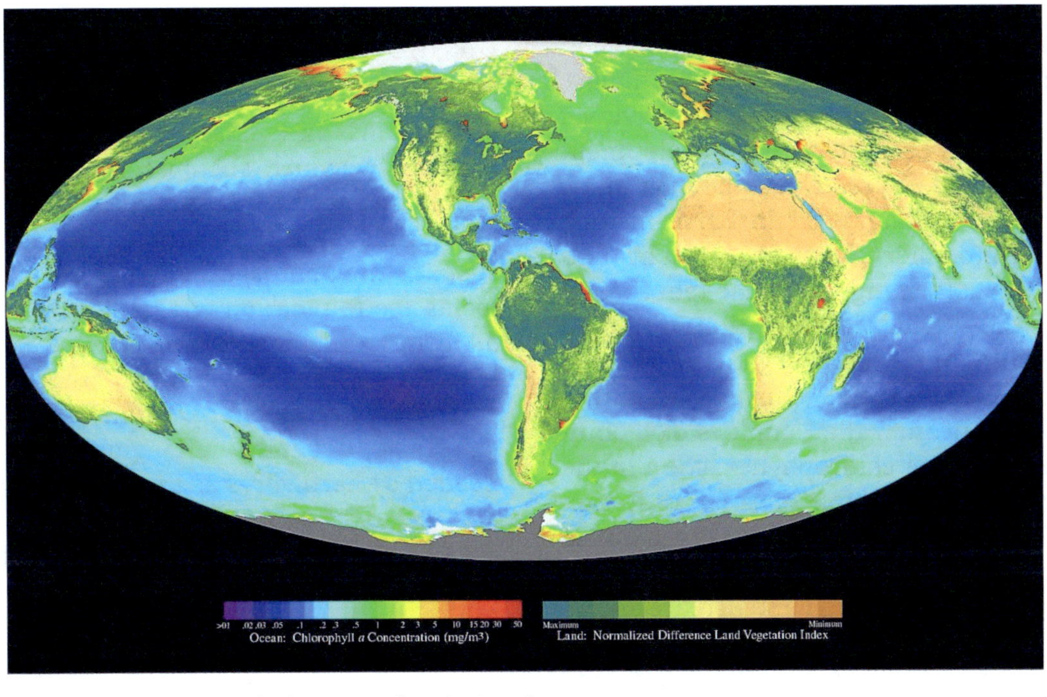

Figure 2.15. The process of photosynthesis in plants.

Source: By Provided by the SeaWiFS Project, Goddard Space Flight Center and ORBIMAGE - http://oceancolor.gsfc.nasa.gov/SeaWiFS/BACKGROUND/Gallery/index.html and from en:Image:Seawifs global biosphere.jpg, Public Domain, https://commons.wikimedia.org/w/index.php?curid=387228

2.4.2. The Significance of Chlorophyll and Light Energy

Chlorophyll is an important molecule in the process of photosynthesis that allows the conversion of light energy into chemical energy. Chlorophyll is a green pigment predominantly found in plants, algae, and cyanobacteria and possesses a strong affinity for absorbing light energy in blue and red colors with the reflection of green light, hence the green color of these organisms. The ability to capture light energy is crucial for photosynthetic organisms, as they can fix organic compounds from water and CO_2 through it, the process that forms the foundation of life on Earth. This can be done with the help of electronic transitions as chlorophyll captures photons. ATP and the reducing agent NADPH can be generated. Chlorophyll molecules have the most important function in the Calvin cycle, where atmospheric CO_2 is fixed to glucose, thereby sustaining energy transfer in the biosphere and forming the basis of the food chain (Michaelian et al., 2014). The evolutionary significance of chlorophyll goes beyond individual organisms to the ecosystem. The development of chlorophyll and its associated photosynthetic processes is a turning point in Earth history, boosting atmospheric oxygen content and markedly transforming the atmosphere and climate of the world. The transition promoted the evolution of aerobic organisms, which transformed ecosystems through enabling species diversification that employs oxygen in respiration. Further, chlorophyll's action in the photochemical dissipation of solar radiation illustrates a thermodynamic need: to maximize the global solar photon rate of dissipation. This autocatalytic process of pigment increase shows a non-equilibrium thermodynamic mandate that guided life evolution itself, illustrating how life adapts to stress and enhances the energy output of the biosphere (Michaelian et al., 2014). However, environmental factors, such as iron availability, critically affect the efficiency of chlorophyll-based photosynthesis. Iron is a crucial component in various protein complexes integral to the photosynthetic pathway, and deficiencies can significantly impair photosynthetic performance. For instance, certain cyanobacteria have adaptive mechanisms to iron starvation, such as the accumulation of the iron stress-induced protein IsiA, which compensates for the decreased level of iron-containing complexes like Photosystem I. Such physiological adaptation also demonstrates the complex interplay between light energy, chlorophyll, and the overall ecological dynamics of freshwater ecosystems (Bibby et al., 2013). Correspondingly, the study of chlorophyll not merely exposes fundamental biochemical functions but also highlights the complex interdependence between organisms and their contexts, illustrating how life continues to improvise under ecological pressures.

2.4.3. The Impact of Photosynthesis on the Ecosystem

Photosynthesis is an anchoring process in ecosystems and exists primarily to be the process by which solar energy is trapped as chemical energy. The process, conducted by plants, algae, and certain bacteria, not only sustains the photosynthetic organisms but also forms the foundation energy source for nearly all forms of life. By creating glucose and oxygen from CO_2 and water, photosynthesis initiates complex webs of energy transformations that supply food chains. The glucose created is needed by herbivores as a carbohydrate source, and the oxygen released is vital for aerobic organisms, including humans. Interestingly, the photosynthetic productivity can structure entire communities as an increase or decrease in its rates can affect biomass yield and energy supply to higher trophic levels, ultimately affecting community processes. In addition, the influence of photosynthesis is not restricted to energy production alone; it plays a significant role in regulating atmospheric gases and enhancing ecological resilience. As photosynthetic organisms absorb CO_2, they help mitigate climate change by sequestering carbon in their biomass and soils. This natural carbon sink can significantly influence global carbon cycles, contributing to climate regulation by stabilizing atmospheric concentrations of greenhouse gases. As rivers and other aquatic ecosystems also participate in carbon cycling with photosynthetic microbes, understanding the same, as highlighted in studies, creates insight into their contribution to the ecological balance. Stable isotope measurements-based technologies, as highlighted in studies (Barth et al., 2011), also enable an understanding of carbon and oxygen cycle interactions, enabling one to perform a detailed assessment of these critical processes. The importance of photosynthesis is brought to broader environmental and evolutionary contexts, particularly as human activity continues to disrupt natural systems. It has been argued in some studies that the Anthropocene—a term given to the current geological epoch characterized by widespread human impact on the geology and ecosystems of the planet—represents a threshold for how energy flows through systems. The marriage of human technological progress with biology makes it difficult for classic paradigms of ecological stability and accommodation. By altering energy in previously unknown ways, human beings are affecting the balance of photosynthesis within terrestrial and aquatic regimes, as described by Alberti et al. (2017). Inasmuch as photosynthesis is still an inherent source of life, its control by human action gives rise to basic questions regarding future ecological sustainability and the evolution of life on Earth (Figure 2.16).

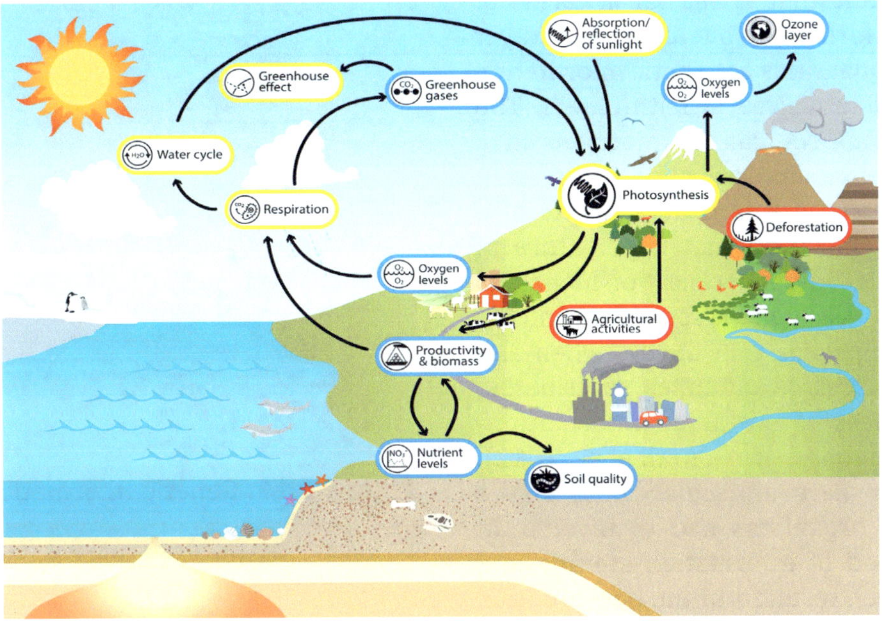

Figure 2.16. The impact of photosynthesis on the ecosystem.

Source: https://ugc.berkeley.edu/wp-content/uploads/2016/03/Photosynthesis_system-model.jpg.

Test Your Knowledge

1. What is the process of photosynthesis in plants?
2. What is the significance of chlorophyll and light energy?

2.5. Genetic Information and DNA

Learning Outcome

1. Structure and function of DNA;
2. The process of DNA replication;
3. The role of RNA in protein synthesis.

Genetic information is central in the search to discover the chemistry of life, especially with regard to the emergence of complexity through molecular interactions. The central issue in this debate is DNA, which packages the hereditary information required by living things to grow, develop and reproduce. It is not only the structure molecule that is special in terms of its double-helix formation, but also in the way it can store huge quantities of genetic information with an incredible degree of faithfulness. The data encoded in DNA can be modified in a context-dependent manner, changing expression and functionality, and this is an example of a hierarchical organization that is a feature of living things. This top-down causation highlights the interplay between genetic sequences and the regulatory mechanisms that dictate when and how genes are expressed, ultimately contributing to the uniqueness of life as an information-driven phenomenon (Cairns-Smith et al., 2012). The jump from mere molecular units to the complex living organisms had to involve an advanced method of information storage and processing, which resulted in the development of DNA as the main genetic material. The earliest stage in the evolution of life, sometimes called the RNA world, depended upon RNA molecules to store genetic information and to catalyze biochemical reactions. Nevertheless, during the evolution of life, a major shift happened with the genetic takeover in which DNA became the central role player in the storage of information. This shift is explained by the higher stability of DNA compared to RNA. The study of error rates during replication highlights why DNA became favored; its lower error rates enable the preservation of genetic information across generations, thus allowing for the accumulation of beneficial mutations and a more stable genome size (Chen et al., 2011). The role of DNA in genetic information goes beyond simple storage; it extends into the level of evolutionary advantages and information transfer fidelity. Of critical importance with this transition was the comparison of RNA and DNA replication, which showed that although replication of RNA was intrinsically less faithful, replication of DNA was more faithful, thus reducing the risks of information degradation.

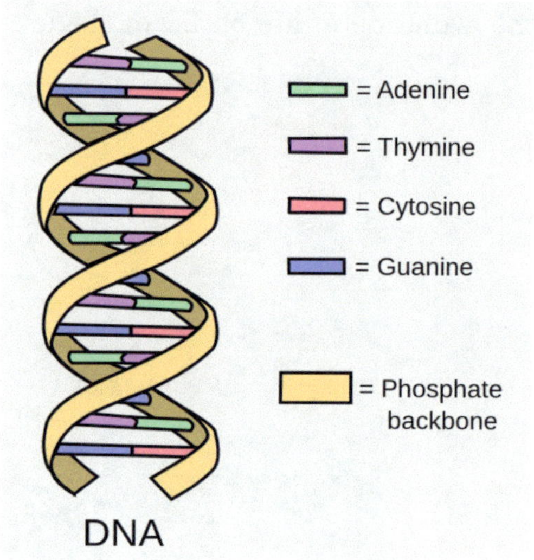

Figure 2.17. Genetic information and DNA.

Source: https://s3-us-west-2.amazonaws.com/courses-images/wp-content/uploads/sites/2318/2016/06/06163522/DNA_simple2.png.

This increased stability allows organisms to store complicated genetic information that can be used in adaptation and survival. In addition, the mechanisms of replication of genetic information provide an example of the delicate waltz of molecular interactions, in which accuracy is the key to the existence of life. The significance of DNA as a reliable repository of genetic data underscores its fundamental role in the chemistry of life, ushering in an era where the richness of biological diversity could flourish (Chen et al., 2011) (Figure 2.17).

2.5.1. Structure and Function of DNA

The deoxyribonucleic acid (DNA) structure is the basis for its role in living organisms, a double helix structure first described by James Watson and Francis Crick. This legendary structure is formed from two strands that are twisted together, stabilized by nitrogenous base pairs: adenine-thymine, cytosine-guanine. The base sequence holds genetic information, being used as blueprints for protein construction and other crucial cellular processes. The base pairing complementarity—adenine to thymine and cytosine to guanine—ensures that DNA can be replicated with very high accuracy since each strand can be employed as a template for the production of a new complementary strand. Structural integrity is crucial since any alteration may produce mutations, which may disrupt essential biological processes.

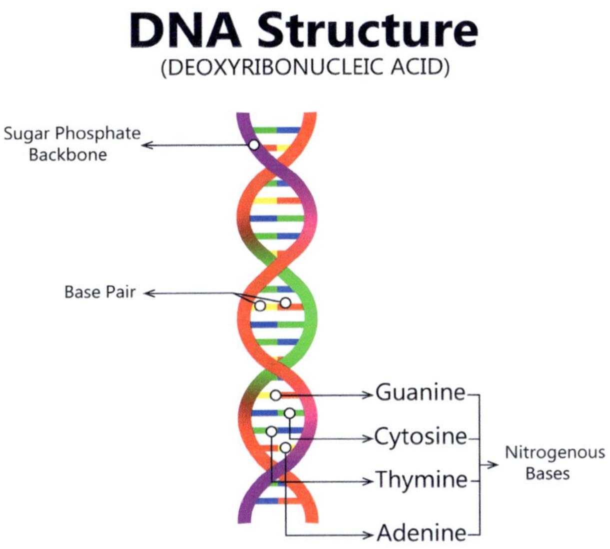

Figure 2.18. DNA structure.

Source: https://external-content.duckduckgo.com/iu/?u=https%3A%2F%2Fstatic.vecteezy.com%2Fsystem%2Fresources%2Fpreviews%2F028%2F627%2F407%2Foriginal%2Fscientific-icon-spiral-of-dna-and-rna-an-illustration-of-the-differences-in-the-structure-of-the-dna-and-rna-molecules-image-poster-structure-rna-and-dna-vector.jpg&f=1&nofb=1&ipt=9d5ae9c047f6649fb009dc6f63a2c33bf129cd17c37bd3088174ef84c86e7c15.

The function of DNA is not limited to the storage of genetic information; it also plays a vital part in regulating cellular activity through gene expression. Each gene in the DNA sequence has specific proteins that perform functions within the cell. Transcription is the initial step of gene expression where messenger RNA (mRNA) is constructed using a template DNA. This mRNA is subsequently translated into proteins at the ribosome. Gene regulation of expression is a complex interaction consisting of various factors, including environmental stimuli and cellular needs. This mechanism of regulation suggests the significance of DNA not only as an information repository for genetic material but also as an active participant in maintaining cellular homeostasis and responding to stimuli, demonstrating its pivotal position in the chemistry of life. Moreover, the evolutionary significance of DNA structure and function cannot be overstated, as it provides the basis for genetic variation and adaptation. Mutations that occur in a DNA sequence of an organism may lead to phenotypic change, with the potential to enhance survival across varied environments. Any genetic variations, whether arising due to environmental pressures outside the organism or as a consequence of replication errors, are imperative driving forces toward evolution. For instance, certain mutations will confer resistance to illness or changing climatic conditions that will affect the overall fitness of a species. This interactive effect of DNA demonstrates its dual function in ensuring vital biological functions and allowing the variability necessary for evolution. Therefore, an understanding of the intricacies of DNA structure and multifunctionality is key to understanding life's basic principles and ongoing adaptation and evolution processes (Laere et al., 2021; Barrera-Martínez et al., 2023) (Figure 2.18).

2.5.2. The Process of DNA Replication

The complex process of replication of DNA lies at the heart of the continuity of life and forms the basis of understanding the chemistry of biological phenomena. It is initiated at specific sites known as origins of replication, where the DNA double helix unwinds and dissociates with the assistance of helicase enzymes. This unwinding gives rise to replication forks, with the nucleotide bases in position to be utilized as templates for new strands of DNA. As the strands are being dissociated, single-strand binding proteins keep the unwound structure together and prevent the strands from re-annealing or degrading. This is a significant step, as it ensures fidelity of the template strands, setting the stage for faithful synthesis of new complementary strands. The correct coordination of these processes not only illustrates the complexity of molecular interactions but also indicates the importance of DNA as the genetic material responsible for heredity. Following the splitting of DNA strands, new strand synthesis takes place as a result of DNA polymerase enzyme action. These enzymes add nucleotides complementary to the uncovered template strands, following rules of base pairing in which adenine pairs with thymine and cytosine pairs with guanine. DNA polymerase is unable to start de novo synthesis or add to a preexisting strand; this must involve the initiation of small RNA primers, which are formed by primase. The presence of primers allows for nucleotide addition by DNA polymerase to start and this results in the continuous synthesis of the leading strand while the lagging strand is synthesized in short sections referred to as Okazaki fragments. This directionality of synthesis mirrors the intrinsic structural properties of DNA as well as the complexity of antiparallel strand alignment. The enzyme-catalyzed assembly of new DNA thus highlights the biochemical principles underlying the continuity of life. The accuracy of DNA synthesis is vital to maintain genetic stability, and numerous proofreading activities lower the error rate during synthesis. DNA polymerase possesses the intrinsic exonuclease activity responsible for removing mispaired nucleotides upon incorporation. The correction of errors is vital as it lowers the overall rate of mutation during replication significantly. In contrast, research has indicated that RNA molecules, as early genetic features had higher error rates in their replication, which may have restricted their evolutionary potential. RNA to DNA evolution into the prevailing genetic material was coupled with increased chemical stability and accuracy of genetic transmission, an event that may have been triggered by its superior capability for error correction (Chen et al., 2011). Studying these replication processes emphasizes the molecular basis of life, showing how stable biochemical systems arise under selective forces over a span of time (Cairns-Smith et al., 2005) (Figure 2.19).

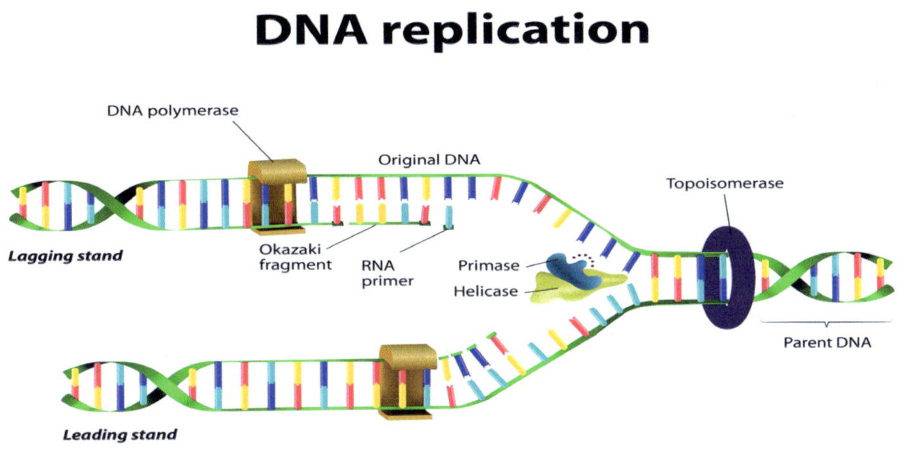

Figure 2.19. DNA replication.

Source: https://d2jx2rerrg6sh3.cloudfront.net/image-handler/picture/2017/5/shutterstock_201778373.jpg.

2.5.3. The Role of RNA in Protein Synthesis

The role of RNA in protein synthesis is the very essence of biochemistry and molecular biology, which illustrates the complex interaction between biological function and genetic information. RNA or ribonucleic acid is the intermediary between DNA and protein and is tasked with the transmission of the genetic information from the nucleus to the cytoplasm, where proteins are being synthesized. The process begins with transcription, wherein messenger RNA (mRNA) is synthesized from a DNA template. The mRNA replica carries the genetic information needed to construct proteins, highlighting RNA's crucial role as an information carrier of biological nature. And so, it is important to learn the mechanisms of RNA in this sense for an appreciation of the broad principles underlying the chemistry of life, particularly specificity and determination in the transmission of hereditary traits from generation to generation. Once mRNA is produced, it then needs to be translated, a complex process where ribosomes decode the mRNA sequence and translate the proteins based on information found there. Transfer RNA (tRNA) also plays a crucial role at this stage as it brings the corresponding amino acids to the ribosome and matches them with the codons in the mRNA sequence. This tRNA anticodons match the mRNA codons is the ultimate example of specificity in biological reactions, where each tRNA is specifically designed to bind with its respective amino acid. The precision of this process is important since errors lead to faulty proteins and, by definition, cellular dysfunction. The dynamic functions of RNA in both transcription and translation, therefore, emphasize its importance in maintaining the integrity of the genetic code as well as proper protein synthesis within the cell environment. Moreover, the communication between RNA and protein synthesis illuminates the evolution of life, as according to the RNA World hypothesis, ancient organisms utilized RNA not only to store genetic material but also as reaction catalysts. According to this perspective, protein synthesis evolved from early RNA-based mechanisms, as RNA molecules progressively developed more complex roles in metabolism and regulation. While evolution made its transition from the RNA World to the Protein World, the specificity of RNA in the propagation of genetic information was essential to the evolution of sophisticated types of life. Therefore, research into the role of RNA in protein synthesis not only illuminates the mechanistic finer points of cellular function but also further reveals the source of life's origin and the evolutionary achievement of biological complexity (Budisa et al., 2019; Griffiths et al., 2017) (Figure 2.20).

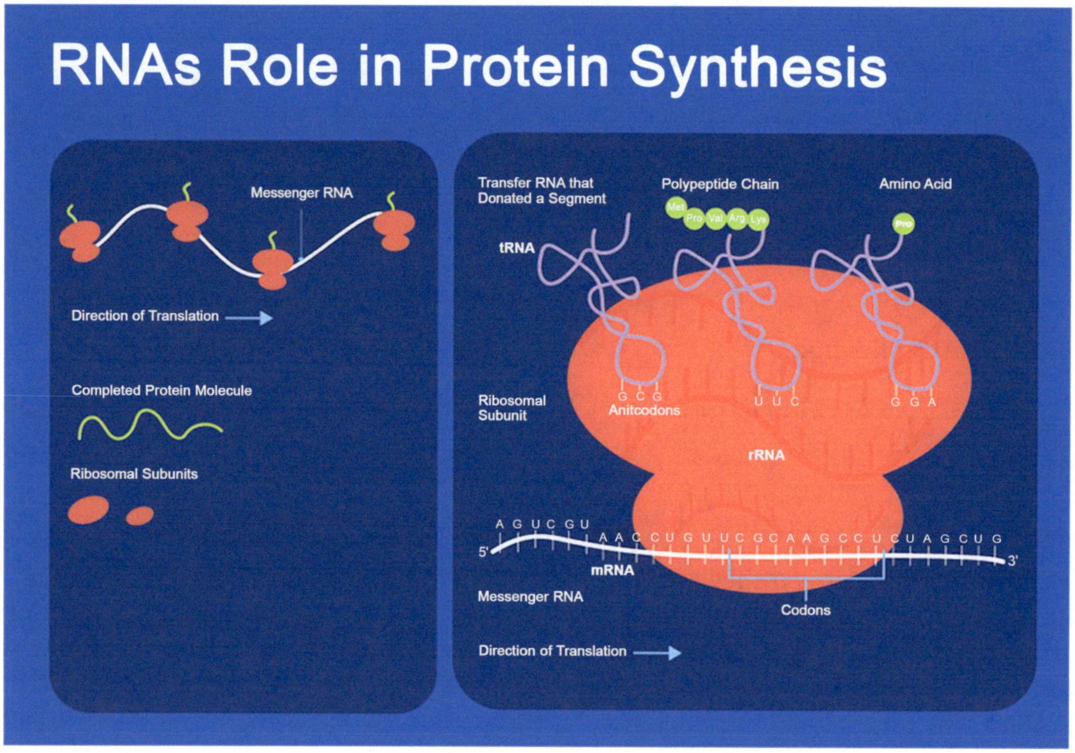

Figure 2.20. The role of RNA in protein synthesis.

Source: https://norgenbiotek.com/sites/default/files/images/blog/240229/Fig3-lrg.jpg.

Test Your Knowledge

1. Explain the structure and function of DNA.
2. What is the process of DNA replication?

2.6. Biochemical Signaling

Learning Outcome

1. Types of signaling molecules;
2. Mechanisms of signal transduction;
3. The importance of signaling in cellular communication.

Cellular communication depends on biochemical signaling, which enables cells to react to different stimuli and ensure homeostasis. This is a complex process whereby the signaling molecules bind to definite receptors, thereby resulting in a sequence of intracellular reactions. The variety of signaling processes, such as autocrine signaling, paracrine signaling and endocrine signaling, demonstrates the versatility and complexity of biochemical networks. As an example, hormonal signaling pathways can control metabolism, whereas local signaling molecules frequently coordinate the response to tissue damage. It is important to learn how these pathways work because their malfunctioning may cause numerous disorders, such as cancer and diabetes. The key to these pathways is the involvement of second messengers, including cyclic AMP and calcium ions, which amplify and transmit the original signal, making sure that the cells react quickly to external stimuli. In addition, the mitochondria are central to cellular signaling in addition to their conventional roles in energy metabolism. Recent studies have emphasized the significance of the mitochondrial calcium uptake in the modulation of many physiologic events, including ATP production and apoptosis. Mitochondrial calcium uniporter complex is an example of how calcium level imbalance can trigger cell death, highlighting the narrow balance that must be achieved to allow cells to work optimally. Such a regulatory process shows that mitochondrial signaling not only plays a role in energy balance but also in cell survival and death pathways. This level of complex interactions shows the importance of mitochondrial health as a disruption in these pathophysiological conditions can occur, further highlighting the connected nature of biochemical signaling and cellular chemistry as a whole. The growing appreciation of these mitochondrial signaling pathways reflects an evolving understanding of life at the molecular level, opening avenues for therapeutic interventions targeting metabolic disorders and degenerative diseases (Quintana et al., 2017). Lastly, the occurrence of stochastic attractors in biochemical signaling networks provides a further example of the complexity of cellular communication. Such attractors are a property of nonlinear networks and may correspond to different states of the cellular level driven by environmental gradients and internal feedback interactions. They provide a dynamic range of responses to stimuli, which makes the cells adjust to varying conditions. The presence of these attractors underscores the co-existence of randomness and organization in biological systems, which results in emergent properties that determine cellular behavior. Such self-organization is actually important in processes such as gene regulation and cellular signaling, where relatively small fluctuations can result in large changes in cellular behavior. As research continues to unravel these complex dynamics, it becomes increasingly clear that understanding biochemical signaling is essential to deciphering the fundamental processes of life itself (Ge et al., 2012) (Figure 2.21).

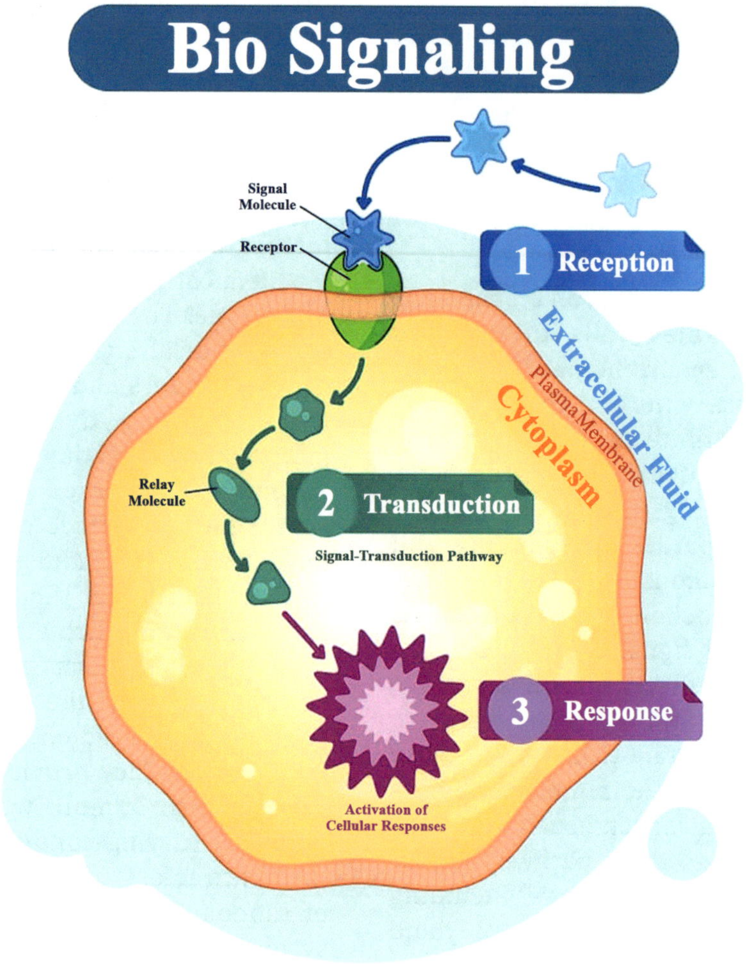

Figure 2.21. Biochemical signaling.

Source: https://biologydictionary.net/wp-content/uploads/2017/05/Cell-Signaling-Pathway.jpg.

2.6.1. Types of Signaling Molecules

In cell communication, signaling molecules play a crucial role to play in mediating various biological processes essential for life. The molecules belong to different classes with different functions and pathways. One of the most common are the hormones, which are secreted from glands and flow through the bloodstream, regulating processes such as growth, metabolism, and homeostasis. In addition, neurotransmitters operate at synapses, conveying information between neurons and modulating mood, pain, and cognition. The diversity of signaling molecules, including growth factors (GFs), cytokines, and small molecules, is an indicator of the complexity of cell-cell interactions and the highly regulated systems permitting communication among numerous cell types. GFs are a classic example of a crucial class of signaling molecules to trigger cellular growth, differentiation, and repair. They are particularly important in regenerative medicine because they can initiate repair mechanisms in tissues. Nevertheless, the therapeutic application of GFs is frequently hampered by rapid clearance from the body as well as a need for high doses to produce desired results. This has led researchers to search for new means of enhancing their effectiveness but diminishing unwanted side effects. For instance, novel material-based approaches have been developed to maintain the bioactivity of GFs at lower concentrations, utilizing extracellular matrix (ECM) interactions that govern their activity and presentation, with important impacts on tissue regeneration and healing (Cantini et al., 2016). Small molecules represent an additional key category of signaling molecules, including an extensive range of compounds with the capacity to control a large number of cellular functions. These molecules are likely to act as second messengers, relaying signals from the cell surface's hormone receptors to internal processes of metabolism. Small molecules have been in vogue not just for what they accomplish in natural signaling but for what could therapeutically be accomplished with them. Researchers are working to screen small molecule collections for hopefuls that will enhance stress resistance and longevity, particularly in human cells (Zhang et al., 2020). This field of investigation has the potential to discover compounds capable of inducing healthy lifespan prolongation by targeting particular stress-response regulators. Generally, the continued investigation of signaling molecules, including GFs and small molecules, is paramount to enhance our understanding in terms of the biochemical determinants of life and develop effective medical therapies.

2.6.2. Mechanisms of Signal Transduction

The complex signal transduction pathways form the basis of cell-to-cell communication and cell response to the outside world. These processes are controlled by a series of biochemical processes triggered by extracellular signals that interact with specific receptors at the cell surface. Upon stimulation, the receptors initiate a cascade of intracellular signaling reactions involving the phosphorylation or dephosphorylation of various proteins. This signaling network plays an important part in the regulation of numerous cellular functions, including growth, differentiation, and apoptosis. By having the full comprehension of these functions, scientists are now in a better position to ascertain how cells react to changes in their environment, which is vital in maintaining homeostasis and overall organismal health. Recent findings have indicated the effect of some agents in modulating signal transduction cascades, focusing particularly on pharmacological effects such as PUVA therapy. Psoralen in combination with PUNTA shows an unusual interaction with cellular components, profoundly altering lipid membrane structures in platelets. In highly complex biochemical interactions, psoralen and ultraviolet A light (PUNTA) increase the order of lipid phases by experiencing covalent adduct formation with fatty acyl chains in phospholipids and subsequently inhibiting membrane recruitment of major kinases such as Akt and Bruton's tyrosine kinase. Such an interruption leads to compromised phosphorylation events required for effective signaling. Examining such perturbations is highly informative regarding how environmental factors might influence cellular communication mechanisms, illuminating the exact balance involved in signal transduction pathways (Compernolle et al., 2016). In addition, the p38-mitogen-activated protein kinase (MAPK) signaling pathway is an excellent model for the interconnectedness of signal transduction, synaptic plasticity, and neurodegenerative disease. Activation of p38 MAPK has been associated with significant cellular processes such as inflammation and reorganization of actin, which are significant for synaptic plasticity and function. Significantly, p38 MAPK binding to substrates such as MAPKAPK

2 (MK2) and tau protein has been shown to influence mechanisms of neuroinflammation and tau hyperphosphorylation. The elucidation of these signal pathways might reveal novel therapeutic targets aimed at reversing the onset of neurodegenerative diseases like Alzheimer's and Parkinson's disease. Thus, the demultiplexing of signal transduction mechanisms at cellular and molecular levels is crucial to facilitate further understanding of complex biological processes as well as the development of targeted therapies (Corrêa et al., 2012) (Figure 2.22).

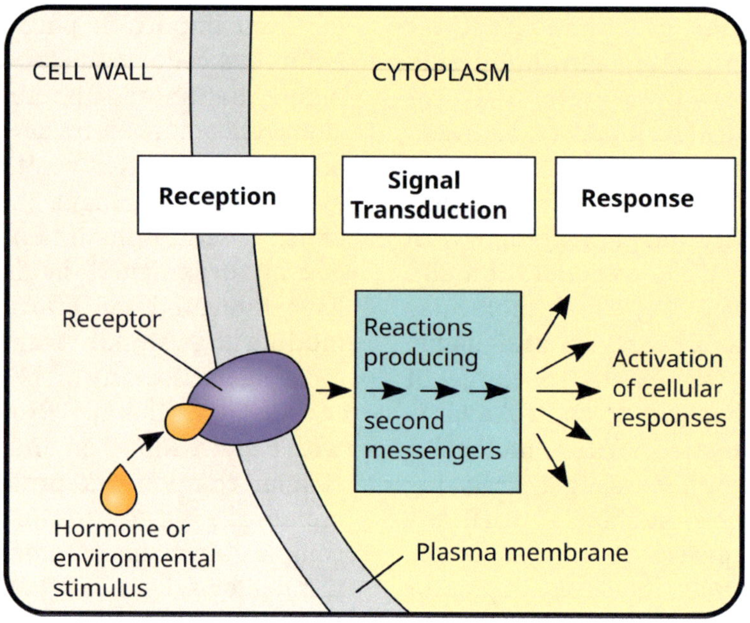

Figure 2.22. Mechanisms of signal transduction.

Source: https://biologydictionary.net/wp-content/uploads/2018/03/Signal-Transduction-Pathways-Model.jpg.

2.6.3. The Importance of Signaling in Cellular Communication

Cell-to-cell communication is the lifecore that involves complex mechanisms of signal transduction that control the varied activities of organisms. These systems of communication are known as cell signaling networks (CSNs) and consist of biochemical pathways that allow cells to respond to stimuli within and outside the cell. CSNs are crucial to learn because they allow basic functions such as growth, differentiation, and immune function. Through signaling, cells can exchange information and modify their behavior according to the surrounding environment, which is vital for homeostasis regulation. Through continued study of CSNs, network modeling and evolution in silico have been suggested as a useful means for unpacking the complexity that is involved within cellular communications systems, which potentially will lead us to have a greater knowledge of biological processes underlying life itself (Decraene et al., 2008). The significance of cellular signaling is not merely a communication issue but contributes significantly to biological evolution. New paradigms in the life definition propose that effective communication is central to the functioning of living organisms. The modern definition of life encompasses cellular self-replication and metabolism, but also emphasizes how communicative interactions drive important processes. These interactions are the basis for how cells respond to their environment, showing the interconnectedness of various cellular processes. Furthermore, recent evidence suggests that these communication networks have evolved more from complex interactions between cells and their ecological partners, such as viruses and RNA networks, rather than from random mutations. This emphasizes the importance of signaling in creating the evolutionary world of life forms (Witzany et al., 2020). In the context of The Chemistry of Life, signaling mechanisms are the coming together of biochemistry and new findings on biological communication. By following the paths taken by cells to receive and send signals, scientists gain important information on how molecular interactions get interpreted as processes of life. For instance, crosstalk—inter-talk between more than one signaling pathway—is how cells adapt dynamically to altering internal and external environments. This adaptability is at the heart of complex biological processes, from immune response to endocrine regulation, and underscores the role of signaling in organismal existence. Deciphering the chemical makeup of these signals and their transmission from cell to cell, we may develop an appreciation of the unimaginable influence that cellular communication wields over the most internal processes of life itself, once again unifying the chemistry that abounds in dynamic living systems.

Test Your Knowledge

1. What are the types of signaling molecules?
2. What are the mechanisms of signal transduction?

Summary

The complexity of interrelationships that constitute the chemistry of life highlights the main principles of biological processes. Whether it is the molecular interactions that underpin cellular processes or the biomechanical systems that underlie organisms, chemistry is the pivot of life on Earth. With ongoing research to decipher the intricacies of the biochemical pathways, it is becoming more and more clear that deciphering these relationships is the key to solving some of the looming global issues, such as environmental degradation and health-related concerns of the population. The persistent threat posed by heavy metals, as highlighted in the literature, poses significant risks to both ecosystems and human populations, necessitating further investigation into their impact on biological systems (Ali et al., 2019). With the increasingly clear connections among chemical stability, ecological balance and human health, there is an urgent need to ensure comprehensive strategies are put in place to alleviate these concerns. In addition, discussion of microbial communities adds depth to the chemistry of life, explaining that even the tiniest of organisms have significant roles to play in biogeochemical cycles. The tremendous diversity of microbial life evident in recent large-scale metagenomic studies suggests that these bacteria and viruses are not incidental inhabitants of their ecological niches but are central players in ecological processes. Notably, the presence of shared functional traits between oceanic microbial communities and the human gut microbiome reveals unexpected connections that could inform health and ecological research (Sunagawa et al., 2015). The identification and utilization of these interactions can help give insights into how a disturbance in these communities could influence the functioning of the ecosystem in general. Therefore, our understanding of the chemistry of life requires interdisciplinary research strategies that integrate approaches of biochemistry, ecology, and environmental science. Last but not least, the necessity of a profound knowledge of the chemistry of life goes beyond academic interest; it has a direct impact on policymaking and the welfare of society. The implications of toxic heavy metals on human health and ecosystems, as discussed, highlight the necessity for active monitoring and remediation strategies to safeguard both human and environmental health (Ali et al., 2019). Also, microbial interactions have further implications that demonstrate that changes in a single element of the ecosystem can propagate across the whole network and eventually affect human health. Considering such complexities, it is important to promote a comprehensive perception of the biochemical basis of life. In the future, interdisciplinary research involving society will be essential in solving these problems to ensure the continuity and prosperity of humans and the diversity of life on the planet.

In exploring the complex relationships of biological systems, the role of chemistry becomes the key to understanding life itself. The arguments surrounding the molecular composition of living organisms emphasize the significance of elements such as carbon, hydrogen, oxygen, and nitrogen. They combine into vital compounds such as carbohydrates, proteins, lipids, and nucleic acids, the very fabric of life. Understanding these biochemical structures not only clarifies metabolic processes but also how organisms respond to their environment. This biochemical structure provides the basis for understanding more sophisticated phenomena, such as the function of enzymes and energy transfer, and it illustrates the relevance of chemistry in cellular functions as well as broader ecological systems. The application of chemistry to practical applications can be seen most readily within the field of pharmacology. The warfarin therapy case study is a perfect illustration of how chemistry is placed into context in clinical practice, demonstrating the significance of drug interaction at the molecular level. It has been noted that students who go through such case studies as this have better pharmacological principle comprehension than students who receive lecture-based education (Birdi et al., 2019). Through combining theoretical understanding with practical application, students come to comprehend the complexities of drug mechanisms and patient care and, in doing so, become improved clinical practitioners. This approach is not only beneficial in fostering a more robust appreciation of chemistry's role in health but also in preparing future practitioners to address more complex challenges of pharmacotherapy. Furthermore, the successful communication of scientific information continues to be an essential role of the chemistry of life. Research shows that researchers who engage in outreach and popularization of their work contribute significantly to academic production as they simultaneously enhance their profile in society (Croissant et al., 2008). Though the contribution of these efforts to a researcher's academic career continues to remain low, their importance in bridging the scientific ideas and the wider society cannot be overstated. The dialogue between scientists and amateurs serves to bridge the gap between complex concepts and common sense, ultimately leading to a greater appreciation of the chemistry of life. The partnership demands that scientists explain their findings well so that the chemistry of life becomes more than just attractive in universities.

KEY TERMS FROM THIS CHAPTER

Acid	Atom	Base
Buffer	Carbohydrate	Compound
Covalent bond	Element	Enzyme
Hydrogen bond	Ionic bond	Lipid
Macromolecule	Molecule	Nucleic acid
pH scale	Polarity	Protein
Solvent	Water (H2O)	

Critical Thinking Questions

1. **What type of bond holds together the atoms within a water molecule?**
   a. Ionic bond
   b. Covalent bond
   c. Hydrogen bond
   d. Metallic bond
2. **Which element is a key component of all organic molecules?**
   a. Oxygen
   b. Nitrogen
   c. Carbon
   d. Hydrogen
3. **Which of the following is NOT a macromolecule essential to life?**
   a. Carbohydrates
   b. Proteins
   c. Water
   d. Nucleic acids
4. **What property of water allows it to dissolve many substances?**
   a. Its polarity
   b. Its density
   c. Its temperature
   d. Its color
5. **What is the primary function of lipids in living organisms?**
   a. Store genetic information
   b. Provide long-term energy storage and insulation
   c. Speed up chemical reactions
   d. Carry oxygen in the blood

Short Answer Questions

1. Explain why water is essential for life, focusing on its unique chemical properties.
2. What does it mean that water is a polar molecule, and how does this polarity influence hydrogen bonding?
3. Describe the concept of pH and explain the significance of acids, bases, and buffers in biological systems.
4. What are macromolecules? List the four main classes of biological macromolecules.
6. How do enzymes affect the rate of chemical reactions in living organisms?

Answers to Critical Thinking Questions

1. (b); 2. (c); 3. (c); 4. (a); 5. (b)

References

1. Ács, B., Bá, C. S., Huba, J. M., Kiss, P., Csermely, P., & Nánási, T., Zoltán. (2008). Ageing as a price of cooperation and complexity: Self-organization of complex systems causes the ageing of constituent networks. https://core.ac.uk/download/pdf/288178.pdf (accessed on 13th June 2025).
2. Alberti, M., Frank, A., & Kleidon, A. (2017). Earth as a hybrid planet–The Anthropocene in an evolutionary Astro biological context. http://arxiv.org/abs/1708.08121 (accessed on 13th June 2025).
3. Braakman, R., & Smith, E. (2013). The compositional and evolutionary logic of metabolism. *Physical biology*, *10*(1), 011001. https://doi.org/10.1088/1478-3975/10/1/011001 (accessed on 13th June 2025).
4. Ali, H., Khan, E., & Ilahi, I. (2019). Environmental chemistry and ecotoxicology of hazardous heavy metals: Environmental persistence, toxicity, and bioaccumulation. *Journal of Chemistry, 2019*, 1–14. https://doi.org/10.1155/2019/6730305 (accessed on 13th June 2025).
5. Zhang, P., Zhai, Y., Cregg, J., Ang, K. K., Arkin, M., & Kenyon, C. (2020). Stress Resistance Screen in a Human Primary Cell Line Identifies Small Molecules That Affect Aging Pathways and Extend *Caenorhabditis elegans*' Lifespan. *G3 (Bethesda, Md.)*, *10*(2), 849–862. https://doi.org/10.1534/g3.119.400618
6. Arce, J., & Betancourt, R. (2001). The design, implementation, and assessment of a new capstone course aimed at science education majors. https://core.ac.uk/download/215487647.pdf (accessed on 13th June 2025).
7. Arese, M., Buscarinu, M. C., Chessa, L., Coarelli, G., et al. (2015). Evidence for detrimental cross interactions between reactive oxygen and nitrogen species in Leber's hereditary optic neuropathy cells. https://core.ac.uk/download/54529895.pdf (accessed on 13th June 2025).
8. Avakyan, S. V., & Baranova, L. A. (2023). Microwave radiation in the environment and its impact on the state of living organisms. *Austin Journal of Infectious Diseases.* https://www.semanticscholar.org/paper/ffdd5962e89912e7116b6a41dae650dcb7d6a16f (accessed on 13th June 2025).
9. Baianu, I.C. and Lozano, P. and Prisecaru, V. I. and Lin, H.C. (2004) *Novel Techniques and Their Applications to Health Foods, Agricultural and Medical Biotechnology: Functional Genomics and Basic Epigenetic Controls in Plant and Animal Cells.* [Preprint](accessed on 13th June 2025).
10. Barrera-Martínez, I., Borrás-Enríquez, A. J., Gonzalez-Escobar, J. L., Moreno-Vilet, L., García-Barrón, S. E., Medina-Moreno, S. A., et al. (2023). Development of a gluten- and lactose-free bakery product with nutritional and functional potential. *Revita Mexicana de Ingeniería Quamina.* https://www.semanticscholar.org/paper/f203fcf73de35aaa8a385c56bbc2a15d269a81d6 (accessed on 13th June 2025).
11. Barth, J. A. C., Freitag, H., Karim, A., et al. (2011). Applications of stable water and carbon isotopes in watershed research: Weathering, carbon cycling, and water balances. https://core.ac.uk/download/12042973.pdf (accessed on 13th June 2025).
12. Braeckman, B. P., Houthoofd, K., & Vanfleteren, J. R. (2009). Intermediary metabolism. *WormBook : the online review of C. elegans biology*, 1–24. https://doi.org/10.1895/wormbook.1.146.1

13. Barthel, J., Eisenberg, B., Corry, B., Doi, M., Engl, H. W., Hyon, Y., et al. (2013). Ions in fluctuating channels: Transistors alive. http://arxiv.org/abs/q-bio/0506016 (accessed on 13th June 2025).

14. Bibby, T. S., Campbell, D. A., Cockshutt, A. M., Fraser, A., et al. (2013). Photo physiological and photosynthetic complex changes during iron starvation in *Synechocystis* sp. PCC 6803 and *Synechococcus elongatus* PCC 7942. [DOI not provided].

15. Birdi, A., Fergus, S., & Riaz, F. (2019). A framework model for a contextualized and integrated warfarin therapy case in a master of pharmacy program. https://core.ac.uk/download/287582262.pdf (accessed on 13th June 2025).

16. Budisa, N., & Kubyshkin, V. (2019). The alanine world model for the development of the amino acid repertoire in protein biosynthesis. https://core.ac.uk/download/287885179.pdf (accessed on 13th June 2025).

17. Cairns-Smith, A. G., Campbell, D. T., Cline, D. B., Davies, P. C. W., Ellis, G. F. R., Gesteland, R. F., et al. (2012). The algorithmic origins of life. http://arxiv.org/abs/1207.4803 (accessed on 13th June 2025).

18. Cairns-Smith, A. G., Mao, C., Winfree, E., Chen, H. L., et al. (2005). Self-replication and evolution of DNA crystals. https://core.ac.uk/download/4888340.pdf (accessed on 13th June 2025).

19. Cantini, M., Cheng, Z. A., Dalby, M. J., García, A., et al. (2016). Material-driven fibronectin assembly for high-efficiency presentation of growth factors. https://core.ac.uk/download/74229857.pdf (accessed on 13th June 2025).

20. Chen, I. A., Gerland, U., Leu, K., Obermayer, B., et al. (2011). The prebiotic evolutionary advantage of transferring genetic information from RNA to DNA. https://core.ac.uk/download/323080491.pdf (accessed on 13th June 2025).

21. Chen, J. G., Crooks, R. M., Seefeldt, L. C., Bren, K. L., Bullock, R. M., Darensbourg, M. Y., Holland, P. L., et al. (2018). Beyond fossil fuel–driven nitrogen transformations. *Science, 360.* https://doi.org/10.1126/science.aar6611 (accessed on 13th June 2025).

22. Compernolle, V., Devloo, R., Feys, H., Sandra, A., et al. (2016). Psoralen and ultraviolet A light treatment directly affects phosphatidylinositol 3-kinase signal transduction by altering plasma membrane packing. https://core.ac.uk/download/74712642.pdf (accessed on 13th June 2025).

23. Corrêa, S. A. L., & Eales, K. L. (2012). The role of p38 MAPK and its substrates in neuronal plasticity and neurodegenerative disease. https://core.ac.uk/download/9257821.pdf (accessed on 13th June 2025).

24. Croissant, Y., Jensen, P., Kreimer, P., Rouquier, J. B., et al. (2008). Scientists who engage with society perform better academically. https://core.ac.uk/download/47891239.pdf (accessed on 13th June 2025).

25. De Laere, B., Crippa, A., Mortezavi, A., Ghysel, C., Rajan, P., Eklund, M., Wyatt, A., et al. (2021). Increased pathway complexity is a prognostic biomarker in metastatic castration-resistant prostate cancer. *Cancers, 13.* https://www.semanticscholar.org/paper/cb643621449f15ef30726f05956860cd1ce7cdbb (accessed on 13th June 2025).

26. Decraene, J., McMullin, B., & Mitchell, G. G. (2008). Modeling and evolving biochemical networks: Insights into communication and computation from the biological domain. https://core.ac.uk/download/11308993.pdf (accessed on 13th June 2025).

27. Freisinger, E., & Sigel, R. K. O. (2019). The bioinorganic periodic table. https://core.ac.uk/download/532357852.pdf (accessed on 13th June 2025).

28. Ge, H., & Qian, H. (2012). Mesoscopic biochemical basis of isogeneic inheritance and canalization: Stochasticity, nonlinearity, and emergent landscape. https://core.ac.uk/download/645587556.pdf (accessed on 13th June 2025).

29. Godoy-Gallardo, M., Eckhard, U., Delgado, L. M., de Roo, J. J. D., Hoyos-Nogués, M., Gil, F. J., & Pérez, R. A. (2021). Antibacterial approaches in tissue engineering using metal ions and nanoparticles: From mechanisms to applications. *Bioactive Materials, 6*, 4470–4490. https://doi.org/10.1016/j.bioactmat.2021.04.033 (accessed on 13th June 2025).

30. Goldford, J. E., Hartman, H., Segrè, D., Smith, T., et al. (2017). Remnants of an ancient metabolism without phosphate. https://open.bu.edu/bitstream/2144/24313/1/Goldford_etal_Cell.pdf (accessed on 13th June 2025).

31. Griffiths, P. E., & Stotz, K. (2017). Biological information, causality and specificity–an intimate relationship. https://core.ac.uk/download/153326381.pdf (accessed on 13th June 2025).

32. Hasanuzzaman, M., Bhuyan, M. H. M. B., Zulfiqar, F., Raza, A., Mohsin, S. M., Al Mahmud, J., Fujita, M., et al. (2020). Reactive oxygen species and antioxidant defense in plants under abiotic stress: Revisiting the crucial role of a universal defense regulator. *Antioxidants, 9*, 681–681. https://doi.org/10.3390/antiox9080681 (accessed on 13th June 2025).

33. Herbert, E. R., Boon, P. I., Burgin, A. J., Neubauer, S. C., Franklin, R. B., Ardón, M., Hopfensperger, K. N., et al. (2015). A global perspective on wetland salinization: Ecological consequences of a growing threat to freshwater wetlands. *Ecosphere, 6*, 1–43. https://doi.org/10.1890/es14-00534.1 (accessed on 13th June 2025).

34. Horowitz, N. H., & Hubbard, J. S. (1974). The origin of life. https://core.ac.uk/download/4869314.pdf (accessed on 13th June 2025).

35. Ignatov, I., & Mosin, O. (2015). Chromatographic separation of deuterium-labeled amino acids, proteins, carbohydrates and lipids isolated from bacterial objects. https://core.ac.uk/download/234697467.pdf (accessed on 13th June 2025).

36. Ignatov, I., & Mosin, O. (2015). Studying the hydrological conditions for origin of first organic forms of life ın hot mineral water with HDO. https://core.ac.uk/download/234697427.pdf (accessed on 13th June 2025).

37. Krenn, M., Ai, Q., Barthel, S., Carson, N., Frei, A., Frey, N. C., Friederich, P., et al. (2022). SELFIES and the future of molecular string representations. *Patterns, 3*, 100588–100588. https://doi.org/10.1016/j.patter.2022.100588 (accessed on 13th June 2025).

38. Leydesdorff, L. (2008). The delineation of nanoscience and nanotechnology in terms of journals and patents: A most recent update. http://arxiv.org/abs/0911.1445 (accessed on 13th June 2025).

39. Li, J., Cao, F., Yin, H., Huang, Z., Lin, Z., Mao, N., Sun, B., et al. (2020). Ferroptosis: Past, present and future. *Cell Death and Disease, 11.* https://doi.org/10.1038/s41419-020-2298-2 (accessed on 13th June 2025).

40. Martins, C. R., Andrade, J. B., & Lopes, W. A. (2013). Solubilities das substances organics. https://www.scielo.br/pdf/qn/v36n8/v36n8a26.pdf (accessed on 13th June 2025).

41. Michaelian, K., & Simeonov, A. (2014). Fundamental molecules of life are pigments which

arose and evolved to dissipate the solar spectrum. http://arxiv.org/abs/1405.4059 (accessed on 13th June 2025).

42. Quintana, A., Raffaello, A., Rasola, A., Tosatto, A., Varadi, A., Rowland, A. A., Tarasov, A. I., et al. (2017). The mitochondrial Ca(2+) uniporter: Structure, function, and pharmacology. https://core.ac.uk/download/129191720.pdf (accessed on 13th June 2025).
43. Rutz, A., Sorokina, M., Galgonek, J., Mietchen, D., Willighagen, E., Gaudry, A., Graham, J. G., et al. (2022). The LOTUS initiative for open knowledge management in natural products research. *eLife, 11.* https://doi.org/10.7554/elife.70780 (accessed on 13th June 2025).
44. Segev, N., & Henderson, R. (2023). Resolution revolution—Seeing the molecules of life with electron cry microscopy. *Volume, 11.* https://www.semanticscholar.org/paper/2ee145cf0a8230ae2644ac9e9a2c68a545ec06eb (accessed on 13th June 2025).
45. Singh, J., Dutta, T., Kim, K. H., Rawat, M., Samddar, P., & Kumar, P. (2018). 'Green' synthesis of metals and their oxide nanoparticles: Applications for environmental remediation. *Journal of Nanobiotechnology, 16.* https://doi.org/10.1186/s12951-018-0408-4 (accessed on 13th June 2025).
46. Steele, K. F. (1998). *Directory of Water Related Courses Offered at Colleges and Universities in Arkansas as of November 1998.* https://core.ac.uk/download/145191276.pdf (accessed on 13th June 2025).
47. Sunagawa, S., Coelho, L. P., Chaffron, S., Kultima, J. R., Labadie, K., Salazar, G., Djahanschiri, B., et al. (2015). Structure and function of the global ocean microbiome. *Science, 348.* https://doi.org/10.1126/science.1261359 (accessed on 13th June 2025).
48. Witzany, G. (2020). What is life? https://core.ac.uk/download/293755771.pdf (accessed on 13th June 2025).

BIOLOGICAL MACROMOLECULES

3

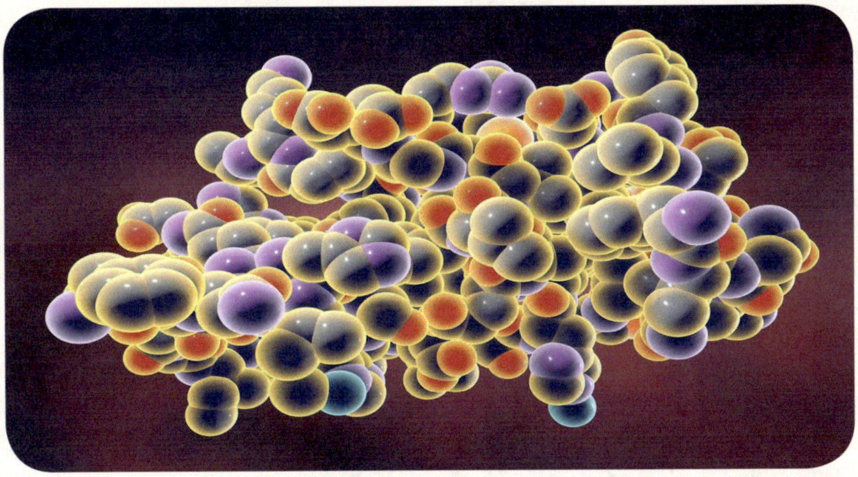

LEARNING OBJECTIVES

After studying this chapter, you will be able to:

1. Define carbohydrates;
2. Discuss about proteins;
3. Describe lipids;
4. Explain about nucleic acids;
5. Focus on interactions between macromolecules;
6. Explore macromolecules in health and disease.

Introduction

The study of biological macromolecules is important in understanding the complexity of life at both the cellular and molecular levels. The macromolecules like proteins, nucleic acids, carbohydrates, and lipids play major roles in various biological processes. The macromolecules not only play the role of structural components of a cell but also that of enzymes to catalyze biochemical reactions, as energy storage molecules, and as genetic information carriers. This complex system of functions leads to the requirement for an intimate knowledge of their structure, properties, and interaction. In addition, the successful incorporation of functional groups into macromolecules enhances their reactivity and allows applications specific to biotechnology and medicine. The coalescence of these factors serves as a basis for exploring the complex dynamics of biological macromolecules and opening up for the tremendous implications of their research and the mechanism underlying their function in living systems. An in-depth study of biological macromolecules reveals the unique positions of these entities, namely their behavior and interaction in biological systems. Macromolecules do not operate as individuals; rather, they are part of dense macromolecular environments with high concentrations of various cellular components. This macromolecular crowding, as described in recent studies, influences diffusion dynamics and biological interactions, leading to non-obvious diffusion behaviors that cannot be predicted with the use of standard models. Syntheses of theoretical models, like Gaussian models and the continuous time random walk (CTRW) model, describe how these phenomena are enacted in cellular behavior. Moreover, post-modification of preexisting reactive particles, as documented in literature, highlights the importance of enhancing macromolecule functionality without compromising their fundamental properties. This union of structure and function sheds light on broader implications for synthetic biology, drug delivery systems, and materials science. In general, the investigation of biological macromolecules is an entry point to the disclosure of underlying biological mechanisms and innovative applications. The roles that such macromolecules play in biological systems are varied, from affecting cellular structure to metabolic processes. Advances in experimental techniques, such as single particle tracking and fluorescence correlation spectroscopy (FCS), have enabled researchers to monitor the complex and often anomalous movements of molecules under crowded conditions. These developments not only enable more understanding of molecular dynamics but also enable the development of new strategies for controlling biological systems. As the science advances, ongoing research continues to bridge gaps in knowledge about structure-property relationships and transport mechanisms within cells. These advances will ultimately pave the way for fundamental innovation in biotechnology, materials science, and drug discovery, once again establishing the core importance of biological macromolecules in medicine and science (Figure 3.1).

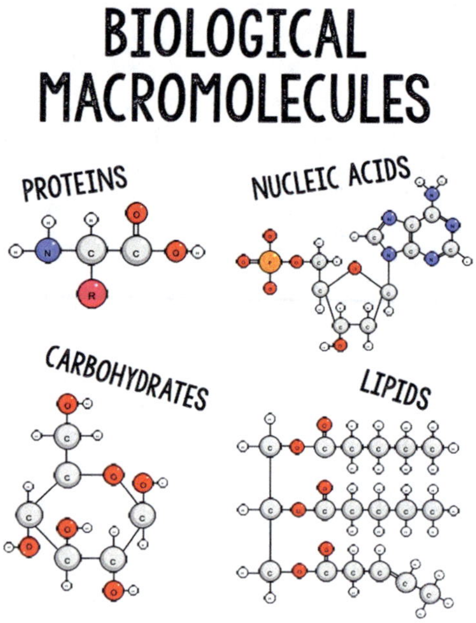

Figure 3.1. Biological macromolecules.

Source: https://i.pinimg.com/474x/66/4c/11/664c11c3ab548a27920c96007a0ab24d.jpg.

Definition of Biological Macromolecules

Biological macromolecules are complex, large molecules that carry the body's cells, tissues, and organs' structure, function, and regulation. Biological macromolecules include proteins, nucleic acids, carbohydrates, and lipids, all with specific and essential roles in living things. Proteins, made up of amino acids linked together by peptide bonds, serve as catalysts in biochemical reactions, enable communication among and within cells, and provide strength in various biological systems. Nucleic acids, DNA and RNA, store and transmit genetic information and regulate protein synthesis, and ensure the continuity of life. Carbohydrates, however, are primarily stores of energy and structural protection factors, while lipids are components of cellular membranes and storage units of energy. Hence, the diverse functions of biological macromolecules further establish their pivotal position in ensuring life and cellular homeostasis. Structure and synthesis of biological macromolecules are greatly dependent on their monomeric genesis. For instance, polymers such as proteins result from translation since ribosomes translate amino acids into polypeptide chains out of mRNA templates. Carbohydrates are synthesized from monosaccharides in reactions such as glycosylation, resulting in the building of starch and cellulose, which are polysaccharides. The biological significance of such macromolecules is emphasized by the hierarchical nature of structure, from humble monomers to highly specialized and functional entities. Highlighted by more recent advances in polymer science, the exploration of macromolecular complexity transcends traditional frontiers, inviting multidisciplinary understanding of how molecular architecture influences biological activity and interaction within cellular environments (Perry et al., 2020). Also, the interplay and interaction among various biological macromolecules form the framework of numerous biochemical processes essential to life. For instance, the complex signal transduction networks in cells largely rely on protein interactions, where enzymes catalyze a specific process and regulatory proteins guide cell response. The coordinated roles of nucleic acids and proteins in gene expression and regulation further demonstrate how macromolecules interact to facilitate complex biological functions. The development of graphical models, such as the systems biology graphical notation (SBGN), becomes imperative in their representation because it involves the accurate description of biochemical processes and facilitates the exchange of effective information among scientific communities (Demir et al., 2010). By illuminating macromolecular interactions, these technologies allow a more concise understanding of the molecular underpinnings of cellular activity, providing a foundation for innovation in everything from biotechnology to synthetic biology.

Importance in Biological Systems

The significance of biological macromolecules to living organisms is that they are essential in their structure and function, and, conversely, regulate most biological processes. Proteins, for instance, are not only structural molecules but also enzymes that catalyze biochemical reactions required for life. Their activity is directly associated with their three-dimensional shape, which is determined by the amino acid sequence. Moreover, the interaction of macromolecules including nucleic acids, lipids, and carbohydrates, facilitates complex cell functions such as signal transduction and storage of energy. Understanding these macromolecules is essential in comprehending the mechanism of operation of cellular systems and how they respond to external stimulation, considering their superordinate role in maintaining the homeostasis of biological organisms. Electrostatic interactions, as defined by macromolecular conformations, play a pivotal role in defining biochemical behaviors necessary within biological systems. Charge distributions on the protein and nucleic acid surfaces, for instance, are key to facilitating interaction with other molecules, allowing molecular recognition and binding processes. Advances in the last few years in modeling charge distributions have contributed to our understanding of their influence on the stability and function of macromolecular structures. Through the study of surface charge interactions with novel methods, researchers can more effectively model the dynamics of biomolecules under physiological conditions. Greater understanding of electrostatics highlights how even minimal variations in charge can create considerable variations in molecular interactions and hence alter cellular processes at a large scale, as indicated by various studies on the relevance of multipole moments to such interactions (Božič 2017). Furthermore, application of computational methods, such as density functional theory (DFT), has revolutionized research in biological macromolecules with the potent tools to study complex interactions. By allowing scientists to be able to model systems with thousands of atoms, DFT facilitates a good representation of the weak interactions that are often decisive in biomolecular assembly and stability. This evolution has rekindled interest in using methods previously established to explore biologically relevant systems, leading to a greater understanding of macromolecular behavior within different environments. As more sophisticated modeling techniques become available, the promise to reveal elaborate subtleties of biological processes grows, in keeping with recent research into the functions of biological macromolecules in global cell processes. The integration of such computational data goes to further enhance our comprehension of the intrinsic functions these macromolecules have in the operation and adaptability of life itself (Kolb et al., 2012) (Figure 3.2).

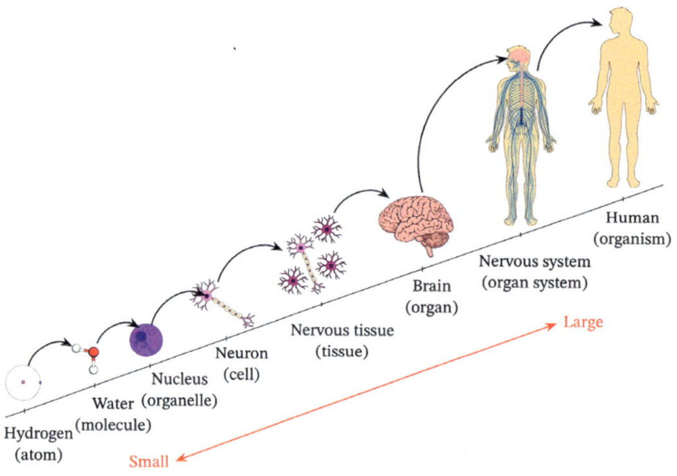

Figure 3.2. Biological macromolecules in living systems.

Source: https://images.nagwa.com/figures/explainers/430187521519/2.svg.

Overview of the Four Main Types

The biological macromolecules are building blocks of life, and there are four major types: carbohydrates, proteins, lipids, and nucleic acids. The different categories have specific roles in biological systems and are involved in different structural, functional, and regulatory processes. Carbohydrates are organic compounds that mainly consist of carbon, hydrogen, and oxygen; they are sources of energy and cell structure. Their monomeric units are called saccharides, which may either make simple sugars or complex polymers, as in the case of glucose and starch, respectively. Carbohydrates play a central role in energy metabolism by way of photosynthesis and cellular respiration, making them ideal to demonstrate the relevance in energy metabolism in sustaining life. Another essential group of macromolecules is proteins, which consist of long chains of amino acids that fold into a particular three-dimensional structure. This folding uniqueness is central to the wide variety of roles that proteins play, including catalyzing biochemical reactions as enzymes, to providing structure in cells. The order of arrangement of the amino acids defines the nature of proteins and their functionality and any minor change can cause a major change in the behavior of the proteins. Here, the development of structures such as the Enzyme Commission number and other standardized nomenclature would be of great help to the researchers to understand the protein interactions. This relevance is reinforced by the SBGN, which aids in representing the complex networks of protein interactions and enzymatic pathways, thereby enhancing our understanding of biochemical dynamics (Demir et al., 2010). Lipids constitute the third type; it is a varied group of hydrophobic molecules that include fats, oils, waxes, and steroids. Their main functions concern the storage of energy, the structure of membranes and signaling. The cell membrane architecture is centrally important to lipids, especially phospholipids, which determine the membrane fluidity and permeability. Complex interplay between lipid constituents frequently governs cellular signaling cascades that are important in cellular communication and homeostasis. Additionally, nucleic acids, comprising DNA and ribonucleic acid (RNA), are vital for the storage and transmission of genetic information. (Sorokin et al., 2009).

3.1. Carbohydrates

Learning Outcome

1. Structure and types of carbohydrates;
2. Functions of carbohydrates in living organisms;
3. Examples of carbohydrates in everyday life.

Carbohydrates are one of the simple forms of biological macromolecules, which are largely required energy sources and structural components in living organisms. They consist of carbon, hydrogen, and oxygen atoms in some combination. They can be categorized as simple sugars or monosaccharides and complex ones , such as oligosaccharides and polysaccharides. Monosaccharides like glucose and fructose are a central part of cellular metabolism and are used as significant energy fuel for energy production by means of processes like glycolysis. Polysaccharides like starch and cellulose also have central functions in storage and structural integrity. Their characteristic conformational properties and branching patterns enable them to perform functions in a range of biological processes, , including cell recognition and signaling reactions, thus showing their significance in metabolic pathways as well as cellular architecture. Moreover, the structural heterogeneity of carbohydrates is noteworthy as it renders them involved in an enormous set of biological interactions. Their ability to form highly branched polysaccharides and glycoproteins allows carbohydrates to carry out cell-cell recognition and communication, which are crucial for immune response and tissue development. Such specificity is derived from variations in monosaccharide composition, glycosidic linkages, and branching that confer the characteristic physical properties of the carbohydrate framework. Additionally, the formation of glycosidic bonds is what leads to the development of disaccharides and longer chain polysaccharides that are often utilized in energy storage and cell structure. These advanced patterns, speaking in terms of the issue at hand with biological macromolecules, are amenable to computational studies that are useful for the identification of structural correlations between these key substances using graph theory techniques (Altschul et al., 2005). Recent investigations have built our understanding of carbohydrates with the biophysics and computational biology methodologies by emphasizing their functions in many biological processes. Investigations into natural surfactant proteins found in certain organisms, , such as tropical frogs, present another facet of carbohydrate function. These proteins, unlike being ubiquitous, are usually complex carbohydrate moieties, which reflect carbohydrate-protein interactions, , such as stability and biocompatibility (Cooper et al., 2017). Strategies to utilize graph theory to study the structures of carbohydrates have proved invaluable in detecting recurring folding motifs and structural motifs among a wide variety of biological macromolecules, enabling a better understanding of their roles in function. This interdisciplinary strategy highlights the totality of carbohydrates in biological systems as a reflection of their complex intertwinements in life at the molecular level (Figure 3.3).

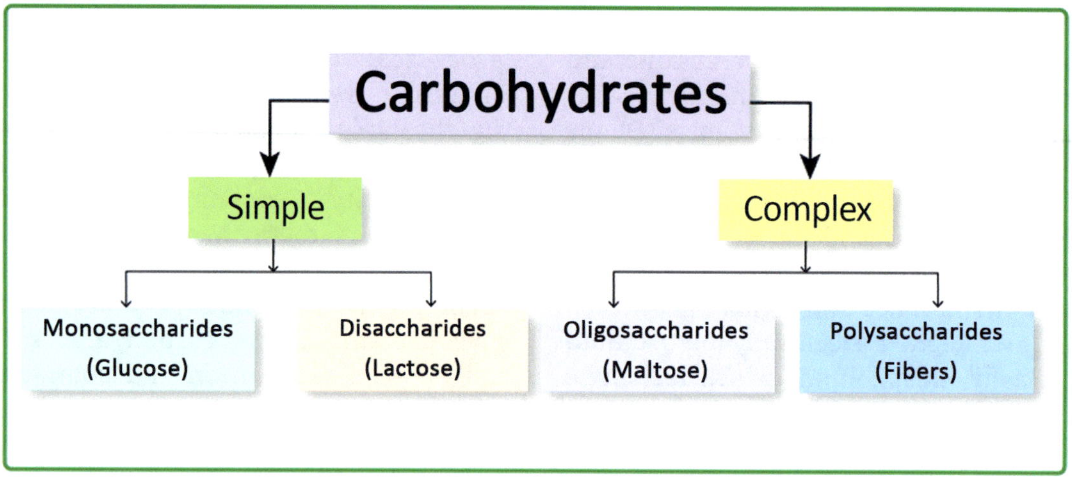

Figure 3.3. Carbohydrates in biological macromolecules.

Source: https://media.geeksforgeeks.org/wp-content/uploads/20240212174616/Classification-of-Carbohydrates.png.

3.1.1. Structure and Types of Carbohydrates

Carbohydrates are one of the four classic biological macromolecules and play an important role in many biological processes. Structurally, carbohydrates contain carbon, hydrogen, and oxygen atoms and are typically found in a 1:2:1 ratio, like the formula CH_2O. By being structured this way, they can create simple sugars, or monosaccharides, which serve as precursors for more complex forms of carbohydrates. Monosaccharides also interact to form glycosidic bonds and form disaccharides, , such as sucrose, or polysaccharides, , such as starch and cellulose. Such structural variability in carbohydrates parallels their diverse functions, from energy storage in animals and plants to structural purposes in bacterial and plant cell walls. It is important to know these subtle structures, not only for biochemistry but also for the origins of life and the process of metabolism in our world as described by Aguirre-Ramírez et al. (2021). Carbohydrates are divided into categories based on size and complexity. Monosaccharides, the simplest, include glucose, fructose, and galactose and are known by the fact that they consist of one sugar unit. Disaccharides consist of two joined monosaccharides, examples being lactose and maltose. At a more general level, polysaccharides , such as glycogen, starch, and cellulose form complex branched or linear sets that are stored as energy as well as providing structural strength. Glycogen, found primarily in animal tissues, is an easily mobilizable energy store, while starch serves a similar purpose for plants. Conversely, cellulose, a major component of plant cell walls, gives strength and hardness. This categorization emphasizes the functional diversity of carbohydrates to be able to easily transition to fulfill different roles in biological systems, as attested by ongoing research work emphasized by Carrère et al. (2017). The biochemical interactions between carbohydrates and other macromolecules also provide insight into their pivotal role in cellular processes. Carbohydrates generally participate in numerous signaling processes and cell recognition pathways, where their individual structure allows specific interaction with lipids and proteins. Protein glycosylation, for instance, can modulate protein folding, stability, and activity, influencing an extensive array of biological processes , including immune response and cell signaling. Additionally, the hydrophilic nature of most carbohydrates enforces their role in cell-to-cell communication and transport across membranes. These interactions not only play a significant role in cell homeostasis but are also accountable for inducing complex physiological responses. The advanced functioning of carbohydrates thus defines their significance as biological macromolecules, indicative of the functioning of structural and functional capabilities in living organisms. This is as presented by Aguirre-Ramírez et al. (2021) (Figure 3.4).

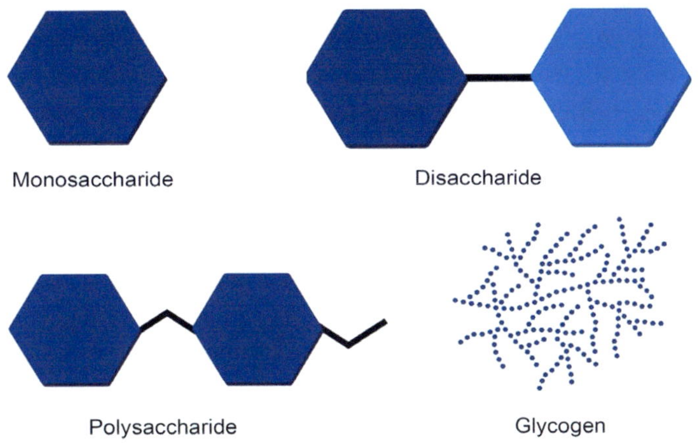

Figure 3.4. Structure and types of carbohydrates.

Source: https://microbenotes.com/wp-content/uploads/2021/11/Types-of-Carbohydrates.jpeg.

3.1.2. Functions of Carbohydrates in Living Organisms

Carbohydrates serve a variety of essential functions in living organisms, the most significant of which is the primary source of energy. Carbohydrates are carbon-oxygen-hydrogen organic compounds that may be classified as simple sugars, complex carbohydrates, or polysaccharides. Glucose, a primary photosynthesis product, is a significant energy source in plants and animals and is the energy needed for cellular processes. When carbohydrates are metabolized, particularly during cellular respiration, organisms convert glucose to generate ATP, an energy currency that is distributed universally throughout cells. Not only does it support metabolic function, but also cellular maintenance and growth of complex multicellular organisms, thereby highlighting how carbohydrates play a significant role in the overall regulation of energy in biological systems. Thus, their energy contribution is a significant consideration in the research of the overall context of biological macromolecules (Aguirre-Ramírez et al., 2021). Apart from the storage of energy, carbohydrates also have significant structural roles in specific organisms. For instance, cellulose is a plant cell wall polysaccharide that provides strength and hardness, enabling plants to maintain their shapes and resist pressures surrounding them. Alternatively, chitin, a polysaccharide containing nitrogen, is the basis of arthropod exoskeletons and fungal cell walls, providing them with structural support in many biological functions. Not only do these structural carbohydrates provide physical strength to organisms, but they also control the processes of growth as well as reproduction. By knowing the functional diversity of carbohydrates, one can envision how vital they are to the maintenance of not only cellular structure but also inter-organismal relationships within ecosystems, with their multifunctionality as a class of biological macromolecules (Aguirre-Ramírez et al., 2021). Carbohydrates are also an important element in cell recognition and signaling. Glycolipids and glycoproteins, the covalently bound carbohydrates to lipids and proteins, respectively, also have significant functions in cell-cell adhesion and communication. These carbohydrate-moiety molecules are used as cell surface recognition sites to facilitate cell-to-environment and cell-to-cell interactions, , including immune response and developmental signals. For example, identification of specific carbohydrates on pathogens by immune cells can trigger protective responses that protect the organism. Besides, the function of carbohydrates in the storage and transmission of genetic information during cell division suggests their importance in short-term and long-term biological process. The multiplicity of functions of carbohydrates is a sign of their basic functions within the overall design of biological macromolecules, attesting to their indispensable nature in sustaining life (Figure 3.5).

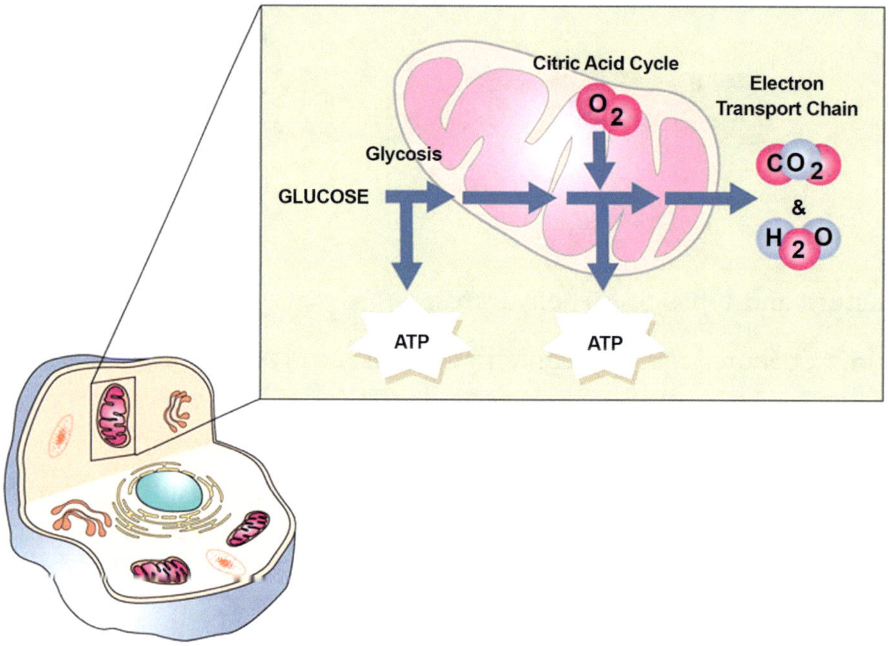

Figure 3.5. Functions of carbohydrates in living organisms.

Source: https://pressbooks-dev.oer.hawaii.edu/humannutrition/wp-content/uploads/sites/10/2017/11/image7-1.jpg.

3.1.3. Examples of Carbohydrates in Everyday Life

Carbohydrates are ubiquitous in our daily life, and play essential roles, not only as a part of human food, but also in many industrial processes. Simple sugars , such as glucose and fructose which are present in fruits, honey among other natural sources, are the most common examples of carbohydrates. These simple sugars are source of fast energy and thus play an important role in body functions like metabolism. In addition, the most common disaccharides that include sucrose, which is mainly obtained from sugar cane and beet are not only useful in making processed food sweet but also play an important role in providing energy. Carbohydrates also go beyond simple nutrition; they are also involved in food preservation, as can be testified by their ability to produce syrups and jams, which prevent the growth of microorganisms because of the high sugar content. Therefore, carbohydrates are versatile, and they have an influence on both health and food technology. Grains, legumes, and vegetables are complex carbohydrates, which are a vital part of a balanced diet. Examples are starches and dietary fiber, essential to keep the digestive system healthy and to control blood sugar levels. Whole grains, beans, and leafy greens are good examples of foods containing such carbohydrates, making them staple foods in most diets around the world. Complex carbohydrates result in a slow supply of glucose into the bloodstream, which supplies energy over a longer period of time and avoids the energy surges that simple sugars cause. It especially applies to people who have to deal with such conditions as diabetes, where the glycemic index of carbs becomes an important factor. Moreover, the fiber content of such foods is beneficial to the health of the gut, as it helps to nurture beneficial gut bacteria, which underscores the broader importance of carbohydrates beyond their role in energy production. Besides their value as a nutrient, carbohydrates are being innovatively used in many areas of science, which reflects their usefulness as biological macromolecules. As an example, new research in polysaccharide studies has led to the discovering of their opportunities in drug development and biomaterials. It has also been demonstrated recently that polysaccharides can be used as a drug delivery system and since they are biocompatible, they can be used to target specific treatments and reduce side effects. This biotechnological advantage paves the way to applications of carbohydrates in regenerative medicine, where they are capable of underpinning tissue engineering and repair. With this convergence of carbohydrate science and technology continuing to grow, current research highlights the topicality of these macromolecules in the solution of contemporary medical problems. The implications of such applications are vast, emphasizing that carbohydrates are not mere dietary components; they are integral to advancements in various scientific domains (Zhang et al., 2019; Aksakal et al., 2021).

Test Your Knowledge

1. Illustrate the structure and types of carbohydrates.
2. What are the functions of carbohydrates in living organisms?

3.2. Proteins

Learning Outcome

1. Amino acids and protein structure;
2. Role of proteins in cellular functions;
3. Enzymes as biological catalysts.

Proteins, the structural elements of all living things, have diverse functions that are essential to life. Comprised of long chains of amino acids, proteins fold into complex three-dimensional structures, which specify their functions in biological systems. The specific amino acid sequence of the proteins ultimately specifies their function and shape, facilitating processes like enzyme catalysis, structural support, and cell signaling. The complexity of protein function is based on the new arrays of 20 unique amino acids that can assemble into nearly infinite numbers of different protein structures. Identification of these structures and functions is vital since they are not only accountable for the integrity of the cells but the complex biochemical reactions that sustain life. Proteins thus illustrate the dynamism and complexity of biological macromolecules in relation to cellular function. Both transcription and translation are the two primary processes involved in protein synthesis, which is required for gene expression. During transcription, messenger RNA (mRNA) is synthesized from DNA, serving as a template for protein synthesis. Subsequently, during translation, ribosomes translate mRNA, which integrates amino acids into a polypeptide chain that folds into a functioning protein. The precision in this biological machinery is incredible; any error in the amino acid sequence can lead to nonfunctional proteins that cause various diseases or abnormal function in organisms. Synthetic biology research now explores how proteins might be designed for more use or novel applications, unlocking the possibility of bridging the gaps between nature and artificial systems. These advances highlight the variety of proteins and their use in biomedical applications, bearing witness to the ongoing relevance of proteins in basic and applied biological sciences. Proteins are applied beyond their natural biological functions, as witnessed in the advancement of synthetic biology and artificial cell design. These artificially created systems implement biomimetic principles to mimic certain characteristics of biological cells, , such as the capacity to synthesize proteins, sort cells, and interface with the environment. As synthetic technologies advance, building blocks like polyoxazolines are becoming increasingly important to create biocompatible materials for drug delivery and biomedical applications. The amphiphilic and nontoxic characteristics of such polymers offer a potential alternative for emerging therapeutic delivery systems. With research progress, deconstructing the interaction of proteins and synthetic materials could make revolutionary biomedical discoveries possible, looking forward to an era where the disparity between natural and synthetic cells becomes much smaller, hence improving drug delivery and therapy interventions (Adams et al., 2007; Ding et al., 2014) (Figure 3.6).

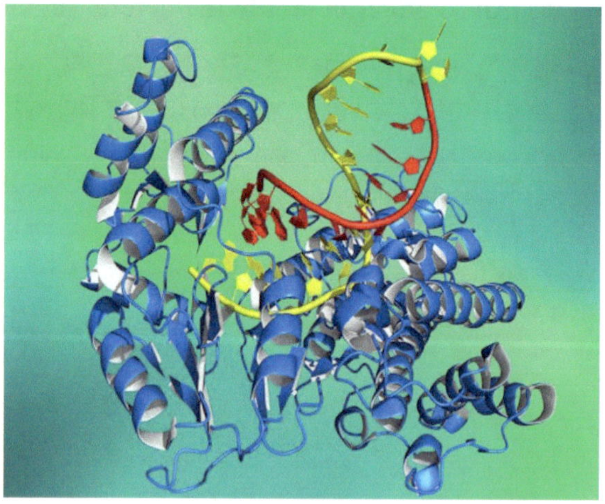

Figure 3.6. Proteins within biological systems.

Source: https://miro.medium.com/v2/resize:fit:764/1*EvXUaXQv3QplyI_P2i-oZw.jpeg.

3.2.1. Amino Acids and Protein Structure

The complex interaction between amino acids and protein structure is fundamental to the understanding of biological macromolecules. Amino acid chains constitute proteins, and amino acids are the humble building blocks responsible for individual structure and function. Amino acids possess varying chemical properties, from their side chain, which determine how these amino acids interact with one another when folding into a functional protein. Moreover, the shape that a protein takes is not a random shape; it is a result of particular interactions between amino acids in the form of hydrogen bonds, ionic bonds, and van der Waals interactions. This brings complexity to the linear relationship between the amino acid sequence and protein three-dimensional structure, thus specifying its biological function and interaction with other molecules. Conformational restraints on the side chains of amino acids are essential protein determinants of function. Studies have revealed that the conformational restraints on side chains introduced by protein structures lead to low entropy in their conformational behavior compared to non-proteinized free amino acids. Occurrence has been accounted for based on a Shannon entropy analysis, and it was demonstrated that proteins hold enormously greater informational content than their individual amino acids. Thus, these limitations not only enhance protein-ligand recognition specificity to other macromolecules but also define how proteins recognize ligands. The impact of this finding is seen in the way proteins evolve to carry out different biological functions, where backbone and side chain geometry are fundamental to protein structural biology (Bojarski et al., 2018). Furthermore, advances in understanding how protein sequence maps onto function have led to new models like the resonant recognition model (RRM). The model interprets amino acid sequences as numerical patterns in relation to a protein's biological function. Computational methods allow researchers to calculate characteristic frequencies that are related to specific protein functions.

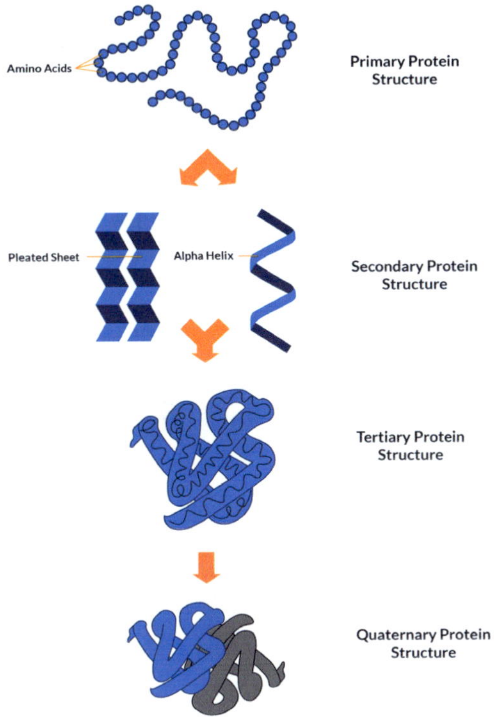

Figure 3.7. Amino acids and protein structure.

Source: https://www.reagent.co.uk/wp-content/uploads/Prot/in/Protein-Structure-1.png.

This approach gives promising insights into how particular amino acid sequences are accountable for the protein's selective binding with other biomolecules. The RRM is also useful in the rational design of peptides with particular biological functions, as exemplified in studies of GFs and oncogenic proteins. Consequently, the examination of the relationship between amino acid sequence and protein function not only enhances our comprehension of biological macromolecules but also opens up avenues for biotechnological exploitation (Cosic et al., 2007) (Figure 3.7).

3.2.2. Role of Proteins in Cellular Functions

Proteins play a key role in numerous cellular processes as the cell's workhorses. They play a central role in nearly every facet of biological function, from metabolism and cell signaling to structural support. Enzyme proteins, for instance, catalyze biochemical reactions, speeding up metabolic pathways that are essential for energy return and the health of the cell. Beyond enzymatic function, proteins also function as signaling molecules, having the capability to convey information within and between cells. This is crucial for cellular functions , such as growth, differentiation, and response to stimuli. The functional versatility of proteins underscores their importance in maintaining life at the cellular level, validating their definition as biological macromolecules that enable the complex choreography of cellular functions. Aside from their involvement in enzymatic and signal transduction functions, proteins are also involved in cellular structure, providing structural support and maintaining cellular integrity.

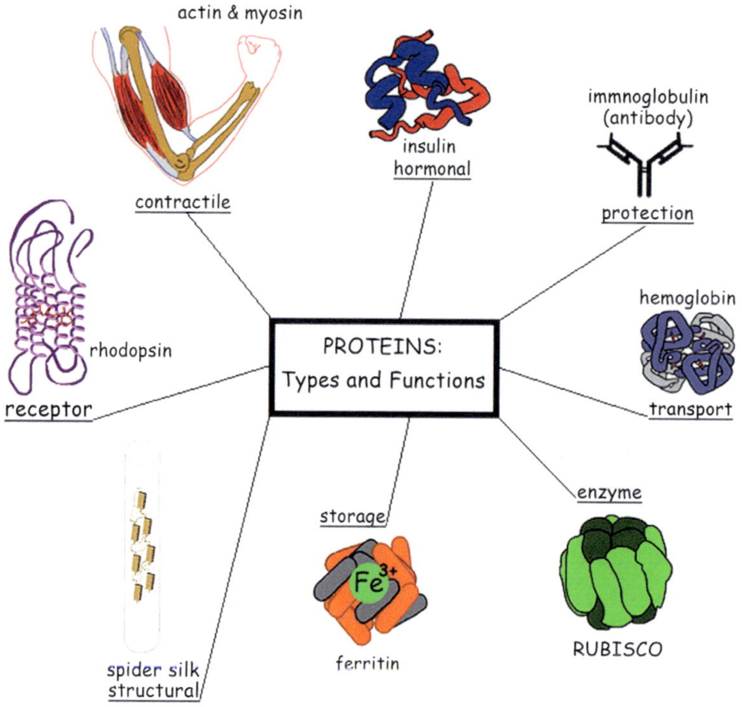

Figure 3.8. Role of proteins in cellular functions.

Source: https://cdn-ccnli.nitrocdn.com/hPbgvskpJmIfLTeYLYGdGpfRfGKbrYYq/assets/images/optimized/rev-f2e7539/alevelbiology.co.uk/wp-content/uploads/2017/11/Functions-of-Proteins.png.

Structural proteins , such as collagen and keratin are involved in the composition of the ECM and the various forms of cell membranes, which is crucial for cell shape and function. Proteins are also a central component of the cytoskeleton, involved in the maintenance of cell shape and motility of cells. Because of the dynamic nature of these structural proteins, cells can react to an enormous number of mechanical stresses and environmental changes. As a reply, the biochemical function of proteins is inseparable from their structural role within the framework of cellular architecture, highlighting their status as biological macromolecules that not just work but also exist as structural life components. The protein-cellular dynamic interaction continues into cellular communication and transport systems, which are crucial for homeostasis. Proteins that are embedded in cell membranes serve as receptors for the perception of and transmission of external signals into the interior of the cells, thereby facilitating the appropriate physiological responses. Proteins are also key to facilitating transport through membranes, and transport proteins regulate the movement of ions, nutrients, and waste products. Selective transport, which is critical to maintaining the cell's internal environment by providing necessary molecules and deterring the accumulation of harmful substances, is regulated by transport proteins. In the case of artificial cells, the design, and construction of artificial systems with comparable protein functions provide promising potential. Not only does this enhance our understanding of biological mechanisms, but it also provides opportunities for new applications in synthetic biology, cellular biology, and biomedical engineering (Braeckmans et al., 2012; Ding et al., 2014) (Figure 3.8).

3.2.3. Enzymes as Biological Catalysts

Enzymes are extremely important biological catalysts in the complex world of biological macromolecules and play a fundamental role in the rate and efficiency of biochemical reactions. These proteins not only make reactions faster but also make sure that metabolic reactions happen under physiological conditions, which makes life as we know it possible. Enzymes are very specific catalysts and most enzymes catalyze a specific type of reaction or work on a specific substrate, hence their significance in ensuring cellular homeostasis. The specific three-dimensional structure of the enzymes enables them to catalyze the reactions between the substrates by reducing the threshold energy needed to activate the reactions and hence speed up the rate of conversion of the substrates to products. Such specificity and efficiency point to the evolutionary importance of enzymes, as a meeting point of the many and varied biochemical processes that support life, as a result of the complex interdependencies among the macromolecules of biological systems. Enzymes catalytic action is further increased by a number of factors, , such as temperature, pH, and substrate concentration. Most enzymes have an optimum temperature where they are most active; below which they may be denatured or become inactive. Likewise, pH influences the ionization of the active site, and accordingly, the ability of the enzyme to react with its substrate in an effective way. According to the existing studies, such a sophisticated interaction of environmental factors not only affects the activity of enzymes but also testifies to the complexity of living systems where enzymes are developed under certain conditions to provide the optimization of metabolic pathways. The adaptability of enzymes to various biochemical environments mirrors the evolutionary processes that enhance survival, suggesting a long-standing coevolution between enzymes and their respective biological systems, which can be conceptualized through paradigms , such as those discussed by Kletzin et al. (2007). Besides, the study of enzymes is not just a biological curiosity; there is a deeper significance to this research, which concerns the possibility of extraterrestrial life. The nature of enzymes, and in particular their ability to adapt to extreme conditions, can tell us about the sort of biological processes that might take place on other planets or moons. As an example, the study of extremophiles, organisms whose enzymes can operate at high temperatures or in other extreme conditions of acidity, can provide useful hints as to the biochemical principles which might apply to life elsewhere in the universe. As articulated in studies involving extraterrestrial exploration, enzymes can serve as markers of potential life, expanding our comprehension of life's resilience and versatility throughout the cosmos (Hoffman et al., 2025). This kind of research not only highlights the biochemical significance of enzymes as catalysts but also points out to their significance as part of the whole picture of biological diversity and the possibility of life elsewhere other than Earth.

Test Your Knowledge

1. What are amino acids and protein structure?
2. What is the role of proteins in cellular functions?

3.3. Lipids

Learning Outcome

1. Types of lipids and their structures;
2. Functions of lipids in the body;
3. Importance of lipids in energy storage.

Lipids are significant because they are one of the major biological macromolecule classes, serving not only as structural components of cell membranes but also as energy reservoirs and signaling molecules. Due to their hydrophobic nature, lipids can pack close together in lipid bilayers to create a boundary that is essential to the compartmentalization of the cell. Arrangement of molecules in this manner is crucial to homeostasis as well as the regulation of the cell interior environment. In addition, lipids are involved in metabolism, being utilized in hormone formation and fat-soluble vitamin absorption. Their diverse structures, extending from simple fatty acids to advanced phospholipids and steroids, point to a wide variety of functions that are performed by the lipids within biological systems. This intricacy highlights the importance of understanding lipid interactions within cellular settings since they may influence various cellular processes and overall organismal health, demonstrating their multifunctional relevance in biological macromolecules (Aguirre-Ramírez et al., 2021). The dynamics of such lipid interaction further illustrate our knowledge of their functionality in biological systems. Especially, these interactions with extracellular stimuli, , such as water, illustrate how lipid functionality is dependent on the external condition. The process of hydration surrounding lipid molecules, , such as interaction between water and lipid surfaces, plays a critical role in stabilizing their structures and promoting their functions in membrane formation. Studies have shown that the character of interacting water molecules with lipids can play a crucial role in influencing their conformational stability and biological activity. When lipid-protein interactions are considered, , such as observed in membrane forms, they become even more crucial, as surfactant-like behavior has a profound impact on maintaining things in a fluid state and allowing biochemical reactions to occur. Therefore, the investigation of the interplay between lipids and hydration and the environment provides significant insights into their role as biological macromolecules (Bridelli et al., 2017). Moreover, the classification of lipids is more than structural categorization since it encompasses their inherent role in health and disease. Perturbations of lipid homeostasis, for instance, are linked with a number of pathologies, , including obesity, cardiovascular ailments, and metabolic disorders. The unique physicochemical characteristics of fatty acids, e.g., saturated, unsaturated, and trans fatty acids, provide an example of how variations in lipid structure can affect health. Furthermore, the role of specific lipids in cell signaling, , such as phosphoinositides in insulin signaling pathways, demonstrates the complex involvement of lipids in cellular functions. By understanding the different types of lipids and biochemical pathways, researchers can plan targeted therapeutic interventions to correct lipid balance and ensure health. Thus, lipids not only have crucial functions as biological macromolecules but are also prime targets for medical treatment and development (Aguirre-Ramírez et al., 2021) (Figure 3.9).

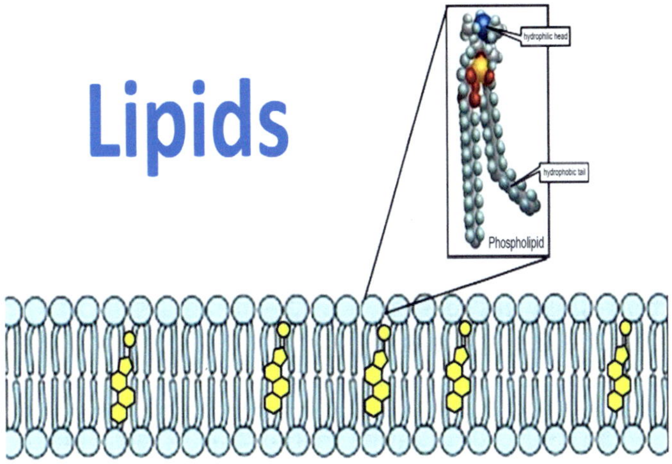

Figure 3.9. Lipids in biological macromolecules.

Source: https://i.ytimg.com/vi/aUIqXB1KQ0s/sddefault.jpg.

3.3.1. Types of Lipids and Their Structures

Lipids are a heterogeneous class of biochemicals that have very significant roles in cellular function and structure and are best characterized by their hydrophobicity. Lipids are classified into various types of lipids, , including triglycerides, phospholipids, sterols, and waxes. Triglycerides, which consist of glycerol and three fatty acid chains, are the form in which energy is stored in adipose tissue. Their compact structure facilitates efficient storage of fat since fatty acids are small and hence facilitate easy storage of fat. Consequently, they become a significant source of energy in the event of metabolic activities. The saturation point of fatty acids also defines the physical properties of triglycerides with regard to fluidity and melting points. The difference is responsible for maintaining membrane integrity and function across different physiological states, thus proving the structural function of lipids in organisms. Phospholipids comprise a very large group of lipids that constitute part of the cell membrane structure. In terms of structure, they consist of two fatty acid tails and a phosphate group attached to a glycerol backbone, which confers on them amphiphilic properties that allow them to spontaneously form bilayers when in solution. The formation of bilayers is significant in cellular compartmentalization since it creates a cellular process barrier. Its hydrophilic head is drawn towards water, while its hydrophobic tails push it away, forming the characteristic lipid bilayer of cellular membranes. In addition, the fluid mosaic model illustrates how these phospholipid bilayers take their position for many proteins to exhibit dynamic interactions vital to membrane function and signaling pathways. This phospholipid structure realization highlights their essential role not just in membrane construction but also in cell-to-cell communication. Sterols, , such as cholesterol, possess specialized yet vital roles in the structure and function of cells. With a multi-ring structure, sterols are integrated into phospholipid bilayers and regulate their fluidity and stability. Cholesterol, in particular, controls membrane properties by preventing fatty acid chains from becoming too closely packed so that membrane fluidity is maintained during temperature fluctuations. This attribute is particularly important in maintaining cell membrane integrity and in allowing proper protein function. Further, the structural variation in sterols affects their roles in the cellular environment, controlling processes , such as membrane trafficking and signaling. The relationship between classes of lipids, such as, such as sterols, phospholipids, and triglycerides—is testimony to the complexity of lipid structures and the implications of functionality in biological organisms, providing insights that are critical to the understanding of activities within cells and overall organism health (Buhleier et al., 2014; Weston et al., 2015) (Figure 3.10).

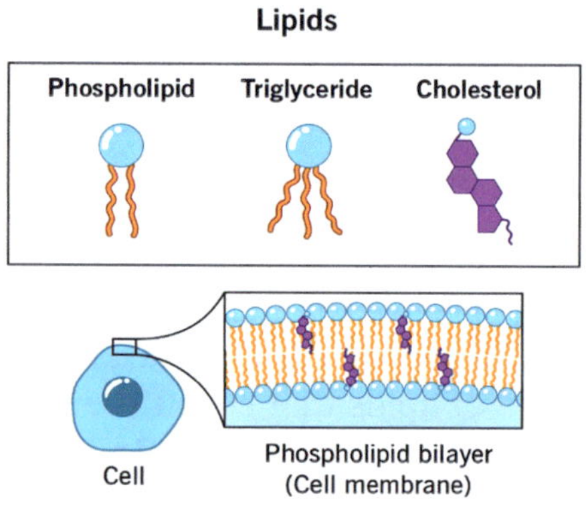

Figure 3.10. Types of lipids and their structures.

Source: https://my.clevelandclinic.org/-/scassets/images/org/health/articles/24425-lipids.

3.3.2. Functions of Lipids in the Body

Lipids serve a variety of crucial functions within the human organism, acting as part of cell structures and as key players in metabolic pathways. Lipids are primarily tasked with the composition of biological membranes, where the phospholipids present a bilayer that serves as a barrier between the interior of the cell and the exterior world. This membrane not only provides cellular integrity but also enables selective molecule transport in and out of the cell. Beyond structural roles, lipids also function as body signals. Some lipids, for instance, serve as hormones and signaling molecules that regulate many physiological processes, , such as growth, metabolism, and immune function, thereby illustrating their numerous roles in human health and homeostasis. In addition to their structural and signaling functions, lipids also function as a vital energy storage system. Adipose tissue, consisting largely of triglycerides, is the body's major store of energy. During caloric deprivation or increased energy demand, lipids are mobilized and hydrolyzed through lipolysis to release fatty acids, which can be employed by a number of tissues, namely muscle and liver cells, to produce energy. Such energy retrieval is superior to carbohydrates and proteins, highlighting the energy efficacy of lipids. Furthermore, energy storage as lipid is necessary for the maintenance of metabolic flexibility in order for the body to adapt to varying energy demands and dietary status. Lipids are also significantly involved in thermoregulation and insulation within the body. Subcutaneous fat layers, composed largely of lipids, protect the body's core temperature from overheating loss in cold temperatures. In addition to holding heat, this adiposity also guards vital organs against mechanical damage. The peculiar character of lipids, , such as their effect on fluidity in cell membranes, extends to affecting the physical attributes of biological systems. Notably, lipid behavior over various conditions can be used to unveil complex biological processes like anesthesia, as exemplified by the interdependence of lipid interactions and anesthetic potency in various research studies (Abelho et al., 2006). In general, such functions accentuate the importance of lipids as biological macromolecules crucial to life (Figure 3.11).

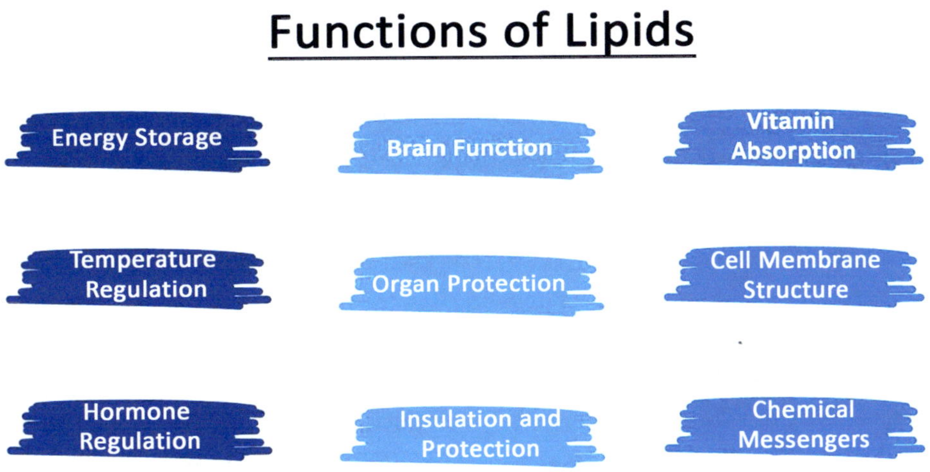

Figure 3.11. Functions of lipids in the body.

Source: https://themasterchemistry.com/storage/2023/07/Functions-of-Lipids-image-1024x597.jpg.

3.3.3. Importance of Lipids in Energy Storage

Lipids are the most important in the energy storage world, being one of the biological macromolecules. Lipids provide a more efficient and compact storage method of energy as opposed to carbohydrates and proteins, which are also used as a source of energy. Their high efficiency is mainly due to the highly reduced composition, which enables them to yield more than twice the energy per gram than carbohydrates. Triglycerides, which are the main storage form of lipids, have hydrocarbon chains that have a considerable amount of energy-dense bonds that give out energy when oxidized. Due to this great energy production, organisms mainly use lipid stores in times of fasting or prolonged physical activity and are thus essential to survival and performance under different environmental conditions. In addition, the structural complexity of lipids also helps them to serve as good stores of energy. The lipids, especially those stored in adipose tissues, are packed in a manner that saves space and at the same time allows rapid access to stored energy. Lipases catalyze the hydrolysis of stored triglycerides when energy is needed, releasing free fatty acids which may then be taken into metabolic processes , including 8-oxidation and the citric acid cycle. This flexibility enables different organisms to effectively mobilize lipid reserves into energy. Also, other metabolic pathways interact with lipid metabolism, which is systemically significant to the homeostasis of energy balance. Lipids also have storage beyond storage; they also act as signaling molecules and can modulate metabolic pathways in response to changes in energy availability, thus coupling energy storage with physiology. The importance of lipids in energy storage assumes a new aspect in the light of developmental biology and nutrition. As an example, resource nutritional quality in early developmental stages of an organism can have a significant effect on metabolic efficiency and memory retention, as indicated by research done on larval stages of parasitic wasps. In this case, quality lipid rich foods stimulate storage of important nutrients , such as triglycerides, which are vital in subsequent performance. Such an interaction depicts that lipids are not only important energy stores but also the basis of ideal growth and cognitive functions across different species. The multifaceted roles of lipids highlight their importance in biological systems, establishing them as fundamental macromolecules essential for energy storage and overall organismal vitality (Abroon et al., 2017; Adams et al., 2007).

Test Your Knowledge

1. What are the types of lipids and their structures?
2. What are the functions of lipids in the body?

3.4. Nucleic Acids

Learning Outcome

1. Structure of DNA and RNA;
2. Functions of nucleic acids in heredity;
3. Role of nucleotides in cellular processes.

Nucleic acids, comprised primarily of DNA and RNA, are essential biological macromolecules responsible for the transmission and storage of genetic information in living things. DNA serves as the blueprint for cellular activity and stores the genetic instructions for the development, functioning, and reproduction of all living organisms that have been found. RNA is, nonetheless, crucial in the transcription of these genetic blueprints to the synthesis of proteins with numerous functions necessary for cellular activity. Awareness of the structure and function of these nucleic acids enlightens us about their evolutionary function, where the nonrandom nature of biomacromolecules reveals a complex association between genetic information and molecular evolution, as hypothetical as it is in studies of the origin of life (Lu et al., 2025). This inherent complexity underscores the importance of nucleic acids in biological organisms, making them a central area of study in biochemistry.

The structural composition of nucleic acids has a highly complex pattern of nucleotides, which constitute the fundamental building units of these macromolecules. A nucleotide consists of a sugar, a phosphate, and a nitrogenous base, of which each contributes to the specific properties that differentiate DNA from RNA. For instance, DNA is double-helical with complementary base pairing that assures accurate replication and transcription. RNA, on the other hand, is typically a single-stranded molecule permitting a vast number of functions, , including messenger RNA (mRNA) that transfers genetic information from DNA to the ribosomes. Moreover, recent breakthroughs in synthetic biology have advanced to create artificial cells equipped with programmable nucleic acids to mimic natural cellular functions, and it is such research that holds the promise of applications of nucleic acid sciences in biomedicine and genetic engineering (Ding et al., 2014). Such a new methodology reflects the adaptability and significance of nucleic acids in natural and artificial systems. As studies go on to de-mystify the complexity of nucleic acids, their role extends beyond mere storage of genetics to the realm of therapies and biotechnological advancements. For example, gene editing technology , such as CRISPR-Cas9 employs nucleic acid sections to edit genomes specifically, demonstrating their revolutionary role in medicine, particularly in the treatment of genetic disorders. Furthermore, the construction of artificial cells with defined nucleic acid components raises questions about the limits of synthetic biology, for example, the minimal elements required for cellular communication and function. They are not only questioning our understanding of life but also forcing the development of new research strategies in biological research. Lastly, the study of nucleic acids illuminates their central role in the design of life's fabric, underscoring the requirement for further research on this dynamic and progressive field (Figure 3.12).

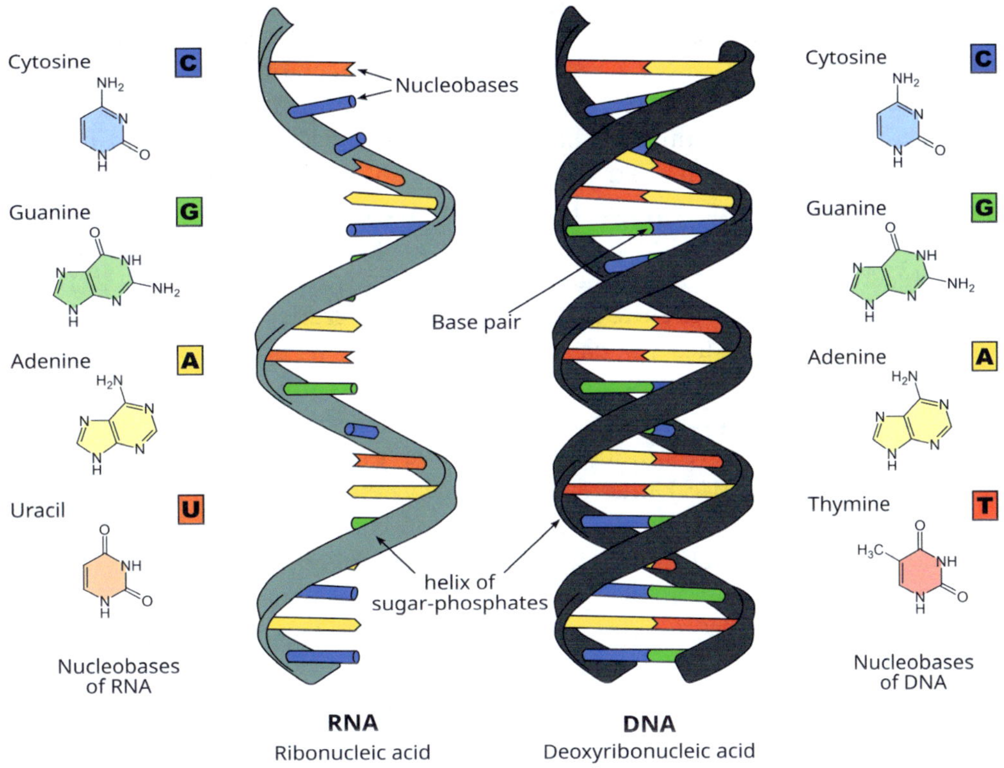

Figure 3.12. Nucleic acids.

Source: https://upload.wikimedia.org/wikipedia/commons/thumb/3/37/Difference_DNA_RNA-EN.svg/1200px-Difference_DNA_RNA-EN.svg.png.

3.4.1. Structure of DNA and RNA

The structure of DNA and RNA is the foundation of how they work as biological macromolecules. DNA, also known as deoxyribonucleic acid, consists of a double-helix structure, which was originally described by James Watson and Frances Crick in the 1950s. The helix structure consists of two strands coiled around a common axis, with the backbone composed of sugar and phosphate units and the inner rungs consisting of paired nitrogenous bases. The unique adenine-thymine and cytosine-guanine pairing, stabilized by hydrogen bonds, is essential to accurate DNA replication and transcription. It not only permits genetic data storage but also triggers repair and conservation machinery to preserve the integrity of heredity from one generation to the next. On the other hand, RNA is single-stranded most of the time and in different shapes and function, e.g., messenger RNA (mRNA), transfer RNA (tRNA), and ribosomal RNA (rRNA). While RNA contains a sugar-phosphate backbone, its sugar is ribose, and uracil replaces thymine as one of the nitrogenous bases. The shape of RNA accommodates a broad range of secondary shapes, which are crucial in its processes of transcription, translation, and enzymatic activity. This versatility is echoed in the design of complex three-dimensional objects, which allow RNA to take an active part in cellular processes. The understanding of RNAs' dynamic functions has increased exponentially with the demonstration of its ability to regulate gene expression and catalyze biochemical conversions. The advancement of structural biology has provided deep insights into the complex architecture of DNA as well as RNA. Techniques , such as X-ray crystallography, which gained popularity early in the 20th century, have played central roles in deciphering these macromolecules' atomic patterns. Curry et al. (2015) illustrate how these analytical instruments have evolved to not only reveal the static structure of DNA and proteins but also the dynamic interactions that underlie biological function. With emphasis on the importance of polymer physics, novel models examine the complexity of DNA and RNA folding, highlighting their importance in biological functioning. As there is growing knowledge about these macromolecules, our ability to design them for biotechnological applications grows also, reiterating their place in the broader realm of molecular biology and genetics (Dhar et al., 2011).

3.4.2. Functions of Nucleic Acids in Heredity

Nucleic acids, which include DNA and RNA, are primary molecules of heredity, specifying the transmission of genetic information from one generation to the next. DNA is a double helix and houses the genetic code that specifies the determination of an organism's characteristics by encoding proteins that play a critical role in biological functions. This permanent genetic information storage allows for precise duplication during cell division so that offspring receive an identical copy of their parent's DNA. As observed in recent studies, nucleic acids not only function as carriers of hereditary information but also as carriers of gene expression and regulation and play a critical role in the growth and functions of all living things (Horowitz et al., 1951). The complex processes of DNA transcription and replication illustrate the complicated machinery of heredity. RNA is involved in a multitude of roles in the expression and transmission of genetic information, acting as a catalyst and messenger in biological processes. Messenger RNA (mRNA), for instance, is replicated from DNA and transports the genetic blueprint to ribosomes, where proteins are constructed. In addition to the role of an intermediary, RNA molecules are involved in gene regulation, an active process of paramount importance to survive environmental changes. New discoveries of non-coding RNAs indicate that the molecules are key players in epigenetic regulation of gene expression and thus influence hereditary characteristics. Furthermore, a greater understanding of RNA's versatility contradicts the traditional genetic inheritance theories, shedding light on diverse means by which nucleic acids contribute to heredity (Witzany, 2017). Such diversity emphasizes nucleic acids more than as bearers of information. The notions of nucleic acids and heredity transcend the biochemical aspects, extending to evolutionary processes as well as biodiversity. As researchers venture into investigating this function further in gene manipulation, it becomes increasingly evident that nucleic acids are in charge of shaping not only individual traits but populations themselves over the long term. By facilitating genetic recombination and adaptation through mechanisms like natural genome editing and horizontal gene transfer (HGT), nucleic acids make genetic diversity for survival in dynamic environments possible. The new approaches characteristic of genetic engineering highlight the utilitarian significance of learning about nucleic acids, outlining their potential to manipulate hereditary traits in various applications , including agriculture, medicine, and conservation. Nucleic acid functions in heredity thus, position nucleic acids at the center of biology, intertwining concepts of stability, variability, and adaptability as essential to life (Figure 3.13).

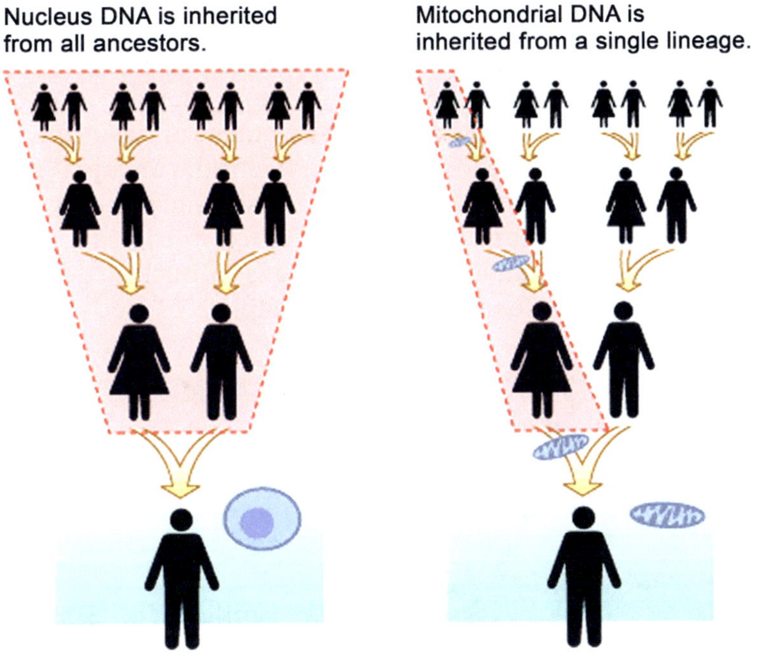

Figure 3.13. Functions of nucleic acids in heredity.

Source: https://d20khd7ddkh5ls.cloudfront.net/mitochondrial_dna_versus_nuclear_dna_3.gif.

3.4.3. Role of Nucleotides in Cellular Processes

Nucleotides serve as essential building blocks of nucleic acids, comprising DNA and RNA, which are vital for genetic information storage and transmission. A nucleotide is a sugar, a phosphate group, and a nitrogenous base, which are the structural backbone of DNA and RNA. These nucleotides combine through a process called polymerization into long chains, which determine the genetic code of organisms. This complex combination of nucleotides carries the information on protein building and cell functioning, which evidences the versatile nature of the nucleotides involvement in cell mechanisms. Moreover, in addition to their role as structural units, nucleotides are also signaling molecules in cells, which regulate many biochemical pathways required to keep the cells at homeostasis and respond to environmental changes. Nucleotides play an important role in the transfer of energy in cells besides their structural functions. ATP, a nucleotide composed of adenosine and three phosphate groups, is often referred to as the energy currency of the cell. The hydrolysis of ATP gives out energy which is utilized in driving a wide variety of cellular processes, , including muscle contraction, active transport processes and enzymatic reactions. This active position of ATP is a representative of the fact that nucleotides are central to bioenergetics and cellular metabolism. The connection between nucleotide metabolism and energy homeostasis highlights the significance of nucleotides beyond their genetic roles and shows a level of complexity to cellular processes. The research of these pathways still contributes to our knowledge of diseases associated with nucleotide metabolism that can result in therapeutic breakthroughs. Furthermore, it is hard to overestimate the evolutionary importance of nucleotides, particularly within the framework of the RNA world hypothesis. According to this theory, early life forms may have depended upon RNA to perform not just genetic information, but also catalytic duties, leading to the transition to contemporary ribonucleic processes. The processes of replication, transcription and translation that are powered by nucleotides are inextricably linked with evolutionary dynamics. Nucleotides could have been important actors in early biological evolution by promoting the formation of cooperative replicator systems, as might have occurred in prebiotic conditions when local communities of nucleotides thrived in mineral pores. Such interactions illustrate how the properties of nucleotides shaped the development of life's complexity, underpinning fundamental concepts in molecular biology and evolutionary theory (Ricardo et al., 2009). Altogether, the wide knowledge of various functions of nucleotides can help us to realize the complicated network of cellular reactions and the secrets of life itself.

Test Your Knowledge

1. What are the functions of nucleic acids in heredity?
2. What are the role of nucleotides in cellular processes?

3.5. Interactions Between Macromolecules

Learning Outcome

1. Protein-carbohydrate interactions;
2. Lipid-protein interactions;
3. Nucleic acid interactions with proteins.

Macromolecular interactions form the cornerstone of the majority of biological activities, and these play multiple structural and functional roles within cells and organisms. The macromolecules, , including proteins, nucleic acids, carbohydrates, and lipids, do not act singularly as a rule but rather interact in complex manners with one another to assist with homeostasis and cellular signaling. For instance, proteins may interact with nucleic acids to regulate the expression of genes and their replication. Similarly, carbohydrates serve as units of recognition for cell-cell communication, allowing cells to communicate and react in response to environmental stimuli. Understanding these interactions is significant to the understanding of mechanisms in cellular processes and organismic responses, which influences biotechnology in medical applications. The function of interaction dynamics between macromolecules is also illustrated in the context of composite materials for biomedical use. Recent studies highlight the fact that protein alloys like silk fibroin blended with fibronectin show enhanced adhesive properties, critical for tissue integration and healing. While being used singularly, silk fibroin is devoid of extensive cell adhesiveness, though it shows enhanced cell attachment and viability when blended with fibronectin, indicating that physicochemical properties of the constituent macromolecules determine biological effects. Not only do these systems of alloys preserve their mechanical integrity under conditions of healing , but they also regulate the interactions that govern cellular behavior, thereby creating the platform for biomaterial breakthroughs in regenerative medicine and tissue engineering (Backman et al., 2017). Besides this, the interaction between complex macromolecular structures can be harnessed for other new applications, , such as the exploration of other biological systems beyond Earth. Findings about macromolecule interactions provide insight into astrobiology, wherein scientists aim to comprehend possible forms of life in extraterrestrial contexts. Through the examination of the stability and versatility of macromolecules in hostile environments, researchers can make educated guesses regarding the biochemical reactions that can support life elsewhere in the universe. Since macromolecules are the very foundations of biological systems, their interactions could be the doorway to identifying biosignatures on other planets as well as to devising life-detection methods. Lastly, understanding these interactions not only makes our planet's complexity of life more comprehensible but also helps guide our search for the existence of life beyond our world (Hoffman et al., 2025) (Figure 3.14).

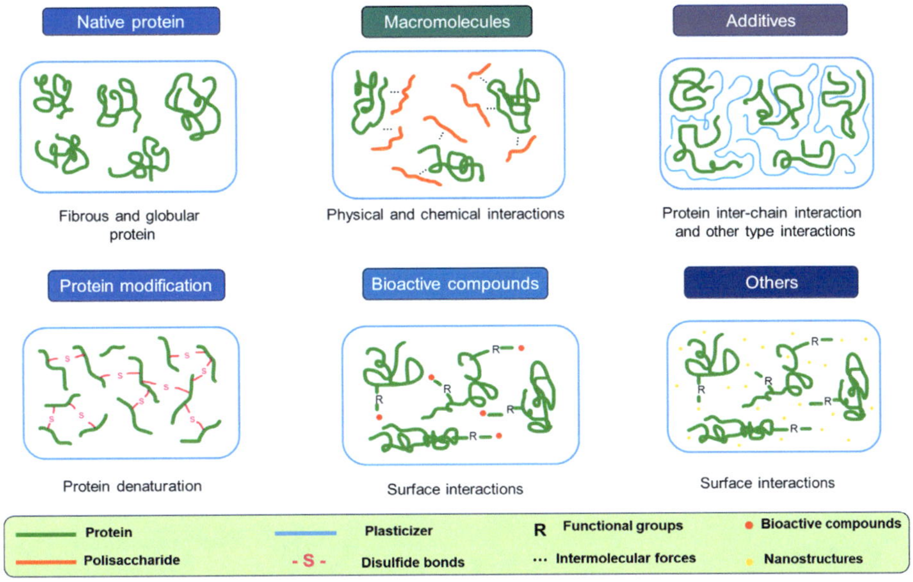

Figure 3.14. Interactions between macromolecules.

Source: https://www.researchgate.net/publication/375516449/figure/fig2/AS:11431281256669045@1719563120955/Interactions-of-proteins-with-other-macromolecule-type.jpg.

3.5.1. Protein-Carbohydrate Interactions

Protein-carbohydrate interaction is of prime importance in the study of biological macromolecules since such interactions are essential in numerous biological processes. Proteins, complex molecules made up of amino acids, are likely to be reliant upon carbohydrates for function, particularly in cell recognition and signal transduction processes. Glycoproteins, carbohydrate groups that are bonded to proteins, account for most physiological reactions, ranging from immune reactions to cell adhesion. The specific binding of proteins to carbohydrates relies heavily on the existence of distinctive sugar moieties, which dictate the way in which these macromolecules interact. This specificity is crucial since it lies at the foundation of vital biological processes, like the recognition of pathogens by immune cells and the progression of disease in cases where such deviations in interaction occur. Protein-carbohydrate interaction biochemistry is very complex, with numerous systems of binding , such as hydrogen bonding, ionic interaction, and van der Waals forces. Diversity in carbohydrate structure offers a series of unique configurations that can be recognized and bound by proteins to form a dynamic interaction controlled by the three-dimensional structure of both entities. This relationship is also more accurately described through the application of graph theory to biological research. Methods , such as subgraph isomorphism and maximum common subgraph isomorphism can reveal the complex structural relationships between proteins and carbohydrates, allowing researchers to gain a deeper understanding of how they co-occur and interact functionally within biological systems (Altschul et al., 2005). This analytical platform provides a powerful platform for describing such interactions at the molecular level, enabling advancements in fields , such as bioinformatics. Moreover, developing work highlights the therapeutic and biomaterial engineering potential of such interactions. For instance, dendritic glycopolymer construction on polyamine scaffolds demonstrates biomedical potential for regulating protein-carbohydrate interactions for drug delivery systems and gene transfection. These glycopolymers leverage the carbohydrate unit's specific binding ability to optimize binding affinity with minimal toxicity compared to traditional polyamine scaffolds (Abbott et al., 2014). All of these advances point to the importance of learning the fundamentals of protein-carbohydrate interactions and how they are used in drug development. Overall, the exploration of these macromolecular interactions is crucial for not only elucidating biological mechanisms but also for applying this knowledge towards solving medicinal challenges (Figure 3.15).

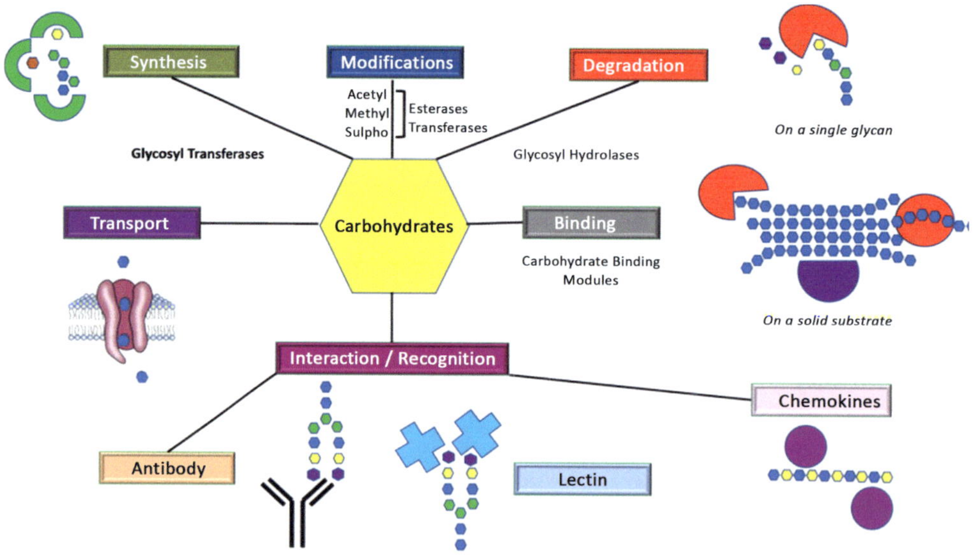

Figure 3.15. Protein-carbohydrate interactions.

Source: https://ars.els-cdn.com/content/image/1-s2.0-S0065231823000033-f03-01-9780323992541.jpg.

3.5.2. Lipid-Protein Interactions

The complex interaction between proteins and lipids is crucial for biological membrane function and is involved in numerous cellular processes , such as signaling, transport, and energy conversion. Interaction among lipids and proteins can substantially affect membrane fluidity and protein conformation, which in turn influence the activity of enzymes and receptors within the membrane. For instance, the fluid mosaic model of membrane structure has it that embedded lipids can be shifted laterally and can have transient facilitating interactions with lipids to maximize function and structural integrity. It is established in research that the lipid environment for a protein can differ, which facilitates modulation of its activity, indicating that coordination of protein signaling cascades by specific lipid domains, i.e., lipid rafts, is imperative.

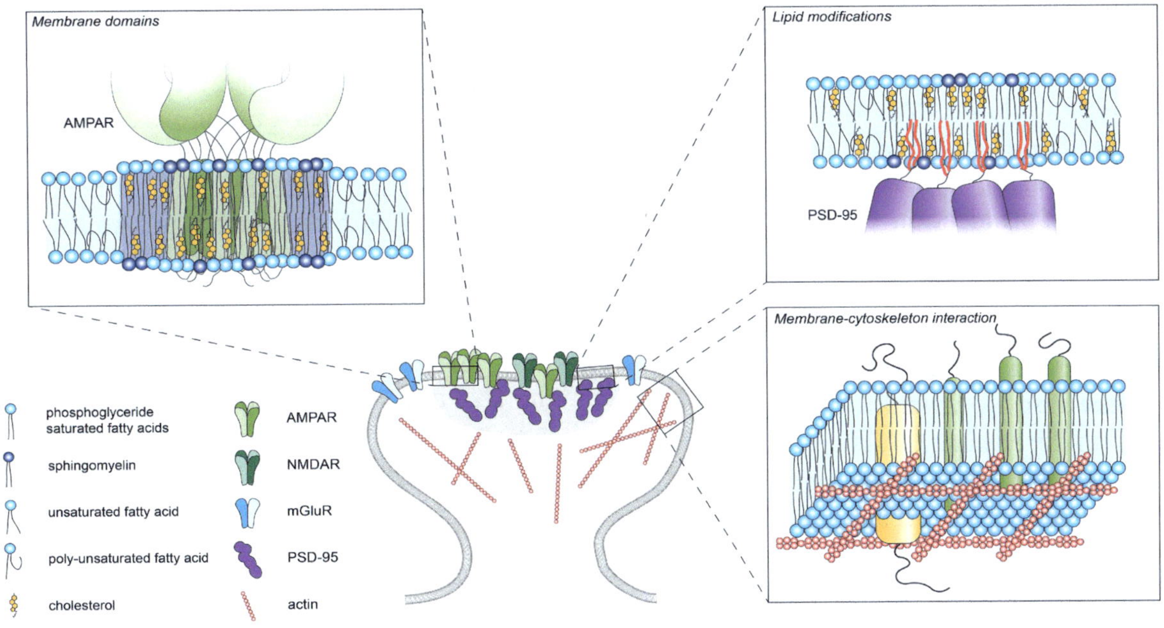

Figure 3.16. Lipid-protein interactions.

Source: https://www.researchgate.net/publication/356649432/figure/fig2/AS:11431281153752498@1682524889712/Lipid-protein-interactions-Schematic-diagram-of-the-different-lipid-protein-interactions.tif.

These complex interactions emphasize the significance of lipid-protein interactions in maintaining cellular homeostasis as well as responding to environmental cues. Advances in molecular modeling techniques have provided insightful information regarding lipid-protein associations' properties, enabling researchers to make educated guesses concerning how some lipid molecules might influence protein function. Computational tools facilitate the simulation of various lipid environments within and around the target proteins and reveal how various contents of lipids may serve to vary protein stability and function. For example, dendrimers, being branched macromolecules, offer a precise structural platform for investigating these interactions in depth. Various studies have highlighted the extent to which changing dendrimer size and surface characteristics can further their use in drug delivery or gene transfer systems, directly related to their lipid interactions.

Understanding these molecular movements not only deepens our insight into biological systems but also promotes the creation of more effective therapeutic agents that exploit lipid-protein interactions for enhanced delivery and efficiency in clinical treatments (Buhleier et al., 2014). Lipid-protein interactions are of concern to nonviral gene delivery systems, particularly in terms of the challenge posed by cellular compartments in gene transfer. Research has indicated that intracellular nanoparticle partitioning within proliferating cells can determine gene delivery efficiency. As an illustration, the intracellular distribution of organelles and foreign macromolecules is altered during the mitotic process, and formulation optimization could lead to better nuclear targeting. Because retroviruses usurp these processes to permit their genetic payload into host cells, nonviral platforms are spurred to utilize similar tactics, leveraging the cellular machinery during division. The development of lipid carriers that can successfully navigate these intracellular dynamics may offer a thrilling alternative to viral vectors, paving the way toward new gene therapy approaches that are safe and effective for application in clinical practice (Braeckmans et al., 2012) (Figure 3.16).

3.5.3. Nucleic Acid Interactions with Proteins

Complex nucleic acid-protein interactions form the basis of numerous biological processes , such as gene regulation, DNA replication and repair mechanisms. These interactions frequently require proteins as critical partners or regulators and such proteins can bind nucleic acids with high specificity and affinity. Nucleic acid-binding proteins have such a variety of functions due to their structural diversity; many different motifs (e.g., zinc fingers, helix-loop-helix domains) mediate specific binding interactions. The study of these binding mechanisms not only explains the intricacies of how a cell works but can also help in the formulation of therapeutic approaches that can be used to target these interactions to treat diseases associated with the failure of gene regulation. As an example, the deregulated nucleic acid-protein binding events can trigger oncogenesis, which highlights the significance of molecular interaction studies to guide the biomedical practice. The sophisticated computational methods have played the key role in the investigation of the nucleic acid-protein interactions at the molecular scale. Molecular modeling, e.g., can be used to predict and visualize binding affinities and interaction dynamics and thus increases our knowledge on these macromolecular interactions. Highly branched macromolecules dendrimers have been widely investigated with respect to their ability to bind to nucleic acids and proteins. Their structural peculiarities can be fine-tuned to better utilize them as drug carriers and therapeutic means. Other experiments have shown that by altering the size and surface properties of dendrimers, it is possible to effect a large change in their behavior, representing a real-world application of computational understanding to biological systems. Such research contributes to the overarching goal of designing innovative carriers capable of precise interactions with nucleic acids and proteins, furthering advancements in drug development (Buhleier et al., 2014). Moreover, the circumstances of the nucleic acids-proteins interaction may also greatly influence the result of such molecular interactions. For example, the phenol emulsion reassociation technique (PERT) illustrates how the environment influences DNA behavior, particularly in terms of adsorption and renaturation. This method allows seeing that interface interactions can tune the efficiency of nucleic acid processes, so heterogeneous systems could offer benefits compared to the classic homogeneous methods in biochemical applications. The results of PERTs demonstrate the importance of physical qualities, like salt concentration and agitation, in motivating nucleic acid-protein dynamics.

This understanding opens new avenues for research on the extraction and manipulation of nucleic acids in various biological contexts, reinforcing the significance of elucidating the complex nature of nucleic acid interactions with proteins (Figure 3.17).

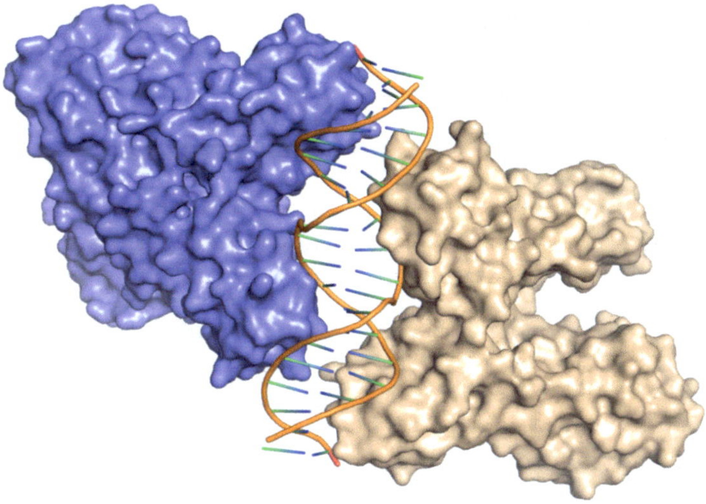

Figure 3.17. Protein–DNA complex showing specific protein–nucleic acid interactions.

Source: https://www.profacgen.com/wp-content/themes/profacgen/img/Protein-Nucleic-Acid-Interactions.png.

Test Your Knowledge

1. What are protein-carbohydrate interactions?
2. What are lipid-protein interactions?

3.6. Macromolecules in Health and Disease

Learning Outcome

1. Role of macromolecules in metabolic disorders;
2. Impact of macromolecule dysfunction on health;
3. Therapeutic applications targeting macromolecules.

In disease and health, macromolecules are at the center of numerous physiological processes and disease mechanisms. Of the major macromolecules, proteins, carbohydrates, lipids, and nucleic acids are crucial for cellular function and structure. Proteins, for instance, are enzymes, structural molecules, and signaling molecules, whereas carbohydrates are cell structural components and energy sources. Their balance and interaction with one another are important in homeostasis. Change in their synthesis, modification, or degradation can lead to disease states, e.g., diabetes, cancer, and neurodegenerative disorders, and thus understanding the role of macromolecules in health and disease states is essential. With research continuing to advance, it is more evident that the examination of these biological macromolecules holds promising frontiers for developing therapeutic interventions and preventive therapies for a multitude of disorders of human health. Chitosan, a biopolymer of chitin origin, has emerged as an important macromolecule in pharmacy and agriculture due to its biocompatibility and biodegradability. Recent studies establish that chitosan oligomers (COs) exhibit superior antifungal activity, making them potential agents for alternative use in sustainable agriculture. The CO silver nanoparticle and propolis combinations have been found to exhibit high antifungal activity against various pathogens, up to 93% inhibition of fungal growth in some cases. The new use of naturally occurring macromolecules indicates their potential to assist not just disease but also health control. By studying eco-friendly alternatives , such as COs to manage phytopathogenic fungi, researchers aim to address the growing concern of chemical agents in agriculture that are harmful to human health and the environment, thus accentuating the interdisciplinary relevance of macromolecules as promoters of sustainable strategies and public well-being (Díez et al., 2017). In the human health context, macromolecules , such as glycosaminoglycans are pivotal to joint health and disease etiology of conditions like osteoarthritis (OA). OA is characterized by cartilage degradation in the joints, which is normally driven by inflammatory cytokines , such as interleukin-1 and tumor necrosis factor-α. Not only do these cytokines drive cartilage degradation acceleration through the depletion of glycosaminoglycans but also suppress chondrocyte viability. Therapeutics, , such as glucocorticoids dexamethasone, have been studied for their ability to counteract these undesirable effects. Dexamethasone has been shown to be useful in improving the cell viability as well as the biosynthesis of glycosaminoglycans in the presence of inflammatory agents, and this indicates that macromolecules can serve both as targets and markers to cure degenerative diseases. This overlap of macromolecules and disease confirms the imperative importance of continued research in biological macromolecules to inform clinical practice and develop new therapeutic strategies (Wallace, 2018).

3.6.1. Role of Macromolecules in Metabolic Disorders

The complex interaction between macromolecules and metabolic disorders reveals fundamental information about cellular and systemic dysfunctions that cause various diseases. Proteins, carbohydrates, lipids, and nucleic acids are the fundamental molecular components of life, and each one of them is very important in maintaining homeostasis. For instance, lipid metabolism disarrangements have been linked to diabetes mellitus and cardiovascular diseases. Lipid dysregulation causes abnormal amounts, which can stockpile toxic metabolic waste products, elevate inflammation and disrupt insulin signaling pathways. These are situations where one witnesses the cascading of effects of macromolecular composition changes leading to dysregulation at the metabolic level. The significance of these interactions is most notable when one considers their importance to therapeutic interventions in averting these disorders. In neurodegenerative diseases, the role of macromolecules becomes further highlighted, particularly with respect to the activation of glial cells and that of inflammatory mediators. It is revealed in research that microglial activation has been strongly associated with metabolic disturbances observed in such conditions as Alzheimer's. Specifically, in a study employing magnetic resonance spectroscopy (MRS), it was possible to demonstrate differential macromolecule levels' responses in the states of health and pathology. The concentrations of metabolites seen with neuroinflammation are also critical markers, and they reflect how macromolecular homeostasis dysregulation can promote disease onset. The increase in the concentrations of macromolecules and lipids observed following lipopolysaccharide (LPS) treatment supports the notion that these molecules do not just reflect metabolic state but also actively participate in the pathophysiological processes underlying neurological disorders (Pardon et al., 2016). The utility of understanding macromolecular function extends beyond diagnosis; they represent pathways of intervention in metabolic disease. Identification of particular macromolecules as biomarkers of a particular disease allows one to design intervention therapies directed toward the root cause of metabolic dysregulation. For instance, research on the alteration in protein expression or lipid profile in a diseased tissue can result in new approaches aimed at reestablishing normal metabolic function. Moreover, the ongoing research into the influence of macromolecules on cellular signaling processes continues to reveal profound revelations regarding the design of pharmacological drugs that will treat metabolic disorders. Lastly, the integration of macromolecular research into clinical practice has the ability to increase patient outcomes in patients suffering from long-term diseases caused by metabolic abnormalities (Wallace, 2018) (Figure 3.18).

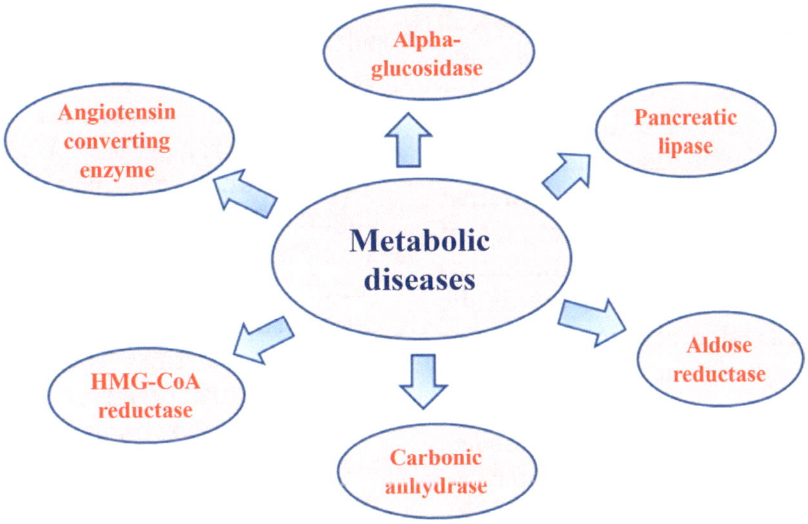

Figure 3.18. Role of macromolecules in metabolic disorders.

Source: https://ars.els-cdn.com/content/image/3-s2.0-B9780128117101000331-f31-03-9780128117101.jpg.

3.6.2. Impact of Macromolecule Dysfunction on Health

Malfunction of biological macromolecules, and more so proteins, nucleic acids, lipids, and carbohydrates, has a strongly impactful influence on health through its interference with fundamental biological processes. Proteins, as enzymes, structural components, and signal molecules, play a vital role in cellular function. Protein misfolding may lead to various diseases, for instance, Alzheimer's and Parkinson's disease, where there is the accumulation of toxic protein that impacts the function of neurons. Such misfolding can arise due to genetic mutation, environmental factors, or general processes of aging. The impact is extensive, for defective protein function can trigger cell degeneration, inflammation, and resulting organ dysfunction, illustrating the tight interconnection between macromolecule integrity and general well-being. Nucleic acids in the DNA and RNA forms are central in storing and transmitting genetic information. Dysfunction of such macromolecules will lead to mutations and abnormal gene expression, resulting in diseases , such as cancer. For instance, external exposures will induce harm to the DNA, which is further exacerbated by long-term inflammation, which also has been found to play a role in age-related diseases as observed in much literature examining the causation and chronicity of chronic inflammatory diseases (Howcroft et al., 2013). If the regulation of gene expression is imbalanced by defective nucleic acids, it can lead to uncontrollable cell proliferation and tumorigenesis. Hence, the integrity of nucleic acids is crucial not just for ensuring gene fidelity but also for preventing the onset of degenerative diseases and retaining overall health. The role of lipids in cell function and structure also points to the impact of dysfunction in macromolecules on health. Excess or deficiency of lipids can lead to metabolic disease, cardiovascular disease, and obesity. For example, fatty acid dysregulation is associated with insulin resistance and type 2 diabetes. In addition, accumulation of certain lipid species can trigger inflammation, which is undesirable for cardiovascular health (Eklund et al., 2003). Carbohydrates, glycoproteins and glycolipids, play roles in cell recognition and the process of signal transduction. Carbohydrate metabolic disorders can lead to diseases , such as diabetes mellitus or glycogen storage diseases, both of which demonstrate the effect of macromolecule malfunction in triggering a chain of disease states. Therefore, maintenance of the functional integrity of lipids and carbohydrates is pivotal in the prevention of metabolic and degenerative diseases, reflecting the quintessential role of biological macromolecules in health defense.

3.6.3. Therapeutic Applications Targeting Macromolecules

Therapeutic use of biological macromolecules is transforming the face of contemporary medicine, especially with regard to drug delivery and immunotherapy. Proteins, nucleic acids, and polysaccharides are macromolecules that are gaining importance as possible means of enhancing therapeutic effects and reducing side effects. Recent advancements have demonstrated that methods like vapor nanobubble (VNB) photoporation can facilitate the efficient delivery of large macromolecules into challenging cell types, , such as primary human $CD4^+$ T cells. This new method that uses gold nanoparticles as sensitizers can be used to introduce macromolecules safely without the cytotoxicity that is commonly experienced with the conventional methods , such as viral transduction or electroporation. Consequently, VNB photoporation not only expands the possibilities for T cell engineering but also holds promise for broader applications in the fields of HIV research and cancer immunotherapies (Braeckmans et al., 2019). Also, the importance of computational modeling in the study and improvement of the characteristics of macromolecules cannot be overestimated. Molecular modeling methods offer useful tools in the design and optimization of therapeutic agents, particularly with regard to dendrimers – highly complex macromolecules which may serve as vehicles to deliver drugs and genetic materials. Using the computational methodologies, researchers can foretell the interaction at the molecular level, which forms the foundation of the development of more efficient drug delivery systems. The investigations of the structure and dynamics of interaction of dendrimers show the way in which the changes in the size and surface characteristics can significantly change their behavior as drug carriers. Such insights are essential for tailoring these macromolecules to meet specific therapeutic needs, thereby enhancing their efficacy in clinical applications (Buhleier et al., 2014). Lastly, therapeutic targeting of macromolecules is not only limited to immune modulation and drug delivery, but also in the creation of diagnostic agents. Application of biological macromolecules in diagnostic technologies takes advantage of their specificity and binding affinity, that can result into enhanced disease detection and monitoring. As an example, engineered antibodies or nucleic acid aptamers may specifically bind disease markers, with increased sensitivity and specificity of diagnostic assays. The continued discovery of multifunctional macromolecules also provides novel routes to simultaneous therapy and diagnostics, or theranostics. The integration of both therapeutic and diagnostic functions of macromolecules has the potential to simplify the handling of patients and therefore result in more personalized treatment regimens that closely match the characteristics of individual patients and eventually enhance patient outcomes in many areas of medical practice.

Test Your Knowledge

1. What is the role of macromolecules in metabolic disorders?
2. What is the impact of macromolecule dysfunction on health?

Summary

An excursion into biological macromolecules shows that these complex structures lie behind the physiological activities of all living organisms. Composed of proteins, nucleic acids, carbohydrates, and lipids, biological macromolecules are not merely static entities but dynamic participants in the ongoing regulation of cellular processes. The codependence of these macromolecules illustrates their engagement with self-organization and homeostasis, which points toward an complex web of interactions sustaining life. Their ability to modify through chemical changes and structural organizations shows the evolutionary significance of these molecules. With populations facing diverse conditions, the functional versatility of biological macromolecules allows them to survive, proving the fundamental postulate that life is an adaptive process governed by molecular complexities. In addition, how biological macromolecules interface with the world can be understood through the free-energy principle (FEP) perspective, in which living systems are characterized by minimizing variational free energy. This is essential to the study of self-organization because it allows us to understand how biological structures emerge and function on gigantic temporal and spatial scales. The FEP assumes that all living systems, from single cells to complex multicellular life, undertake processes that reduce disorder and maximize their biochemical interactions with the environment. Because living organisms hold implicit models of their niches, their structural adaptations and behavior become manifestations of a universal mechanism for maintaining life and enhancing complexity. This perspective places added richness on our understanding of the dynamic roles taken by macromolecules in maintaining life, demonstrating continuous causal interplay between form and function. Finally, recent advances in genetic engineering and genome editing reveal immense insight into the capabilities that are intrinsic to biological macromolecules, particularly in genetic variety and plasticity. The classical notions of genetic engineering, which were used to portray genes as fixed blocks of code, are being altered by the detection of naturally occurring genome-editing mechanisms in other organisms. The accuracy with which some viruses and RNA molecules perform genome editing on command by adaptive imperatives subverts previous paradigms based on mutation as the dominant mechanism for creating genetic diversity. This knowledge change has implications not just for artificial genome editing, but for an enhanced vision of the plasticity and complexity of biological systems as systems. The biology of macromolecules therefore, takes us towards a more systemic vision of life, in which the ability to adapt and self-organize is key to evolutionary success and biological innovation.

It is significant in biological macromolecular science to understand their diverse functions and interactions within the cell context. The macromolecules in question are proteins, nucleic acids, carbohydrates, and lipids, each of which contributes uniquely to living organisms' biology. Proteins, for instance, play essential roles in catalysis, signaling, and structural support, while nucleic acids , such as DNA and RNA, are essential to genetic information storage and its transmission. Not only do carbohydrates offer energy, but also as structural components in cell membranes, and lipids are essential in the formation of biological barriers. The delicate balance and interaction between these macromolecules are the foundation of the complexity of biological systems, which renders them vital in both health and disease. Recent advances in the research of biological macromolecules further enhanced our appreciation of their dynamic nature, particularly in relation to polymer translocation via nanometer-scale pores and channels in membranes. These investigations have enabled researchers to observe the dynamics of single macromolecules in real time at a previously unimaginable resolution. This has led to meaningful theoretical progress in non-equilibrium dynamics that are at the heart of understanding how such molecules behave under different environmental conditions. For instance, the translocation of polymers determines their folding and unfolding dynamics, which are vital to their function. These findings not only broaden our understanding of macromolecular interactions but also present avenues for potential biotechnological applications that leverage such dynamic processes (Barkema et al., 2013). Moreover, investigations on the conformational changes of biological macromolecules under varied conditions, , such as temperature, have yielded an enhanced perspective of their structural dynamics. A work on ultrafast DNA ensemble unfolding illustrates that the conditions of the environment exert a profound influence on macromolecular behavior, representing complex order-disorder transitions. Particularly, studies employing molecular dynamics simulations have demonstrated that processes like unzipping and strand dissociation are non-linear in their course, demonstrating the intricacy of macromolecular conformations. These findings form the basis for insights into the mechanism of biological processes and the influence of mutations on nucleic acid structure, thus testifying to the significance of biological macromolecules in molecular biology (Lin et al., 2009).

KEY TERMS FROM THIS CHAPTER

Amino acids	Carbohydrates	Disaccharides
DNA (deoxyribonucleic acid)	Enzymes	Fatty acids
Lipids	Monosaccharides	Nucleic acids
Peptide bonds	Phospholipids	Polysaccharides
Primary structure	Proteins	Quaternary structure
RNA (ribonucleic acid)	Secondary structure	Steroids

CRITICAL THINKING QUESTIONS

1. **Which biological macromolecule is primarily responsible for storing genetic information?**
   a. Carbohydrates
   b. Proteins
   c. Nucleic acids
   d. Lipids
2. **Which macromolecule is made up of amino acids?**
   a. Carbohydrates
   b. Proteins
   c. Lipids
   d. Nucleic acids
3. **Which of the following is a carbohydrate?**
   a. DNA
   b. Glucose
   c. Fatty acid
   d. Enzyme
4. **Lipids are mostly composed of:**
   a. Sugars and amino acids
   b. Glycerol and fatty acids
   c. Nucleotides
   d. Monosaccharides
5. **Which macromolecule acts as enzymes to speed up chemical reactions?**
   a. Carbohydrates
   b. Proteins
   c. Lipids
   d. Nucleic acids

Short Answer Questions

1. What are biological macromolecules? Why are they important to living organisms?
2. Name the four major types of biological macromolecules.
3. What are the monomers (building blocks) of carbohydrates?
4. Describe the difference between simple sugars (monosaccharides) and complex carbohydrates (polysaccharides).
5. What elements are primarily found in lipids?
6. What is the role of nucleic acids in cells? Name the two main types.

Answers to Critical Thinking Questions

1. (c); 2. (b); 3. (b); 4. (b); 5. (b)

References

1. Abbott, L., Agarwal, R., Agashe, S., Agrawal, A., Albersheim, P., et al. (2014). Dendritic glycopolymers based on dendritic polyamine scaffolds: View on their synthetic approaches, characteristics and potential for biomedical applications. https://core.ac.uk/download/71974968.pdf (accessed on 13th June 2025).
2. Abelho, M., Jackson, A. D., Boheim, P., Cantor, C., et al. (2006). The thermodynamics of general anesthesia. https://core.ac.uk/download/44146229.pdf (accessed on 13th June 2025).
3. Abroon, P., Alford, L., Ashouri, A., Farahani, M., et al. (2017). Early life nutritional quality effects on adult memory retention in a parasitic wasp. https://core.ac.uk/download/83960681.pdf (accessed on 13th June 2025).
4. Adams, N., & Schubert, U. S. (2007). *Poly(2-oxazolines)* in biological and biomedical application contexts. https://core.ac.uk/download/1301686.pdf (accessed on 13th June 2025).
5. Pardon, M. C., Yanez Lopez, M., Yuchun, D., Marjańska, M., Prior, M., Brignell, C., Parhizkar, S., Agostini, A., Bai, L., Auer, D. P., & Faas, H. M. (2016). Magnetic Resonance Spectroscopy discriminates the response to microglial stimulation of wild type and Alzheimer's disease models. *Scientific reports*, *6*, 19880. https://doi.org/10.1038/srep19880 (accessed on 13th June 2025).
6. Aguirre-Ramírez, M., Banat, I. M., Diaz De Rienzo, M. A., Silva-Jimenez, H., et al. (2021). Surfactants: Physicochemical interactions with biological macromolecules. https://core.ac.uk/download/394999208.pdf (accessed on 13th June 2025).
7. Aksakal, R., Badi, N., Du Prez, F., Mertens, B., et al. (2021). Applications of discrete synthetic macromolecules in life and materials science: Recent and future trends. https://core.ac.uk/download/416426770.pdf (accessed on 13th June 2025).
8. Altschul, S., Artymiuk, P. J., Barnard, J. M., et al. (2005). Graph theoretic methods for the analysis of structural relationships in biological macromolecules. https://core.ac.uk/download/51878.pdf (accessed on 13th June 2025).
9. Howcroft, T. K., Campisi, J., Louis, G. B., Smith, M. T., Wise, B., Wyss-Coray, T., Augustine, A. D., McElhancy, J. E., Kohanski, R., & Sierra, F. (2013). The role of inflammation in age-related disease. *Aging*, *5*(1), 84–93. https://doi.org/10.18632/aging.100531 (accessed on 13th June 2025).
10. Backman, D., Rim, G., Jacobsen, M. M., Li, H., et al. (2017). Silk-fibronectin protein alloy fibers support cell adhesion and viability as a high strength, matrix fiber analogue. https://core.ac.uk/download/215931762.pdf (accessed on 13th June 2025).
11. Badcock, P., Constant, A., Friston, K., Hesp, C., et al. (2025). A multi-scale view of the emergent complexity of life: A free-energy proposal. https://core.ac.uk/download/158370660.pdf (accessed on 13th June 2025).
12. Barkema, G. T., Kolomeisky, A. B., & Panja, D. (2013). Through the eye of the needle: Recent advances in understanding biopolymer translocation. http://arxiv.org/abs/1308.4712 (accessed on 13th June 2025).
13. Bojarski, A., Nowak, M., & Testa, B. (2018). Conformational constraints on side chains in protein residues increase their information content. https://core.ac.uk/download/159155717.pdf (accessed on 13th June 2025).

14. Božič, A. L. (2017). From discrete to continuous description of spherical surface charge distributions. http://arxiv.org/abs/1711.02854 (accessed on 13th June 2025).

15. Braeckmans, K., Brans, T., De Smedt, S., Fraire, J. C., et al. (2019). Gold nanoparticle-mediated photoporation enables delivery of macromolecules over a wide range of molecular weights in human CD4+ T cells. https://core.ac.uk/download/237011703.pdf (accessed on 13th June 2025).

16. Braeckmans, K., De Smedt, S., Rejman, J., Remaut, K., et al. (2012). Intracellular partitioning of cell organelles and extraneous nanoparticles during mitosis. https://core.ac.uk/download/55801726.pdf (accessed on 13th June 2025).

17. Bridelli, M. G. (2017). Fourier transform infrared spectroscopy in the study of hydrated biological macromolecules. https://core.ac.uk/download/322429298.pdf (accessed on 13th June 2025).

18. Buhleier, W., Carbone, M., Florindo, H., Hess, M., Klajnert, B., Silva, L., et al. (2014). Molecular modeling to study dendrimers for biomedical applications. https://core.ac.uk/download/29849184.pdf (accessed on 13th June 2025).

19. Carrère, H., Ferrer Martí, I., Garfi, M., Solé Bundó, M., et al. (2017). Enhancement of microalgae anaerobic digestion by thermo-alkaline pretreatment with lime (CaO). https://core.ac.uk/download/87654010.pdf (accessed on 13th June 2025).

20. Cooper, A., Kennedy, M. W., Smith, B. O., Vance, D., et al. (2017). Frog foams and natural protein surfactants. https://core.ac.uk/download/74232128.pdf (accessed on 13th June 2025).

21. Cosic, I., & Pirogova, E. (2007). Bioactive peptide design using the resonant recognition model. https://core.ac.uk/download/pdf/15611515.pdf (accessed on 13th June 2025).

22. Curry, S. (2015). Structural biology: A century-long journey into an unseen world. https://core.ac.uk/download/77010710.pdf (accessed on 13th June 2025).

23. Demir, E., Schreiber, F., Mi, H., Le Novere, N., & Moodie, S. L. (2010). Systems biology graphical notation: Process description language level 1. https://core.ac.uk/download/pdf/289411.pdf (accessed on 13th June 2025).

24. Dhar, A., Elcock, A., Grosberg, A., Minton, A., Pisliakov, A., et al. (2011). Capturing the essence of folding and functions of biomolecules using coarse-grained models. http://arxiv.org/abs/1109.5970 (accessed on 13th June 2025).

25. Díez Casero, J. J., Flores Pacheco, J. A., Martín García, J., Martín Gil, J., et al. (2017). Potential control of forest diseases by solutions of chitosan oligomers, propolis and nano silver. https://core.ac.uk/download/322906396.pdf (accessed on 13th June 2025).

26. Ding, Y., Tan, C., & Wu, F. (2014). Synthetic biology: A bridge between artificial and natural cells. https://core.ac.uk/download/323068426.pdf (accessed on 13th June 2025).

27. Eklund, G. (2003). Cadmium in newborns. https://core.ac.uk/download/11694662.pdf (accessed on 13th June 2025).

28. Franosch, T., & Höfling, F. (2013). Anomalous transport in the crowded world of biological cells. http://arxiv.org/abs/1301.6990 (accessed on 13th June 2025).

29. Gruber, A., Klinger, D., & Navarro, L. (2019). Reactive precursor particles as synthetic platform for the generation of functional nanoparticles, nanogels, and microgels. https://core.ac.uk/download/288114298.pdf (accessed on 13th June 2025).

30. Gu, L., Kholodovych, V., Martin, A. A. T., Moghe, P., et al. (2015). Carbohydrate-derived amphiphilic macromolecules: A biophysical structural characterization and analysis of binding behaviors to model membranes. https://core.ac.uk/download/323067811.pdf (accessed on 13th June 2025).
31. Hoffman, D. B., & Kontaratos, A. N. (2025). Biological generalizations and the search for extraterrestrial life. https://core.ac.uk/download/pdf/80660612.pdf (accessed on 13th June 2025).
32. Horowitz, N. H., & Mitchell, H. K. (1951). Biochemical genetics. https://core.ac.uk/download/4869357.pdf (accessed on 13th June 2025).
33. Kletzin, A., Anbar, A. D., Alberts, B., Pullman, B., Duve, C. de, Myers, C. R., et al. (2007). Bootstrapping the energy flow in the beginning of life. https://core.ac.uk/download/pdf/15453002.pdf (accessed on 13th June 2025).
34. Kolb, B., & Thonhauser, T. (2012). Molecular biology at the quantum level: Can modern density functional theory forge the path? http://arxiv.org/abs/1207.1075 (accessed on 13th June 2025).
35. Lin, M. M., Shorokhov, D., & Zewail, A. H. (2009). Structural ultrafast dynamics of macromolecules: Diffraction of free DNA and effect of hydration. https://core.ac.uk/download/4883691.pdf (accessed on 13th June 2025).
36. Liu, Y., Miao, Y., Min, W., Qian, X., et al. (2019). Spectral tracing of deuterium for imaging glucose metabolism. https://core.ac.uk/download/323065665.pdf (accessed on 13th June 2025).
37. Lu, T. S. (2025). The origin of biological macromolecules on the earth: The hypothesis of inorganic template. https://core.ac.uk/download/pdf/42876014.pdf (accessed on 13th June 2025).
38. Perry, S. L., & Sing, C. E. (2020). 100th anniversary of macromolecular science viewpoint: Opportunities in the physics of sequence-defined polymers. https://core.ac.uk/download/286729732.pdf (accessed on 13th June 2025).
39. Ricardo, A., Guerrier-Takada, C., Biondi, E., Gallori, E., Szathmáry, E., et al. (2009). The origin of life: Chemical evolution of a metabolic system in a mineral honeycomb? https://core.ac.uk/download/11855055.pdf (accessed on 13th June 2025).
40. Sorokin, A., Schreiber, F., Mi, H., Le Novere, N., & Moodie, S. L. (2009). Systems biology graphical notation: Process description language level 1. https://core.ac.uk/download/pdf/288790.pdf (accessed on 13th June 2025).
41. Wallace, Chelsey. Boston University ProQuest Dissertations & Theses, 2018. 10786747. (accessed on 13th June 2025).
42. Witzany, G. (2017). Artificial and natural genetic information processing. https://core.ac.uk/download/186332662.pdf (accessed on 13th June 2025).
43. Zhang, J. (2019). *BIOC 321: Elements of Biochemistry–A Peer Review of Teaching Project Benchmark Portfolio.* https://core.ac.uk/download/215200862.pdf (accessed on 13th June 2025).

CELL STRUCTURE AND FUNCTION

4

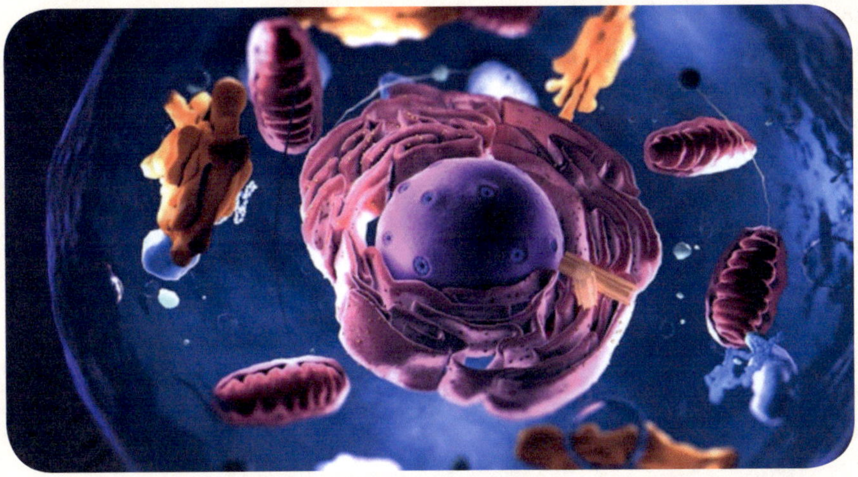

LEARNING OBJECTIVES

After studying this chapter, you will be able to:

1. Define the cell theory;
2. Explain about prokaryotic cells;
3. Discuss eukaryotic cells;
4. Illustrate the cell membrane structure;
5. Explore cellular metabolism;
6. Describe cell communication.

Introduction

Cell structure and function form the basis of the examination of all biological systems. The cell, the basic unit of life, harbors a wealth of structures that confer functions crucial to the maintenance and reproduction of living organisms. These structures, , including the nucleus, mitochondria, and ribosomes, operate in concert such that they aid various cellular processes , such as energy production, storage of genetic information, and protein synthesis. Understanding how these building blocks are interfaced and operate together to generate general cellular function provides a foundation to study more complex biological phenomena like tissue formation and organ system function. Structure-function interaction within a cell serves to highlight the significance of cellular biology because understanding obtained there can provide insight into a path towards medical and biotechnological innovation. A detailed description of cell structure also involves delving into the molecular components that facilitate cellular processes. Proteins are particularly at the heart of cellular operations, performing multifunctional roles by binding to nucleic acids and other biomolecules. The role and effectiveness of proteins are intimately connected with their three-dimensional forms, which are defined by amino acid sequences that constitute them. For example, the ability of a protein to catalyze biochemical reactions or transport molecules across membranes is determined by its precise configuration. The information gained from the examination of protein structures and functions is supplemented further by structural bioinformatics techniques that make use of computational strategies to analyze and foresee these structures. This allows researchers to model cellular environments and predict protein interactions, ultimately clarifying the molecular basis of life itself (Abeln et al., 2023). Also, the research on cell structure goes beyond cellular component descriptions; it embraces knowledge of dynamic processes that maintain life. In cell biology, thermodynamics, molecular simulations, and bioinformatics converge to inform us about the behavior and functioning of cells under fluctuating physiological conditions. Quantification, prediction, and simulation of protein structure are some of the methods that reflect the interconnected nature of cellular units and their functions in cellular dynamics. With this multi-faceted perspective of cells, researchers can learn about the multifarious interactions that define life at the molecular level. Lastly, a full introduction to cell structure and function not only forms an academic foundation but also accentuates the importance of cellular research for innovations in health and disease control (Abeln et al., 2023) (Figure 4.1).

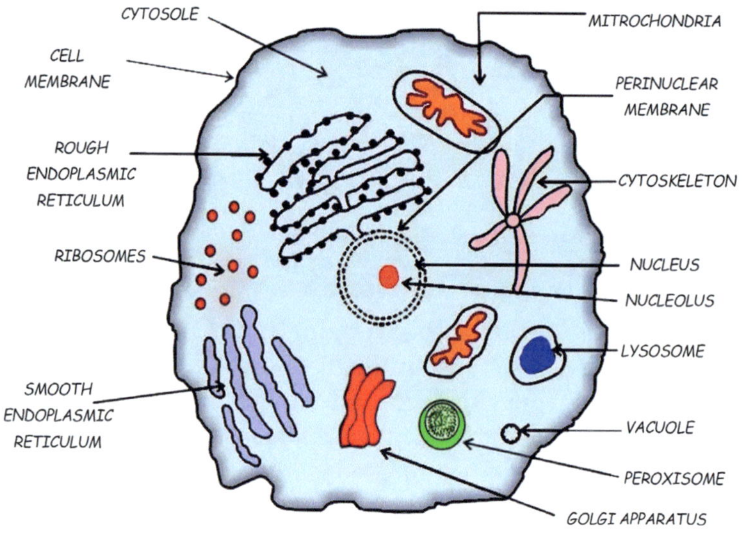

Figure 4.1. Cell structure and function.

Source: https://medicosage.com/wp-content/uploads/2021/11/CELL-STRUCTURE.webp.

Definition of a Cell

In biology, the cell is the simple building unit of life, the building block for all living organisms. A cell is both functional and structural as the smallest unit capable of performing the functions of life. Each cell is an complex system with a variety of organelles that carry out specialized functions required to ensure the survival of the organisms. For instance, cells in multicellular organisms may differentiate to carry out unique functions, contributing to functionality and complexity at the higher level. The concept here underscores cellular organization, indicating that even the most primitive forms of life consist of one or more cells, which are quite capable of processing duties , such as metabolism, energy conversion, and waste removal, laying the groundwork on which all biological operations take place. The complex cell structure is not a rigid scaffolding; it is actively involved in the cells' function as well as their interaction with the environment. For example, lipid bilayer cellular membranes play a significant role in regulating the movement of material into and out of the cell, homeostasis, and signaling to adjacent cells. In addition, organelles , such as the mitochondria, the cell's powerhouses, convert nutrients to energy, and the ER and Golgi apparatus have key functions in protein synthesis and modification. It is central to the structure and function of cell biology because it enables researchers to determine how alterations in cellular components can lead to malfunctioning, which can be manifested in many diseases. Hence, what one learns about the cell's definition is much more than recognizing its existence; it entails the understanding of its role in wider biological environments. Also, the implications of cellular structure extend far and encompass other fields like medicine and genetics. For instance, dysfunction and structural abnormalities in cells can cause disease states , such as primary ciliary dyskinesia (PCD), a genetic disorder resulting in abnormal ciliary structure that impairs mucociliary clearance and health (Noone, 2001). Of most significance in cellular pathology is this convergence of function and structure, especially when looking at conditions caused by pathogens. The study of how parasites like *Plasmodium falciparum* invade host cells to acquire infection represents a paradigm for strategic coordination among cellular constituents (Angrisano et al., 2012). By deciphering these complex interactions, researchers can develop targeted therapies that are meant to correct or minimize cellular pathology. Thus, an informed definition of a cell must reflect its cellular components, their functions, and the broader importance to health and disease (Figure 4.2).

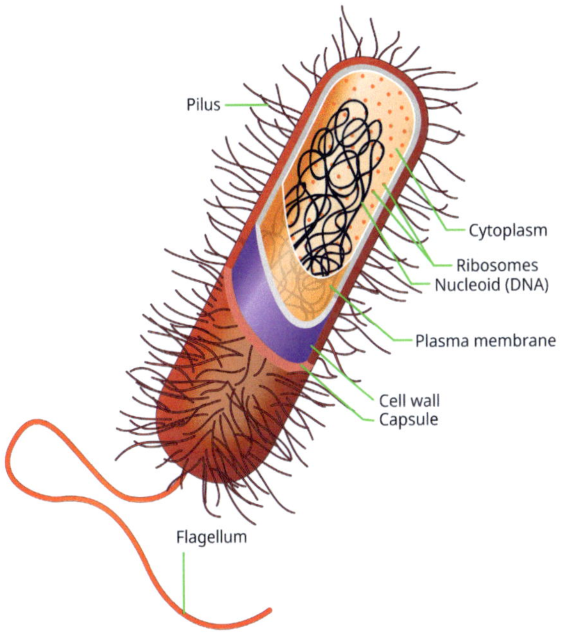

Figure 4.2. Definition of a cell.

Source: https://upload.wikimedia.org/wikipedia/commons/thumb/c/c5/Prokaryote_cell.svg/330px-Prokaryote_cell.svg.png.

Importance of Studying Cell Structure and Function

Cellular structure and function is a keystone pillar for the study of biology at multiple levels. Cells are the basic units of life, containing the processes that direct organismal behavior and evolution. A careful examination of these structures makes it easier to comprehend how complicated processes are permitted by specific components. For instance, the complex cell component architecture of membranes and organelles accounts for how proteins and genetic material engage and generate physiologic systems. All structural elements are important in facilitating cellular processes to yield a nested hierarchy in which macroscopic properties emerge from atomic-scale interactions. This hierarchical structure connects evolutionary biology with physical sciences and presents a paradigm for how organisms endure and adapt to their environments. By considering the complex relationships between form and function, researchers are in a better position to understand mechanisms of evolution and development in organisms as well as to justify the study of cell structure and function. Cell membranes, which enclose and protect cells, are important in controlling the movement of substances into and out of the cell, controlling eventually many physiological processes. Central to this function are the ion channels, which are found localized in such membranes and act as gatekeepers to allow ions , such as Na^+, K^+, Ca^{++}, and Cl^- to flow through the barrier. The dynamical operation of these channels between open and closed positions in response to stimuli demonstrates the sophisticated mechanism under cellular communication and signaling. Just as transistors in a computer control the flow of electricity, ion channels control the same in biological organisms, passing electrical signals needed for muscle contraction and brain functions. Understanding such mechanisms is crucial in identifying the physiological basis of life since it finally validates the importance of studying cell structure and cell function as part of a complete picture of biology (Barthel et al., 2013). Interplay between cell function and structure extends beyond isolated components, with a focus on conducting research on integrated systems. Cellular processes are generally interrelated and regulated by processes that create emergent properties only observable if the cell is studied as a whole. Based on studies, differences in channel structures affect how they respond to stimuli and impact cellular behavior overall. This interdependence is a reflection of the need for an integrated understanding of biology, from comprehending genetic coding to understanding the functioning of proteins within systems. Secondly, cells are not static units; they reproduce, undergo evolution, and transform, enabling life to be perpetuated generation after generation. An exhaustive comprehension of the way cells exist in the ecosystem forms the foundation for future progress in medicine, biotechnology, and ecology. For this reason, cell structure and function are crucial to the scientific advancement and solving contemporary biological issues (Eisenberg et al., 2018).

4.1. The Cell Theory

Learning Outcome

1. Historical background of cell theory;
2. Key components of cell theory;
3. Implications of cell theory in biology.

The Cell Theory is a key concept in the foundations of our knowledge of the biological sciences and is stated as all living things are made of cells, the cell is the fundamental unit of life and all cells are created as a result of other existing cells. This framework not only divides organisms into single-celled and multi-celled organisms but also highlights the fact that the cellular structure is universal to all the different forms of life. This paradigm has allowed scientists to create a unifying concept that crosses cut across taxonomy, allowing the connection of organisms as diverse as prokaryotic bacteria to complex eukaryotic plants and animals. The application of this theory goes as far as the comprehension of the cellular functions, where it gives an insight into how the different components of the cell interrelate to maintain the life processes. With the understanding that performance with regard to functioning may be broad among cells, there is a greater realization of cellular specialization and plasticity in specific environmental conditions. The Cell Theory also points out the amazing structure and functionality of cellular structures, besides establishing the cell as a basic unit of life. All the cells contain specialized parts, like nucleus, mitochondria, and ribosomes, which work together to direct life's processes. This compartmentalization enables cells to perform complicated biochemical reactions effectively improving their overall functioning. Moreover, studies of cellular structure have shown that the relationship between structure and function is not incidental but, on the contrary, it is crucial to the health and viability of organisms. As an example, the dynamic property of the plasma membranes can show how the communication between cells and the intake of nutrients can be largely affected by the alterations in the structure. This relationship is important to understand in the development of biotechnological applications and therapeutic interventions to target cellular dysfunction, which causes diseases. Besides, the very fact that we seek answers about the interaction of cells emphasizes the significance of cooperative activity, which is represented in evolutionary models of cooperation. The example of the diffusion of public goods between cells explains how cooperative behavior may evolve, with the help of spatial structures, which promote nutrient sharing. These processes rely on more complex cellular interactions, in which colony shape affects resource retention and use. The discoveries, in their turn, not only explain the evolutionary benefits of cooperation between microorganisms but also chime in with the ideas of the Cell Theory due to the emphasis they put on the interdependence of cellular processes. The inability of researchers to capture these interactions with complex models notwithstanding, the necessity of combining both theoretical and empirical studies emerges. Furthering our knowledge of cooperation between and among cells will be a great addition to our quest of cellular mechanisms that will further add to our understanding of cell structure and functionality (Figure 4.3).

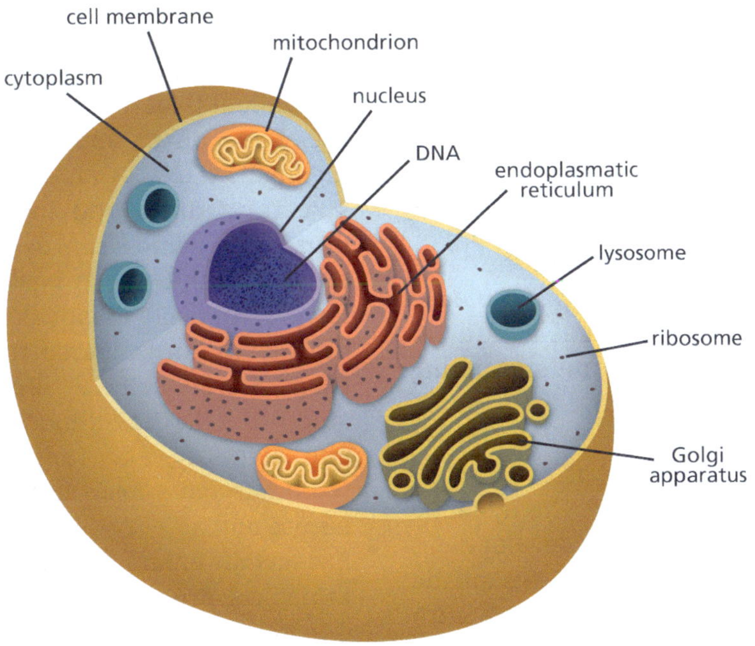

Figure 4.3. The cell theory.

Source: https://images.saymedia-content.com/.image/t_share/MTc2NDY0OTk0NzYwNjY0Mjc3/cell-cell-theory-nucleuscell-wall-and-cytoplasmic-organelles.png.

4.1.1. Historical Background of Cell Theory

The cell theory background is an important milestone in the history of biology that paved the way for our modern knowledge of cellular structure and function. The earliest notion of cells dates back to the 17th century, with innovators , such as Robert Hooke conducting experiments with crude microscopes. Hooke's discovery of the cells of cork in 1665 was momentous, as he was the first to describe the microscopic chambers that he saw as cells. Hooke's initial exploration of viewing cells resulted in more study of the nature of living organisms and their makeup. By the time the early 19th century arrived, work by scientists like Matthias Schleiden and Theodor Schwann resulted in developing the first two principles of cell theory: that all living things consist of cells and that the cell is the fundamental unit of life and is composed of basic blocks from which complex biological systems are constructed. The contributions of Louis Pasteur and Rudolf Virchow helped solidify the position of cell theory as a foundation of biological sciences. Pasteur's germ theory of disease in the 1860s reaffirmed the view that microorganisms made up of cells are the key to understanding disease processes, hence connecting cellular structures with health and disease. At the same time, Virchow's statement of *omnis cellula e cellula*, that all cells arise from cells already in existence, gave a fundamental understanding regarding cell reproduction and continuity. Altogether, these clarifications of fact did not just form precedent principles of cell theory but went beyond them to capture the dynamic aspect of biological processes. This development represents a transition towards more systematic investigation of life at the cellular level, with implications spreading across numerous scientific disciplines. Additionally, technological advancement in microscopy methods during the late 19th and early 20th centuries further advanced the investigation of cells. As new staining methods were developed and improved optical equipment became available, scientists could not only see the outer structure of the cell but also the inner workings, , such as the nucleus and organelles. These observations were the basis for the third pillar of cell theory, that cells arise from cell division. As scientists like Walther Flemming described mitosis, the relationship between cell function and structure became increasingly evident. In following the history of cell theory, it can be seen that the interaction between technology and theory, as exemplified through the development of optics and microscopy, has made it possible to understand life at its most basic level, opening the doors to future studies in genetics and cellular biology (Costa & Frezza, 2014; Valparaiso University, 1969).

4.1.2. Key Components of Cell Theory

The fundamental laws of cell theory have been critical in shaping our understanding of biology and the position of living organisms. The first fundamental factor asserts that all living things are made up of one or more cells. This law highlights the fact that cells are the fundamental structural units and organizational units of organisms, which can be either unicellular or multicellular. By positing that cells are the basis for life, this rule establishes a coherent idea in biology linking diverse forms of life through their shared cellular form. The microscopic study of cells has illustrated their complexity and diversity, and it is for this reason that scientists have been investigating the specific nature and activity of different cell types, which has further enhanced information about general biological processes and the history of life on our planet. The second cellular premise states that the cell is the unit of life. This hypothesis negates previous assumptions that more complex organisms may form from non-cellular forms or spontaneous generation. By organizing life in cellular structure, this factor points out that all biological activities occur within cells. Consequently, the examination of cellular processes like metabolism, growth, and reproduction has become core to the research in biology. In realizing that cells are modules of life, research on cellular structure and function has been pushed in various fields like genetics and physiology. Consequently, this principle has also enabled the development of advanced technologies that enable scientists to see deeper into cellular functions and thus unravel the complexities of living systems, a manifestation of the interactive relationship between form and function at the cellular level. The final component of cell theory extends the concept that all cells emerge from existing precells through cell division. This term is significant to the explanation of life continuity because cells that are new are formed to replace those that die off, facilitating the growth and maintenance of homeostasis in organisms. The term also illuminates how genetic information is transmitted from a parent generation to offspring, with a focus on the importance of cellular processes like mitosis and meiosis. Cellular reproduction not only explains the development of the organisms but also serves in evolution by making genetic differences feasible. Additionally, this aspect of cell theory forms the foundation of the biotechnology sector, whereby understanding cellular reproduction opens up avenues in medical therapies and genetic manipulation, indicating the utilization of the theory in the existing scientific advancement (Figure 4.4).

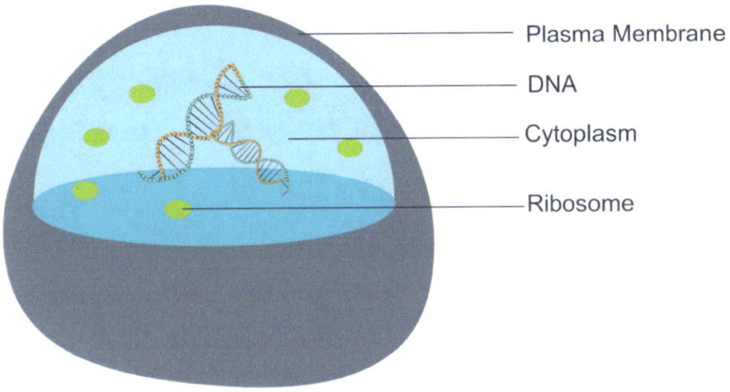

Figure 4.4. Components of cell theory.

Source: https://pressbooks.bccampus.ca/humanbiology053/wp-content/uploads/sites/1414/2021/06/Basic-Components-of-a-cell.png.

4.1.3. Implications of Cell Theory in Biology

The cell theory implications of biology go far beyond simple classification; they constitute the underlying principle of all biological sciences. Cell theory holds that living things consist of cells and that cells are the units of structure and function in living things. This view has profound implications for our conception of biological phenomena, as it emphasizes the significance of cellular interactions and the microenvironment. For instance, molecular transport processes in more recent research indicate cellular constituents can exhibit nonergodic behavior, something which is central to describing how cells maintain dynamic functions and adapt to changing circumstances (Lapeyre et al., 2015). By an appreciation that cells are more than passive containers for biological machinery, but rather active participants in the regulation of themselves, scientists can gain a deeper understanding of cellular organization and how it influences organismal health and behavior in general. In addition, the cell theory's emphasis on cellular structure provides the potential to explain the specificity of cellular function. All cell types, ranging from neurons to immune cells, have certain structural features that enable it to perform specialized functions necessary for the survival of the organisms. As an example, the unique morphology of neurons enables rapid communication through the nervous system, while the plastic structure of immune cells allows them to effectively recognize and respond to pathogens (Barreiro et al., 2017). An understanding of these specialized functions highlights the complex interrelationship between cell structure and biological function, supporting the notion that alterations in cellular architecture can lead to dysfunction and disease, , including cancer. Cell theory is, in this sense not only a unifying theory for biological research but also assists in the comprehension of mechanisms of cellular specialization. Finally, the implications of cell theory also echo in medical and biotechnology fields, facilitating advances in targeted therapy and regenerative medicine. The paradigm that all diseases stem from cellular abnormalities empowers researchers to be able to design interventions for treating the underlying cellular abnormalities. For instance, stem cell innovations apply the regenerative abilities of some cell types to unleash the mysteries of how cell theory-derived knowledge can lead to breakthrough treatments (Lapeyre et al., 2015). Furthermore, an understanding of cellular signaling and interaction pathways has led to advances in targeted drug delivery systems and the efficacy of pharmacological interventions. The larger significance of cell theory thus resides not only in its capacity to explain but also in its capacity to transform human health and therapeutic interventions in contemporary biology.

Test Your Knowledge

1. What are the key components of cell theory?
2. What are the implications of cell theory in biology?

4.2. Prokaryotic Cells

Learning Outcome

1. Characteristics of prokaryotic cells;
2. Differences between prokaryotic and eukaryotic cells;
3. Examples of prokaryotic organisms.

Even the simplest organisms , such as prokaryotic cells display an amazing complexity in their structure and functioning. These cells have traditionally been defined by the absence of a defined nucleus and organelles, but confound the idea of simplicity with their complex organizational principles. The key features of prokaryotic cells are a plasma membrane, cytoplasm, ribosomes, and genetic material, which is commonly presented as a single circular chromosome. More recent research has shown that prokaryotic cells too have a variety of cytoskeleton proteins, long considered the preserve of eukaryotes, and so do have a more complex evolutionary lineage than had been realized by early microbiologists. The implications of these findings suggest that the basic framework for cellular organization and function may have originated much earlier in evolutionary development than previously recognized, reshaping our understanding of cell biology and its foundational elements (Douglas, 2010). In addition, the versatility of prokaryotic cells is supported by the fact that they can construct structures (e.g., biofilms and microbial consortia) and it is a dynamic interaction between the individual cells and the community structures. In a feedback process, the hierarchy of both genetic and community structures affect the rate of proliferation of both, thus optimizing the biological growth at several levels. This complex interrelation demonstrates the way prokaryotic cells could change their shape and behavior to fit the environmental challenges and promote resilience and survival across different ecosystems. The evolutionary benefits that bacteria have in changing conditions are seen in essential survival mechanisms, like the accumulation of intracellular storage polymers. These polymers not only serve as energy reserves but also facilitate adaptations, enabling prokaryotic cells to thrive in their often inhospitable habitats (Douglas, 2010). One can also understand the importance of prokaryotic cells in terms of their evolutionary implications. Being among the most fundamental elements of life, they provide a clue into the nature of the formation of cellular organization and the development of multicelled organisms. Stanier and van Niel's demarcation of prokaryotes and eukaryotes formed the basis of perceiving biological diversity. But the identification of common features, , including the existence of cytoskeleton components and metabolic storage processes, causes one to reconsider the evolutionary TOL on Earth. It is this newfound view that emphasizes that prokaryotes are not simple organisms but complex organisms that are critical to the ecological balance and biochemical processes by which life can be sustained. In light of such revelations, prokaryotic cells are increasingly recognized as critical players in our planet's ecosystems and warrant further exploration to fully appreciate their multifaceted roles in cellular structure and function (Berlanga & Guerrero, 2010) (Figure 4.5).

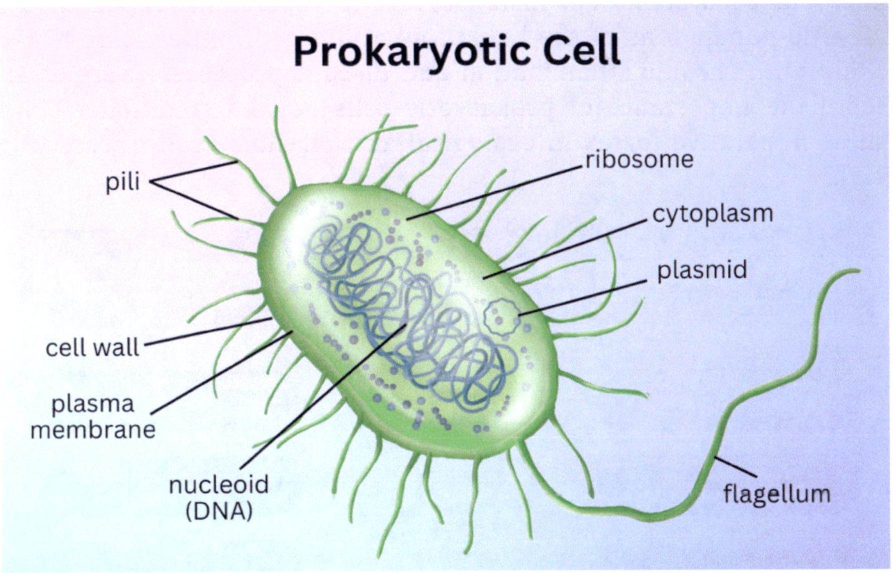

Figure 4.5. Prokaryotic cell diagram.

Source: https://sciencenotes.org/wp-content/uploads/2023/06/prokaryotic-cell-diagram.png.

4.2.1. Characteristics of Prokaryotic Cells

Prokaryotic cells are among the simplest and oldest living organisms on Earth, varying by a set of characteristic features that determine their structure and functions. To begin with, these cells lack a particular nucleus, which distinguishes more complex eukaryotic cells. Genetic information in prokaryotes is coiled into one circular chromosome located within an area designated as the nucleoid. The organization facilitates efficient replication and expression of genes. Prokaryotic cells can also include plasmids—small circular DNA pieces having the ability to transfer favorable traits , such as resistance to antibiotics. These adaptations thus facilitate their existence in various environments and make prokaryotes an important player in ecological communities and human health. Yet another feature of prokaryotic cells is their relatively simpler cellular composition. In contrast to eukaryotic cells, prokaryotes lack membrane-bound organelles; instead, cellular functions occur in the cytoplasm, where protein synthesis is carried out by ribosomes. This reduced complexity usually leads to fast rates of division, where prokaryotes can divide rapidly under optimal conditions. Also, their cell walls, primarily composed of peptidoglycan, provide shape and protection against environmental stresses, , including changes in osmotic pressure. The structural difference in cell walls also distinguishes prokaryotes as Gram-positive or Gram-negative, having significant implications on their susceptibility to antibiotics and immune response. Understanding the ecology of the structural components enhances our understanding of prokaryote adaptability and plasticity in diverse habitats. The metabolic specialization of prokaryotic cells enables them to occupy a wide range of ecological niches. Their ability to metabolize a variety of substrates, from organic molecules, gases, and minerals, indicates their evolution and ecological versatility. Some prokaryotes are autotrophs and can synthesize food by photosynthesis or chemosynthesis, while others are heterotrophic and utilize organic substrates for energy. Studies have shown that prokaryotic populations can be influenced to a great extent by environmental factors, which led to the concept of microbial ecosystems. For instance, recent research has highlighted the importance of specific prokaryotic populations of the human microbiome in influencing health, for instance, in their relationship with chronic inflammation and disease processes (Berg et al., 2020). These findings underscore the importance of prokaryotic cells beyond traditional biological contexts, positioning them as imperative forces in ecological and physiological processes (Shrestha et al., 2017) (Figure 4.6).

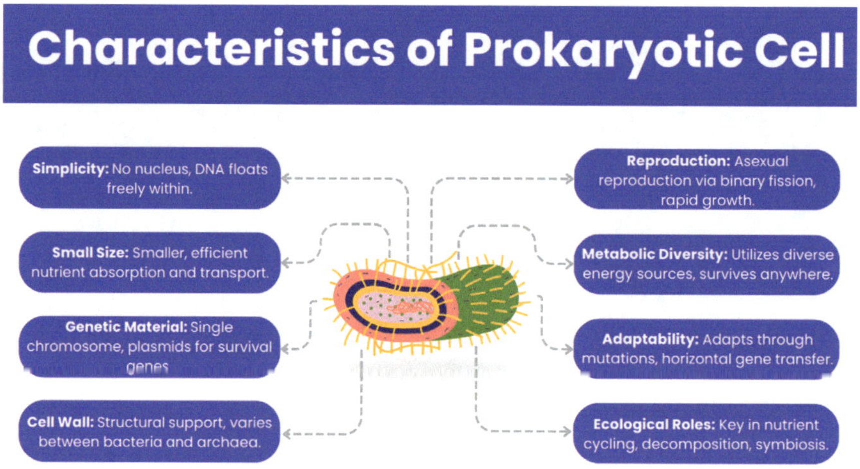

Figure 4.6. Characteristics of prokaryotic cells.

Source: https://images.examples.com/wp-content/uploads/2024/03/Characteristics-of-Prokaryotic-Cell.png.

4.2.2. Differences Between Prokaryotic and Eukaryotic Cells

The differences between prokaryotic and eukaryotic cells are foundational in thinking about cell structure and function. Prokaryotic cells, , such as archaea and bacteria, are smaller and simpler compared to eukaryotic cells. One of the most identifying traits may be that prokaryotic cells do not have a nucleus; instead, their genetic material resides in a nucleoid area, a non-membrane-bound structure within the cell. This structural simplicity is further enhanced by other traits , such as the lack of membrane-bound organelles, which makes them very different from eukaryotic cells. Eukaryotic cells possess a well-established nucleus where their genetic material resides and an enormous array of organelles , such as mitochondria and the ER, that provide compartmentalization of biochemical reactions and make cellular efficiency possible. Both the number and organization of chromosomes are unique in prokaryotic and eukaryotic cells. Prokaryotic cells typically possess a single circular chromosome, but eukaryotic cells have multiple linear chromosomes. Histone proteins also make eukaryotic cell DNA packaging more complex since these proteins package and coil the DNA into a structure referred to as chromatin. The structural division allows for more advanced regulation of gene expression in eukaryotes, which results in the development of higher-order biological systems and multicellularity. Further, eukaryotic cells possess sexual reproduction through meiosis, a component of genetic diversity—something that is advantageous to rapidly changing environments. Genetic flexibility is not as potent in prokaryotes, which asexually reproduce through binary fission under normal conditions. That said, prokaryotes also possess their own form of genetic exchange in the form of HGT, which works to render them adaptable and successful evolutionarily (Witzany, 2008).

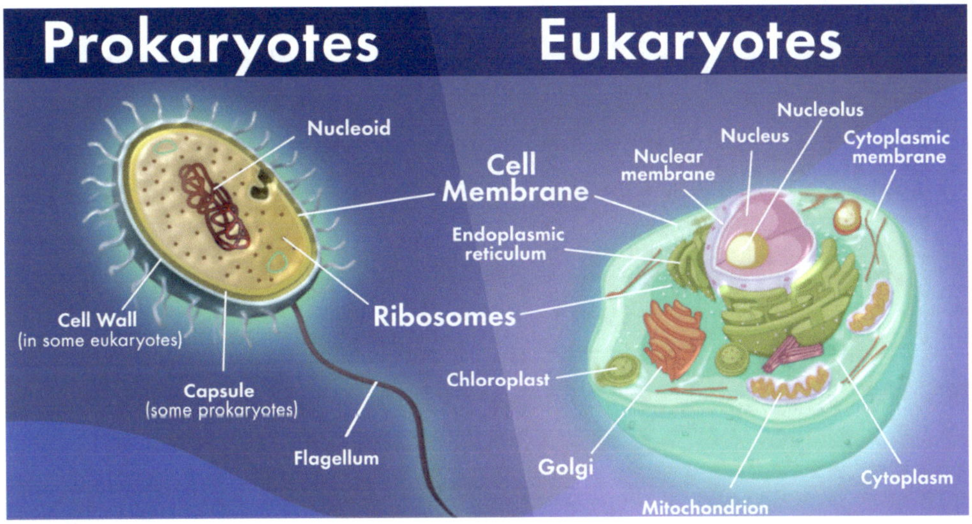

Figure 4.7. Differences between prokaryotic and eukaryotic cells.

Source: https://cdn.hswstatic.com/gif/Prokaryotesvseukaryotes.jpg.

Also, the metabolic abilities of prokaryotic and eukaryotic cells reflect their distinct structural make-up. Prokaryotes can have a variety of metabolic paths, getting accustomed to a variety of environments and showing extreme flexibility in energy production, often employing anaerobic mechanisms under unfavorable conditions. Eukaryotic cells developed specialized organelles, , such as chloroplasts in plant cells to carry out photosynthesis and mitochondria for aerobic respiration, making it possible to produce energy more efficiently within a complex cell architecture. Their common evolutionary past is underscored by their contrasting features, which have allowed them to occupy different ecological niches for over billions of years. Interestingly, a few features of eukaryotic cell structure, , such as telomeres, are of viral origins dating back before the advent of eukaryotic cells, indicating a fascinating evolutionary history where prokaryotic relatives played a pivotal role in founding higher eukaryotic life (Witzany, 2007) (Figure 4.7).

4.2.3. Examples of Prokaryotic Organisms

Prokaryotic cells, dominated by bacteria and archaea, present a wide range of diversity in the structural and functional features. Bacteria, ubiquitous in the environment, can be classified into various genera and species based on their morphology and metabolic processes. Escherichia coli, a ubiquitous intestinal inhabitant of man, is an indispensable model system in molecular biology and exhibits the simple biological processes typical of most prokaryotes. A critical example is Streptococcus pneumoniae, a bacterium responsible for pneumonia and other illnesses, and one that highlights the biomedical significance of prokaryotic diversity. Not only are the microbes crucial to human disease and health, but they are also kept in check by and perform ecological balance, nutrient cycling, and biogeochemical processes and are therefore vital for scientific investigation and environmental functions.

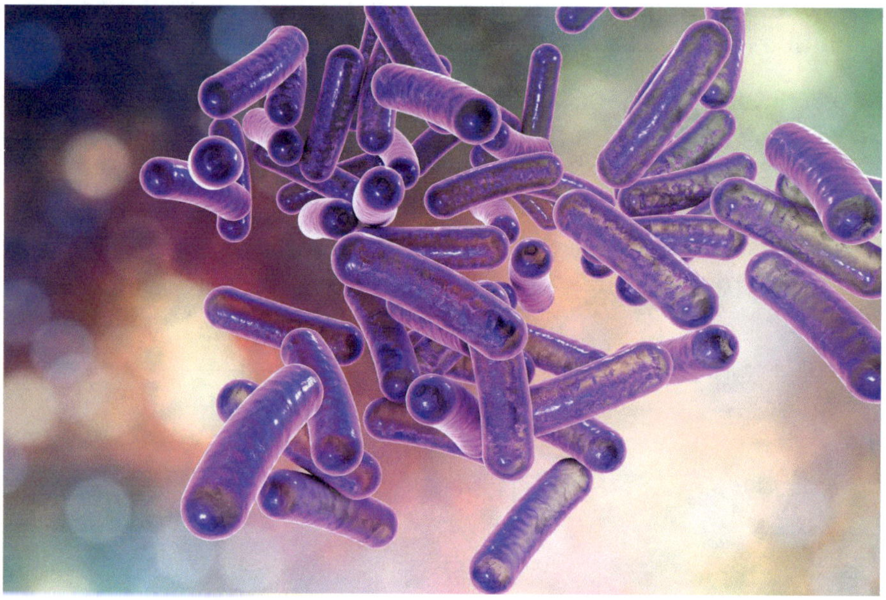

Figure 4.8. Illustration of Shigella bacteria, a genus of gram-negative bacteria known to cause gastrointestinal infections.

Source: https://www.thoughtco.com/thmb/IMXlkIr1Wk2bawFAus81a77MTjc=/1500x0/filters:no_upscale():max_bytes(150000):strip_icc()/shigella-bacteria--illustration-758308491-5a02252f9e9427003c1759be.jpg.

In contrast to bacteria, archaea represent a distinct class of prokaryotic life with unique biochemical and genetic features that set them apart. Normally living in extreme conditions, like hot springs and salt lakes, *Halobacterium salinarum* and related organisms demonstrate high-level adaptations to extreme environments, emphasizing the staggering range of prokaryotic life. Archaea membranes, composed of ether-linked lipids, are the reason behind their ability to survive harsh conditions, compared to the ester-linked lipids found in bacterial membranes. In addition to influencing their survival, this structural feature also provides clues for early evolutionary processes, connecting the transition from prokaryotes to more complex eukaryotic cells. Studies on such extremophiles establish the versatility and evolutionary function of prokaryotic organisms in life in general on Earth. Prokaryote interactions with the environment reveal crucial characteristics of cellular function and organization, particularly through the lens of HGT, a process underlying their evolution. Recent research reveals that DNA repair mechanisms, long thought to be exclusive to eukaryotic cells, also occur in prokaryotic cells, as evidenced by the discovery of a non-homologous end-joining (NHEJ) process in certain bacteria. These findings not only prove the cellular activities of prokaryotic cells to be more sophisticated than projected but also challenge traditional assumptions of prokaryotic and eukaryotic classifications. Additionally, the social competence of bacteria enables them to form complex social structures, which are comparable to multicellular ones, where they regulate behavior and interaction essential for effective survival in different niches. This expansive treatment of prokaryotic life again reaffirms their vital role in the biological spectrum, with significant implications for inferring cellular processes in all living beings (Bowater et al., 2006; Witzany, 2008) (Figure 4.8).

Test Your Knowledge

1. What are the characteristics of prokaryotic cells?
2. Differences between prokaryotic and eukaryotic cells?

4.3. Eukaryotic Cells

Learning Outcome

1. Structure of eukaryotic cells;
2. Types of eukaryotic cells (plant vs. animal);
3. Functions of organelles in eukaryotic cells.

Characteristic of eukaryotic cells is that they are complex in structure, containing membrane-bound organelles that carry out specific functions necessary for the survival and proliferation of the cell. Among these organelles, the nucleus plays a pivotal role, housing the genetic material (DNA) and regulating gene expression. This compartmentalization enables more complex control of cellular processes than in prokaryotic cells. Moreover, the eukaryotic cells possess specialized organelles like the ER and Golgi apparatus that aid in the synthesis and modification of proteins and lipids. The work of these organelles is interconnected in such a way that the cell can be at homeostasis and appropriately respond to different internal and external signals, which highlights the complexity of coordination that characterizes the eukaryotic cellular functioning and promotes overall biological processes. Besides being structurally complex, eukaryotic cells are functionally versatile, thus able to perform several processes essential to life. An example of this can be seen in the function of ribonucleases, especially T2 ribonucleases, which are preserved among eukaryotic organisms. The enzymes play an important role in the control of a variety of cellular processes, , including plant self-incompatibility and the human immune response regulation. It is worth mentioning that the studies have revealed that the RNaseT2 proteins influence the neurological development and tumor growth, indicating the catalytic and catalytic-independent functions in the preservation of cellular health. The understanding of these enzymes, especially through studies conducted on proteins like Rny1 in *Saccharomyces cerevisiae*, reveals the multifaceted nature of eukaryotic cellular functions, which are often intertwined with critical physiological outcomes essential for organismal development and health (Luhtala et al., 2012). In addition, eukaryotic cell functioning is important in disease processes. For instance, the mitochondrial tumor suppressor 1 (MTUS1) gene illustrates how specific genetic variations, particularly nonsynonymous single-nucleotide polymorphisms (nsSNPs), can substantially influence cellular behavior. Mutations in genes like MTUS1 have been observed to be associated with a variety of cancers and other disorders because such mutations may result in the destabilization of the finely balanced process of cell growth and cell proliferation normally regulated by the tumor suppressor genes. The overall study of these nsSNPs is a crucial initiative in the direction of interpreting the functional consequence of the genetic variation to generate focused therapeutic strategies. Ultimately, exploring the complex relationships between eukaryotic cell structure and function not only advances our understanding of cellular biology but also informs strategies for disease prevention and treatment (Rozario et al., 2021) (Figure 4.9).

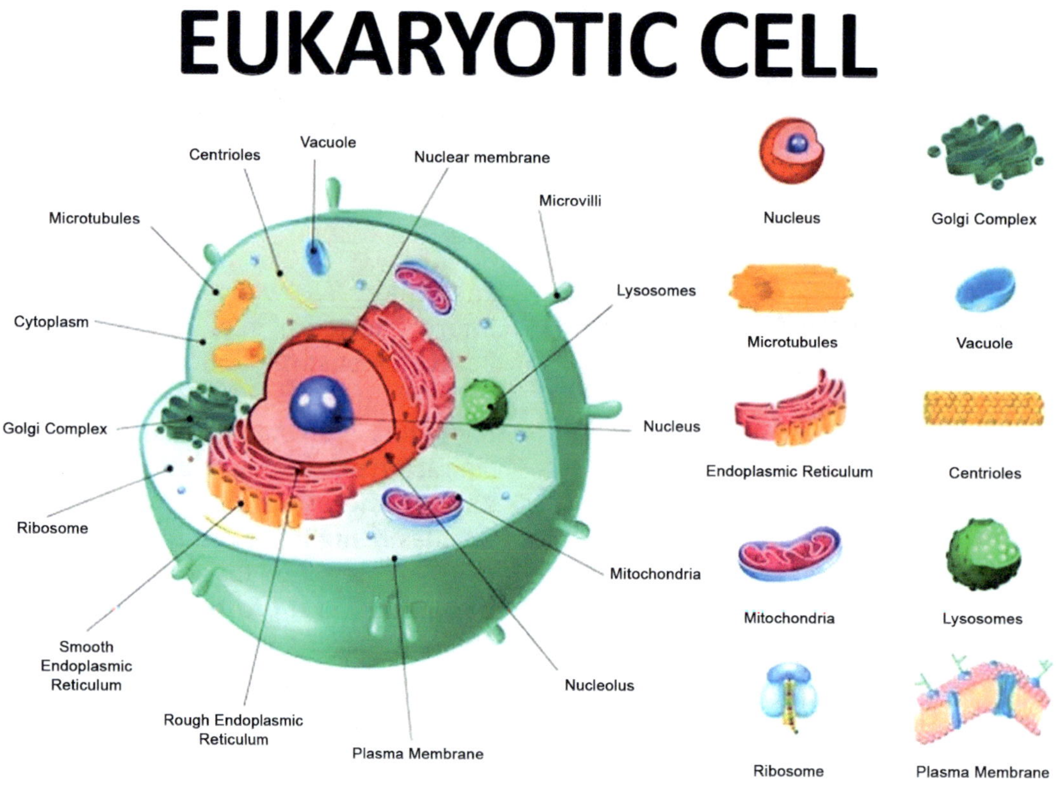

Figure 4.9. Diagram of an eukaryotic cell.

Source: https://media.geeksforgeeks.org/wp-content/uploads/20240402104838/Eukaryotic-Cell.webp.

4.3.1. Structure of Eukaryotic Cells

The cellular organization of eukaryotes is especially complex, which indicates a highly evolved organization typical of these organisms. Eukaryotic cells are characterized by the presence of a membrane-bound nucleus, which holds the genetic material organized in chromosomes. This is in contrast to the absence of a true nucleus in prokaryotic cells. Inside the nucleus, processes , such as transcription occur, allowing for the synthesis of messenger RNA that will be translated into proteins afterwards. In addition, eukaryotic cells consist of a variety of specialized organelles with various purposes that sum up to the cellular function. Organelles like the mitochondria and the ER work in harmony to assist in producing energy and synthesizing proteins, demonstrating the efficiency of eukaryotic cells in maintaining life processes with the provision of complex multicellular organization. At the core of understanding the structure of eukaryotic cells is how membrane systems, particularly those of the secretory pathway, function.

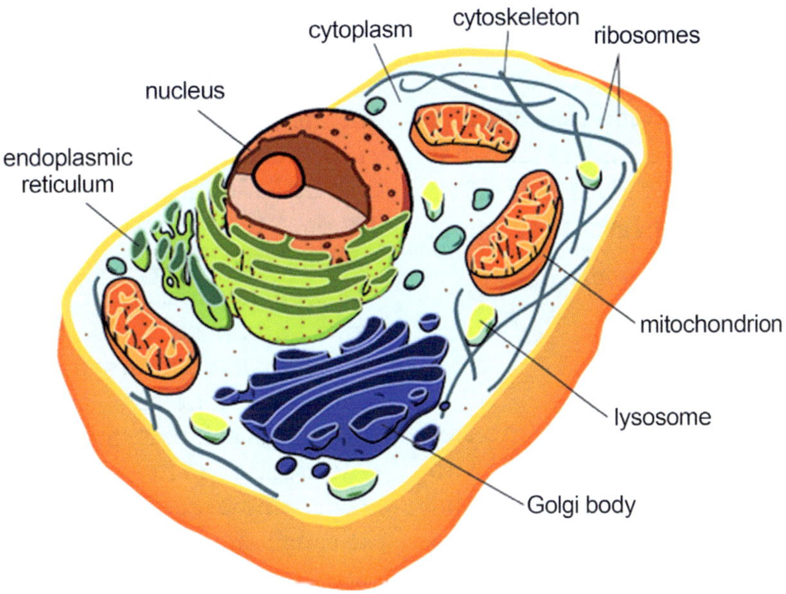

Figure 4.10. Structure of eukaryotic cells.

Source: https://media1.shmoop.com/images/biology/biobook_eukaryotes_graphik_2.png.

Proteins that are synthesized in the ER are exported through the Golgi apparatus before they are inserted into the plasma membrane. In this context, receptor expression-enhancing proteins (REEPs) serve the function of vital adapters that assist in the trafficking and localization of integral membrane proteins (IMPs), thereby ensuring their proper delivery to functional sites. Such processes are vital for cellular homeostasis and function, particularly in signaling and communication in the organism. In addition, this regulatory mechanism is not only for certain proteins but is extended across a range of cellular processes and signifies the importance of membrane dynamics and structure in the functioning of the cells, justifying the fact that structure has a highly significant role in influencing function in eukaryotic cells (Angelotti, 2022). Eukaryotic cells also illustrate the importance of genetic regulation in maintaining the integrity of cellular function. MTUS1 is an example of how genetic processes can influence tumorigenesis and cell growth. Research demonstrated that MTUS1 downregulation is associated with many cancers and has provided valuable information on cell mechanisms of regulation of growth. Examination of this gene's nsSNPs has revealed potential harmful impacts on protein structure and function, which further contribute to the regulation of cellular behavior in a tumor context. The ability to discern deleterious nsSNPs from neutral ones with the help of bioinformatics enhances our understanding of cellular pathology and has implications for personalized medicine and treatment strategies. Thus, the examination of structural and functional properties of eukaryotic cells reveals a complicated correlation between cell structure and the regulation of genes and establishes that a cell's structure is not only a part of its function but also its overall health (Rozario et al., 2021) (Figure 4.10).

4.3.2. Types of Eukaryotic Cells (Plant vs. Animal)

Eukaryotic cells are generally divided into two types: plant cells and animal cells, both with unique structural features and functions. Plant cells are distinguished by the presence of a hard cell wall made of cellulose that assists in giving support and protection. Aside from the cell wall, plant cells contain chloroplasts, which are specialized organelles that enable photosynthesis by collecting sunlight and converting it into chemical energy. This unique feature highlights the key difference between plant and animal cells, in that the latter lack chloroplasts and obtain energy from the metabolism of organic compounds. Besides this, plant cells typically possess large vacuoles for storage and turgor pressure in the center, while animal cells possess small vacuoles. These structural differences indicate the adaptation of plant and animal cells to their environment and mode of living. Another significant difference between plant and animal cells lies in the composition of the ECM. The ECM in animal cells consists of a complex mixture of proteins and carbohydrates that serves to facilitate cell signaling, adhesion, and communication. The matrix is tasked with creating a dynamic interaction among cells to enable tissues to develop and regenerate. Conversely, plant cells contain a rigid extracellular structure due to their cell walls, which serves not only a mechanical function but also participates importantly in cell communication through plasmodesmata; minute channels between neighboring plant cells. The presence of the channels allows water, nutrients, and signaling molecules to pass through, which supports coordination in plant tissue. These differences in the structure and function of ECM illustrate how eukaryotic cells have evolved to meet the particular demands of their environments, which impact their general physiology and function within multicellular organisms. Additionally, the discrepancy in cellular processes between plant and animal cells also indicates their diverse biological functions. For instance, during cell division, plant cells exhibit a unique process known as cell plate formation that is essential in the creation of new cell walls between daughter cells. Animal cells, however, undergo cytokinesis through the development of a cleavage furrow as a result of cytoskeleton contraction. These include variations in organelle structure, particularly in the ER and Golgi apparatus, both of which display alternative distributions and functions in the two cell types. This variation indicates the sophisticated adaptations that have been developed during evolution, with each eukaryotic line having developed sophisticated mechanisms to utilize its ecological niche. A comparative analysis of these cell characteristics makes our understanding of cell structure and function, and of the evolutionary relationships that unite the plant and animal kingdoms within the broad framework of multicellular eukaryotic life, deeper (Panstruga et al., 2023; Naish & Henderson, 2024) (Figure 4.11).

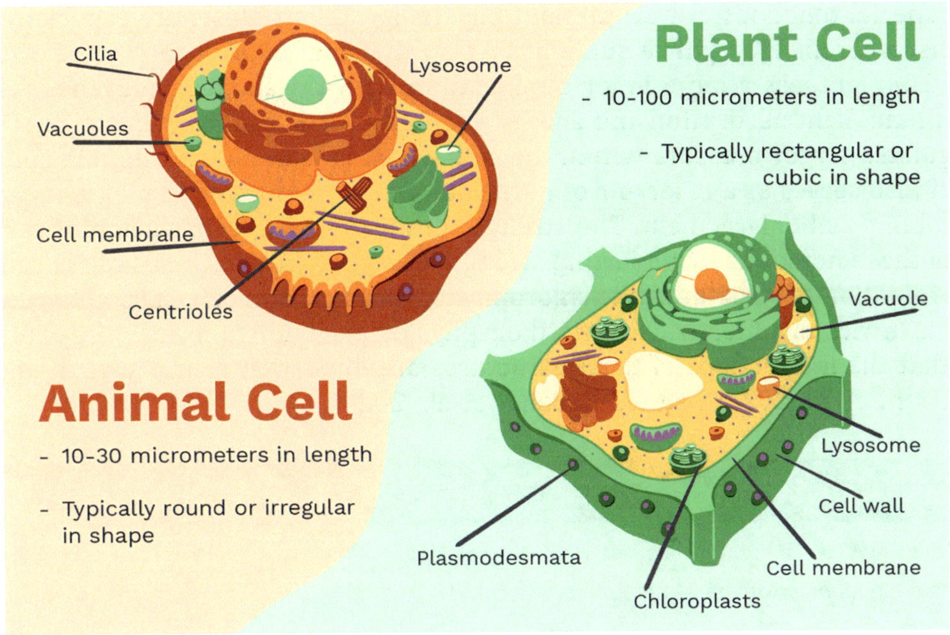

Figure 4.11. Types of eukaryotic cells (plant vs. animal).

Source: https://www.thoughtco.com/thmb/KC4a3LJ9Lvarsw6QaZ69z-tsKEU=/1500x0/filters:no_upscale():max_bytes(150000):strip_icc()/animal-cells-vs-plant-cells-373375_final-5b462d7fc9e77c00375014f1.png.

4.3.3. Functions of Organelles in Eukaryotic Cells

Of all the organelles found in eukaryotic cells, one of the most important is the nucleus, which serves as the cellular activity control center. It houses the genetic information as DNA, which is packaged into chromosomes. The nuclear envelope, a double membrane, surrounds the DNA and regulates molecular traffic in and out of the nucleus through nuclear pores. This selective permeability is significant in the process of protein synthesis because it allows the mRNA to exit the nucleus after transcription, and it keeps DNA from being exposed inside. The nucleolus, which is located within the nucleus, also plays a role in ribosome biogenesis, where the ribosomal RNA (rRNA) is synthesized and combined with proteins to form the subunits that will be required for protein synthesis. Thus, the nucleus not only carries the genetic blueprint of the organism but also the central location in the total operation and regulation of the cell. Another crucial group of organelles that exist in eukaryotic cells is the mitochondria, which have been loosely referred to as the energy factories of the cell. These cells are mainly responsible for transferring ATP into a usable form of energy through the process of cellular respiration. The structure of mitochondria is also well-suited for this function, with an outer and inner membrane, where electron transport occurs in an extensive fold known as cristae. The matrix, the inner space inside the inner membrane, contains enzymes that catalyze the Krebs cycle, which leads to the synthesis of ATP. Importantly, mitochondria also possess their own DNA, suggesting endosymbiotic origin, and are implicated in metabolic process, apoptosis, and cell signaling. Thus, mitochondria exemplify the dynamic relationship between structure and function in eukaryotic cells as primary energy producers for cellular functions. Plant cells contain organelles , such as chloroplasts, cell walls, and central vacuoles that perform special functions critical to plant physiology. Chloroplasts are especially adapted for photosynthesis, wherein they utilize sunlight to trap CO_2 and water into glucose with oxygen as a byproduct. They possess a large inner organization with thylakoid membranes folded up in a granum for optimal light absorption and energy generation. The central vacuole, another essential organelle, regulates turgor pressure within the cell, which is vital for the support and structure of plant tissue. It also serves as a reservoir of nutrients, waste products, and pigments and facilitates growth by enabling cell enlargement. The strength of the cell wall, constructed of cellulose fibers, also facilitates these functions by imparting strength and protection against stress to the cell. This organelle coordination of structures like chloroplasts, vacuoles, and cell walls illustrates how plant cell structure is intricately intertwined with their physiological functions because extensive studies have shown that the interactions of these structures are important to the growth and survival of plants (Kheyrodin et al., 2022; Fuertes-Rabanal et al., 2024).

Test Your Knowledge

1. What are the types of eukaryotic cells (plant vs. animal)?
2. What are the functions of organelles in eukaryotic cells?

4.4. Cell Membrane Structure

Learning Outcome

1. Composition of the cell membrane;
2. Functions of the cell membrane;
3. The fluid mosaic model.

The cell membrane or plasma membrane is a simple building block of cell structure that is vital to enable cellular communication with the external world and to give cellular integrity. It is primarily made up of a phospholipid bilayer, in which hydrophilic heads facing the aqueous inside and outside environments and the hydrophobic tails facing inward in semi-permeable fashion. It is this bilayer morphology that facilitates selective permeability, with required molecules entering and waste substances exiting the cell. Further, within this bilayer are found various proteins—integral and peripheral—playing structural, transport, and signaling functions. With this understanding of the complexity of the cell membrane, one would appreciate its significance in keeping cellular functions and responding to environmental stimuli, thus affecting the overall health and behavior of the organism. Besides the fundamental phospholipid and protein content, cell membrane also includes cholesterol, which affects the fluidity and stability of the membrane. Cholesterol molecules interleave themselves between the phospholipids, keeping them from packing too tightly and thus allowing an optimal level of fluidity for membrane to function. This fluidity is crucial for vesicle formation and the function of membrane proteins. Significantly, alterations in membrane structure entail severe disruptions to cellular function; alterations in cholesterol levels, for example, have an effect on signaling cascades and cell-cell communication. The dynamic nature of the cell membrane also enables the development of specialized regions, , such as lipid rafts, which play a key role in organizing proteins and lipids necessary for signal transduction and membrane trafficking. Recent studies have also introduced other complexities in the organization of cell membranes, which include the participation of heat shock proteins and cavin proteins in its process. For instance, the small heat shock protein Hsp12 stabilizes lipid membranes by making them more resilient under stress conditions, , such as those faced by the pathogenic fungus *Ustilago maydis*. The observation that Hsp12 can bind to lipid membranes and inhibit protein aggregation contributes to the body of evidence regarding the structural organization of membranes and their correlation with cellular survival under pathological conditions (Mitra et al., 2023). Independently, cavin proteins, being the key players in caveolae formation, have been implicated in being essential regulators of cellular functions , such as endocytosis and signal transduction (Yang et al., 2022). The explanation of such functions adds richness to our understanding of the function of membrane proteins contribute not only to the provision of structural integrity, but also to the regulation of vital cellular activity, demonstrating the intimate relationship between cell membrane structure and function (Figure 4.12).

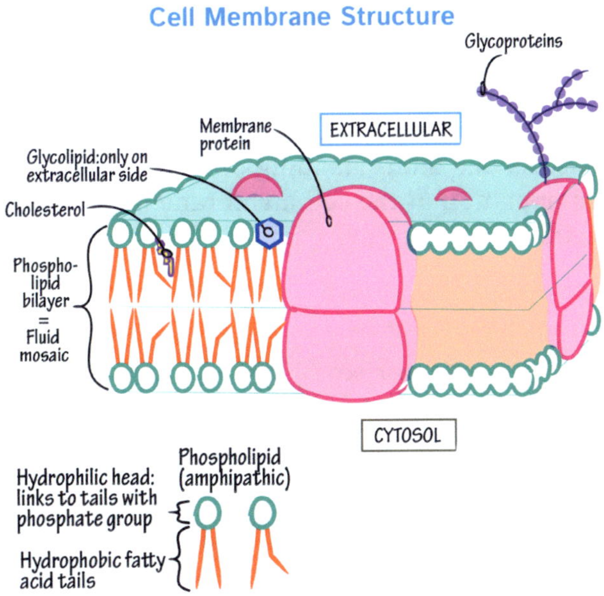

Figure 4.12. Cell membrane structure.

Source: https://d1j63owfs0b5j3.cloudfront.net/term/images/cell-membrane-structure-3082.png.

4.4.1. Composition of the Cell Membrane

Also known as the plasma membrane, the cell membrane is a fundamental boundary that distinguishes the internal cell environment and the external surroundings. This semi-porous membrane is mainly made up of a phospholipid bilayer, in which the phospholipids are aligned so that their hydrophobic tails are pointing towards the interior and their hydrophilic heads are towards the aqueous surroundings. This organization is essential not only to provide the structural integrity of the cell but also to enable the selective permeability of the cell so that the specific molecules can get in or out of the cell whenever necessary. This bilayer is home to proteins that are critical to several cellular processes, , such as cellular signaling, transport, and ECM interaction, and thus the overall functionality and versatility of the cell in regard to changes in the environment. Incorporation of proteins in the lipid bilayer forms a mosaic model that underlines the dynamics of the cell membrane. These membrane proteins may further be classified into integral and peripheral membrane proteins; the former being membrane-spanning, whereas the latter being connected to the membrane surface. The integral proteins can also act as channels or transporters, to facilitate the passage of ions and polar molecules through the hydrophobic core of the bilayer. Conversely, the peripheral proteins are usually involved in signaling or are enzymes. Carbohydrate chains are also found attached to proteins (glycoproteins) and lipids (glycolipids) that project above the extracellular surface and play a central role in cell recognition and communication. This biological membrane is even made more fluid by the incorporation of cholesterol, which stabilizes the membrane structure at different temperatures, making sure that the fluid mosaic model is really representative of the versatile nature of cell membrane functionality. The cell membrane is not only involved in structural integrity, but it is also involved in homeostasis in the cell as well as communication with other cells. Having different receptors on the surface of the membrane enables the cells to detect any changes in the surroundings and react on them. As an example, the binding of extracellular signals to these receptors can trigger a series of intracellular reactions changing the cellular processes, , including gene expression or metabolism. In addition, endocytosis and exocytosis, which facilitate the absorption of larger molecules and substance secretion, respectively, depend on the cell membrane. As our understanding of membrane dynamics expands, it becomes increasingly clear that the composition of the cell membrane has profound implications for overall cellular function and interaction with the environment (Alí et al., 2018; Joel et al., 2020) (Figure 4.13).

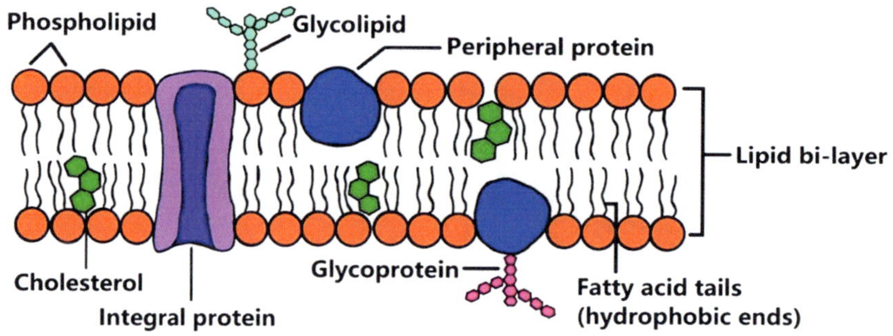

Figure 4.13. Composition of the cell membrane.

Source: https://external-content.duckduckgo.com/iu/?u=https%3A%2F%2Fsciencequery.com%2Fwp-content%2Fuploads%2F2021%2F02%2Fcomponents-of-cell-memebrane.jpg&f=1&nofb=1&ipt=8245b9ae6d4a7033cf1902c136f8503c9a51114b111286ab11ed60357e69770e.

4.4.2. Functions of the Cell Membrane

The cell membrane, a fundamental component of all living cells, serves a number of vital functions that contribute towards cellular homeostasis and integrity. Comprising primarily a bilayer of phospholipids, the membrane creates a semi-permeable barrier across which material movement is regulated. This permeability is essential to the preservation of the cell's internal environment, permitting essential nutrients to enter the cell and removing waste products. The proteins that are integral to the membrane perform a variety of functions, like transport, signaling, and structural support. The cell membrane thus not only protects the cell contents but also controls the dynamic interaction between the cell and the external world to offer communication and coordination among other cells surrounding it. The cell membrane plays a crucial role in cell communication and signaling as well. Membrane surface receptor proteins bind to specific ligands, which trigger a series of intracellular reactions that impact multiple cellular processes. Binding of hormones or neurotransmitters to their respective receptors, for instance, can trigger metabolic pathways or alter gene expression. In addition, the fluid mosaic model of the membrane introduces the mobility of the lipids and proteins, through which rapid responses to environmental stimuli can be triggered. These exchange processes emphasize the importance of the cell membrane in facilitating not only structural support but also dynamic communication, which is essential for cellular adaptation and function across varying physiological conditions.

In addition to the communicative and structural functions, the cell membrane is also integral to facilitating transport processes vital for cellular function. Various transport proteins, , including channels and carriers, facilitate the passive or active transport of ions and molecules across the membrane. For instance, carrier proteins facilitate the transport of glucose into the cell via a concentration gradient, while active transport processes, e.g., ATP-driven pumps, move ions against their gradients to maintain the necessary ion concentrations in the cell. Furthermore, it has been shown through the study of bacterial membranes that mechanisms of transport are far more complex , including mechanisms like those in *Sphingomonas* sp. strain A1, that use dynamic structures to accumulate certain molecules. These instances show that there is far more to the function of the cell membrane than just encapsulation and that it performs an essential function in the metabolism, communication, and cellular homeostasis (Huijuan et al., 2019; Murata et al., 2022) (Figure 4.14).

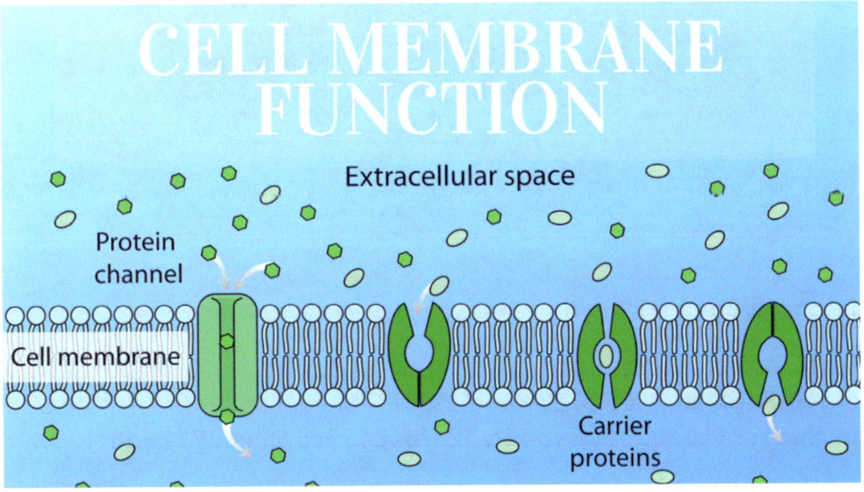

Figure 4.14. Functions of the cell membrane.

Source: https://external-content.duckduckgo.com/iu/?u=https%3A%2F%2Fsciencetrends.com%2Fwp-content%2Fuploads%2F2018%2F05%2Fe3214b48-cell-membrane-function-1536x864.png&f=1&nofb=1&ipt=3776dd2d01fe15e448e2767d6d9964e086016f84422cc9957858ef7746e9a414.

4.4.3. The Fluid Mosaic Model

The fluid mosaic model is a fundamental conceptual theory for describing biological membrane structural and functional dynamics. It suggests that cell membranes are not fixed structures but fluid membranes made up of a phospholipid bilayer containing embedded proteins, cholesterol, and carbohydrates. The fluidity of the membrane is essential for several physiological processes, , such as the transport of lipids and proteins, facilitating signaling pathways and nutrient transport. The mosaic character refers to the multitude of various proteins in the membrane, each with specific functions, , such as receptors, enzymes, and transporters. This organization allows membranes to quickly respond to environmental changes, maintaining homeostasis within the cell but also permitting interaction with the external world, necessary for cell function and survival. Cholesterol plays a critical role in membrane fluidity and integrity as well, providing stability to the fluid mosaic model. Cholesterol molecules actually insert between phospholipids and affect the packing of the lipid tails and thus the permeability and flexibility of the membrane. Existing research highlights cholesterol's position in lipid rafts, specialized membrane microdomains that allow for the clustering of signaling molecules, enhancing their affiliation and functions. For example, Alzheimer's disease amyloid-β peptide deposition has been directly linked to membrane lipids, and particularly to cholesterol, thereby illustrating the models use in the explanation of disease causation. Through its impact on membrane composition, cholesterol alters protein-protein interaction in the fluid mosaic domain, and consequently, signaling pathways of cells and is involved in a wide array of pathologies, e.g., neurodegenerative disorders (Fabiani & Antollini, 2019).

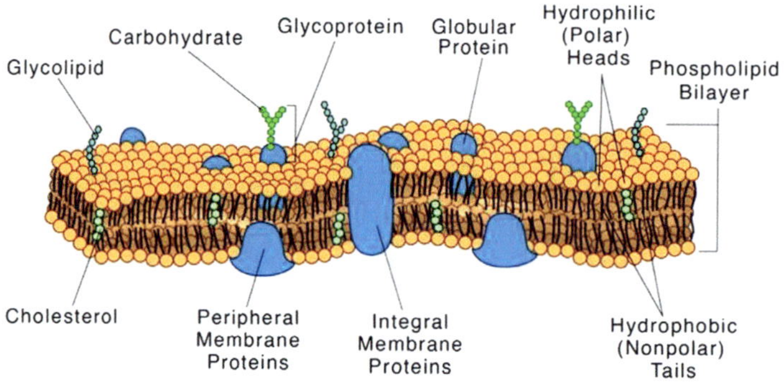

Figure 4.15. Fluid mosaic model.

Source: https://external-content.duckduckgo.com/iu/?u=https%3A%2F%2Fwww.sciencefacts.net%2Fwp-content%2Fuploads%2F2022%2F11%2FFluid-Mosaic-Model-768x528.jpg&f=1&nofb=1&ipt=5791e904ac4a21eb9c1a02b1064342fed25f84c435c5bfc06a7176d8f470207e.

The fluid mosaic model is not merely applicable to structural understanding but also gets activated with cellular signaling mechanisms. Receptor clustering and membrane localization are required for initiating cascades of signal transduction. One such example is the vascular endothelial growth factor (VEGF) signal transduction, where VEGF receptors tend to be localized in areas of the membrane , such as caveolae. In such domains, clustering of receptors allows for dimerization upon ligand binding, which is a prerequisite for effective signal transmission. This clustering behavior highlights the complex interaction between embedded proteins and membranes that helps in modulating cellular response. Processes of such mechanisms allow researchers to examine therapeutic avenues for interfering with membrane processes and shed light on cellular behavior in varying physiological states (Chen et al., 2013). Hence, the fluid mosaic model highlights the complex interaction between structure and function of membranes that is necessary to maintain cellular homeostasis (Figure 4.15).

Test Your Knowledge

1. What are the functions of the cell membrane?
2. What is the fluid mosaic model?

4.5. Cellular Metabolism

Learning Outcome

1. Overview of metabolic processes;
2. Role of enzymes in metabolism;
3. Energy production in cells.

The complex processes of cellular metabolism form the foundation of life at the cellular level and serve the dual purpose of energy generation and the synthesis of cellular constituents needed. At the center of metabolism are biochemical pathways through which cells can use nutrients to generate energy through catabolic reactions, through glycolysis and the Krebs cycle. These pathways are tightly controlled by various enzymes that allow the conversion of substrates into useful products. Moreover, energy produced in the form of ATP plays a crucial role in propelling cellular activities , such as muscle contraction, active transport through a membrane, and biosynthesis of macromolecules. Understanding such metabolic pathways not only recognizes the critical processes that sustain cellular life but also sheds light on potential interventions in metabolic diseases, giving hints on how metabolic pathways may be manipulated to develop therapeutic remedies. The role of cellular structures, particularly the cell membrane, in regulating metabolism cannot be underemphasized because they have a critical function in maintaining homeostasis within the cell. The cell membrane potential influences metabolic processes profoundly; it sets up a gradient upon which metabolites and ions can be transported. This is regulated by the membrane proteins, which render the membrane selectively permeable. As Kundenko et al. (2024) demonstrate, complex interactions between these proteins and external stimuli, including changes in electromagnetic fields, may be able to alter cellular metabolic processes. These findings about membrane dynamics will be able to inform future research on how to therapeutically regulate cellular metabolism and offer pathways for the reversal of dysregulated metabolic states by regulation of membrane function. Studies on membrane proteins illustrate their essential roles in numerous metabolic processes, , including signal transduction and the coordination of cellular response to environmental stimuli. The movement of metabolites and ions is credited to membrane proteins, which facilitate the movement necessary for maximum cellular metabolism. From biology to bioengineering, studies have demonstrated that understanding such proteins' structural and functional attributes can advance multiple fields considerably, as presented by Tosaka & Kamiya (2023). By providing mechanisms to study these proteins, scientists are capable of unraveling their functions and interactions within cellular metabolism. This knowledge not only enhances our appreciation of basic biological function but also informs drug and treatment strategy development aimed at the restoration of normal cellular function compromised in metabolic disease. Such an interface of cellular structure and metabolism is then a target for ongoing investigation by scientists (Figure 4.16).

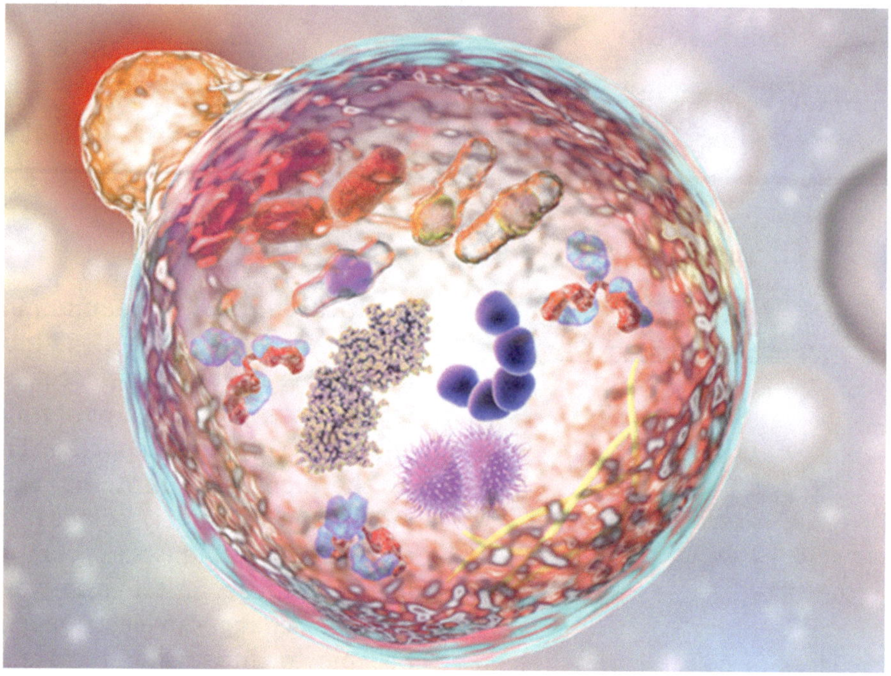

Figure 4.16. Cellular metabolism.

Source: https://www.amsbio.com/wp//wp-content/uploads/2022/05/metabolism-featured-image-correct-size.png.

4.5.1. Overview of Metabolic Processes

The metabolic network is highly complex and is instrumental in ensuring cellular homeostasis, which regulates all the biochemical reactions that ensure the survival of organisms. The two types of processes central to metabolism are catabolism and anabolism. Catabolic pathways involve the degradation of complex molecules to simpler ones, with the release of energy, which can be utilized by the cell, and anabolic pathways consume energy to construct larger macromolecules out of smaller building blocks. Such a balance between catabolism and anabolism is an example of how cells regulate energy flow and molecular content in response to intracellular and extracellular signals. And these processes do not exist in a vacuum either, but operate in highly regulated networks, which make cells highly adaptive to changing environments, and which underline the relationship between structure and function on cellular and molecular scales, which is the basis of life form itself. The numerous cellular structures facilitate metabolic processes, and the mitochondria are commonly termed as the powerhouses of the cell. These organelles are primarily involved in aerobic respiration, a catabolic process that converts biochemical energy from nutrients into ATP. Moreover, mitochondria are a key point in the whole metabolic network, as they not only affect energy synthesis but also control the metabolic pathways by the intermediates formed in different reactions. Such interdependence of metabolic pathways depicts the efficiency of cellular organization in maximizing energy conservation and use. Also, the interconnection between the functioning of the mitochondria and other cell systems, , such as the ER and the plasma membrane, points to the importance of the comprehensive nature of the knowledge of metabolism in the framework of the cell structure and functioning. New findings on the metabolic processes have given rise to new findings on unravelling the mysteries of cellular communications and adaptations via epigenetics. These mechanisms highlight the potential of environmental inputs to change metabolic activity without changing genetic sequences and thus offer a dynamic way for organisms to adapt to environmental stressors. It has been recently demonstrated that metabolic motivations are closely connected with epigenetic control, which is observed in the phenomena of how some traits can be passed through generations without being encoded in direct genetic information. These results imply that metabolism can act as a carrier of epigenetic information and result in transgenerational modifications that further outline the importance of metabolism in evolutionary biology. Thus, a comprehensive overview of metabolic processes not only elucidates cell function but also opens pathways to understanding broader biological implications, , including adaptation and evolution itself, as outlined by the implications of the Informational Model of Consciousness (Gaiseanu et al., 2019) and the relevance of multi-scale modeling in biological systems (Hasenauer et al., 2015).

4.5.2. Role of Enzymes in Metabolism

Enzymes are of vital importance in cellular metabolism as they function as biological catalysts that facilitate chemical reactions to be faster, thus enabling organisms to sustain important processes of life more efficiently. Enzymes are mostly proteins, but some catalytic RNA molecules called ribozymes also occur. Enzymes reduce the amount of activation energy needed for reactions in cells and thus make rates of reaction very high, which is important in metabolic pathways like glycolysis and the citric acid cycle. Several enzymes fall into these pathways, and each of the enzymes catalyzes a specific chemical conversion, thus ensuring that metabolic reactions occur in a controlled and regulated manner. The regulation and specificity are critical for the maintenance of cellular homeostasis and to enable the bidirectional flow of metabolic substrates and products, thus enabling the cells to possess adequate energy and building blocks for biosynthetic processes. The importance of the regulation of enzymes is exemplified by recent advances in their mechanisms, , such as allosteric regulation, covalent modification, and feedback inhibition. For instance, in response to metabolic intermediates, some enzymes are conformationally modified to rise or fall in their activity and thus regulate the metabolic flow commensurate with the energy needs of the cell. Besides, fluctuations in enzyme expression levels are also capable of significantly influencing metabolic pathways, , such as changes in nutrient status or stress. Interestingly, SCFAs, which are generated through fiber fermentation within the intestine, have been shown to be involved in enzyme regulation processes, influencing transcriptional adaptation among varying tissues by modulating the function of histone-modifying enzymes to impact gene expression with regard to metabolic well-being (Bart et al., 2021). These dynamic interactions emphasize the complex web of metabolic regulation at the enzyme level and its broader significance in organismal physiology. Furthermore, integration with advanced databases , such as KEGG extends our understanding even further into the role of enzymes in metabolism via a more global understanding of metabolic pathways and interactions. KEGG combines genomic data with functional explanation and illustrates the role of enzymes in complex cellular processes using a system of functional orthology (Kanehisa et al., 2016). This allows researchers to map how enzymatic activity contributes to metabolic networks, which can facilitate the identification of novel therapeutic targets for metabolic disorders. Thus, the knowledge of the role played by enzymes not only expands our comprehension of cellular metabolism but also dictates biotechnological developments and drug research on controlling these processes for therapeutic health outcomes. The analysis of enzymes thus remains a fundamental component of cell biology, and the discovery of significant correlations between enzyme function, regulation of metabolism, and cellular overall health (Figure 4.17).

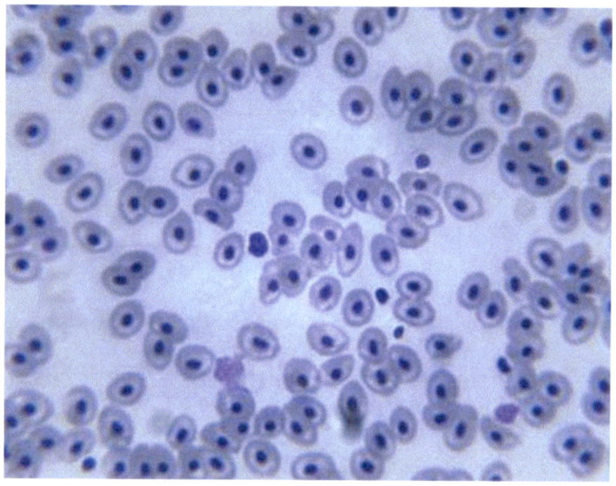

Figure 4.17. Role of enzymes in metabolism.

Source: https://www.sciencing.com/sciencing/role-enzymes-metabolism-6390077/AA003849_XS.jpg.

4.5.3. Energy Production in Cells

Cell energy production is basically linked with the integrity and functionality of cell structure, in particular the mitochondria, popularly known as the power plant of the cell. The organelles yield oxidative phosphorylation, by which ATP, the primary energy currency of biological systems, is generated. This process happens through ETCs in the inner mitochondrial membrane, where redox reactions cause proton pumping and the creation of an electrochemical gradient. The flow of protons returning into the mitochondrial matrix through ATP synthase allows for ATP synthesis. Efficiency of the process is crucial for cellular function, as ATP not only energizes metabolic reactions but also controls cellular functions , such as growth, division, and response to environmental stimuli. Disrupted energy production may result in cellular malfunction and is behind many diseases, , including neurodegenerative diseases. The link between energy generation and cellular health emphasizes the role of enzymes and proteins in mitochondrial activity. For example, studies have identified the involvement of oxidative stress in energy metabolism and how aberrant protein interactions lead to impaired ATP production. Studies have shown that disruption of core cellular functions, , such as those that interfere with molecular pathways in mitochondrial bioenergetics, can be involved in the pathology of diseases like Alzheimer's. Cellular dysfunction in energy production is conveyed by decreased ATP content and altered mitochondrial morphology that can produce catastrophic outcomes on cellular viability. The proteomics tells us how cellular components operate in concert, and how the complex balance must be maintained to maintain energy homeostasis. These results illustrate the sophistication of energy cell dynamics and the consequences for overall cell health and function, particularly in the context of disease pathology (Swomley et al., 2014). Furthermore, the mechanisms that ensure mitochondrial integrity under conditions of cellular stress are crucial in guaranteeing energy supply, particularly under conditions where cell survival is at risk, , such as under apoptosis. Experiments with single proteins, e.g., α1-microglobulin (A1M), have indicated that they possess protective roles against oxidative stress under conditions of cell death. A1M displays a high capacity to bind mitochondrial components, specifically the ETC complex I, thereby enhancing the conservation of ATP production during stress situations. This action is important since it not only maintains the energy production during programmed cell death but also decreases oxidative damage in nearby healthy cells. The role of the proteins indicates the intricacy of adaptations cells implement to maintain mitochondrial integrity, illustrating the profound interdependence of energy metabolism, cell structure, and function in preserving cellular homeostasis (Fellman et al., 2013) (Figure 4.18).

Figure 4.18. Energy production in cells.

Source: https://snworksceo.imgix.net/jhn/bde75d89-2f5d-4be6-90f3-a2547d05bda4.sized-1000x1000.png?w=1000.

Test Your Knowledge

1. What is the role of enzymes in metabolism?
2. Illustrate energy production in cells?

4.6. Cell Communication

Learning Outcome

1. Mechanisms of cell signaling;
2. Importance of cell communication in multicellular organisms;
3. Examples of signaling pathways.

The complex process of cell communication is the core behind the maintenance of homeostasis and activation of various physiological reactions in multicellular life forms. Cell communication occurs through a number of mechanisms , including direct cell-to-cell contact, chemical communication, and the transmission of messages via the ECM. All these interactions rely heavily on IMPs, which allow cells to communicate with each other and their environment. They allow for the movement of nutrients, the removal of waste products, and passing of signals, which are essential for cell function and coordination among cells in tissues. Polypeptide mutations and environmental exposure can, however, inhibit these protein functions, leading to disrupted pathways and compromised cell communication. Clarification of IMP structure and dynamics is thus essential to comprehend processes of cell communication, which may have severe implications for treatment development to reinstate cellular function under disease conditions where communication is perturbed (Majeed et al.,

2021). The ECM is another dominant modulator of cell communication in tissues. Being a dynamic protein network, the ECM structurally supports cells as well as facilitates biochemical signaling that modulates cellular behavior. For example, perineuronal nets (PNNs) assembly in the ECM regulates neuron-to-neuron interactions and thereby synaptic transmission and neuronal network activity. Such spatiotemporal regulation of cell-to-cell communication is particularly important in psychiatric illness, where stress-induced remodeling of ECM components can disrupt the fine balance of neurotransmission. Evidence of PNN alteration and its correlation with parvalbumin-positive interneurons demonstrates the highly complex relationship between ECM composition and neuronal health. These interactions need to be grasped to manage the cellular and molecular mechanisms of mood disorders and other pathologies associated with impaired cell communication (Spijker et al., 2020).

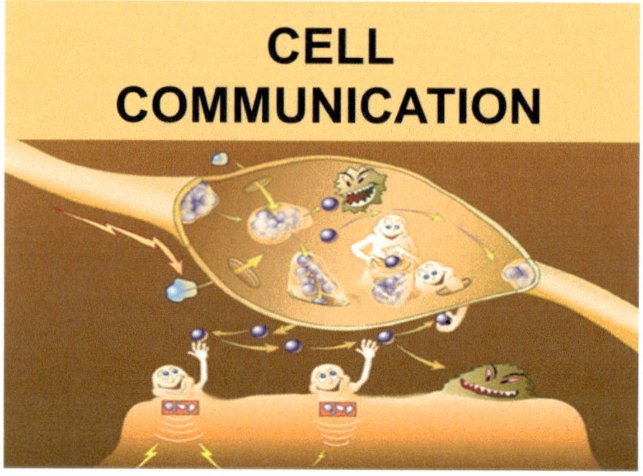

Figure 4.19. Cell communication.

Source: https://quizizz.com/_media/quizzes/01271704-d9e3-4196-84e4-4b0a46d6aa79.

Generally, cell communication processes are multifaceted and critical to the overall function of cells and tissues. Cells communicate and process essential information for coordinated cellular responses through IMPs and ECM components. Dysfunctions in these routes of communication can have severe physiological consequences, reflecting the relevance of understanding IMPs and ECM in the context of disease and health. Through investigations of the molecular mechanisms of cell communication, scientists can further investigate the dynamics of cellular interactions and develop novel therapeutic options. This information is important, not only to advance elementary biological understanding but also to address clinical issues , such as degenerative disease, cancer, and psychiatric disorder, in which communication networks are generally disrupted (Majeed et al., 2021; Spijker et al., 2020) (Figure 4.19).

4.6.1. Mechanisms of Cell Signaling

Cell signaling pathways form a core basis of cellular interactions and functioning in multicellular organisms. These signaling pathways help the communication between the cells, making them adapt to their environment. There are many different types of signaling modalities , such as autocrine, paracrine, endocrine, and juxtacrine, that add to the diverse nature of cellular communication. Autocrine signaling: Cells can also respond to signals produced by themselves, which enhances local responses. Conversely, paracrine signaling implies the neighboring cells reacting to the released signaling molecules, generating local effects. Endocrine signaling has a longer range, where hormones are transported by the blood to other distant organs. Juxtacrine signaling is a direct cell-cell contact-dependent signaling that demonstrates the significance of cell-to-cell interactions in many physiological processes. Such diversity of signaling pathways is an illustration of the simplicity and dependence of cellular processes, which are the main determinants of tissue formation and homeostasis in the whole organism. The complex systems of intracellular signaling networks tend to act in synergy with cell-cell communication strategies to stabilize homeostasis. To take a few examples, in the repair of tissue or in immune reactions, cells communicate using a variety of signaling molecules to make sure that there is a vigorous response to an external signal. An important feature of these pathways is the ability to exert time-dependent modulation of responses, which is important in fine-tuning physiological outcomes. Recent studies highlight a framework that characterizes these intracellular networks as black boxes, assessing input-output relationships without needing extensive details on intermediary steps (Altschuler et al., 2018).

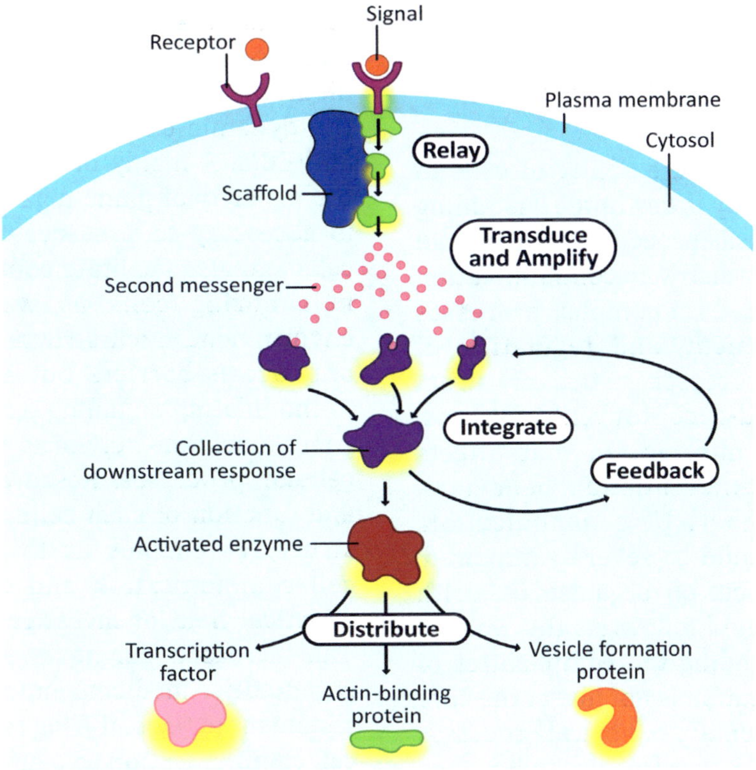

Figure 4.20. Mechanisms of cell signaling.

Source: https://open.oregonstate.education/app/uploads/sites/178/2023/11/Figure-07-03.png.

This method has elucidated processes like synchronization and delay in response time and provided information about the timing and dynamics of cytokine secretion which are major players in the communication between immune cells. It is through such incorporation of aspects that we can begin to comprehend how ordered cellular interactions required to sustain life are. Furthermore, cell signaling principles are not restricted to instant physiological effects, and instead, they are highly influential in developmental processes that enable adequate organismal growth and adaptation. Homeostasis is not a dead state rather, it is a dynamic process involving the interaction of biological systems that change with time, combining the information of different cellular interactions. This perspective positions homeostasis as a diachronic mechanism; one that transcends time and fosters the evolution of physiologic structures and functions (Torday, 2015). In this respect, cell signaling mechanisms play a central role in embryogenesis, physiological control and tissue repair, which shows the scale-free universality of these principles in various biological systems. Therefore, deciphering the intricacies of cell signaling not only adds to our wealth of cellular and molecular biological knowledge but also has prospective use in therapeutic interventions and regenerative medicine (Figure 4.20).

4.6.2. Importance of Cell Communication in Multicellular Organisms

Within multicellular life, the ability of cells to communicate effectively is key to the sustaining of numerous biological processes and overall organismal homeostasis. Cell communication is inherently coupled with complex processes of growth, development, and regulation of physiology. With processes , such as direct contact and ligand-receptor interactions, cells share critical information that affects specialization and differentiation behaviors. Without these channels of communication, cellular processes would be severely impaired, and detrimental effects on organismal health and development would follow. As the work is structured, understanding the complexities of cell-to-cell communication is not just at the core of deciphering rudimentary biological processes but also of exploring the therapeutic implications regarding diseases caused by communication failures, ultimately illuminating the importance of cell communication in sustaining life in multicellular organisms (Su et al., 2024). The diverse array of signaling pathways relating to cell communication is crucial in tissue homeostasis as well as environmental regulation. Cells employ a variety of signaling molecules to transmit information over distances, facilitating responses required for survival in a changing environment. One such example is hormonal signals regulating physiological responses required for development and metabolism, while immune cell-to-cell communication plays a critical role in responding to infection with pathogens and remaining healthy. The complexity of these signaling networks such that aberrations would lead to severe dysfunctions and diseases emphasizes the extent to which accurate communication among cells is important. Research has been focused on elucidating these pathways since their identification can result in new treatments and interventions for diseases related to malfunctioning signaling mechanisms (Su et al., 2024). Also, the cell membrane and ECM structural basis reveal the dynamic character of the cell membrane and ECM. A highly ordered cell wall, found in the majority of plant, algal, and microbial cells, is necessary to preserve the integrity of the cells and also facilitate contact with immediate neighboring cells as well as the outside environment. Such structural components not only act as barriers but also actively engage in modulating signaling activities and forming communication networks that are integral to cellular processes. Research on the structure and function of such cellular and extracellular structures enables us to understand cell-to-cell communication and cellular adaptation, a critical field of investigation in cell biology. This increased insight can foster innovations in regenerative medicine and tissue engineering, a testament to the enduring relevance of exploring cell communication as a matter of cell structure and function.

4.6.3. Examples of Signaling Pathways

Signaling pathways are crucial for the maintenance of cellular function and permit cells to talk with one another inside and between cells. They commonly involve a series of molecular interactions that carry signals from extracellular stimuli into intracellular responses. One good example is the Notch signaling pathway, which plays a critical role in cell differentiation and development. Specifically in mammalian cells, synthetic genetic programs derived from Notch pathways have been established for precise regulation of cell-cell communication. This approach regulates cell adhesion through differential expression of cadherin family proteins and leads to the formation of multilayered spheroids in cell cultures. As research goes on, such artificial pathways enable one to study complex cellular processes and hold potential in applications in regenerative medicine, in which control of tissue architecture is vital for successful outcomes in tissue repair and engineering (Lam et al., 2022). Another significant signaling pathway is seen in the fruit fly, *Drosophila melanogaster*, which is an omniscient model utilized to examine hematopoiesis and immune response mechanisms. In Drosophila, the lymph gland is a hematopoietic organ that offers a stem cell niche-like microenvironment similar to that of vertebrates. The evolutionarily conserved signaling pathways and transcription factors (TFs) that control blood cell fate reinforce the similarity in function between species and the convergent evolution. Experiments conducted on Drosophila lymph gland revealed the intricacy of cell-to-cell communication networks that reveal different regulatory schemes for governing hematopoietic stem cell behavior. These findings not only contribute to increasing our understanding of Drosophila biology but also generate valuable information concerning vertebrate stem cell regulation and its potential clinical application (Kharrat et al., 2022). The study of signaling pathways extends to the model organisms and ventures into synthetic biology, where constructed systems are engineered to regulate cellular functions for specific aims. Scholars have developed a computational approach that allows for the design and prediction of contact-dependent CSNs and their roles in cell adhesion. Scholars can accurately predict developmental pathways of cells by systematically studying parameterized genetic programs. This novel technique enables the creation of novel cellular structures that have not previously been documented, , such as multi-layered structures that replicate complex tissue organization. The synergy between computational design and synthetic biology holds significant potential to propel forward the field of regenerative medicine approaches and enhance our understanding of cellular dynamics in the framework of tissue formation and repair (Lam et al., 2022).

Test Your Knowledge

1. What is the mechanism of cell signaling
2. What is the importance of cell communication in multicellular organisms

Summary

The function and structure of cells, therefore, it is evident that the complex design of cells is at the core of the series of processes that allow for life to exist. From the uncomplicated prokaryotic cell to more advanced eukaryotic cells, knowing cellular architecture illuminates how various organelles and the cell membrane collaborate to maintain homeostasis and normal cellular function. The gene expression control mechanisms, which were first noted by Jacob and Monod, convey the dynamic nature of cells, showing how external stimuli can cause extreme changes within the cell environment. These findings emphasize the importance of cellular regulation, vital for adaptability and survival under fluctuating conditions, as noted in the extended conclusions of decades of studies (Cavalli et al., 2011). Apart from that, cell biology has also been enhanced by more sophisticated modeling approaches like the hybrid cellular automata (HCA). The model enables researchers to study biological phenomena at different scales and map complex interactions discovered in diverse cellular microenvironments. HCA and theirpotential applications, particularly in cancer research, testify to their role in helping determine the complex dynamics of cell growth and regulation (Agur et al., 2004). Through the isolation of the interaction between space and time effects on cellular behavior, HCA is a significant step in our ability to predict and manipulate cellular outcomes and thus transform our approach to examining cell structure and function on a larger scale. In general, the deductions that can be made from the complex interaction between cell function and structure extend far beyond the mere observation of cellular anatomy. They encompass a strong appreciation of the regulatory mechanisms involved and the approaches employed to study such mechanisms. Ongoing technology and methodological advances ensure that next-generation research will certainly tell us much more about cellular systems, potentially leading to revolutionary findings in medicine and biology. And as our knowledge deepens, there is a need to realize that all cellular components, from the nucleus to the cytoskeleton, play a vital role, not just for the survival of one cell but also in the overall tapestry of life itself. Emphasizing the complexity and sophistication of cell-to-cell interactions only serves to underscore the necessity of more research in this field.

The study of cell structure and function is the most important area in the study of biology because it forms the basis of understanding how organisms function in a biochemical and physiological perspective. The cell is constructed in a very complex way so as to accommodate all its functions, which include the production of energy and transmission of genes. Mitochondria, ER, and the Golgi apparatus organelles work together to ensure the crucial processes are completed effectively. Each structure has specialized roles—mitochondria serve as the powerhouse that generates ATP, while the ER is involved in protein and lipid synthesis. Therefore, the relationship between these elements plays a key role in the maintenance of cellular homeostasis, enabling cells to change and react to the variations in the environment, which is fundamental to the survival of both unicellular and multicellular organisms. The effectiveness of cell functioning usually depends on the regulatory processes of gene expression. TFs are key players in eukaryotic cells, which ensure the coordination of the timely expression of genes in response to internal and external stimuli. The chromatin structure and epigenetic modifications are some factors that influence gene expression and form a dynamic environment where genes are regulated in terms of activity. Recent studies employing methods like dynamic regulatory events miner (DREM) illustrate how coordinated transcription can lead to synchronized cellular responses, as observed in TGF-β treated MCF10A cells. Such investigations highlight the importance of understanding how genes interact within an complex network, thereby emphasizing the mutual dependence of cellular structure and function in achieving organismal homeostasis and adaptation to varying conditions (Bai et al., 2019). Besides transcriptional coordination, analysis of parameter interactions in models also demonstrates the complexity of cellular processes. As an example, the idea of active subspaces enables scientists to reduce models with many variables to lower-dimensional representations, improving our knowledge of cellular output in different conditions. This approach shows that a one-diode solar cell model has the advantage of a one-dimensional active subspace, showing how dimensionality reduction can simplify studies of cellular responses. By defining the parameters that have a great impact on the cellular behavior, scientists would be able to conduct more directed studies whose predictive value would be applicable to the cell functionality. Ultimately, recognizing the interplay between cellular structure and function, alongside effective modeling techniques, fosters a comprehensive understanding of biological systems (Campanelli et al., 2015).

KEY TERMS FROM THIS CHAPTER

Amino acids	Carbohydrates	Disaccharides
DNA (deoxyribonucleic acid)	Enzymes	Fatty acids
Lipids	Monosaccharides	Nucleic acids
Peptide bonds	Phospholipids	Polysaccharides
Primary structure	Proteins	Quaternary structure
RNA (ribonucleic acid)	Secondary structure	Steroids

Critical Thinking Questions

1. **What is the basic unit of life?**
   a. Organ
   b. Tissue
   c. Cell
   d. Organism
2. **Which organelle is known as the "powerhouse of the cell"?**
   a. Nucleus
   b. Mitochondrion
   c. Ribosome
   d. Golgi apparatus
3. **Which part of the cell controls cell activities and contains genetic material?**
   a. Cytoplasm
   b. Nucleus
   c. Cell membrane
   d. Lysosome
4. **What is the function of the cell membrane?**
   a. Produce energy
   b. Control what enters and leaves the cell
   c. Store nutrients
   d. Make proteins
5. **Which organelle is responsible for protein synthesis?**
   a. Ribosome
   b. Chloroplast
   c. Lysosome
   d. Vacuole

Short Answer Questions

1. What is the cell theory? List its three main principles.
2. What are the main differences between prokaryotic and eukaryotic cells?
3. Name the three main parts of a cell and their general functions.
4. What is the function of the cell membrane?
5. Describe the role of the nucleus in a eukaryotic cell.

Answers to Critical Thinking Questions

1. (c); 2. (b); 3. (b); 4. (b); 5. (a)

References

1. Abeln, S., Feenstra, K. A., Gavaldá-Garciá, J., Hoekstra, A. G., et al. (2023). Introduction to protein structure. http://arxiv.org/abs/2307.02169 (accessed on 13th June 2025).
2. Abeln, S., van Dijk, E., Feenstra, K. A., Gavaldá-Garciá, J., et al. (2023). Introduction to protein structure. https://core.ac.uk/download/577889947.pdf (accessed on 13th June 2025).
3. Agur, Z., Alarcón, T., Maini, P. K., Marron, P., et al. (2004). The use of hybrid cellular automaton models for improving cancer therapy. In *Proceedings, Cellular Automata: 6th International Conference on Cellular Automata for Research and Industry, ACRI 2004, Amsterdam, The Netherlands*, eds P. M. A. Sloot, B. Chopard, A. G. Hoekstra. https://core.ac.uk/download/pdf/96784.pdf (accessed on 13th June 2025).
4. Alí Alejo, T., Matamoros, M., & Guerra, G. A. (2018). A proteomic atlas of the African swine fever virus particle. *Journal of Virology, 92.* https://doi.org/10.1128/jvi.01293-18 (accessed on 13th June 2025).
5. Allen, B., Gore, J., & Nowak, M. A. (2013). Spatial dilemmas of diffusible public goods. *eLife, 2.* https://doi.org/10.7554/elife.01169 (accessed on 13th June 2025).
6. Altschuler, S. J., Thurley, K., & Wu, L. F. (2018). Modeling cell-to-cell communication networks using response-time distributions. https://core.ac.uk/download/323079318.pdf (accessed on 13th June 2025).
7. Angelotti, T. (2022). Exploring the eukaryotic Yip and REEP/Yop superfamily of membrane-shaping adapter proteins (MSAPs): A cacophony or harmony of structure and function? *Frontiers in Molecular Biosciences, 9.* https://www.semanticscholar.org/paper/3b3a2ef6681 4923b9533d12096e4aceee72e015f (accessed on 13th June 2025).
8. Angrisano, F., Baum, J., Dekiwadia, C., Gout, A., et al. (2012). Subcompartmentalisation of proteins in the rhoptries correlates with ordered events of erythrocyte invasion by the blood stage malaria parasite. https://core.ac.uk/download/162205513.pdf (accessed on 13th June 2025).
9. Fabiani, C., & Antollini, S. S. (2019). Alzheimer's Disease as a Membrane Disorder: Spatial Cross-Talk Among Beta-Amyloid Peptides, Nicotinic Acetylcholine Receptors and Lipid Rafts. *Frontiers in cellular neuroscience, 13*, 309. https://doi.org/10.3389/fncel.2019.00309(accessed on 13th June 2025).
10. Bai, F., Chen, H., Li, R., Taft, R., et al. (2019). Spatial clustering and common regulatory elements correlate with coordinated gene expression. https://core.ac.uk/download/225561466.pdf (accessed on 13th June 2025).
11. Barreiro, A. K., & Ly, C. (2017). When do correlations increase with firing rates in recurrent networks? https://core.ac.uk/download/84402991.pdf (accessed on 13th June 2025).
12. Barthel, J., Eisenberg, B., Corry, B., Doi, M., Engl, H. W., Hyon, Y., et al. (2013). Ions in fluctuating channels: Transistors alive. http://arxiv.org/abs/q-bio/0506016 (accessed on 13th June 2025).
13. Berg, G., Rybakova, D., Fischer, D., Cernava, T., Champomier Vergès, M. C., Charles, T. C., Chen, X., et al. (2020). Microbiome definition re-visited: Old concepts and new challenges. *Microbiome, 8.* https://doi.org/10.1186/s40168-020-00875-0 (accessed on 13th June 2025).
14. Berlanga, M., & Guerrero, R. (2010). The hidden side of the prokaryotic cell: Rediscovering the

microbial world. https://core.ac.uk/download/159084328.pdf (accessed on 13th June 2025).

15. Bowater, R., & Doherty, A. J. (2006). Making ends meet: Repairing breaks in bacterial DNA by non-homologous end-joining. https://core.ac.uk/download/2728184.pdf (accessed on 13th June 2025).
16. https://core.ac.uk/download/pdf/82278911.pdf Swomley, A. M., Förster, S., Keeney, J. T., Triplett, J., Zhang, Z., Sultana, R., & Butterfield, D. A. (2014). Abeta, oxidative stress in Alzheimer disease: evidence based on proteomics studies. *Biochimica et biophysica acta, 1842*(8), 1248–1257. https://doi.org/10.1016/j.bbadis.2013.09.015 (accessed on 13th June 2025).
17. Caldwell, D. E. (2010). The calculative nature of microbial biofilms and bio aggregates. https://core.ac.uk/download/159084571.pdf (accessed on 13th June 2025).
18. Campanelli, M., Constantine, P. G., & Zaharatos, B. (2015). Discovering an active subspace in a single-diode solar cell model. http://arxiv.org/abs/1406.7607 (accessed on 13th June 2025).
19. Cavalli, G. (2011). From linear genes to epigenetic inheritance of three-dimensional epigenomes. https://core.ac.uk/download/30436052.pdf (accessed on 13th June 2025).
20. Chen, Y., Edwards, J. S., Halász, Á. M., Short, B., et al. (2013). The impact of high-density receptor clusters on VEGF signaling. http://arxiv.org/abs/1309.0868 (accessed on 13th June 2025).
21. Eisenberg, B. (2018). Asking biological questions of physical systems: The device approach to emergent properties. http://arxiv.org/abs/1801.05452 (accessed on 13th June 2025).
22. Fatichi, S., Leuzinger, S., & Körner, C. (2013). Moving beyond photosynthesis: From carbon source to sink-driven vegetation modeling. *New Phytologist, 201*, 1086–1095. https://doi.org/10.1111/nph.12614 (accessed on 13th June 2025).
23. Fellman, V., Gram, M., Kotarsky, H., Leanderson, P., et al. (2013). The radical-binding lipocalin A1M binds to a Complex I subunit and protects mitochondrial structure and function. https://core.ac.uk/download/83785446.pdf (accessed on 13th June 2025).
24. Fuertes-Rabanal, M., Rebaque, D., Largo-Gosens, A., Encina, A., & Mélida, H. (2024). Cell walls, a comparative view of the composition of cell surfaces of plants, algae and microorganisms. *Journal of Experimental Botany.* https://www.semanticscholar.org/paper/235e8356c7365f710d60c5642e43e632ca063d0e (accessed on 13th June 2025).
25. Gaiseanu, F. (2019). Epigenetic information–body interaction and information-assisted evolution from the perspective of the informational model of consciousness. https://core.ac.uk/download/287611914.pdf (accessed on 13th June 2025).
26. Gale, N., Heath, G., Cameron, E., Rashid, S. F., & Redwood, S. (2013). Using the framework method for the analysis of qualitative data in multi-disciplinary health research. *BMC Medical Research Methodology, 13.* https://doi.org/10.1186/1471-2288-13-117 (accessed on 13th June 2025).
27. Hasenauer, J., Hross, S., Jagiella, N., Theis, F. J., et al. (2015). Data-driven modelling of biological multi-scale processes. http://arxiv.org/abs/1506.06392 (accessed on 13th June 2025).
28. Kanehisa, M., Furumichi, M., Tanabe, M., Sato, Y., & Morishima, K. (2016). KEGG: New perspectives on genomes, pathways, diseases and drugs. *Nucleic Acids Research, 45*, D353–D361. https://doi.org/10.1093/nar/gkw1092 (accessed on 13th June 2025).
29. Kharrat, B., Csordás, G., & Honti, V. (2022). Peeling back the layers of lymph gland structure and regulation. *International Journal of Molecular Sciences, 23.* https://www.semanticscholar.

org/paper/6aa2ffa6c931cca09d1410aad9424ebfe26b425e (accessed on 13th June 2025).

30. Kheyrodin, H., Jami, R., & Rehman, F. (2022). Cellular structure and molecular functions of plants, animals, bacteria, and viruses. *Cellular, Molecular and Biomedical Reports.* https://www.semanticscholar.org/paper/57c818c94829c5b1de8ab4962a125074430349e8 (accessed on 13th June 2025).

31. Kundenko, M., Rudenko, A., & Mardziavko, V. (2024). Model of studying the membrane potential in the interaction of cell structure with an electromagnetic field. *2024 IEEE 42nd International Conference on Electronics and Nanotechnology (ELNANO).* https://www.semanticscholar.org/paper/cf8d5f4863852d8a1781790800873e44d89c5d44 (accessed on 13th June 2025).

32. Lam, C., Saluja, S., Courcoubetis, G., Yu, D., Chung, C., Courte, J., & Morsut, L. (2022). Parameterized computational framework for the description and design of genetic circuits of morphogenesis based on contact-dependent signaling and changes in cell–cell adhesion. *ACS Synthetic Biology.* https://www.semanticscholar.org/paper/d4c2f92c4055785250a688496020eb7a0b91d7f2 (accessed on 13th June 2025).

33. Lapeyre Jr., G. J., Lewenstein, M., Manzo, C., Massignan, P., et al. (2015). Weak ergodicity breaking of receptor motion in living cells stemming from random diffusivity. https://core.ac.uk/download/41778549.pdf (accessed on 13th June 2025).

34. Liu, Z., Zhao, Z., Peng, B., Duan, X., & Huang, Y. (2020). Beyond extended surfaces: Understanding the oxygen reduction reaction on nano catalysts. *Journal of the American Chemical Society, 142*, 17812–17827. https://doi.org/10.1021/jacs.0c07696 (accessed on 13th June 2025).

35. Luhtala, N., & Parker, R. (2012). Structure-function analysis of Rny1 in tRNA cleavage and growth inhibition. *PLoS ONE, 7.* https://www.semanticscholar.org/paper/9be93d56547376afc963394acc37e4ee64c1dd18 (accessed on 13th June 2025).

36. Lv, H., Ren, D., Yan, W., Wang, Y., Liu, H., & Shen, M. (2019). Linoleic acid inhibits *Lactobacillus* activity by destroying cell membrane and affecting normal metabolism. *Journal of the Science of Food and Agriculture.* https://www.semanticscholar.org/paper/4fe106b8b86595e4db09e97d60144275765a3d45 (accessed on 13th June 2025).

37. Majeed, S., Ahmad, A., Sehar, U., & Georgieva, E. R. (2021). Lipid membrane mimetics in functional and structural studies of integral membrane proteins. *Membranes, 11*, 685–685. https://doi.org/10.3390/membranes11090685 (accessed on 13th June 2025).

38. Mitra, A., Bhakta, K., Kar, A., Roy, A., Mohid, S. A., Ghosh, A., & Ghosh, A. (2023). Insight into the biochemical and cell biological function of an intrinsically unstructured heat shock protein, Hsp12 of *Ustilago maydis. Molecular Plant Pathology, 24*, 1063–1077. https://www.semanticscholar.org/paper/8e822edd0d5156f18e0c07d8384be7a33d9034e3 (accessed on 13th June 2025).

39. Murata, K., Kawai, S., & Hashimoto, W. (2022). Bacteria with a mouth: Discovery and new insights into cell surface structure and macromolecule transport. *Proceedings of the Japan Academy. Series B, Physical and Biological Sciences, 98*, 529–552. https://www.semanticscholar.org/paper/1fb15832c9f1635fb0cdbb80ca138df4e31ed9cb (accessed on 13th June 2025).

40. Naish, M., & Henderson, I. R. (2024). The structure, function, and evolution of plant centromeres. *Genome Research, 34*, 161–178. https://www.semanticscholar.org/paper/f9538989be542cfdee90586460cb55f4eb094b87 (accessed on 13th June 2025).

41. Noone P. G. (2001). Primary ciliary dyskinesia: a report from ATS 2001, May 18–23, San Francisco. *Respiratory Research, 2*(5), E001. https://doi.org/10.1186/rr75 (accessed on 13th June 2025).

42. Panstruga, R., Antonin, W., & Lichius, A. (2023). Looking outside the box: A comparative cross-kingdom view on the cell biology of the three major lineages of eukaryotic multicellular life. *Cellular and Molecular Life Sciences: CMLS, 80.* https://www.semanticscholar.org/paper/bdcde4d778d817c125d9936a1228ad387ff9afcf (accessed on 13th June 2025).

43. Parvez, S. M. M., & Babu, D. (2022). Comprehensive overview of plant cell structure and functions: Analysis of organelles, roles, and their interactions in plant physiology. *Australian Herbal Insight.* https://www.semanticscholar.org/paper/002a00e3a9ef48582e30b6c886342 0eda6abb2ca (accessed on 13th June 2025).

44. Riley, J. S., & Tait, S. W. G. (2020). Mitochondrial DNA in inflammation and immunity. *EMBO Reports, 21.* https://doi.org/10.15252/embr.201949799 (accessed on 13th June 2025).

45. Rozario, L., Sharker, T., & Akter Nila, T. (2021). *In silico* analysis of deleterious SNPs of human *MTUS1* gene and their impacts on subsequent protein structure and function. *PLoS ONE, 16.* https://www.semanticscholar.org/paper/623838f3564570143ffe74b6820a7a291a4aa88b (accessed on 13th June 2025).

46. Shrestha, E., White, J. R., Yu, S. H., Kulaç, İ., Ertunç, O., De Marzo, A. M., Yegnasubramanian, S., et al. (2017). Profiling the urinary microbiome in men with positive versus negative biopsies for prostate cancer. *The Journal of Urology, 199*, 161–171. https://doi.org/10.1016/j.juro.2017.08.001 (accessed on 13th June 2025).

47. Spijker, S., Koskinen, M. K., & Riga, D. (2020). Incubation of depression: ECM assembly and parvalbumin interneurons after stress. *Neuroscience & Biobehavioral Reviews, 118*, 65–79. https://doi.org/10.1016/j.neubiorev.2020.07.015 (accessed on 13th June 2025).

48. Su, J., Song, Y., Zhu, Z., Huang, X., Fan, J., Qiao, J., & Mao, F. (2024). Cell–cell communication: New insights and clinical implications. *Signal Transduction and Targeted Therapy, 9.* https://www.semanticscholar.org/paper/66951f323694740b1a975607a9b538ed69f917a4 (accessed on 13th June 2025).

49. Torday J. S. (2015). Homeostasis as the Mechanism of Evolution. *Biology, 4*(3), 573–590. https://doi.org/10.3390/biology4030573(accessed on 13th June 2025).

50. Tosaka, T., & Kamiya, K. (2023). Function investigations and applications of membrane proteins on artificial lipid membranes. *International Journal of Molecular Sciences, 24.* https://www.semanticscholar.org/paper/31a3b952e2c27ef33ab8bbc847e2c8ae55592781 (accessed on 13th June 2025).

51. Valparaiso University. (1969). *Graduate Catalog, 1969–1970.* https://core.ac.uk/download/144553297.pdf (accessed on 13th June 2025).

52. van der Hee, B., & Wells, J. M. (2021). Microbial regulation of host physiology by short-chain fatty acids. *Trends in Microbiology, 29*, 700–712. https://doi.org/10.1016/j.tim.2021.02.001 (accessed on 13th June 2025).

53. Witzany, G. (2007). Telomeres in evolution and development from biosemiotics perspective. https://core.ac.uk/download/pdf/287482.pdf (accessed on 13th June 2025).

54. Witzany, G. (2008). Bio-communication of bacteria and its evolutionary interrelations to

natural genome editing competences of viruses. https://core.ac.uk/download/pdf/287751.pdf (accessed on 13th June 2025).

55. Yang, J., Huang, J., Wang, H., Liu, Y., Tang, Y., Lin, C., Zhou, Q., et al. (2022). Expression of the Cavin family in childhood leukemia and its implications in subtype diagnosis and prognosis evaluation. *Frontiers in Pediatrics, 10.* https://www.semanticscholar.org/paper/e6e6ddadd10534a4fc076bbdeb4db06d0574e7d4 (accessed on 13th June 2025).

MEMBRANES AND TRANSPORT

5

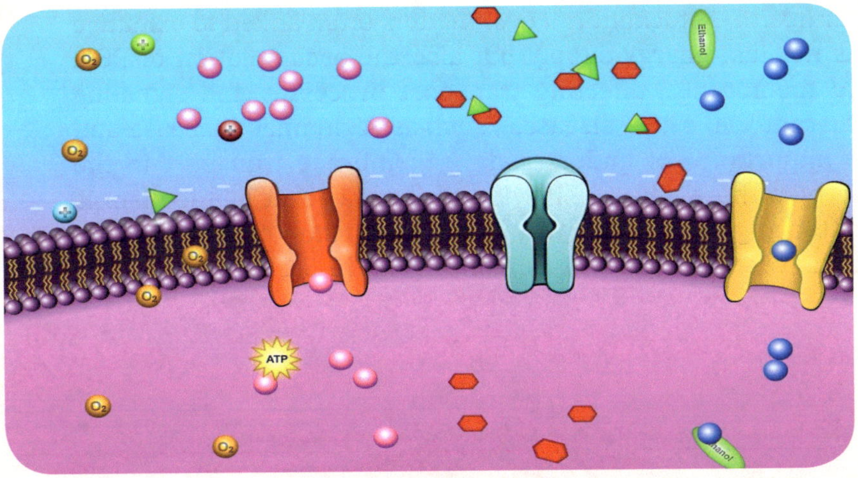

LEARNING OBJECTIVES

After studying this chapter, you will be able to:

1. Structure of biological membranes;
2. Types of membrane transport;
3. Passive transport;
4. Active transport;
5. Bulk transport;
6. Membrane transport in health and disease.

Introduction

The complex forms and functions of cellular membranes are at the forefront of the understanding of biological mechanisms of transport. The very center of cellular life, membranes are permeable barriers that selectively regulate the movement of substances in and out of cells. This selectively permeable attribute is of utmost importance to the maintenance of homeostasis to bring important nutrients in and to expel waste products. They are made up mostly of phospholipids, in a bilayer that is a fluid and dynamic environment. This structure not only provides structural integrity, but also various transport processes ranging from passive diffusion to active transport. The enzymes embedded within these membranes catalyze the reactions required for membrane function. The significance of membranes is supported by their role in cellular communication and signaling, which is necessary for coordinating physiological reactions in multicellular organisms. Membranes and transport are significant enough to cross the realm of metabolic processes, particularly with regard to conditions like Type 2 Diabetes

Mellitus (T2DM). T2DM is an outcome of a complex interplay between a defective insulin signaling pathway and the complexity of cellular uptake mechanisms. The insulin secretion and production by pancreatic β-cells are closely associated with cell membrane function. The role of membrane integrity and function of transport proteins can be severely disrupted as a result of allowing entry of glucose into target tissues by insulin. This emphasizes the necessity of the study of membrane transport mechanisms not only structurally but also from a pathophysiological perspective. For instance, knowing the mechanism of transporters in the membrane helps shed important light on the altered biochemical pathways in T2DM and ultimately guides therapeutic intervention. In addition, oxidative stress is a significant threat to the integrity of membranes and transport mechanisms, particularly in neurodegenerative disorders. As highlighted by research, the central nervous system (CNS) relies significantly on the effective functioning of mitochondria and membrane dynamics for ATP production and cellular health. Mitochondrial dysfunction, often linked to oxidative stress, creates a disruption that affects not just energy supply but also the accumulation of ROS. Excessive ROS can be harmful to lipid membranes, affecting transport processes and signaling pathways in neurons. Ultimately, this cascade can cause diseases , such as Alzheimer's disease and Parkinson's disease, underscoring the importance of understanding membrane transport in the context of oxidative damage. It does so by a thorough examination of the way that metabolism is integrated with membranes and how attacking membranes would be able to stop oxidative stress from interfering with cellular function, as well as transport dynamics (Figure 5.1).

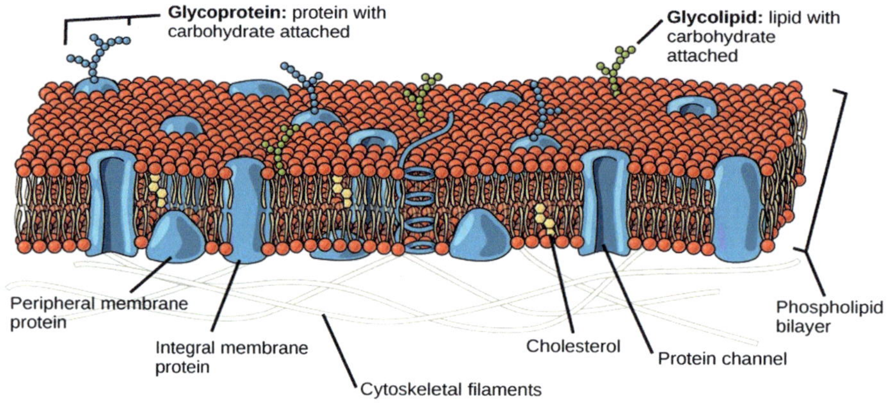

Figure 5.1. Membranes and transport.

Source: https://rwu.pressbooks.pub/app/uploads/sites/29/2020/07/image2-5.jpeg.

Definition of Biological Membranes

Biological membranes are basic structures that delimit cellular compartments and hence establish the interface between a cell's internal and external environment. They are primarily made up of a phospholipid bilayer decorated with proteins, cholesterol, and carbohydrates that are accountable for their fluidity and function. This particular structure allows membranes to perform vital functions, , including acting as selective barriers in controlling the passage of substances in and out of cells. This structure is described by the fluid mosaic model, with an emphasis on the dynamic nature of the composition of membranes, which can change and reorganize in ways that accommodate varying cellular functions. In the study of membranes, it is essential to find out how their structure determines their function, giving an outline of studying critical processes , such as transport, signaling, and energy transduction in cellular biology. On a deeper level, biological membranes play a role in complex interactions that decide cellular behavior. They are controlled by proteins spanning the membrane or binding to its surface, facilitating communication and transportation of molecules. Membrane proteins may be functioning as receptors, enzymes, or transport channels, each carrying out particular tasks that are essential to homeostasis within the cell. An example would be the selective permeability of membranes, which is a vital feature enabling cells to maintain their internal environment irrespective of the changing conditions around them. At the center of this function lies a process known as vesiculation, where vesicles are created to transport material in the cell, and from and to the external space. The binding between these proteins and lipid moieties signifies the complexity and adaptability of biological membranes to the physiological needs of cells. Besides, improvements in computational models, , such as the BioBeta model, have pushed the frontiers of biological membrane modeling and interaction further. These models enable protein-level and membrane-level interactions to be investigated while offering an overview of cellular processes. Using bigraphical representations, researchers are in a position to define the dynamic nature of biological systems more accurately. Furthermore, these structures offer a method of formal justification for higher-order membrane processes in terms of lower-level protein interactions, , such as phagocytosis and vesiculation. Such an interdisciplinary examination of biological membranes not only enhances our understanding of cellular dynamics but also helps in formulating strategies for addressing complex biological problems, thereby illuminating the central role membranes play in transport and cellular function as a whole (Grohmann et al., 2009) (Figure 5.2).

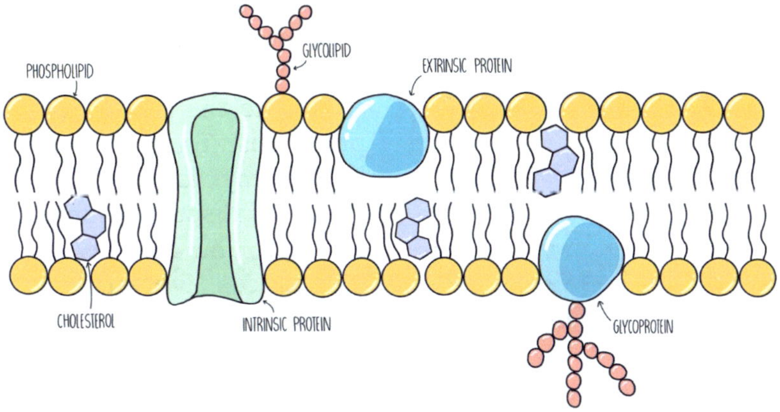

Figure 5.2. Biological membranes.

Source: https://images.squarespace-cdn.com/content/v1/5c5aed8434c4e20e953d6011/1586786436596-HX8BCCQ1KMHMG3ADL5DV/IMG_1121.jpeg.

Importance of Membranes in Cellular Function

Cell membranes are important for maintaining homeostasis in cells by serving as active barriers to regulate the exchange of substances in and out of the intracellular environment. The phospholipid bilayer cell membrane framework, stabilized by various proteins, carbohydrates, and cholesterol, enables selective permeability since vital nutrients can pass through while the unwanted substances are kept out. This facilitated transport is critical to cellular survival and functioning as it is accountable for the creation and upholding of the precise internal environment conducive to biochemical reactions. Secondly, incorporation of membrane proteins lies at the heart of signal transduction by which cells can react to the external world through receptor-mediated processes. The relationship between membrane composition and function is evident in a number of cellular processes , including metabolism, cell growth, and apoptosis, thereby indicating the importance of the membranes to facilitate rudimentary cellular functions.

In addition to selective transport, membranes are also very critical in intracellular organization and the operation of organelles. The interactions between the membrane and the cytoskeleton, particularly with the system of microtubules, are central to the positioning of the organelles and membrane trafficking. The complex interaction between cytoskeletal elements and membranes has lately been highlighted, with microtubules influencing not only the physical distribution of organelles but also sorting of cargo and directional transport inside the cell. This interaction between the microtubule system and membranes coordinates the distribution of cellular components efficiently and their activities accurately. As the interactions continue to be the focus of extensive research, it has become increasingly clear that the integrity of the cellular membrane is also very much associated with optimal operation of cellular machinery, necessary for normal function as well as for stress adaptations to the environment (Stephens et al., 2012). Moreover, evolutionary innovations in the composition of membranes, particularly the incorporation of cholesterol, allowed cells to adapt to varying environmental conditions. Higher oxygen levels in the environment caused difficulties for early organisms, necessitating cellular respiration and oxygen regulation improvements. Cholesterol, a critical component as a lipid, facilitates the clustering of lipid rafts, unique membrane regions involved in signal transduction and modulation of cellular functions. For instance, such cholesterol-rich caveolae are microdomains allowing hypoxia adaptation and oxygen diffusion regulation across membranes. These membrane adaptations illustrate the complex interplay among cell function and external limitations, substantiating that membranes are more than mere structural features but engaged participants in the regulation of vital physiological processes (Zuniga-Hertz & Patel, 2019). Thus, the function of membranes is more than mere compartmentalization as prime movers in adapting to and thriving in diverse biological environments.

Overview of Transport Mechanisms

Transport processes through biological membranes are fundamental to the establishment and maintenance of cellular homeostasis as well as to the execution of many physiological processes. Transport may be classified in a broad way as passive and active transport. Passive transport is dependent on concentration gradient and involves the free movement of substances without using energy. Simple diffusion, facilitated diffusion, and osmosis are some examples of passive transport, in which the molecules pass through the membrane through protein channels or directly, according to their size and polarity. These processes play an important role in the transport of gases, nutrients and waste products and it reaffirms the significance of selective permeability that is characteristic of biological membranes, such that vital molecules can get in and unwanted ones get out of the cell comparatively readily. Conversely, active transport systems are energy-coupled processes that allow cells to transport molecules opposite to their concentration gradient. This is usually accomplished by the hydrolysis of ATP, which powers proteins referred to as pumps, , such as the sodium-potassium pump, to maintain ion gradients that are essential to cellular functioning. Active transport Active transport plays an especially important role in regulating the internal environment of cells, allowing them to concentrate necessary nutrients or ions that might be present in lower concentrations externally.

The membrane transport systems thus form dynamic equilibria, which is essential to the cellular integrity and response to environmental stimuli, making them complex in biological systems. Among the more specialized transport mechanisms is forward osmosis (FO), a process that capitalizes on osmotic pressure differences between two solutions across a semipermeable membrane. In FO, the water is concentrated by passing a dilute feed solution to a more concentrated draw solution. The process allows the purification and concentration of solutes without the addition of external energy source, a factor that significantly limits fouling propensities on membranes, a problem that has bedeviled numerous applications. FO's functionality has illuminated its potential within various fields, , including wastewater treatment and resource recovery systems like FO membrane bioreactors (FO-MBR) that join molecular transport dynamics with technological applications. As research into FO continues, it highlights the evolving landscape of membrane transport technologies and their adaptive applications in addressing contemporary environmental and biological challenges (Oliveira et al., 2010; Arslan et al., 2018) (Figure 5.3).

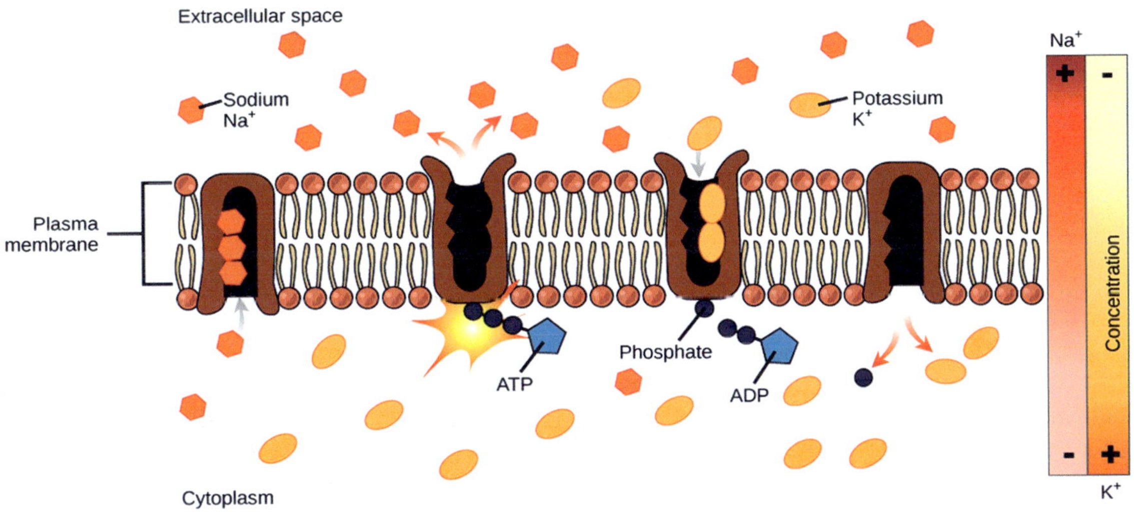

Figure 5.3. Transport mechanisms.

Source: https://open.oregonstate.education/app/uploads/sites/157/2021/02/0308_Sodium_Potassium_Pump.jpg.

5.1. Structure of Biological Membranes

Learning Outcome

1. Phospholipid bilayer composition;
2. Role of proteins in membrane structure;
3. Fluid mosaic model of membrane organization.

Biological membranes are basic structures that separate the internal and external environments of cells and support a range of transport processes that contribute to the cell's function. The membrane's basic structure consists primarily of a bilayer of lipids with dominant phospholipids that provide fluidity and structure. The stratified nature of this arrangement allows for inserting proteins that provide transport, signaling, and structural capabilities. Lipid structure, which is influenced by environmental temperature, also plays a significant role in determining membrane properties, , such as permeability and fluidity. Current evidence highlights the dynamic nature of membrane components, and research demonstrates that their organization is not rigid but rather a highly regulated and mobile process that is subject to temperature and protein-protein interactions.

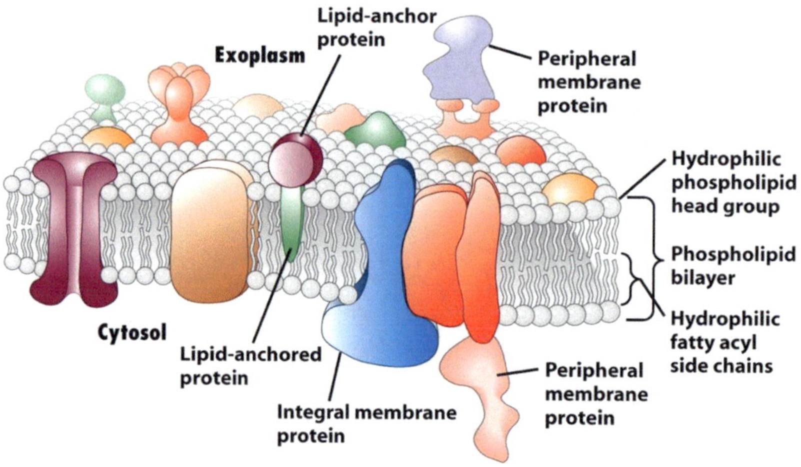

Figure 5.4. Structure of biological membranes.

Source: https://blogger.googleusercontent.com/img/b/R29vZ2xl/AVvXsEgSYeOxdSU5k9JMF5CTfv85uRl0_KsF-epAWnxSU9dK8wYwPvwDuYoiGnAaKJnf18AuckCq0-23i7kGHH2Eesb2-OzD_nK3FwkBT1kEcLmgU4ATDW_cglsvebWH9a6WDDiugf-50VPqLM1U/s1600/Picture1.jpg.

This sophistication is particularly relevant in the maintenance of homeostasis and in responding to physiological change and emphasizes the sophisticated role of membrane structure in biological function. Inherent within the function of biological membranes are membrane proteins, which may be classified either as integral or peripheral based on their relationship to the lipid bilayer. Integral proteins are inserted into the hydrophobic interior of the membrane, typically acting as channels or transporters, transferring ions and molecules across the membrane. Selective permeability is required for cellular processes , such as nutrient absorption and waste expulsion. Peripheral proteins are loosely attached to the surface of membranes and participate in signaling pathways and cell shape maintenance. Along with it, there is new insight into the dynamics of some receptors, , such as

DC-SIGN, which shows that the proteins experience nonergodic subdiffusion within the membrane, meaning a complex interaction between molecular mobility and structure in facilitating transport (Lapeyre et al., 2015). This dynamic enhances our insight into cellular function and breaks from previous assumptions of membrane composition as static. Evolutionary adaptation of biological membranes is also crucial, as seen in the role that cholesterol plays in regulating membrane fluidity and function. Cholesterol, a necessary lipid, contributes to both membrane structural stability as well as regulation of permeability to gases , such as oxygen. The rise in atmospheric oxygen concentration necessitated the development of complex membrane structures and signaling pathways to regulate oxygen efficiently (Zuniga-Hertz & Patel, 2019). Caveolae, cholesterol-enriched membrane microdomains, are centers for signaling pathways and play critical roles in the regulation of oxygen diffusion. This association is an illustration of how membrane structure adapts to the needs of more sophisticated organisms to initiate proper physiological responses to stimuli in the environment. Accordingly, research into these adaptations sheds light on the complex relationship between membrane structure and function and provides useful insights into the physiological processes of cellular transport and organism survival (Figure 5.4).

5.1.1. Phospholipid Bilayer Composition

The organization of the phospholipid bilayer is responsible for cellular membrane function, modulating permeability, fluidity, and protein interactions. Phospholipids, possessing hydrophilic heads and hydrophobic tails, assemble into a bilayer, which is impermeable to the flow of ions and polar molecules. Compartmentalization of cellular processes is made feasible by such a structural organization, thereby maintaining internal and external environments separate. Furthermore, inclusion of other lipids, i.e., cholesterol, can affect bilayer properties through stabilization of the membrane structure and affecting its fluidity. Cholesterol, in particular, is responsible for the creation of microdomains within the membrane that contribute significantly to the cell's signaling and transport functions. Selective permeability of the phospholipid bilayer finally keeps us alive by permitting the diffusion of required nutrients and waste materials while protecting the cell. Other than structural concepts, the specific composition of phospholipids within the bilayer significantly influences membrane dynamics and functionality. For instance, cardiolipin, a phospholipid that is specifically found in mitochondrial membranes, has been found to impact membrane curvature and organelle structure. Recent findings indicate that localized pH alterations initiate tubulation in cardiolipin-rich membranes, which demonstrates the degree to which subtle variations in environmental conditions can initiate radical structural changes. These are generalizable to cardiolipin; phosphatidylglycerol is another phospholipid with similar behavior in acidified conditions. The capacity of these lipids to sense environmental signals is the membrane plasticity and the demonstration of the complex relationship between lipid composition and biology. This plasticity is a requirement for cellular respiration and the regulation of metabolic processes across various physiological conditions. The interaction of various lipid constituents of the bilayer also has evolutionary significance for complex cellular systems. As organisms evolved with the advancement of their respiratory systems, adaptations in membrane lipid composition also made oxygen management and transport more efficient. For example, cholesterol-enriched membrane regions called caveolae function as platforms for cell signaling and participate in the regulation of oxygen diffusion. These cholesterol-laden structures are related to the selective forces that have been placed by increasing atmospheric oxygen tensions, where membranes needed to become central sites for the modulation of physiological response to hypoxia. Through breaking the code of the evolutionary significance of lipids, researchers can learn about cellular adaptation to environments in terms of involved phospholipid-complex interactions and compositions. This appreciation also serves to underscore the fundamental significance of the phospholipid bilayer in the general scheme of membranes and transport, since it is the basis for many physiological functions required for life (Figure 5.5).

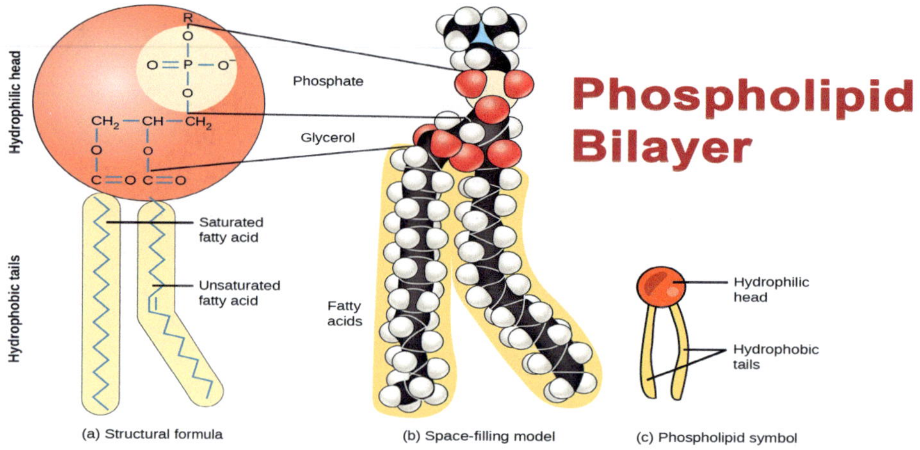

Figure 5.5. Phospholipid bilayer composition.

Source: https://microbenotes.com/wp-content/uploads/2022/09/Phospholipid-Bilayer.jpg.

5.1.2. Role of Proteins in Membrane Structure

Proteins have a function in membrane structure as essential scaffolds for containing and demarcating cell boundaries. IMPs, in particular, span the lipid bilayer, generating a framework of structure that maintains membrane integrity. Proteins possess discrete domains that engage with the hydrophobic center of the lipid bilayer with favorable interactions to generate stability but allow necessary fluidity. Here, transport proteins play a critical role in regulating the movement of ions, metabolites, and other substrates across the membrane. The selective permeability of membranes is primarily because of these proteins, emphasizing their importance in securing homeostasis within cells and biochemical balance. By facilitating selective transport, IMPs are the focal point of a variety of physiological processes, further underlining their architectural and functional roles in membrane biology. Their dynamic character is also highlighted by their interactions with the biomolecules in contact with them, which influence structural stability and activity. Membrane proteins are not static; their correlated molecular motion is essential to their functioning. Previous studies have indicated that the hydration shell surrounding membrane proteins is also crucial in controlling such dynamics, ranging from transport efficiency to the elasticity of the membrane itself. Novel technologies, , such as inelastic neutron scattering, have been central to interrogating such molecular mechanisms and providing insights on how protein-protein interactions and lipid content affect functional properties. Understanding these dynamics is fundamental because it indicates that structure in the membrane is not merely a passive barrier but an active participant that can respond to many cellular requirements and extrinsic stimuli (Caillé et al., 2008). In addition to channeling and transporting, membrane proteins partake in complex signaling pathways that establish cellular responses. Their location in the lipid bilayer allows them to structure into crucial signaling complexes to effectively communicate with the extracellular environment and intracellular machinery. For example, proteins , such as receptors and enzymes guarantee the delivery of signals that initiate critical cellular processes, , including growth and metabolism. Membrane transport proteins also play a vital role in defense mechanisms against antibiotics in microbial membranes, where they enable selective exclusion of toxins to ensure survival (Santha et al., 2016). These multiple functions underscore the importance of proteins, aside from their structural contribution in membranes, in coordinating the myriad cellular processes essential to life and illustrate their evolutionary significance in membrane dynamics and transport functions (Figure 5.6).

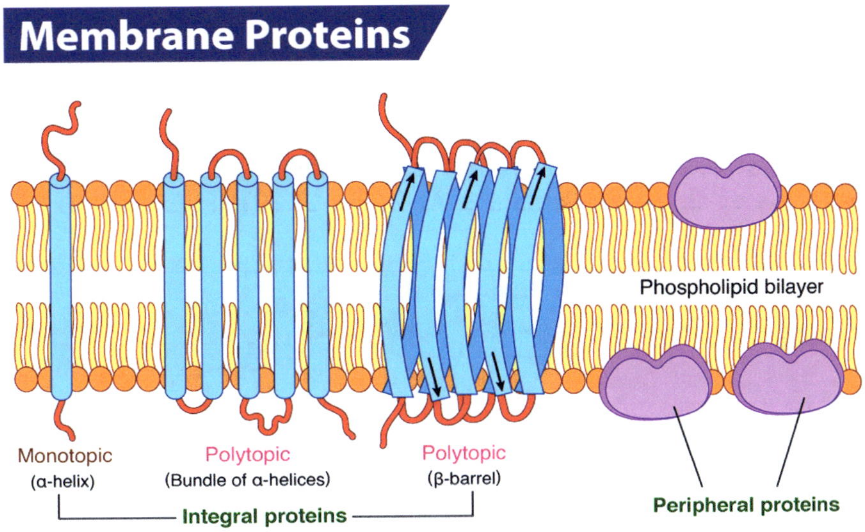

Figure 5.6. Role of proteins in membrane structure.

Source: https://www.sciencefacts.net/wp-content/uploads/2021/12/Membrane-Proteins.jpg.

5.1.3. Fluid Mosaic Model of Membrane Organization

The fluid mosaic model of membrane organization provides a complete conceptualization of biological membrane structure and functioning. This model predicts that biological membranes are dynamic systems in which the different lipid and protein components are distributed laterally in a fluid matrix, similar to a mosaic. This organization provides some flexibility that permits proteins to float in or to move across the bilayer, a requirement in membrane functions involved in transport, signal transduction, and cell recognition. The model also highlights the fluidity of membranes, thus explaining how this property facilitates cellular functions, , such as fast translocation of molecules and the existence of lipid rafts, which are microdomains enriched with specific lipids and proteins. This fluidity is important in ensuring cellular integrity in enabling interactions that ensure appropriate physiological functioning. The fluid mosaic model has implications that are beyond the simple structure of the membrane into the realms of complex cellular interactions and processes. As an example, asymmetrical lipid distribution, i.e., enrichment of certain classes of lipids in either the inner or outer leaflet of the membrane, is important in cellular signaling and communication. , such asymmetry is part of the cell response to external stimuli and the coordination of activities inside the cell. Moreover, cholesterol presence in the membrane makes it fluid and stable, which also promotes several processes, , such as membrane fusion or vesicle transport. This dynamic of the membrane components enables the cells to recruit proteins involved in the signaling pathways, enabling the cells to respond to the internal and external changes effectively, thus maintaining homeostasis. The importance of lipid-protein interactions has recently been pointed out as a key factor in the functioning of the membrane in a given pathological condition, i.e., neurodegenerative diseases. In particular, changes in lipid composition, especially those of cholesterol and sphingolipids, have been connected to the pathological aspects of such diseases as Alzheimer's. The aggregation of amyloid peptides which influence the integrity and functioning of the membranes, can be influenced by aberrations in these lipid milieus. The correlation between the characteristics of membranes and disease conditions underlines the topicality of the fluid mosaic model in terms of its ability to determine how membrane dynamics affect the wellbeing of cells, thereby providing possible points of therapeutic interventions. By elucidating the interactions between lipids and membrane proteins, researchers can better comprehend the mechanisms underlying neurodegeneration, leading to innovative treatment strategies aimed at correcting these lipid imbalances and restoring proper cellular function (Bellen et al., 2005; Fabiani & Antollini, 2019) (Figure 5.7).

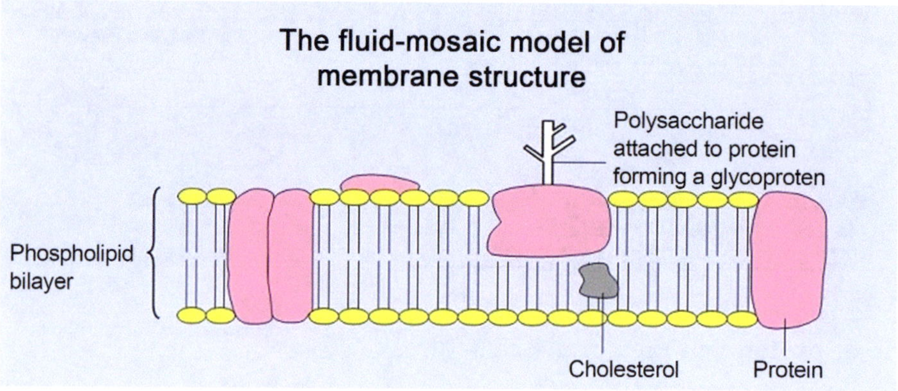

Figure 5.7. Fluid mosaic model of membrane organization.

Source: https://www.totalassignmenthelp.com/uploads/fluid-mosaic-model.webp.

Test Your Knowledge

1. What is the phospholipid bilayer composition?
2. What are the roles of proteins in membrane structure?

5.2. Types of Membrane Transport

Learning Outcome

1. Passive transport mechanisms;
2. Active transport mechanisms;
3. Bulk transport processes.

The transport of substances across cellular membranes is at the core of sustaining homeostasis and supporting all cell processes. In general, membrane transport could be classified into two types: passive and active transport. Passive transport relies on spontaneous diffusion of molecules along their concentration gradient, without requiring any input of energy. Illustrations include simple diffusion, by which small nonpolar molecules can directly pass through the lipid bilayer, and facilitated diffusion, which uses specialized membrane proteins to help the transport of larger or polar molecules, like glucose, through the membrane. The process demonstrates the fundamental principle that cells will attempt to equal internal and external concentrations, leading to dynamic exchange of materials for the support of life.

The efficiency of these processes is at times enhanced by structural alterations within the membrane, highlighting the subtle relationship between membrane structure and transport processes. Active transport, however, is an energy-requiring process that moves substances opposite to their concentration gradient and hence requires cellular energy, often in the form of ATP. This process is necessary for the inward transport of required nutrients and the outward export of waste products, allowing cells to control their internal environment in a state of disequilibrium relative to the external world. One such example is the sodium-potassium pump's contribution to neuronal transmission by actively pumping sodium ions out of and potassium ions into the cell, establishing electrochemical gradients essential for the transmission of nerve impulses. Additionally, several transporters and pumps, , such as those associated with the endocytic and exocytic pathways, are necessary for the bulk transport of large molecules. The incorporation and regulation of these active mechanisms reveal a complex framework that is crucial for cellular function and overall physiological balance. The research has also disentangled the role of membrane structure in facilitating transport processes, particularly in the interaction between membranes and the cytoskeleton.

The coordination of membrane dynamics with the microtubule network supports efficient trafficking and organization of organelles in the cell. Cooperation between them is crucial for sorting and delivering some cargoes between different membrane compartments, an essential aspect of cellular organization and signaling. Specialist membrane domains, , such as cholesterol-rich caveolae, have been identified as key sites for sorting and integration of signals, which affect transport selectivity and efficiency. Membrane evolutionary changes, which reflect environmental requirement changes, such as, such as the increase in atmospheric oxygen levels—also reveal how transport mechanisms evolved to attain cellular efficiency and sustain complex life forms. Thus, understanding the form of membrane transport not only accounts for cellular function but is also amenable to explanation in broader biological and evolutionary processes, like the role that cholesterol plays in oxygen management in advanced respiratory systems (Zuniga-Hertz & Patel, 2019) and the correlation of membrane motors for enforcing transport (Stephens et al., 2012) (Figure 5.8).

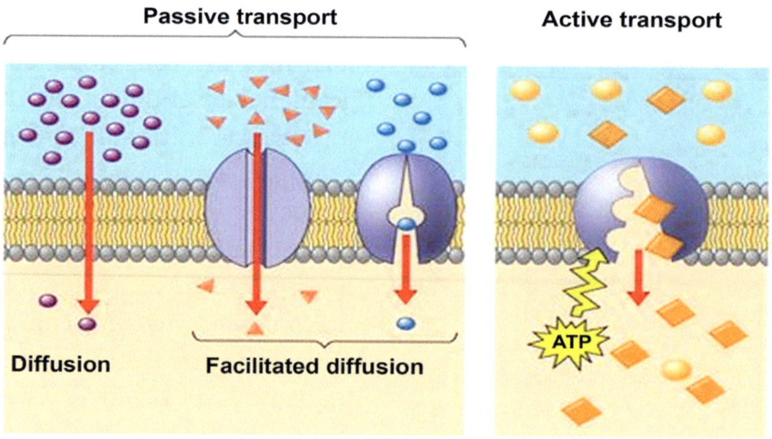

Figure 5.8. Types of membrane transport.

Source: https://www.mrgscience.com/uploads/2/0/7/9/20796234/transport-types_orig.jpg.

5.2.1. Passive Transport Mechanisms

The concept of passive transport mechanisms forms the core of studying how things are transported across biological membranes without using the direct expenditure of energy. The transport is primarily conducted in the direction of the concentration gradient, enabling molecules to pass from areas of increased to areas of decreased concentration. Simple diffusion, facilitated diffusion, and osmosis are the modes of passive transport, each with its unique mechanism and conditions. Free diffusion allows small nonpolar molecules , such as oxygen and CO_2 to pass through the lipid bilayer directly, while facilitated diffusion uses membrane proteins to assist in transferring large or polar molecules, , such as glucose and ions, through the membrane more effectively. This fundamental disparity calls for biological systems to regulate solute levels as part of homeostasis to ensure necessary nutrients and ions are available to support cellular activity. Although passive transport tends to occur with minimal interference, various factors might influence efficiency and directionality. Temperature, for instance, has a direct effect on molecular movement; higher temperatures tend to increase kinetic energy and thereby enhance diffusion across membranes. Furthermore, the membrane itself is permeable to a great extent; lipid solubility, for example, determines the ease with which substances traverse the bilayer. Highly permeable membranes exhibit extreme variations and can facilitate the quick transport of necessary solutes. Furthermore, various external factors, for example, electromagnetic fields, can change the transport processes. For instance, experiments have shown that ion transport and nutrient uptake can be significantly influenced by exposure to particular electromagnetic frequencies, which demonstrates that environmental factors can shape effective mechanisms of passive transport (Stange et al., 1995). In biochemistry and pharmacology, knowledge of how passive transport works is critical when developing effective drug-delivery systems. Compounds categorized under biopharmaceutics classification system (BCS), Class III, are highly soluble but poorly permeable across biological membranes and thus suffer from bioavailability. Most efforts at drug permeation enhancement involve time-old strategies , such as the application of prodrugs or peptides as permeation enhancers. More contemporary strategies involving methodologies of nanosizing and hybridization are presented to overcome the barriers posed by the limitation of passive transport (Dave et al., 2016). By optimizing these mechanisms, scientists seek to make therapy more successful and pharmaceuticals more effective. Thus, research into passive transport remains a core area of research with far-reaching importance for biological systems and also for therapeutic treatments (Figure 5.9).

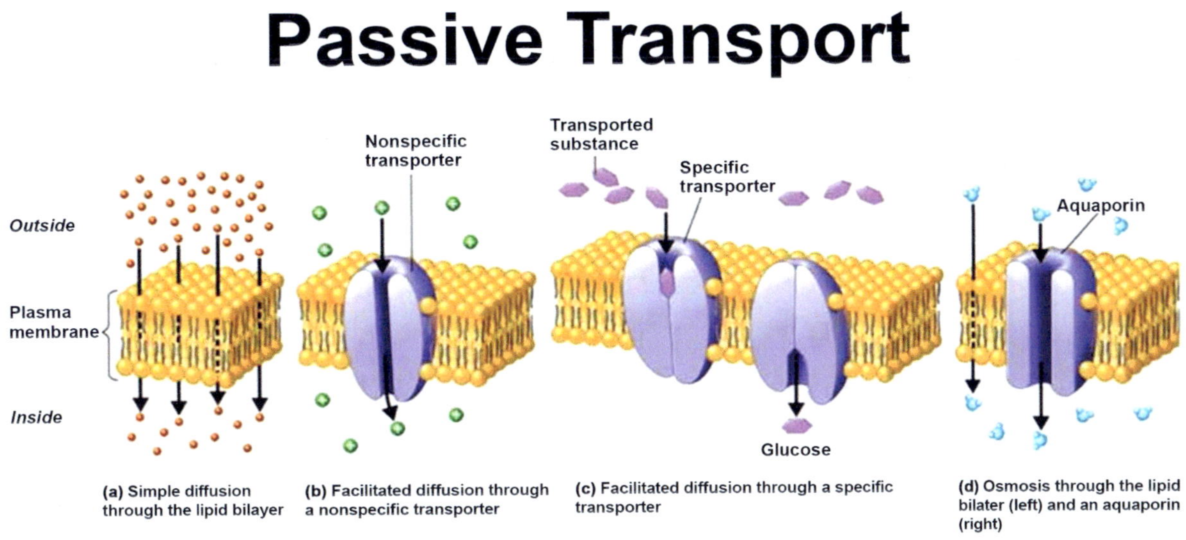

Figure 5.9. Passive transport mechanisms.

Source: https://microbenotes.com/wp-content/uploads/2019/11/Passive-Transport.jpg.

5.2.2. Active Transport Mechanisms

Active transport mechanisms are a component of cellular physiology, enabling ions and molecules to be transported against their concentration gradients. Unlike passive transport, a diffusion-dependent process, active transport relies on energy, typically derived from the hydrolysis of ATP or electrochemical gradients. Different transport proteins facilitate this energy-requiring process, which may be classified as primary and secondary transporters. Primary active transport involves direct utilization of ATP, , such as the Na^+/K^+ ATPase, to maintain the necessary ion gradients that allow cellular homeostasis. Secondary active transport utilizes the energy provided by primary transporters to transport other molecules, , such as glucose co-transport with sodium ions. Understanding these processes is relevant to knowing how cells maintain their internal environments (homeostasis) and interact with the environment. The intricacy of active transport is illustrated in the function of G-protein-coupled receptors (GPCRs), the axis of cellular signaling cascades. Recent research has identified that GPCRs , such as the β1-adrenergic receptor are capable of producing signals both from the plasma membrane and intracellular locations like the Golgi apparatus. This process of signaling nicely reflects the intricacy of membrane dynamics during active transport as it is receptor-specific and encompasses the transport of ligands through facilitated mechanisms like the organic cation transporter 3 (OCT3) in the interest of a speedy response. Such findings reflect the functional selectivity achieved by ideas of location bias in GPCR activity, with novel therapeutic intervention possibilities for these fundamental receptors (Conti et al., 2017). In addition, active transport mechanisms significantly influence metabolic processes, especially in pathologic states , such as hyperglycemia. During such states, disruption of erythrocyte membrane integrity can alter the oxygen binding capacity of hemoglobin, a factor essential to maintain cellular respiration. During conditions of blood glucose elevation, there are several biochemical changes noted that are observed, , such as glycosylation of membrane proteins, which influences the efficiency of active transport functions. The lipid composition changes, i.e., decreased phospholipid components and higher diacylglycerol levels, have the potential to destabilize the erythrocyte membrane, which may lead to potential hemolysis and impaired oxygenation. These observations emphasize the significance of active transport pathways, membrane migration, and overall cellular health, which necessitates further investigation in therapeutic uses (Bourdon & Eveillard, 2017) (Figure 5.10).

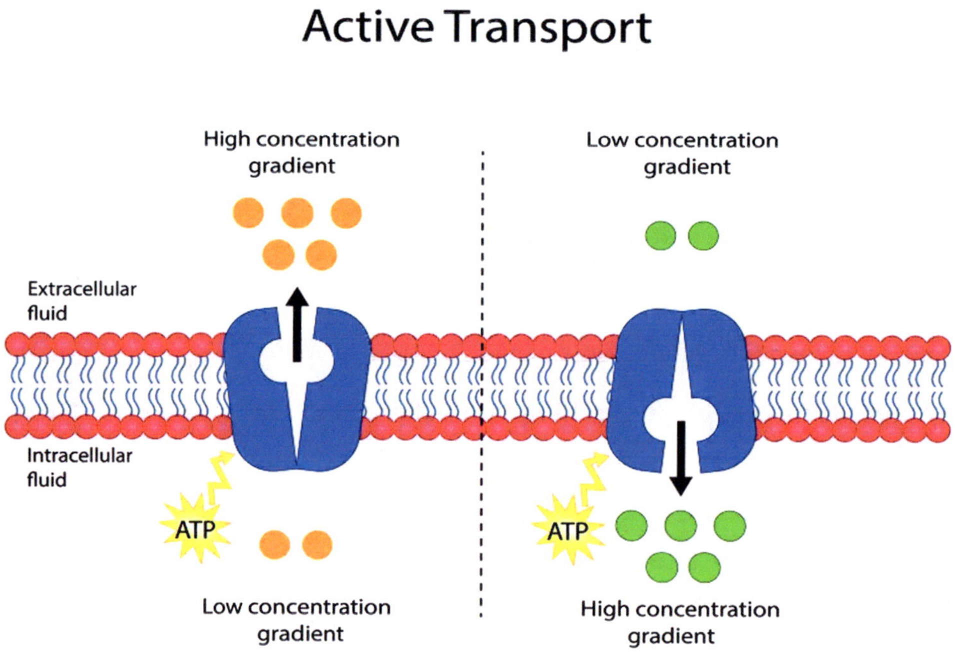

Figure 5.10. Active transport membrane.

Source: https://thumbs.dreamstime.com/b/active-transport-design-289162461.jpg.

5.2.3. Bulk Transport Processes

The study of the mechanisms of bulk transport is essential to the explanation of the way cells can sustain homeostasis and communicate. They include the transporting of bulk amounts of molecules across membranes, primarily utilizing endocytosis and exocytosis. Endocytosis is the process by which cells take up substances into the cell by enwrapping them in a membrane-bound vesicle and then transports macromolecules that are not soluble in the lipid bilayer directly. The process plays an important role in nutrient absorption and receptor-mediated signaling, which are the basis of several cellular processes. Through such mechanisms, cells can effectively regulate their internal environment as well as adaptively respond to external stimuli, making bulk transport largely significant in preserving cellular integrity and functionality. Exocytosis enables cells to secrete substances into the extracellular space. This is the most important process when it comes to the secretion of hormones, neurotransmitters as well as other proteins that are required in the communication between different cells. Exocytosis, like endocytosis, deals with membrane-bound vesicles that merge with the plasma membrane and discharge their contents out of the cell. This interaction is not only crucial in cellular communication but also in the maintenance of the ECM as well as the regulation of local environments. As suggested in recent research, the coordination of these transport processes depends heavily on the spatial arrangement of proteins and their interactions, underscoring the complex relationship between membrane transport and cellular signaling pathways, which can result in fluctuations and gradients critical for cellular function (Kholodenko et al., 2002). More recent work has revealed further details of the complications of membrane transport, especially in the part played by GTPases , such as ARF1 in the generation of tubular carriers in transport reactions. These are long, slender carriers that do not depend on conventional coat proteins but rather depend on certain proteins to mediate transport functions that are both anterograde and retrograde. By utilizing techniques , such as CRISPR/Cas9 and super-resolution nanoscopy, researchers have established that the GTP-bound form of ARF1 is essential for the stability and formation of these carriers (Baddeley et al., 2017). This finding is an illustration of the role of bulk transport processes in the interaction at the Golgi apparatus, which once again emphasizes the importance of transport mechanisms in the cell in the correct distribution of cell components and the preservation of the functional structure of the cell (Figure 5.11).

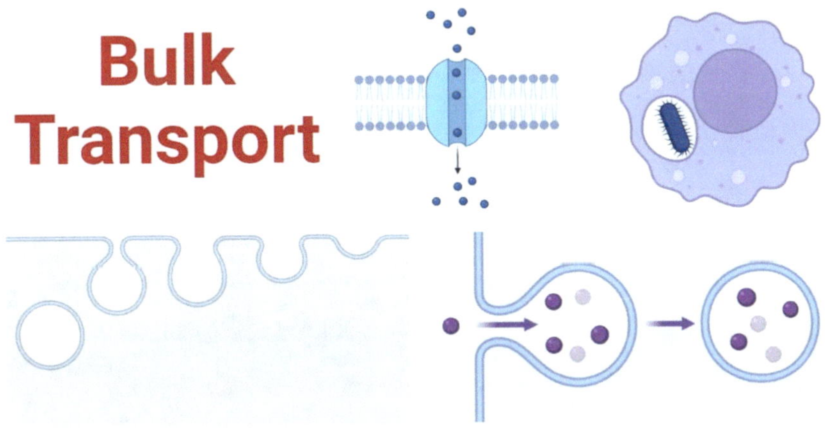

Figure 5.11. Bulk transport processes.

Source: https://microbenotes.com/wp-content/uploads/2023/05/Bulk-Transport.jpg.

Test Your Knowledge

1. What is the passive transport mechanism?
2. What is the active transport mechanism?

5.3. Passive Transport

Learning Outcome

1. Diffusion and its significance;
2. Facilitated diffusion through channel proteins;
3. Osmosis and water movement.

Passive transport is a simple process by which substances are transported across biological membranes without the expenditure of metabolic energy. The process relies on the internal kinetic energy of the molecules to drive them along their concentration gradient from an area of high concentration to low concentration. Examples of passive transport include simple diffusion, facilitated diffusion, and osmosis. During simple diffusion, small and nonpolar molecules like CO_2 and oxygen readily pass through cell membranes. Facilitated diffusion, on the other hand, includes transport proteins that assist the passage of polar or larger molecules, which provides the demonstration of the cell's regulation of transport through specialized channels. The efficacy of passive transport is particularly important in cellular homeostasis because it allows for the selective exchange of material pertinent to cellular function. In the case of cellular environments, passive transport phenomena correct for the addition of differing classes of molecules and their dynamic interactions with membranes. For instance, while small hydrophobic molecules can passively diffuse into lipid bilayers with great ease, charged and polar molecules usually rely on specialized carrier proteins to facilitate their entry and exit. The latter is particularly well-illustrated in ion and other hydrophilic material transport. Some recent research has also made clear how certain endogenous ligands utilize available transport pathways to reach intracellular compartments.

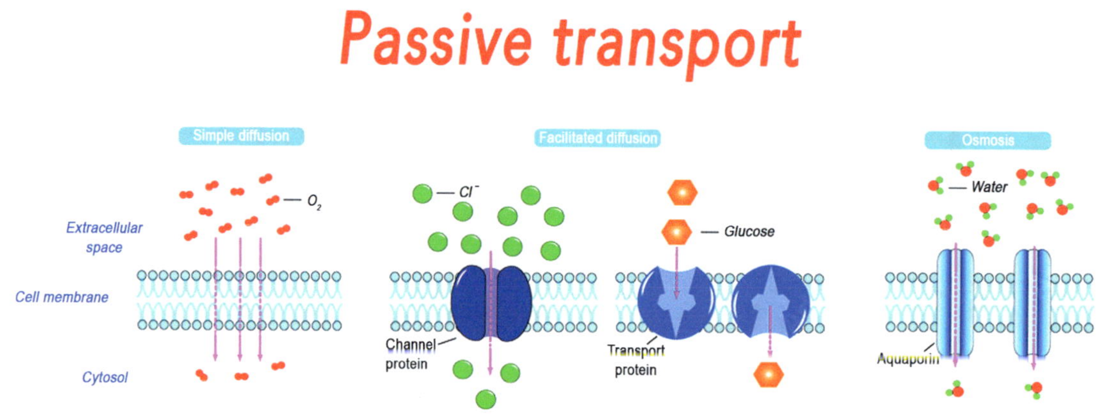

Figure 5.12. Passive transport.

Source: https://as1.ftcdn.net/jpg/04/35/33/20/1000_F_435332019_pR8D2cWIaB6FmJvEG4LzU5C7mf1lSe78.jpg.

For example, the β 1-adrenergic receptor is brought into a reservoir of Golgi apparatus-resident receptors by facilitated transport, requiring OCT3 for hydrophilic ligands but allowing hydrophobic drugs to be passively diffused (Conti et al., 2017). This opens up the subtle facades of passive transport and how it is engaged in cellular signaling. Furthermore, the implications of passive transport extend beyond the simple movement of molecules; they are crucial to the maintenance of diverse cell structure and function. By allowing for accurate sorting of lipids and proteins into specialized organelles, passive transport allows for essential cell functions localized in particular areas of the membrane. In photoreceptor cells, for example, dynamic differentiation of protein populations relies on a balance of protein-membrane attraction and obstructed diffusion so that cellular components are transported to their destinations in an optimal way (Aidan et al., 2019). This precision of transport not only emphasizes the importance of passive mechanisms in cellular form but also emphasizes the potential for such phenomena to be harnessed as targets for therapeutic innovation in disorders where transport is disrupted. Appreciation of the intricacies of passive transport thereby is at the very core of the advancement of knowledge within cell biology and within the area of therapeutic innovation (Figure 5.12).

5.3.1. Diffusion and Its Significance

Diffusion is a fundamental process that regulates a vast array of biological functions, particularly in relation to cell membranes and transport mechanisms. Diffusion can be simply defined as the free movement of molecules from areas of higher concentration to areas of lower concentration, driven by particles' intrinsic kinetic energy.

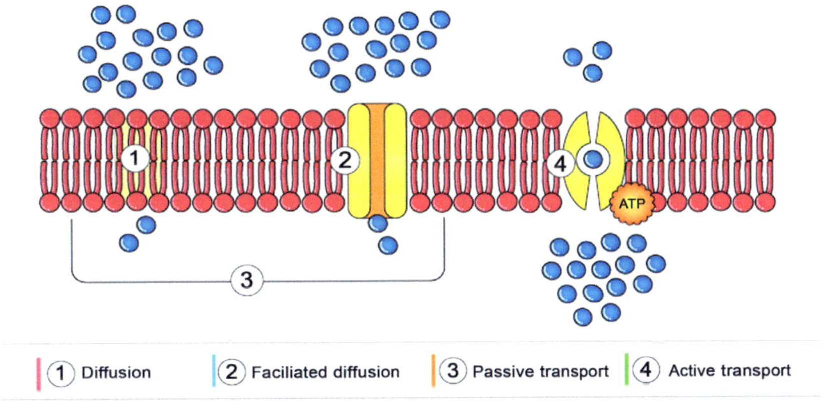

Figure 5.13. Diffusion.

Source: https://cdn1.byjus.com/wp-content/uploads/2018/11/diffusion-means-of-transport.png.

This takes place in a critical process that ensures cells maintain homeostasis, allowing the passage of required substances like nutrients, gases, and wastes. Efficiency during diffusion depends on several factors influencing membrane permeability, molecular weight, and temperature. Because cells are situated in semi-permeable membranes, understanding such dynamics is at the heart of grasping not only nutrient uptake, but also cell-to-environmental communication and response (Finkelstein et al., 2017). Diffusion is not mere substance transport; rather, it is at the heart of numerous physiological processes. For instance, oxygen diffusion from alveoli to the blood is essential for gas exchange to be effective, for aerobic tissue respiration to be possible. Disruption of this process can lead to respiratory impairment and cellular dysfunction. Diffusion is also critical in signaling pathways. For example, the crossing of signaling molecules across membranes can initiate the response that controls cellular behavior, growth, and adaptation. The mechanisms of diffusion account for how important it is for the space and time organization of various cellular functions and therefore why cellular biology is most significant in membrane and transport system study (Finkelstein et al., 2017). Additionally, in disease scenarios, the application of diffusion is even more consequential. Consider healing of a wound; collagen fiber alignment and organization can influence the degree to which nutrients and signal molecules diffuse to the wound. Models that examine diffusion in relation to deposition here tell us about healing kinetics and processes in tissue repair. It is possible with multiple manipulations of fibroblast migration and growth factor availability to predict scarring and outcome of tissue regeneration. Thus, diffusion not only is the determining process for cellular actions but also influences therapeutic modalities directed toward enhancing recovery and delineating all manner of disease states. Continued exploration of the field of diffusion continues to shed new light on its pivotal role in health and disease, further ensuring its relevance in the field of membranes and transport (Figure 5.13).

5.3.2. Facilitated Diffusion Through Channel Proteins

The physiological mechanisms of facilitated diffusion through channel proteins are a focal aspect of cellular transport in lipid membranes. As molecules , such as ions and polar substances do not typically diffuse straight across lipid bilayers due to their hydrophobic nature, channel proteins provide a selective pathway for their transport. These IMPs facilitate selective transport of single substances, thereby providing homeostasis and enabling cells to respond optimally to varying environmental stimuli. The formation of these aqueous pores facilitates the rapid transportation of ions, which is essential in electrical signaling within neurons and muscle contraction. Thus, the occurrence and regulation of channel proteins are critical to the preservation of cellular environments favoring life and sensitive to external change, proving the requirement of channel proteins for membrane dynamics. Channel proteins may be classified based on their mechanism of gating as being voltage-gated, ligand-gated, or mechanically gated. Voltage-gated channels are opened by changes in membrane potential, allowing ions , such as Na^+ and K^+ to pass across membranes during action potentials. Ligand-gated channels, on the other hand, are opened by binding specific molecules, functionally linking extracellular stimuli to intracellular responses. Mechanically gated channels, however, are opened by physical stimulation, , including stretch or pressure. This classification highlights the flexibility and diversity of channel proteins in facilitated diffusion. The efficiency of these channels ensures rapid, controlled transport, emphasizing the intricacy of biological membranes and their function in mediating basic processes. Channel proteins are thus facilitators of transport, as well as integral elements of cellular regulation and signaling. Besides, the integration of cholesterol and other components of the membrane can elucidate the functioning of channel proteins in more complex systems. Cholesterol, a necessary cellular component of the membrane, dictates fluidity and function of such proteins and optimizes their role in the facilitation of ion transport. As discussed as well in membrane adaptation research, structural development of caveolae represents the manner by which local lipid milieus may affect the channel proteins' functioning. Such structural interaction is crucial as caveolae are associated with cellular signaling pathways capable of regulating processes like oxygen diffusion, creating a feedback loop required for maintaining physiological conditions in the face of changing conditions. The function of facilitated diffusion through channel proteins is one chapter in a grander story of membrane structure and function, where these elements cooperate synergistically to facilitate life at the cellular level and respond to specific metabolic demands, as documented in current studies on membranes and transport (Kassai et al., 2015; Zuniga-Hertz & Patel, 2019) (Figure 5.14).

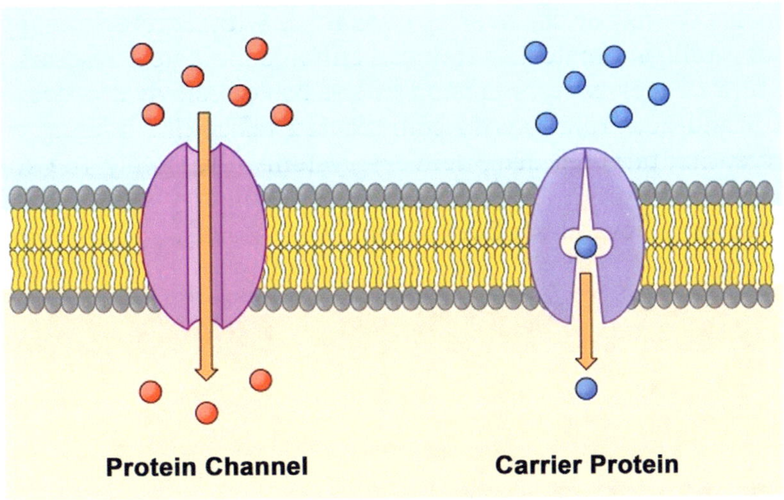

Figure 5.14. Facilitated diffusion through channel proteins.

Source: https://old-ib.bioninja.com.au/_Media/facilitated-diffusion_med.jpeg.

5.3.3. Osmosis and Water Movement

Osmosis plays an important role in the homeostasis of cells because it allows the transfer of water across semipermeable membranes. This passive transport process can be facilitated when water molecules flow or are transported from the region of low solute concentration to the region of high solute concentration, and this process will lead to the equal concentration of solutes on the two sides of the membrane. The knowledge of osmosis dynamics is not just a theoretical issue; it gives an idea about the essential biological processes and organism survival. As an example, our understanding of the water permeation through cellular membranes has undergone a significant boost by the identification of aquaporins, which are proteins acting as water channels. Despite the long history of osmotic research, gaps remain in fully comprehending the molecular underpinnings that dictate osmotic permeability and the roles that these channels play in cellular environments and responses to varying osmotic conditions (Alleva et al., 2020). The effectiveness of membranes, in regard to water movement, is one of the main aspects affected by the principles of osmosis. The recent developments in the field of membrane technology have signified the prospects of using special materials that are developed to maximize the effectiveness of water filtration and desalination. , such as an example is the nanopore membranes, which have extraordinary properties of effectively rejecting ions yet permit water to permeate at a high rate. This two-in-one property is of great importance to both biological applications and engineering applications, including water purification. These membranes, as demonstrated by molecular dynamics simulations, do not only increase water transport but also explain processes that form the basis of salt rejection. Such innovations are essential in addressing global water scarcity, wherein advanced membranes could provide sustainable solutions for accessing clean water, thereby bridging the gap between natural processes and human-engineered systems (Maiti et al., 2022). Besides, the applications of osmosis also reach many other spheres , such as biotechnology, environmental science, and medicine. With the help of the phenomena of osmosis and membrane transport, scientists are working on the new methodology to create more efficient therapeutic interventions and biotechnological tools. Water transport based on osmotic gradient is vital in the hydration of cells, absorption of nutrients and the elimination of wastes. Moreover, the overlap of osmosis with nanotechnology has paved the way to the development of intelligent materials that can emulate biological systems. Such achievements not only contribute to a deeper comprehension of the basic biological processes but also lead to the development of brand-new solutions to complicated tasks, like making crops more resilient in a dry climate or creating targeted drug delivery systems. Ultimately, the study of osmosis and water movement remains a vibrant and essential field of inquiry within the broader context of membranes and transport (Alleva et al., 2020; Maiti et al., 2022).

Test Your Knowledge

1. What is diffusion and its significance?
2. What is Osmosis and water movement?

5.4. Active Transport

Learning Outcome

1. Definition and energy requirements;
2. Examples of primary active transport;
3. Secondary active transport and its mechanisms.

Active transport is dependent upon a significant biological mechanism through which the molecules are transported through the cellular membrane against the gradient of concentration, utilizing energy that is typically supplied through the hydrolysis of ATP. It is a vital mechanism for cellular homeostasis and supplying necessary ions and nutrients in adequate concentrations to sustain cellular processes. One of the more well-characterized examples of active transport is the sodium-potassium pump (Na^+/K^+ ATPase), which actively moves sodium ions out of the cell and also brings potassium ions in. It serves as the foundation for much cellular activity, including the generation of membrane potential and cell volume regulation. By this mechanism, cells can maintain their internal environment constant, permitting the optimal conditions for biochemical reactions that support life, e.g., in hepatocytes, where cellular volume can be regulated even in the presence of inhibitors like ouabain, which blocks the Na^+/K^+ pump (Morgante et al., 2015). The dynamic of active transport is also aided by the employment of specialized transport proteins and channels for molecule movement. The transport mechanisms, , such as ABC transporters and ion channels, offer avenues for the uptake of vitamins, minerals, and other substances vital to cell function. One example of a high-level system of active transport is the process through which retinol binding protein (RBP) transports vitamin A into cells. RBP serves as a carrier molecule, exporting vitamin A in a controlled form using the STRA6 receptor on cell surfaces. It illustrates the specificity of transport mechanisms but highlights the need for active transport in the delivery of nutrients and metabolic regulation throughout the organism. As evident here, both the vitamin A transport mechanism and the sodium-potassium pump show how active transport plays a vital role in various cellular processes and physiological functions (Kassai et al., 2015). Active transport mechanisms are also not passive mechanisms along cell membranes; they show the evolutionary wisdom in homeostasis and cell adaptation. Cells developed various strategies for active transport so that they could respond to various environmental conditions. For instance, hepatocytes possess vesicular systems to control their volume by sequestration of water and ions in intracellular vesicles so that they can operate under adverse conditions. Such systems are crucial not only for the survival of individual cells but also for the physiological stability of tissues and organs as a whole. An understanding of the complex interplay among multiple active transport systems is what guides potential therapies, particularly in diseases where transport deficiencies are pathogenic. Therefore, the study of active transport is not purely rudimentary biology but also has intersections with medicine and pharmacology as we explore its application in disease treatment and human health augmentation (Figure 5.15).

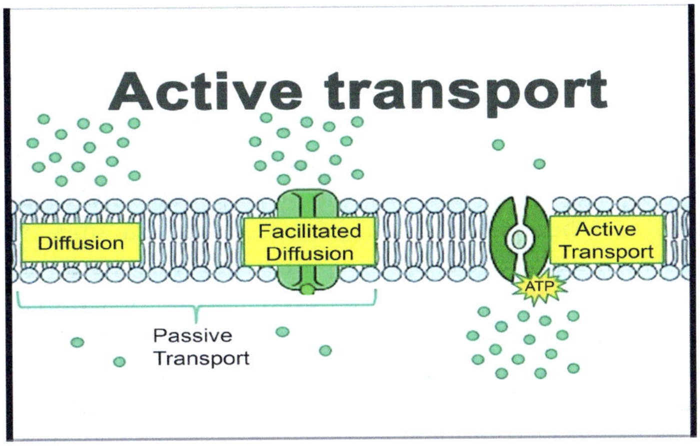

Figure 5.15. Active transport.

Source: https://i.ytimg.com/vi/EmPfjfmw_08/sddefault.jpg.

5.4.1. Definition and Energy Requirements

Understanding the definition of transport mechanisms in biological membranes involves understanding the energy. Transport across the membrane is categorized into passive and active transport. Passive transport relies on the flow of molecules along their concentration gradient, eliminating cellular expenditure of energy. Diffusion and facilitated diffusion are examples of passive transport, where material spontaneously flows without the need for ATP assistance. Conversely, active transport is against a concentration gradient and always requires ATP or other forms of energy. This is significant because it upholds the fundamental principles of cellular homeostasis and nutrient uptake and emphasizes the tightrope that cells must walk between energy cost and metabolic function. Membrane transport energy requirements become more complex when considering specific nutrient uptake, as seen in plant biology. For example, the uptake of significant ions like phosphate demonstrates active and passive processes. Studies have indicated that phosphate-deficient plants exhibit altered kinetic values for phosphate uptake with elevated energy requirements in taking up the essential nutrient in stressed conditions. Of interest are phosphate uptake mechanisms demonstrating distinctions in kinetic characteristics, , such as Km and Vmax, depending on phosphorus nutrition of plants. These interactions not only enhance the comprehension of transport mechanisms but also demonstrate the dependence on energy for effective utilization of nutrients, further directing agricultural approaches that are best able to serve plant growth under various environmental stresses (Clark et al., 1997). As cellular mechanisms developed, the modifications in the structure of membranes have contributed significantly to formulating the energy components of transport. The development of cholesterol-rich domains, , such as caveolae, shows how membrane composition can predispose transport efficiency under altering environmental stress, for example, increased atmospheric oxygen tension. The role of cholesterol as a membrane-stabilizing lipid helps with the regulation of gas diffusion, , such as oxygen, to meet cellular energy demands. Caveolae are also engaged in localized oxygen sensing and regulation of transport activities through the generation of distinctive micro-environments for signaling and adaptation. This evolutionary perspective emphasizes the continuous play of energy requirements and membrane adjustment in sustaining cellular activities in multicelled animals (Zuniga-Hertz & Patel, 2019).

5.4.2. Examples of Primary Active Transport

The concept of primary active transport is foundational in comprehending cellular function within the context of membranes and transport. It necessitates the application of ATP directly to drive the pumping of ions or molecules against their concentration gradient, thereby creating the required gradients essential for cellular homeostasis. One such familiar example is the sodium-potassium pump (Na^+/K^+ ATPase), which actively transports potassium ions inside the cell and sodium ions outside the cell. This mechanism is essential in different physiological processes, including nerve impulse transmission and muscle contraction. This process is fueled by the energy liberated during ATP hydrolysis, thus allowing cells to achieve their resting membrane potential and regulate volume. Cells would not be able to sustain their ionic balance without effective primary active transport mechanisms like the Na^+/K^+ ATPase, leading to disrupted metabolic activities and cellular processes. Besides the sodium-potassium pump, another key case of primary active transport is the calcium pump (Ca^{2+} ATPase), used in muscle contraction and the release of neurotransmitters. This pump, driven by ATP, expels calcium ions from cells or sequesters them in the sarcoplasmic reticulum and resulting in a reduced intracellular concentration of calcium. Regulation is essential to muscle physiology since elevated concentrations of calcium lead to contraction. Interestingly, the functioning of calcium pumps is especially important in neuron signaling because it controls the supply of calcium for synaptic transmission and hence influences inter-neuronal communication. The coupling of ATP hydrolysis with calcium ion transport is only one aspect in which the principal active transport is integrated with cellular signaling mechanisms and muscle activity and emphasizing its functionality in physiological processes and cell health in general. Primary active transport is not just analyzed in mammalian systems but occurs in other organisms, for instance, in fungi, where transport mechanisms are analogous and function to preserve cellular well-being. Fungal cells, for instance, rely on various transporters that use ATP to absorb essential nutrients like glucose and ions for growth and metabolism. In particular, much of our understanding of these processes has emerged from model organisms, , such as *Saccharomyces cerevisiae* and *Schizosaccharomyces pombe*, which have been instrumental in elucidating transport mechanisms within the context of nutrient assimilation and cell division. By exploring the mechanisms through which fungal cells regulate their internal environment through active transport, scientists unveil elementary principles of cell physiology, including the recycling of cellular material, further highlighting the importance of primary active transport in cellular growth and function (Boswell et al., 2017). Not only does this study deepen our insight into transport mechanisms, but it also broadens the range of implications for biotechnology and health sciences (Figure 5.16).

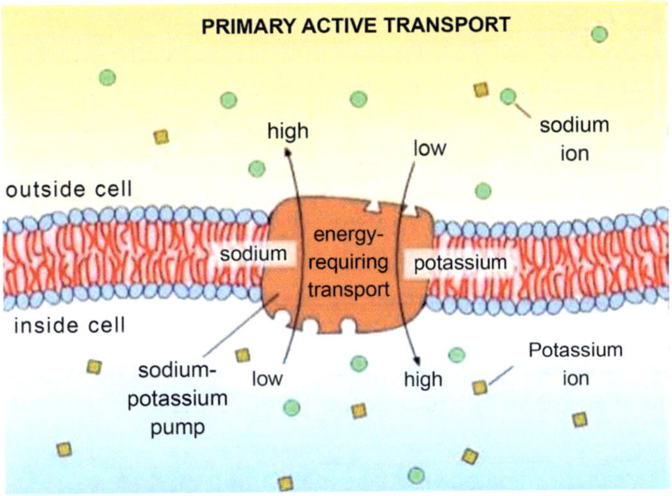

Figure 5.16. Primary active transport.

Source: https://i.ytimg.com/vi/-1evc75Z0G0/hqdefault.jpg?sqp=-oaymwEmCOADEOgC8quKqQMa8AEB-AHKBIACsAOKAgwIABABGGUgWihIMA8=&rs=AOn4CLCCph5iJeXlPg8l6FW_dtCXR53fBg.

5.4.3. Secondary Active Transport and Its Mechanisms

The knowledge of secondary active transport is of importance in the explanation of how cells can transport substances against their concentration gradient, with the use of the energy stored in the electrochemical gradients created by the primary active transport. In contrast to primary active transport, where ATP is directly used to move molecules, in secondary active transport, the energy in the form of gradients of primary transport is used. This mechanism typically involves symporters and antiporters. Symporters move two kinds of molecules in the same direction, whereas antiporters move one type of molecule in exchange for another in opposite directions. Such a subtle dynamics indicates the significance of membrane proteins as cellular homeostasis and nutrient uptake controllers. As an example, sodium-glucose symporter utilizes the sodium gradient generated by Na^{+}/K^{+} ATPase to actively glucose absorption in the intestines, which illustrates the importance of secondary active transport in nutrient absorption processes that support the cellular functionality and energy metabolism. A closer look at the biochemical processes involved in the secondary active transport shows the complexity of interaction between transporters and cellular signaling pathways. The cellular membrane and molecular interactions play a central role in controlling the transport systems. This movement of ions in and out generates an electrochemical gradient, which is central to many physiological processes, including neural transmission and muscle contraction. For example, in Vicia faba (broad bean) root-tip cells, experiments have shown that varying electromagnetic fields can influence the uptake of amino acids, showcasing the impact of external signals on transport mechanisms. The appearance of pinocytotic vesicles on the exposed cells demonstrates a flexibility of the transport mechanisms, which enables the absorption of nutrients. These results not only point to the dynamic nature of the secondary active transport but also to its possible vulnerability to the external environmental factors, which is why the continued research of such processes in the wider scope of the cell biology seems to be necessary. Besides the absorption of nutrients, secondary active transport plays an essential role in the balance of ions and signaling cascades in other cell types. The capability of cells to control the ion concentrations on the different sides of membranes is of utmost importance to such processes as osmoregulation and signal transduction. As an example, the sodium-calcium exchanger uses the sodium gradient to extracellularly expel calcium ions, which is important in the functioning of the heart and nerves. On the same note, alterations caused by, e.g., electromagnetic fields can influence fundamental cellular functions related to secondary transport processes. Without effective secondary transport mechanisms, cells could have difficulty in creating suitable internal environments and this can result in dysfunction or disease. Therefore, the detailed knowledge of the secondary active transport mechanisms is not only significant to cellular physiology but also provides insight into the possible therapeutic interventions of the metabolic disorders and diseases of the transport pathways, which makes its significance in health and disease conditions alike (Figure 5.17).

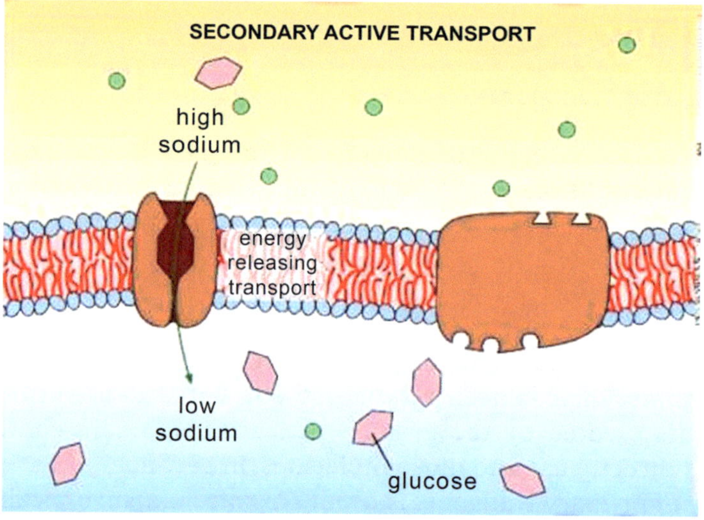

Figure 5.17. Secondary active transport and its mechanisms.

Source: https://i.ytimg.com/vi/nYC3_3hb54Q/hqdefault.jpg.

Test Your Knowledge

1. What are the examples of primary active transport?
2. What is secondary active transport and its mechanisms?

5.5. Bulk Transport

Learning Outcome

1. Endocytosis and its types;
2. Exocytosis and cellular secretion;
3. Role of vesicles in bulk transport.

It is significant to know the bulk transport mechanisms within the aspect of membrane technology, where solute passage is by a number of transport phenomena. The mechanisms can be grouped based on operation, e.g., sieving action and affinity or charge variation-based diffusion. Passage through a membrane is normally via particular processes, like Knudsen diffusion in narrow pores or viscous flow in wide channels. Specifically, each mechanism plays a significant role in rendering the transport process as a whole effective and efficient. Moreover, the properties of the membrane, , such as crystallinity and surface charge, have a great influence on diffusion rate and resulting flux rates across these barriers, hence influencing separation process performance in industrial applications (Smolders et al., 1991). For bulk transport, concentration polarization effects are of particular interest. The effect arises if the solute concentration in the vicinity of the membrane surface differs from the solute concentration in the bulk solution. Such gradients may have a profound impact on the transport rate and, consequently, the efficiency of the membrane process. The mass transfer coefficient, determining the extent of concentration polarization, is irrevocably connected with the diffusion coefficient. Therefore, boundary layer behavior and diffusion during the membrane formation process need to be comprehensively understood to maximize membrane technology. The parameters are dealt with to escape the adverse effects of concentration polarization, thus enhancing the overall performance of separation applications (Smolders et al., 1991). The study of proton exchange membranes like Nafion enhances the comprehension of bulk transport in membranes, particularly in fuel cell technologies. This polymer possesses unique structural features that arise from the demarcation of hydrophobic and hydrophilic regions. In hydration, Nafion creates water channels within which protons migrate in the form of hydronium ions. Water sorption and desorption behavior of the membrane matrix are critical in establishing transport properties, especially with varying temperatures. Observations of ice formation in thermal cycling highlight the role of temperature in membrane proton conductivity and membrane hydration. This refined understanding of water transport at low temperatures not only deepens membrane science but also sheds light on how to improve proton transport in energy conversion technology (Figure 5.18).

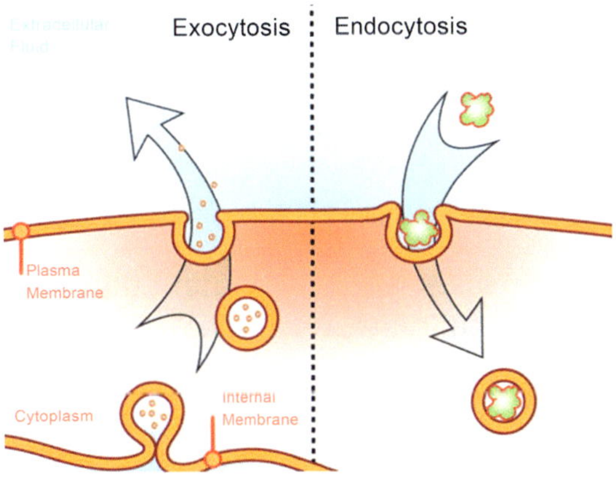

Figure 5.18. Diagram of bulk transport.

Source: https://external-content.duckduckgo.com/iu/?u=https%3A%2F%2Fo.quizlet.com%2FZ3l7Tkhxf3WM5hbKTY7QUw_b.jpg&f=1&nofb=1&ipt=711fe2a0dc375487f1c654ec0d273b15285a948f7d50a22e926ba3135c74e500.

5.5.1. Endocytosis and Its Types

Having knowledge of endocytosis is important for the understanding of the interaction between cells and the environment, particularly through the internalization of substances. Endocytosis is a process the internalization of extracellular material by surrounding it with the cell membrane, which invaginates and forms vesicles. Various types of endocytosis exist, which tend to fall under the umbrella of phagocytosis, pinocytosis, and receptor-mediated endocytosis. Phagocytosis, or the eating of cells, is how large particles, , such as pathogens or cellular debris, become engulfed in large membrane-bound vesicles called phagosomes. Phagocytosis is an important part of the immune response, enabling the removal of disease-causing substances from the body and maintenance of tissue homeostasis.

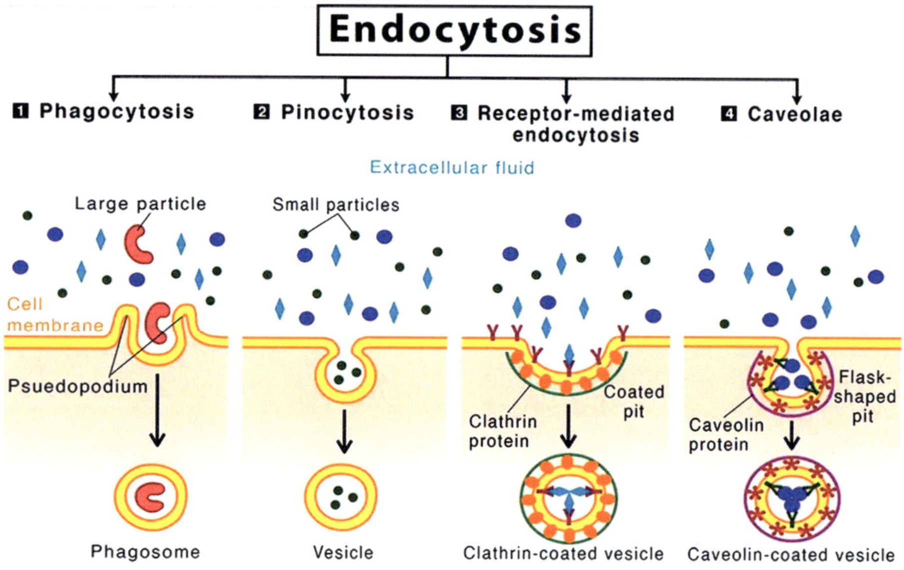

Figure 5.19. Endocytosis.

Source: https://external-content.duckduckgo.com/iu/?u=https%3A%2F%2Fwww.sciencefacts.net%2Fwp-content%2Fuploads%2F2020%2F08%2FEndocytosis-768x545.jpg&f=1&nofb=1&ipt=062ccfc559cb7806df778e31786a985a454dbe3b00072e9dfc7da8f9f6bb1ddde.

Therefore, understanding these processes emphasizes their role in disease resistance and cell health. Pinocytosis, or cell drinking, is another vital endocytic mechanism whereby cells can ingest fluid and small solutes. Initially characterized as a non-selective form of endocytosis, pinocytosis has been described to involve more sophisticated regulation than initially supposed. Unlike phagocytosis, the ingestion of large particles, pinocytosis permits the monitoring of the outside environment effectively by ingesting extracellular fluid. This mechanism is essential for nutrient uptake so cells can obtain biomolecules required of them in an ever-changing environment. Additionally, newer studies further highlight how pinocytosis is regulated by specialized cellular machinery, with important roles played by specific signaling pathways and cytoskeletal arrangements for this form of endocytosis to function. The communication between pinocytosis and nutritional supply highlights its importance in cellular function and metabolism. Receptor-mediated endocytosis is a highly selective subset of the endocytic activities involving membrane receptor-ligand binding. It not only enhances the selectivity of cell uptake but also allows for more efficient internalization of material necessary for cell function. All classes of molecules, including hormones, nutrients, and pathogens, use this pathway to gain entry into the cells. For instance, insulin and iron uptake are facilitated by receptor functions that ensure such critical biomolecules are delivered to cells with utmost efficacy. Furthermore, emerging research places in the limelight further pathways exploited by pathogens to take over this endocytic pathway, bypassing cellular countermeasures. The examination of such mechanisms reveals avenues for therapeutic delivery system improvement, particularly in cases where increased accessibility of drugs is paramount. Lastly, the diverse modalities of endocytosis highlight its significance in regular cellular function and disease (Medina-Kauwe et al., 2007; Bayat et al., 2011) (Figure 5.19).

5.5.2. Exocytosis and Cellular Secretion

Exocytosis is a cellular process that is vital in the release of substances from a cell through the fusion of vesicles with the plasma membrane. It is a vital mechanism for numerous physiological processes, including hormone, neurotransmitter, and enzyme release, which are necessary for the attainment of homeostasis and the facilitation of cell-to-cell communication. In membrane and transport contexts, the lipid bilayer organization of the plasma membranes lies at the center of exocytosis, as it regulates the dynamics of membrane merging and protects the cell integrity during the process. The involvement of specific proteins, , such as SNAREs, is crucial for specific targeting and fusion of vesicles, which indicates the complexity of exocytotic pathways in eukaryotic cells where coordination of various proteins is responsible for effective cell secretion of cell products. Regulated cell volume is finely linked with exocytosis, demonstrating how complex cellular processes are interconnected. The exocytic pathway assists in cell volume regulation by exporting ions and water from the cell and thereby diminishing the effect of osmotic variations. In particular, the hepatocytes are capable of exporting excess water and replenishing ion balances even when the Na^+/K^+-pump has been inhibited by ouabain, suggesting that there must be additional mechanisms at play in volume regulation in cells. Research has established that hepatocytes utilize a vesicular pathway to sequester and release ions and water in an efficient manner, underlining the importance of exocytosis not only in secretion but also in the maintenance of cellular homeostasis under fluctuating extracellular environments. Exocytosis is thus an inherent component of both excretory functions and the overall regulation of cellular milieus, confirming its pivotal position in cell biology (Morgante et al., 2015).

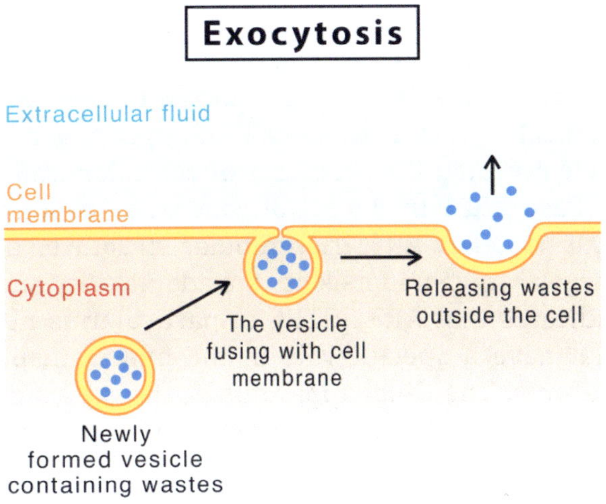

Figure 5.20. Exocytosis.

Source: https://external-content.duckduckgo.com/iu/?u=https%3A%2F%2Fwww.sciencefacts.net%2Fwp-content%2Fuploads%2F2020%2F07%2FExocytosis.jpg&f=1&nofb=1&ipt=a91bbb85d48a02e40a8096261bf8af4cb9fbcfecfee7f654ce7d4e2db424cc0c.

The functional impact of exocytosis extends to numerous cellular processes and influences both immune responses and tissue morphogenesis. Exocytosis, for instance, is essential to the secretion of signaling molecules , such as cytokines, which coordinate immune responses through the communication between immune cells. In addition, understanding the molecular mechanisms of exocytosis, particularly the roles of proteins , such as syntaxins, is essential to the unraveling of these processes. Syntaxin studies and their interaction show their necessity in the process of membrane fusion during exocytosis in most organisms, including yeast, where evidence has determined redundancy between certain syntaxins while highlighting the necessity of certain isoforms in various cellular contexts (Öyen et al., 2003). These findings are vital in further elaborating our knowledge of cellular secretion mechanisms and their applications in biomedicine, particularly in the treatment of diseases linked to exocytotic dysfunctions (Figure 5.20).

5.5.3. Role of Vesicles in Bulk Transport

Vesicles play a crucial role in bulk transport within cells, allowing proteins, lipids, and other macromolecules to be moved from one organelle to another, as well as to the plasma membrane. The small, membrane-bounded compartments are necessary to have dynamic cellular compartmentalization. They are involved in both anterograde transport, which transports materials from the ER to the Golgi apparatus, and in retrograde transport, which transports materials from the Golgi to the ER or further to lysosomes. Direction is critical for the organization and operation of cells. Besides, the specificity of vesicular transport is also regulated by specific protein coats, , such as COPI and clathrin, that aid in the cargo molecule selection and packaging so that only the correct molecules are transported to their intended destinations. Recent advances in microscopy, particularly those reliant on CRISPR/Cas9 gene-editing technologies and super-resolution microscopy, have revealed the processes of vesicular transport. Above all, research has characterized the small GTPase ARF1 in the act of generation of long, tubular carriers independent of the classical Golgi coat proteins. ARF1-rich tubular structures are capable of mediating anterograde and retrograde cargos, which underscores the adaptability and diversity of vesicular transport processes. The study indicates that ARF1[GTP], in particular, is necessary for the formation of such tubular networks that have a specific role in the bulk transport system, branching off from the classical vesicle features. The Golgi apparatus itself plays a crucial role in understanding the vesicular transport, as it serves as a central processing and sorting station for cargo imported from the ER. Despite significant advances in our knowledge of Golgi structure and function over the past century, many questions regarding its operational mechanisms remain unanswered. The history of research into the Golgi has seen ongoing debates about its roles and organization, which have contributed to the development of contemporary models of vesicular transport. Effective communication between other cell compartments and the Golgi is controlled by vesicles integrated closely into the organelles function to properly modify macromolecules and deliver them to their destination. The complexity of this process signifies the importance of vesicles in bulk transport within the overall framework of membranous interaction in the cell.

Test Your Knowledge

1. What is endocytosis and its types?
2. What is the role of vesicles in bulk transport?

5.6. Membrane Transport in Health and Disease

Learning Outcome

1. Importance of transport in nutrient absorption;
2. Role of membrane transport in drug delivery;
3. Implications of transport dysfunction in diseases.

Membrane transport is vital to the maintenance of cellular homeostasis, particularly in pathology , such as hypertension, which has emerged as a serious public health issue. Sodium (Na^+) reabsorption, a vital process controlled by the kidneys, is particularly responsive to dietary consumption of salt. Excessive intake of salt results in the development of salt-sensitive hypertension, which is linked with abnormal Na^+ handling in the nephron. Specifically, the thick ascending limb of the loop of Henle (TAL) has been identified as a crucial location where Na^+ reabsorption is impaired under conditions of high salt. Dysfunction in this segment can exacerbate sodium retention, leading to elevated blood pressure and associated complications, all of which act together to cause millions of deaths globally each year, highlighting the necessity for specific therapeutic interventions directed at this nephron segment (Dominiczak et al., 2017). In neurology, membrane transport systems are also accountable for neurodegenerative illness like neurodegeneration with brain iron accumulation (NBIA). NBIA is a group of conditions in which there are accumulations of iron within the brain that lead to hindrances in proper cellular function and result in a range of motor dysfunction. Iron overload is directly associated with the abnormal functioning of transport proteins involved in iron homeostasis, , such as ceruloplasmin and ferritin light chain. Mutations in genes encoding such transporters may lead to iron dyshomeostasis that encourages cellular conditions , such as oxidative stress and impaired autophagy. Understanding how these transport mechanisms intersect with other cell pathways is pivotal to building targeted therapies that could minimize symptoms or slow disease in affected patients (Arber et al., 2016). The intersection of membrane transport mechanisms in health and disease highlights the fine balance required for cellular function and systemic health. Transport process disruption has the potential to cascade into an avalanche of pathological events, as seen with hypertension and neurodegenerative disease. Therefore, developing a greater insight into these transport processes not only refines knowledge regarding disease pathophysiology but also opens the way to novel therapeutic strategies. By focusing on single transport proteins and their roles in cellular context, one can identify targets for intervention. This emphasis is essential, for a greater knowledge of membrane movements could result in novel treatments that can abrogate the effects of disease associated with malfunctioning transport, greatly enhancing long-term patient outcomes.

5.6.1. Importance of Transport in Nutrient Absorption

The transport of nutrients is hugely dependent on cell membrane transport mechanisms as barriers to control the entry of required nutrients into cells. In the intestines, for instance, transport proteins selectively mediate the transport of macromolecules and micronutrients from digested food. These transport proteins are organically specific molecules, ions, and water, and serve to maintain cells in homeostasis while effectively ingesting the necessary nutrients for energy and growth. Interestingly, deficiencies in these transport systems can yield many metabolic disorders, and this says a lot about their integral role in nutrient regulation. As demonstrated in metabolic tests, efficient nutrient uptake through independent pathways of transportation not only facilitates the gain of energy but also regulates complex inter-organ communication essential for the maintenance of overall metabolic health and adaptation to changing nutrition in the diet (Bauer et al., 2019). Furthermore, adaptation of the intestinal epithelium upon feeding further indicates the role of transport in the uptake of nutrients. In such snakes as the Burmese python, morphological adjustment after feeding demonstrates a remarkable ability for maximizing nutrient uptake. The specific example of a rise in enterocyte size and elongation of microvilli enhances the absorptive surface area, allowing for more efficient nutrient transport from the gut lumen into the circulatory system. Research suggests that such an adaptive response is triggered rapidly following feeding, with significant villus elongation and enterocyte expansion occurring in days (Aldewachi et al., 2006). Such quick adaptation capacity not only maximizes nutrient uptake under conditions of food availability but also points to the crucial interaction between structural adaptation mechanisms and transport mechanisms within the intestinal context. In addition to structural adjustment, nutrient-absorbed gut signaling pathways play a crucial role in the regulation of energy homeostasis. For instance, enteroendocrine cell-released hormones , such as Bursicon α in *Drosophila* demonstrate how nutrient absorption can influence hormonal signal transduction as well as metabolic responses within the organism. The signaling evoked by the availability of nutrients in the blood has important impacts on systemic metabolism. These include energy utilization and storage. Dysregulation of such systems results in metabolic dysregulation, justifying the decisive role of transport not only for absorption but for maintaining inter-organ communication (Bauer et al., 2019). Hence, recognizing the intricacies of transport in the context of nutrient uptake is necessary to have an idea about overall metabolic processes and corresponding therapeutic possibilities in disorders (Figure 5.21).

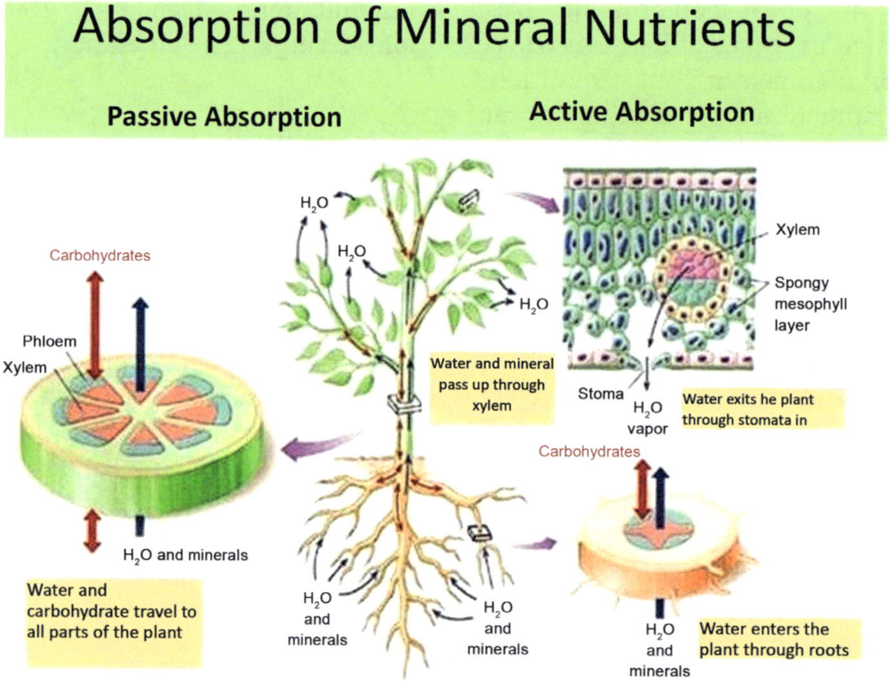

Figure 5.21. Importance of transport in nutrient absorption.

Source: https://www.iasexpress.net/wp-content/uploads/2023/06/image-83.png.

5.6.2. Role of Membrane Transport in Drug Delivery

Successful delivery of drugs relies heavily on the processes of membrane transport, through which drugs are transported across cell membranes. Biological membranes are selectively permeable barriers that regulate the movement of substances, , such as drugs, into and out of the cell. Awareness of the various transport phenomena, passive diffusion, facilitated diffusion, and active transport, is useful to maximize drug delivery systems. These procedures not only determine the bioavailability of a drug but also impact its pharmacokinetic and pharmacodynamic profiles. More especially for hydrophilic or large drugs, membrane permeability is a critical consideration because limited absorption can compromise therapeutic effectiveness. Enhancements in membrane transport concepts can therefore significantly promote drug absorption and hence facilitate better therapeutic effects with reduced side effects. Recent research has identified extracellular vesicles (EVs) as novel drug delivery vehicles by taking advantage of their natural role in intercellular communication. EVs are membrane-enclosed vesicles that can encapsulate various bioactive molecules and transfer them to target cells.

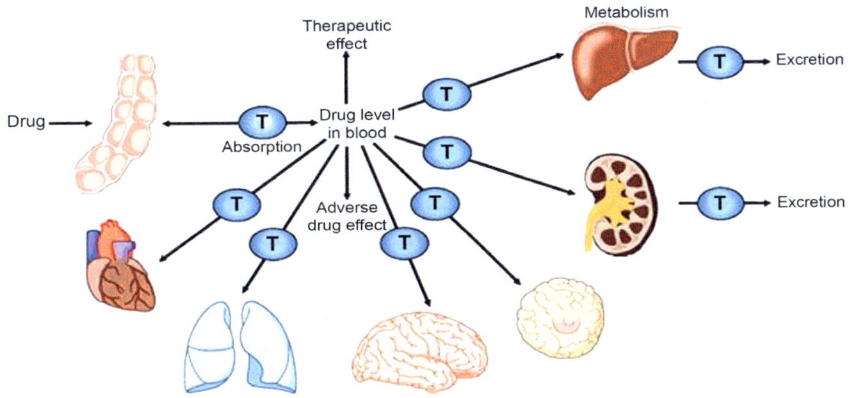

Figure 5.22. Role of membrane transport in drug delivery.

Source: https://basicmedicalkey.com/wp content/uploads/2016/07/00070.jpeg.

The utilization of plant-derived EVs is a hopeful approach, given that these vesicles exhibited beneficial properties in plant physiology as well as human health. Through the exploitation of natural crosstalk between plant EVs and mammalian systems, scientists are investigating them as vectors for therapeutic compounds, emphasizing their function in alternative protein secretion routes. This novel approach could provide a remedy for encapsulating drugs that exhibit poor membrane permeability, thus overcoming some of the most critical challenges in drug delivery and bioavailability (de Lousa et al., 2022). The BCS plays a critical role in guiding scientists in the improvement of drug delivery processes, particularly in the case of Class III molecules, which are highly soluble but poorly permeable. The classification brings to the limelight that the effectiveness of a compound could be compromised by its ability to cross biological membranes. Accordingly, various approaches have been researched to enhance the permeability of such drugs, ranging from the development of prodrugs, permeation enhancers, and advanced methods like nanoencapsulation. Current innovation proposes that hybrid methodologies combining traditional and modern approaches offer effective solutions to overcome permeability challenges. Through overcoming the barriers to drug absorption, these approaches can greatly enhance the therapeutic potential of Class III molecules, ultimately paving the way for better therapeutic interventions (Dave et al., 2016) (Figure 5.22).

5.6.3. Implications of Transport Dysfunction in Diseases

Transport dysfunction forms a critical component of the pathophysiology of many diseases, led in particular by neurodegenerative disorders. Mitochondria and their dynamic functions like fusion, fission, and transport form the very essence of neuronal homeostasis. Pathology of these mitochondrial functions can lead to catastrophic effects, as in the instance of Charcot-Marie-Tooth disease type 2A and dominant optic atrophy that are due to mitochondrial fusion defects. These conditions demonstrate that dysfunction in transport also affects energy metabolism, as well as reflects on the general integrity of neurons to induce progressive degeneration and cell death. Furthermore, systemic derangement of the transport processes may exacerbate neurodegenerative diseases, where the inability to effectively control mitochondrial dynamics renders the neurons susceptible, and targeted interventions would be critical to potentially restore these processes and stop disease progression (Alexander et al., 2009). In addition to mitochondrial transport malfunction, dysfunction of membrane transporters' capacity for management of neurotoxic aggregates is a dire danger in neurodegenerative diseases like Alzheimer's and Parkinson's. Accumulation of amyloid-beta and α-synuclein in these conditions comes with an inherent correlate of abnormal transport in cellular membranes, leading to mitochondrial damage and additional neuronal loss. Interestingly, studies have suggested that mitochondrial phospholipid membranes are destabilized by contact with such aggregates, and this compromises mitochondrial function, triggering a cascade of bioenergetic failure. The dysfunctions connect cellular transport systems to the neuron's bioenergetics, and it underscores the need to explore how therapeutic agents, like polyphenolic compounds, might protect mitochondria against aggregate-induced injury. The interrelated nature of transport dysfunction and neuronal health underscores the complexity of disease processes and potential for targeting (Caruana & Vassallo, 2014). The scope of transport dysfunction extends beyond neurodegenerative disease, as it infiltrates other diseases , such as metabolic syndromes and cardiovascular disease. In these diseases, disturbance of transport functions has the potential to lead to extensive metabolic dysregulation, as seen in diseases characterized by insulin resistance. Transporters of membranes that are responsible for glucose and lipid uptake can be impaired, and this results in impaired cellular function and heightened susceptibility to the occurrence of long-term complications. Likewise, in cardiovascular disease, transport abnormality in cholesterol and fatty acid uptake is responsible for atherosclerosis and myocardial infarction. Such instances bring down the extensive role of transport mechanisms towards cell health and the pathogenesis of disease. A general understanding of these dysfunctions can lead the way to new therapeutic strategies for the restoration of normal transport processes, thereby enhancing health outcomes in various diseases.

Test Your Knowledge

1. What is the importance of transport in nutrient absorption?
2. What are the implications of transport dysfunction in diseases?

Summary

The dynamic and multifaceted nature of transport processes and membranes is at the heart of homeostasis and cellular function. Cellular membranes are not only barriers but also permit the selective passage of various molecules, controlled tightly by specialized transport proteins. The technological advancement making carrier function analysis possible, like that developed in the study of auxin transporters, points out the significance of effective transport systems in maintaining growth and development compatible cell environments. By studying the actions of proteins like PIN-FORMED (PIN) and their role in auxin homeostasis, we get an idea of the broader importance of transport mechanisms involved in cell signaling pathways and metabolic processes (Barbez et al., 2013). Even in other biological systems like wound healing and tissue repair, the relationship between transport processes and membrane dynamics is shown. Experiments that probed collagen deposition and fibroblast interaction with the ECM gave insight into mechanical and biochemical cues that influence cellular activity during healing. The multiscale model of collagen orientation and fibroblast activity shows the complexity of tissue regeneration and how transport and membrane-related phenomena are inseparably interdependent with physiological response. This relationship highlights the imperative of a unified comprehension of these processes to direct therapeutic action in the promotion of wound healing and tissue repair (Dallon et al., 2005). Ultimately, the study of membranes and transport processes has promising potential for innovation in areas of biotechnology and medicine. As researchers are drawn deeper into the machinations of such systems, there is potential for the development of purpose-built interventions that can drive transport processes toward desired outcomes. Either through manipulation of the hormonal signal transduction pathways or through the engineering of biomaterials with comparable native tissue properties, there will be an intimate knowledge of these cell systems that will unlock new treatments and technologies. Thus, continued studies into membrane transport and dynamics are not only academically relevant but also carry significant implications in solving pragmatic problems in disease and health management.

Membrane detail and transport knowledge is vital in most fields of scientific and industrial use. Critical aspects of membrane function reveal that they are present as selective barriers which manage the movement of material from ions through to large molecules. Selectivity is vital in biological systems such that cells can maintain homeostasis as well as facilitate key processes , such as the intake of nutrients and excretion of waste products. In industrial processes, membrane technology development can be expected to make energy usage more efficient and environmentally responsible, particularly in the generation of energy and resource conservation. The potential for engineering membranes with specific properties expands their applications, as seen in laboratory work on oxygen separation and the emission of toxic fumes. Such a case in point is the development of dense ceramic oxygen separation membranes, an example where simple principles meet practical applications. These membranes utilize high-temperature, effective rapid oxygen-ion conducting ceramics, which address crucial challenges faced in natural gas conversion and oxy-fuel combustion processes. By effectively fractionating oxygen from air, the membranes reduce emissions of nitrogen oxides (NOx) and improve sequestration of CO_2, making it possible to have cleaner processes. Dual-phase composite studies, , such as gadolinia doped ceria and strontium-doped lanthanum cobaltite or silver, highlight the importance of combining different materials to increase membrane conductivity and performance and overall performance in real applications (Seeharaj et al., 2010). Furthermore, characterization of these membranes confirms that they can achieve balancing ionic and electronic conductivity, which is important in their operation. Advanced techniques, , such as secondary ion mass spectrometry, have been employed to study oxygen tracer diffusion and surface exchange kinetics in the composite materials. Results of large oxygen diffusivity in the solid state and better mechanical strength are promising for greater efficiency in oxygen separation processes. With continuing research, these membranes not only have significant promise for industrial applications but also imply broader implications for emerging energy and environmental technologies, highlighting the strategic importance of membranes for revolutionary scientific breakthroughs (Peña et al., 2017).

KEY TERMS FROM THIS CHAPTER

Active transport	Bulk transport	Carrier proteins
Channel proteins	Cholesterol (in membranes)	Diffusion
Electrochemical gradient	Endocytosis	Exocytosis
Facilitated diffusion	Fluid mosaic model	Glycoproteins
Integral proteins	Osmosis	Peripheral proteins
Phospholipid bilayer	ocytosis	Selective permeability

CRITICAL THINKING QUESTIONS

1. **What is the main function of the cell membrane?**
   a. Provide energy
   b. Control what enters and leaves the cell
   c. Store genetic material
   d. Produce proteins
2. **The cell membrane is primarily made up of:**
   a. Carbohydrates
   b. Proteins
   c. Phospholipids
   d. Nucleic acids
3. **Which process involves the movement of molecules from an area of high concentration to low concentration?**
   a. Active transport
   b. Diffusion
   c. Endocytosis
   d. Exocytosis
4. **What requires energy to move substances across the cell membrane against their concentration gradient?**
   a. Passive transport
   b. Osmosis
   c. Active transport
   d. Diffusion
5. **Which of the following is a type of passive transport specifically involving water?**
   a. Diffusion
   b. Endocytosis
   c. Osmosis
   d. Phagocytosis

Short Answer Questions

1. What is the basic structure of the cell membrane?
2. Explain the fluid mosaic model of the cell membrane.
3. What roles do proteins play in the cell membrane?
4. Define selective permeability in terms of the cell membrane.
5. What is the difference between passive and active transport?
6. Describe diffusion and give an example of a molecule that moves by diffusion.

Answers to Critical Thinking Questions

1. (b); 2. (c); 3. (b); 4. (c); 5. (c)

References

1. Arber, C. E., Li, A., Houlden, H., & Wray, S. (2016). Review: Insights into molecular mechanisms of disease in neurodegeneration with brain iron accumulation: unifying theories. *Neuropathology and applied neurobiology*, *42*(3), 220–241. https://doi.org/10.1111/nan.12242
2. Aldewachi, H., Carey, J. B., Carey, M., Cramp, G., Helmstetter, C., Diamond, M. E., Diaz, J. M., et al. (2006). Postprandial morphological response of the intestinal epithelium of the Burmese python (*Python molurus*). http://arxiv.org/abs/q-bio/0611019 (accessed on 13th June 2025).
3. Alexander, C., Baloyannis, S., Banchs, I., Bossy-Wetzel, E., Brockmann, K., Cassidy-Stone, A., Chang, D. T., et al. (2009). Mitochondrial dynamics–fusion, fission, movement, and mitophagy–in neurodegenerative diseases. https://core.ac.uk/download/4879584.pdf (accessed on 13th June 2025).
4. Alleva, K. E., Amodeo, G., & Chara, O. (2020). Aquaporins: Another piece in the osmotic puzzle. https://core.ac.uk/download/322675728.pdf (accessed on 13th June 2025).
5. Angelova, M. I., Arroyo, M., Bitbol, A. F., Fournier, J. B., et al. (2014). Interplay of packing and flip-flop in local bilayer deformation: How phosphatidylglycerol could rescue mitochondrial function in a cardiolipin-deficient yeast mutant. https://core.ac.uk/download/41779995.pdf (accessed on 13th June 2025).
6. Antollini, S. S., & Fabiani, C. (2019). Alzheimer's disease as a membrane disorder: Spatial cross-talk among beta-amyloid peptides, nicotinic acetylcholine receptors and lipid rafts. https://core.ac.uk/download/287882156.pdf (accessed on 13th June 2025).
7. Arslan, S., Eyvaz, M., İmer, D., Koyuncu, I., et al. (2018). Forward osmosis membranes–A review: Part II. https://core.ac.uk/download/322435369.pdf (accessed on 13th June 2025).
8. Baddeley, D., Bewersdorf, J., Bottanelli, F., Erdmann, R., et al. (2017). A novel physiological role for ARF1 in the formation of bidirectional tubules from the Golgi. https://core.ac.uk/download/323064987.pdf (accessed on 13th June 2025).
9. Barbez, E., Friml, J., Kleine-Vehn, J., Laňková, M., et al. (2013). Single-cell-based system to monitor carrier-driven cellular auxin homeostasis. https://core.ac.uk/download/55845457.pdf (accessed on 13th June 2025).
10. Bartolini, P., Deme, B., Petrillo, C., Plazanet, M., et al. (2013). Water in a polymeric electrolyte membrane: Sorption/desorption and freezing phenomena. http://arxiv.org/abs/1305.0437 (accessed on 13th June 2025).
11. Bauer, C., Cordero, J. B., Kruspig, B., Maddocks, O. D. K., et al. (2019). A neuronal relay mediates a nutrient-responsive gut/fat body axis regulating energy homeostasis in adult *Drosophila*. https://core.ac.uk/download/161513999.pdf (accessed on 13th June 2025).
12. Bayat, A., Jintapattanakit, A., Lin, A., Kotzé, A. F., Kotzé, A. F., Kotzé, A. F., Kotzé, A. F., et al. (2011). Uptake and transport of novel amphiphilic polyelectrolyte-insulin nanocomplexes by Caco-2 cells–towards oral insulin. https://core.ac.uk/download/9553230.pdf (accessed on 13th June 2025).
13. Bellen, H. J., Callaerts, P., Dermaut, B., Kania, A., et al. (2005). Aberrant lysosomal carbohydrate storage accompanies endocytic defects and neurodegeneration in *Drosophila* benchwarmer. https://core.ac.uk/download/55755360.pdf (accessed on 13th June 2025).

14. Boswell, J. S., Carlile, M. J., Daniel, J. W., Deacon, J. W., Deak, T., Falconer, J., Feldmann, H., et al. (2017). Introduction to fungal physiology. https://core.ac.uk/download/228177974.pdf (accessed on 13th June 2025).

15. Bourdon, J. C., Gromova, N. V., Grunyushkin, I. P., Klenova, E. M., et al. (2017). Physical and chemical processes and the morpho functional characteristics of human erythrocytes in hyperglycemia. https://core.ac.uk/download/141567522.pdf (accessed on 13th June 2025).

16. Brown, A. I., Besharse, J., Brangwynne, C. P., Koslover, E. F., Goehring, N. W., Hagins, W. A., Hu, J., et al. (2019). Drive, filter, and stick: A protein sorting conspiracy in photoreceptors. https://core.ac.uk/download/323066927.pdf (accessed on 13th June 2025).

17. Caillé, A., König, S., Lipowsky, R., & Rheinstädter, M. C. (2008). Collective molecular dynamics in proteins and membranes. https://core.ac.uk/download/pdf/81791364.pdf (accessed on 13th June 2025).

18. Caruana, M., & Vassallo, N. (2014). Select polyphenols protect mitochondria against amyloid aggregates in Alzheimer and Parkinson diseases. https://core.ac.uk/download/46602598.pdf (accessed on 13th June 2025).

19. Clark, G. T. (1997). *Phosphate Absorption by Arabidopsis Thaliana: The Effects of Phosphorus Nutritional Status* (Master's thesis, Massey University, Palmerston North, New Zealand). https://core.ac.uk/download/157568896.pdf (accessed on 13th June 2025).

20. Conti, M., Huang, B., Irannejad, R., Mika, D., et al. (2017). Functional selectivity of GPCR-directed drug action through location bias. https://core.ac.uk/download/213442723.pdf (accessed on 13th June 2025).

21. Dallon, J. C., Maini, P. K., & Sherratt, J. A. (2005). A multiscale model for collagen alignment in wound healing. https://core.ac.uk/download/pdf/96773.pdf (accessed on 13th June 2025).

22. Dave, V. S., Gupta, D., Gupta, S. V., Nguyen, T., et al. (2016). Current and evolving approaches for improving the oral permeability of BCS Class III or analogous molecules. https://core.ac.uk/download/81665712.pdf (accessed on 13th June 2025).

23. de Marcos Lousa, C., Eldahshoury, M. K., & Farley, J. T. (2022). Unconventional secretion of plant extracellular vesicles and their benefits to human health: A mini review. https://core.ac.uk/download/580952407.pdf (accessed on 13th June 2025).

24. Dominiczak, A., Ferreri, N. R., & Graham, L. A. (2017). The role of renal transporters and novel regulatory interactions in the TAL that control blood pressure. https://core.ac.uk/download/82893296.pdf (accessed on 13th June 2025).

25. Farquhar, M. G., & Palade, G. E. (1998). The Golgi apparatus: 100 years of progress and controversy. https://core.ac.uk/download/323062257.pdf (accessed on 13th June 2025).

26. Finkelstein, A., Flamholz, A., Nenninger, A., Pockels, A., Schmidt, A., Upadhyaya, A., Dupuy, A. D., et al. (2017). Membranes by the numbers. https://core.ac.uk/download/211380558.pdf (accessed on 13th June 2025).

27. Galicia-García, U., Benito-Vicente, A., Jebari, S., Larrea-Sebal, A., Siddiqi, H., Uribe, K. B., Ostolaza, H., et al. (2020). Pathophysiology of type 2 diabetes mellitus. *International Journal of Molecular Sciences, 21*, 6275–6275. https://doi.org/10.3390/ijms21176275 (accessed on 13th June 2025).

28. Grohmann, D., Ciobanu, G., Bacci, G., & Miculan, M. (2009). A framework for protein and membrane interactions. https://core.ac.uk/download/53347931.pdf (accessed on 13th June 2025).
29. Grohmann, D., Ciobanu, G., Bacci, G., & Miculan, M. (2009). Biographical models for protein and membrane interactions. https://core.ac.uk/download/53328652.pdf (accessed on 13th June 2025).
30. Kassai, M., Kawaguchi, R., Sun, H., Ter-Stepanian, M., et al. (2015). Vitamin A transport mechanism of the multitransmembrane cell-surface receptor STRA6. https://core.ac.uk/download/323078464.pdf (accessed on 13th June 2025).
31. Kholodenko, B. N., Peletier, M. A., & Westerhoff, H. V. (2002). Control of spatially heterogeneous and time-varying cellular reaction networks: A new summation law. https://core.ac.uk/download/301661846.pdf (accessed on 13th June 2025).
32. Lapeyre Jr., G. J., Lewenstein, M., Manzo, C., Massignan, P., et al. (2015). Weak ergodicity breaking of receptor motion in living cells stemming from random diffusivity. https://core.ac.uk/download/41778549.pdf (accessed on 13th June 2025).
33. Maiti, P. K., Moid, M., Naskar, S., Sahoo, S., et al. (2022). Ultra-high permeable phenome nanotube membranes for water desalination. https://core.ac.uk/download/560387358.pdf (accessed on 13th June 2025).
34. Medina-Kauwe, L. K. (2007). "Alternative" endocytic mechanisms exploited by pathogens: New avenues for therapeutic delivery? https://core.ac.uk/download/323077467.pdf (accessed on 13th June 2025).
35. Morgante, E., Van Rossum, G. D., Russo, A., RUSSO, G., et al. (2015). Ouabain-induced cytoplasmic vesicles and their role in cell volume maintenance. https://core.ac.uk/download/54521390.pdf (accessed on 13th June 2025).
36. Oliveira, S. de, & Saldanha, C. (2010). An overview about erythrocyte membrane. https://core.ac.uk/download/12426151.pdf (accessed on 13th June 2025).
37. Öyen, M. (2003). Functional studies of plasma membrane syntaxis in yeast. https://core.ac.uk/download/11694668.pdf (accessed on 13th June 2025).
38. Zuniga-Hertz JP and Patel HH (2019) The Evolution of Cholesterol-Rich Membrane in Oxygen Adaption: The Respiratory System as a Model. Front. Physiol. 10:1340. doi: 10.3389/fphys.2019.01340 (accessed on 13th June 2025).
39. Peña Villamil, D. M., & Torres Londoño, N. P. (2017). Deployable arches based on regular polygon geometry. https://core.ac.uk/download/81581757.pdf (accessed on 13th June 2025).
40. Santha, R. P., & Kumar, A. (2016). Structure, function and regulation of integral membrane transport proteins. https://core.ac.uk/download/141671480.pdf (accessed on 13th June 2025).
41. Seeharaj, P. (2010). Mixed-conducting LSC/CGO and Ag/CGO composites for passive oxygen separation membranes. https://core.ac.uk/download/1588692.pdf (accessed on 13th June 2025).
42. Singh, A., Kukreti, R., Saso, L., & Kukreti, S. (2019). Oxidative stress: A key modulator in neurodegenerative diseases. *Molecules, 24*, 1583–1583. https://doi.org/10.3390/molecules24081583 (accessed on 13th June 2025).

43. Smolders, C. A., van den Berg, G. B., et al. (1991). Diffusional phenomena in membrane separation processes. https://core.ac.uk/download/pdf/11452626.pdf (accessed on 13th June 2025).
44. Stange, B. C. (1995). *The Structural and Functional Effects of Electromagnetic Fields on The Plasma Membrane of Vicia Faba, The Broad Bean: A Thesis Dissertation Presented in Partial Fulfilment of The Requirements for the Degree of Master of Science, Plant Biology at Massey University.* https://core.ac.uk/download/162616944.pdf (accessed on 13th June 2025).
45. Stephens, D. J. (2012). Functional coupling of microtubules to membranes–Implications for membrane structure and dynamics. https://core.ac.uk/download/29026529.pdf (accessed on 13th June 2025).

ENERGY, ENZYMES, AND METABOLISM

6

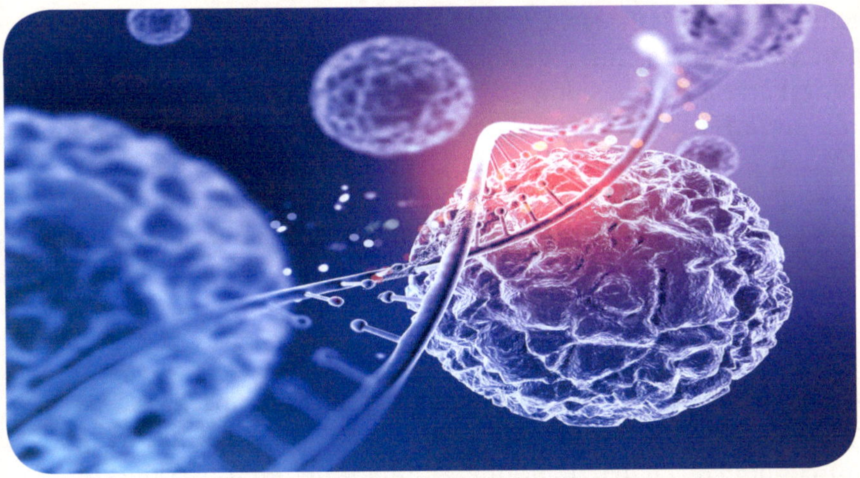

Introduction

The complex interplay of energy, enzymes, and metabolism forms the template of biological activities in all living organisms. Underlying these interactions is the idea of bioenergetics that explores how cells make use of energy to sustain life processes. Enzymes as biological catalysts have a vital role in accelerating the amount of chemical reactions needed for metabolism. Their ability to decrease the activation energy of reactions allows metabolic pathways to proceed under the ambient conditions of the cellular environment effectively. Also, their regulation is crucial, as it ensures that metabolic processes are responsive to the fluctuating demands of the organism and therefore maintain homeostasis and equilibrium in energy. The dynamic image of enzyme function and energy processes has deep significance for fields like bioengineering, ecology, and medicine, where the regulation of such systems could lead to new technologies and therapeutic interventions. The understanding of metabolic pathways and enzyme kinetics underlies the mechanisms by which organisms adapt to their environment and optimize

LEARNING OBJECTIVES

After studying this chapter, you will be able to:

1. Define the concept of energy;
2. Explain enzymes: catalysts of metabolism;
3. Describe metabolic pathways;
4. Describe energy transfer in cells;
5. Explore enzyme regulation and inhibition;
6. Focus on the metabolism and homeostasis.

energy utilization. Some enzymes have complex regulatory mechanisms, closely controlling metabolic reaction rates depending on internal and external conditions. For instance, in denitrification reactions in hyporheic zone sediment, one study discovered significant time lags in the response of enzymes to a shift in substrate concentration, emphasizing the complex interactions underlying metabolic activity (Chen et al., 2017). This temporal consideration emphasizes the importance of a robust grasp of microbial community traits and their corresponding enzyme functionalities, which are predictive of biogeochemical processes in the environment. With metabolic fluxes varying due to environmental changes, the predominance of enzymes becomes more apparent, emphasizing their role in ecological stability and resilience. This interconnectedness of enzymes, energy, and metabolism is true not only on an organismal level but also in larger ecological and evolutionary levels. While energy flows through ecosystems, enzymes play a role in converting energy from one form into another, facilitating nutrient cycling, and funding a range of biological processes. Furthermore, the evolution of enzyme systems illuminates the ability of organisms to adapt in response to various environmental stresses. The complex interplay of transcriptional and translational processes only makes this situation more complex, as emerging models illustrate correlations between enzyme concentration and metabolic flux responses, forming the basis for predictive biogeochemical modeling (Amigo et al., 2011). The models present useful tools to scientists looking to uncover and control the complex dynamics of ecosystems, opening doors to new breakthroughs in environmental sustainability and management.

Definition of Energy in Biological Systems

The concept of energy in living organisms is at the forefront of understanding how life operates on a cellular and systemic basis. Energy is typically described as the potential to perform work, allowing for processes of growth, reproduction, and homeostasis in living things. Biologically, energy comes in many forms that are each of vital importance to metabolic functions. For instance, chemical energy in macromolecules like lipids and carbohydrates is transferred by metabolic pathways to ATP, which is referred to as the cell's energy currency. It is vital for energizing biochemical reactions sustaining basic cellular functions, accentuating dynamic mutual interaction between energy dissipation and metabolism in living organisms. The relationship between enzymes and energy is a key aspect of metabolism. Enzymes, being biological catalysts, accelerate the conversion of substrates to products and reduce the activation energy required for reactions to happen significantly. Such efficacy is crucial in facilitating the sustenance of the various metabolic pathways through which energy and matter are converted in living organisms. Enzyme activity is usually influenced by factors like temperature, pH, and the concentration of the substrate or inhibitor, which may impact the overall energy dynamics of a cell. The understanding of this relationship enhances our knowledge on how cells adapt to different energy demands, particularly in pathophysiologic conditions where alterations in enzyme activity can lead to metabolic diseases and emphasizes the pivotal role played by enzymes in energy conversion processes in biological systems (Peyton et al., 2013). In addition, as we delve further into the complexities of energy definitions in biology, a systemic appreciation necessitates an integrative approach. Different areas of biological study, including molecular biology, ecology, and cognitive biology, all offer unique insights into how energy operates in various systems. For instance, systems biology offers an environment in which the connectivity of energy exchange can be modeled, allowing a systems view of metabolic pathways and their systemic impact (Nagarjuna et al., 2004). This is increasingly pertinent as we think about intersections of metabolism with more general biological and cognitive processes, moving towards a more integrated view of life processes. As such, the biological definition of energy is more than a question of measurement or transduction but rather a consideration of the fundamental principles that underlie life itself and provide sustenance to the subtle interrelationships that permit the existence of biological diversity and function (Figure 6.1).

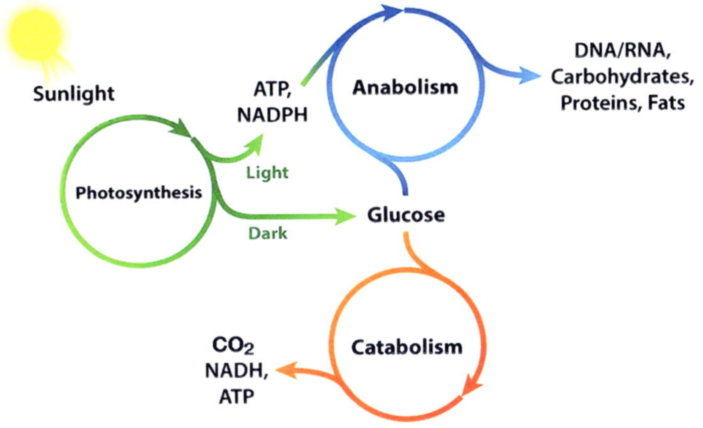

Figure 6.1. Energy in biological systems.

Source: https://openoregon.pressbooks.pub/app/uploads/sites/220/2021/02/Biochemistry_Page_413_Image_0003.jpg.

Importance of Enzymes in Metabolic Processes

The complex role of enzymes in metabolic processes cannot be overlooked because they are such critical catalysts that enable biochemical reactions for life. Enzymes lower the activation energy of a reaction, and in so doing, hasten metabolic processes significantly. For instance, in cellular respiration, specific enzymes are responsible for converting glucose to usable energy, which signifies their critical role in ATP generation, the currency of cellular energy. Without enzymes, metabolic reactions would go on too slowly to sustain life. Enzyme activity must also be controlled; there are many conditions , such as pH, temperature, and substrate concentration, that can vary their effectiveness significantly. This control mechanism allows organisms to alter their metabolic processes in response to changing internal and external stimuli to regulate efficiency and balance in biochemical systems. Not only do enzymes facilitate catabolic processes, where large molecules are broken down to yield energy, but they also help considerably with anabolic processes, which construct complex molecules from simpler molecules. This is beautifully illustrated through the folate cycle, where some of the enzymes participate in the metabolism of amino acids, facilitating the biosynthesis of significant cellular components. Research indicates that immune responses result in specific patterns of enzyme expression that direct the metabolic flux toward survival-providing pathways upon infection. As an example, the cooperative action of TFs , such as Dif and FOXO in response to immune stimulation influences the expression of enzymes involved in the production of serine and glycine, demonstrating immunity and metabolism. These enzyme-mediated metabolic adjustments are examples of how organisms optimize energy expenditure under stress, ultimately impacting health. The function of enzymes extends beyond the individual metabolic process since they operate in concert within complex pathways that underlie cellular activity. Insights from research into enzymes in the context of metabolic reactions within a variety of organisms, including model systems *Drosophila melanogaster*, illustrate how evolutionary innovations drive metabolic plasticity. These adaptations are particularly marked in stressful environments, , such as the polar regions, where microorganisms have maximized enzymatic processes to survive. With the development of the omics technologies, these adaptations have been highlighted at the molecular level with emphasis on the potential applications of understanding enzymatic functions to biotechnology and ecological research. By drawing on these findings, scientists can develop novel mechanisms for problem-solving because of climate change and the stability of ecosystems, ultimately demonstrating the vast importance of enzymes in ensuring metabolic processes that are vital in ensuring life on earth (Figure 6.2).

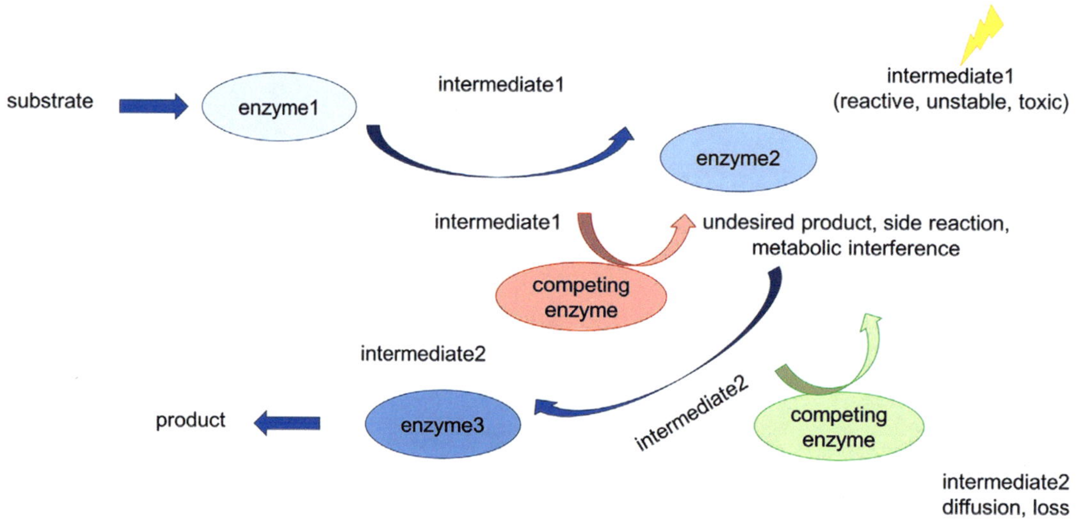

Figure 6.2. Enzymes in metabolic processes.

Source: Pröschel M, Detsch R, Boccaccini AR and Sonnewald U (2015) Engineering of metabolic pathways by artificial enzyme channels. Front. Bioeng. Biotechnol. 3:168. doi: 10.3389/fbioe.2015.00168

Overview of Metabolism and Its Significance

Metabolism is the complex biochemical activities whereby organisms convert food to energy and necessary cellular components. Metabolism can be generally classified under two categories: catabolism and anabolism. Catabolism refers to the breakdown of complex molecules , such as carbohydrates and lipids, which yield energy that is harnessed by the cell to carry out various functions. In contrast, anabolism harnesses this energy for constructing necessary molecules necessary for growth, maintenance, and repair. This dualism highlights metabolisms role in achieving homeostasis and facilitating life's variety of processes. The appreciation of metabolism not only points to its critical function in preserving organismal health but also to an understanding of a range of disease states, in which metabolic dysregulation may be causal to conditions , such as obesity and diabetes. Importantly, metabolism impacts the effectiveness with which energy is utilized, thus the overall physical vitality and endurance. The control of metabolic processes is crucially supported by enzymes, the biological catalysts. Enzymes catalyze biochemical reactions to enable necessary changes to occur under physiological conditions. They are influenced by a variety of factors, including pH, temperature, and substrate concentration, which reflects their critical contribution to effective metabolism. For instance, glycolytic enzymes facilitate the breakdown of glucose, while others, including those in the citric acid cycle, are required for the synthesis of ATP. Enzymes have a function beyond catalysis; they are central to the regulation of processes that dictate energy storage and expenditure. As outlined in many studies, enzyme activity disturbances are to blame for initiating metabolic disorders, further indicating the importance of enzymes in maintaining not just personal health but also the overall metabolic balance of ecosystems like human physiology (Fonteh et al., 2018). Energy balance can also be understood by the theory of metabolizable energy (ME), or rather an indicator of the energy being provided to an organism after digestion and absorption. Studies that have analyzed dietary intake have shown that variations in ME values can determine variations in energy prediction models, asserting the need for a better understanding of nutrient digestibility and utilization (Zou et al., 2007). These results are important across areas from nutrition science to health economics since they influence anything from personal dietary recommendations to health policy. The significance of metabolism, therefore, cannot be overemphasized since it embraces and influences developmental and health destiny across life. Understanding these complex pathways and paradigms of energy is required for the planning of useful strategies for the preservation of health and avoidance of metabolic disorders, making metabolism an essential aspect of scientific inquiry and health management.

6.1. The Concept of Energy

Learning Outcome

1. Types of energy relevant to biological systems;
2. Laws of thermodynamics and their implications;
3. Role of ATP as the energy currency of the cell.

The concept of energy is a central doctrine of comprehending life processes, most notably in biochemistry and metabolism. Energy is the ability to do work, and in biology, it plays a crucial role in facilitating many reactions that sustain life. Enzymes, the biological catalysts, play a critical role in these processes through the lowering of activation energy required for reactions to proceed. This dynamic interplay between energy and enzymatic function places the complexity of metabolic pathways into context. By virtue of maximum energy efficiency, enzymes enable organisms to reach peak metabolic yield to provide a consistent supply of energy for growth, reproduction, and homeostasis. The delicate interplay between energy availability and enzymatic activity, therefore, is essential to the survival of living organisms. Furthermore, the energy-enzyme relationship is not merely mechanical in nature; it is a manifestation of deep biological interdependence that expands across numerous scientific disciplines. To appreciate this complexity, there is a necessity to consider the broader implications of energy dynamics within biological systems. As debated in contemporary biological theory, there exists a model that integrates molecular biology, evolutionary dynamics, and ecological systems, it proposes that energy flow is an integrator of special importance for these disciplines. Semantic view of scientific theories suggests that current energy models in physics are possibly too simple to totally account for the biological processes that take place because of energy interaction. This idea includes a call for expanded awareness of the energy role, one admitting the constraints of traditional models and requesting innovative approaches to investigate biological reality with consideration of physical as well as chemical bases of life (Nagarjuna et al., 2004). In real biological systems, enzyme kinetics and resulting dynamics of energy transfer may be significantly slowed based on a variety of factors. Previous studies, , such as the dynamics of denitrification, show that the response of enzymes is time-lagged to changes in substrate concentration. Using the trait-based models, scientists have begun unveiling the complexity of these interactions and enzymatic responses being potentially under the control of transcription processes rather than substrate availability. This paradigm revolution not only propels understanding of metabolic processes but also emphasizes the significance of energy regulation at a microbial level. The observations unveiled through such research point to the predictive significance of such studies, making it possible for scientists to model biogeochemical processes more effectively by integrating the complex dynamics of energy, enzymes, and metabolism (Chen et al., 2017).

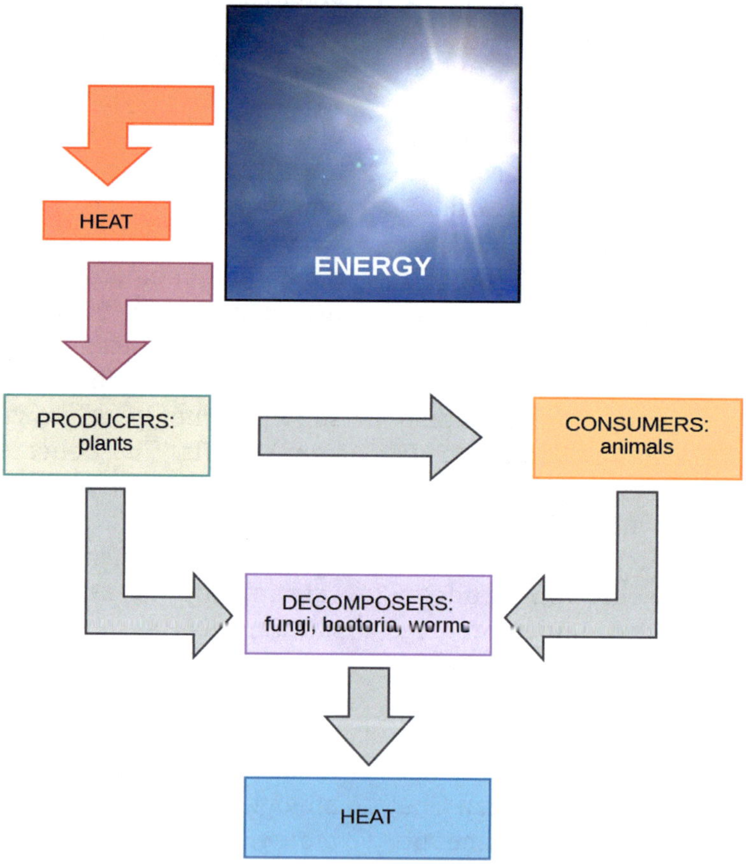

Figure 6.3. Biological energy.

Source: https://opentextbc.ca/biology/wp-content/uploads/sites/96/2015/08/Figure_04_01_012.jpg.

6.1.1. Types of Energy Relevant to Biological Systems

It is helpful to be familiar with the types of energy that are relevant to biological systems to understand the fundamental processes underlying metabolism and enzymatic activity. Biological systems are dependent primarily on chemical energy, acquired from molecular bonds to fuel growth and homeostasis. Cellular respiration, demonstrating the breakdown of glucose, is an example of this conversion, where cells access and release energy sequestered in chemical bonds. Besides, ATP is the cells' "energy currency," where different biochemical reactions require energy input. This intricacy in interactions among various forms of energy accentuates the importance of energy transformations in keeping life going and supports the assertion that biosystems are structured according to efficient utilization of energy, as attested by metabolic pathways and enzyme control, which optimize energy yield and function. In addition to chemical energy, biological organisms also rely on light energy, particularly in the case of photosynthesis.

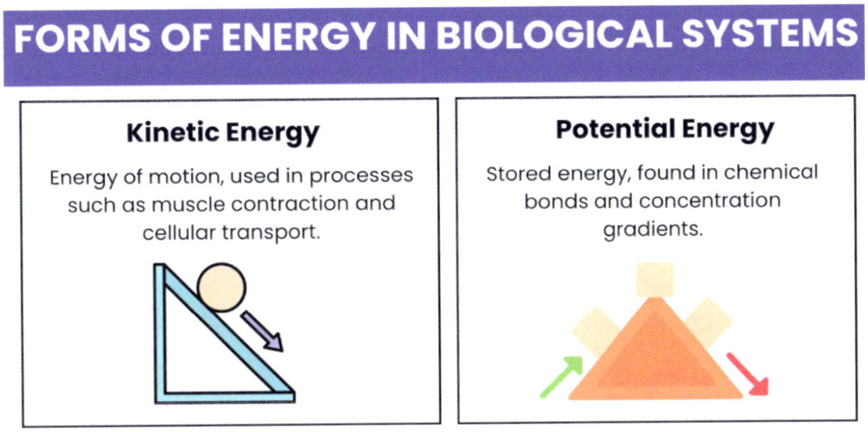

Figure 6.4. Types of energy relevant to biological systems.

Source: https://images.examples.com/wp-content/uploads/2024/07/Forms-of-Energy-in-Biological-Systems.png.

Phototrophic organisms, i.e., plants and certain bacteria, utilize light energy as chemical energy by creating glucose out of CO_2 and water. Apart from providing a large energy source for autotrophs, this supports heterotrophic organisms indirectly through food webs. The efficiency and optimality of light energy capture and conversion in photosynthesis also demonstrate the dynamical nature of biological energy systems. As noted above, the modularity of metabolic architecture allows for the optimal use of light energy, implying that metabolic networks are designed with intrinsic flexibility to adapt to varying conditions in the environment and thus points to the interdependence of energy forms in biological systems (Braakman & Smith, 2013Braakman & Smith, 2013). Furthermore, the use of thermal energy, less typical as it is to light and chemical energy, is also critical to the functioning of biological systems. Thermoregulation by endothermic organisms is perhaps the best case of adaptation to thermal energy in the maintenance of metabolic rate and enzymatic activity. The effect of temperature fluctuation on the kinetics of enzymes further underlines the interaction between energy and metabolism. Enzymes operate optimally at given temperature ranges, and deviations beyond these can result in reduced efficiency or denaturation. Through homeostasis, organisms manage to maintain their enzymic processes unaffected, as emphasized in research into how metabolic processes respond to ambient thermal conditions (Barbara et al., 2010). The different types of energy that can be used with biological systems collectively account for a complex and efficient functioning required for the sustenance of life (Figure 6.4).

6.1.2. Laws of Thermodynamics and Their Implications

The laws of thermodynamics are fundamental principles that govern energy conversions within biological processes, particularly in metabolism and enzyme-catalyzed reactions. The first law, or energy conservation law, is that energy cannot be created or destroyed but can only be changed from one form to another. This is a fundamental principle to apply when describing how cells generate energy in the form of useful energy through metabolic pathways. For instance, when cells perform cellular respiration, chemical energy in glucose is converted to ATP, the currency of the cell's energy. Such mechanisms demonstrate the interdependence of energy and biological function and how cells maintain their dynamic equilibrium through complex metabolic pathways regulated by thermodynamic principles. The second law of thermodynamics places entropy as the quantification of the natural direction that systems will move in the direction of disorder. As far as metabolism, this implies that energy transfer within the cell is inherently inefficient because some energy is constantly lost in the form of heat, and therefore it leads to increased entropy. This has enormous implications on enzyme function; enzymes catalyze reactions to decrease activation energy, thereby facilitating faster attainment of equilibrium states as well as ensuring energy efficiency. Enzyme cost minimization models in experiments demonstrate that the interaction between enzyme supply and energy can influence metabolic efficiency. Based on these thermodynamic constraints, scientists can create improved models for metabolic pathways to account for energy yield along with the inherent costs of enzyme production (Bar-Even et al., 2016). Furthermore, using economic theory to enrich thermodynamic principles provides insight into metabolic fluxes within cellular systems. Metabolic fluxes are not only influenced by physical constraints but also by their economic relevance as well, balancing enzyme cost versus metabolic benefit. Economic flux analysis as a principle recognizes that low flux pathways are energetically favorable and can efficiently allocate resources within the cell. With the integration of thermodynamic and economic perspectives, metabolic engineering can better maximize pathways for target products while minimizing resource consumption. This unified view not only sheds light on our comprehension of enzyme dynamics and control of metabolism but also offers a basis for advancements in synthetic biology and bioengineering, whereby precise modifications can enhance the function and efficiency of enzymes (Liebermeister et al., 2014).

6.1.3. Role of ATP as the Energy Currency of the Cell

ATP is usually referred to as the cell's primary energy currency, as it is involved in funding most of the biochemical processes. Essentially, ATP contains a nitrogenous base (adenine), a five-carbon sugar (ribose), and three phosphate groups bonded together in a chain. The phosphate group bonds are high-energy bonds that have energy that can be easily released when hydrolysis is done, as this provides the energy required for cellular processes. The release of energy serves a two-pronged purpose—providing power for the basic metabolic processes and enabling muscle contraction, signal transduction, and biosynthesis. Thus, ATP not only serves as an all-purpose energy carrier but also as a central necessity for cellular integrity and function in the complex process of metabolism and enzyme activity. The versatility of ATP is evident in its ability to participate in a variety of biochemical reactions, namely those catalyzed by enzymes. Biological catalysts, simply referred to as enzymes, catalyze reactions that would otherwise occur at an imperceptibly slow rate under normal conditions. Coupling ATP hydrolysis with other enzymatic reactions, cells utilize the released energy to drive processes that would be thermodynamically unfavorable under nonstandard conditions. This co-operation is essential to mechanisms , such as active transport, where ATP is spent to move ions in a direction opposite to their concentration gradient, and biosynthesis, where ATP-driven reactions construct elaborate macromolecules from simple building blocks. The seamless coordination of ATP and enzymes underscores the importance of ATP not only as an inert carrier of energy but as a purposeful catalyst of metabolic reactions, reflecting the character of cellular processes and structure. Besides, ATP production and reformation are crucial to sustain its energy currency role. Oxidative phosphorylation, which occurs in the mitochondria and transforms food molecule energy into ATP, is the principal process by which ATP is produced. Other mechanisms, , such as substrate-level phosphorylation, also contribute to ATP production in glycolysis and the citric acid cycle. ATP regeneration is required for cellular energy homeostasis, as energy utilized in numerous metabolic processes far exceeds the readily available energy within the cell. This cycle of ATP generation and utilization resembles a very advanced control of energy in the cell, maintaining the dynamic equilibrium of metabolic reactions, which are the foundations of life. Thus, ATP not only illuminates the relationship between energy, enzymes, and metabolism but also is an important aspect of cell biology deserving further research (Figure 6.5).

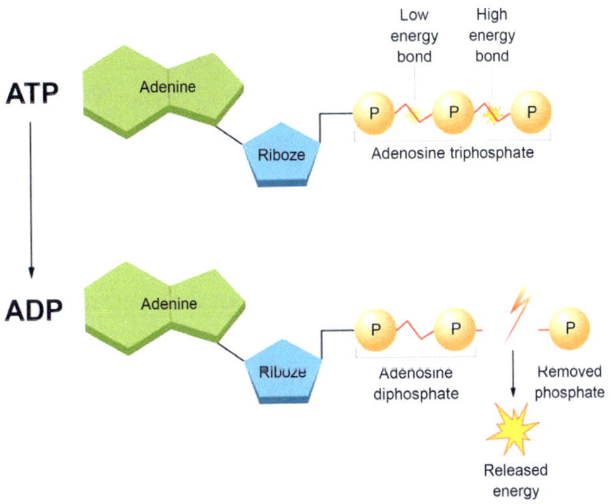

Figure 6.5. ATP as the energy currency of the cell.

Source: https://thumbs.dreamstime.com/z/atp-energy-currency-cell-atp-adenosine-triphosphate-called-energy-currency-cell-organic-241465102.jpg.

Test Your Knowledge

1. What are the types of energy relevant to biological systems?
2. What is the role of ATP as the energy currency of the cell?

6.2. Enzymes: Catalysts of Metabolism

Learning Outcome

1. Structure and function of enzymes;
2. Mechanisms of enzyme action;
3. Factors affecting enzyme activity.

Enzymes are simple catalysts of metabolic reactions, accelerating biochemical reactions essential for living. By lowering the activation energy required for reactions, enzymes facilitate cell functions at rates compatible with the demands of living organisms. Such catalytic activity is important since many metabolic reactions are sluggish by nature in the absence of enzymes. Enzymes are also very specific, catalyzing some reactions out of many potential substrates. This specificity gives metabolic pathways an adjustment and an efficient response to the changing energetic demands of cells. So, the role of enzymes is not only to make reactions happen, but also plays a role in being part of the control and economy of metabolism, allowing cells to adapt dynamically to environmental cues while maintaining homeostasis. Their integration into metabolic networks is an equivalent nested complexity that finds expression in modularity and hierarchy. Metabolic processes, as opposed to isolated chemical reactions, are not independent but are part of a system in which the activity of one enzyme can affect the activity of others within the same pathway or between branching pathways. This kind of architecture enables compartmentalization of reactions and metabolic modules' definition, which can be reused and reused in various contexts. The components of such modules will often have conserved functions as well as structure, which is a sign of evolutionary stability that has withstood the test of time. When metabolic processes became more sophisticated, they incorporated responses to the evolution of the environment, illustrating how enzymes mirror and guide metabolic evolution by being very flexible and specific in function and thus are catalysts of metabolic architecture itself (Braakman & Smith, 2013). The recent forays into elucidating the origin of metabolic pathways have shed new light on the feasibility of primitive metabolisms that may have existed before the advent of modern enzymatic systems. Studies have shown that there is a possibility that early metabolic reactions would have taken place independently of phosphates but through alternative mechanisms involving thioesters and iron-sulfur clusters, an ancient form of biochemistry (Goldford et al., 2017).

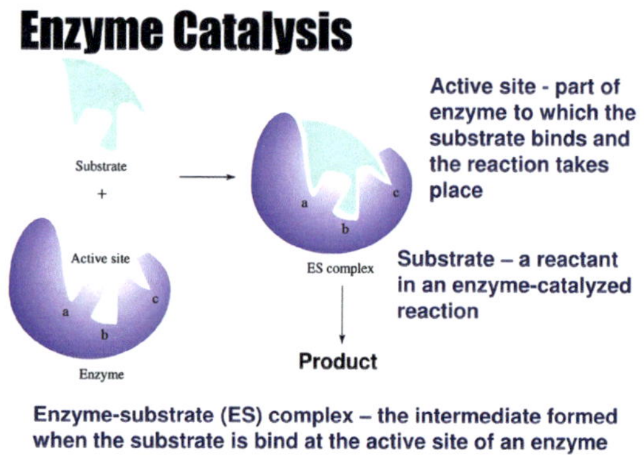

Figure 6.6. Enzyme: a biological catalyst.

Source: https://external-content.duckduckgo.com/iu/?u=https%3A%2F%2Fimage2.slideserve.com%2F3680619%2Fenzyme-catalysis-l.jpg&f=1&nofb=1&ipt=8ec9eb382156bf8748e699b2dc23174466385df5c8b6f7d40f649fc1e055beea.

This would suggest that such metabolic fossils could hold the secret to tracing the evolutionary origins of metabolism and the evolution of enzymic activities. Understanding these earlier metabolic processes illuminates the modern understanding of how enzymes have fine-tuned the harnessing of energy over time, continuously optimizing how energy is effectively harnessed in various biochemical environments. Not only does such analysis enrich the theoretical understanding of enzyme function, but it also underscores the complex interdependence of enzymatic activity, energy regulation, and the evolutionary history of life's metabolic machines (Figure 6.6).

6.2.1. Structure and Function of Enzymes

Enzymes are biological catalysts that are critical in facilitating many biochemical reactions necessary for life. Enzyme function is very optimized in their structure; enzymes are usually globular proteins consisting of long chains of amino acids that get folded into specific three-dimensional structures. The folding causes the creation of an active site, a specific area that will interact with specific substrates to form enzyme-substrate complexes. The selectivity of enzymes is a consequence of the specific interactions between molecular structures, in which even small variations have drastic impacts on catalytic efficiency.

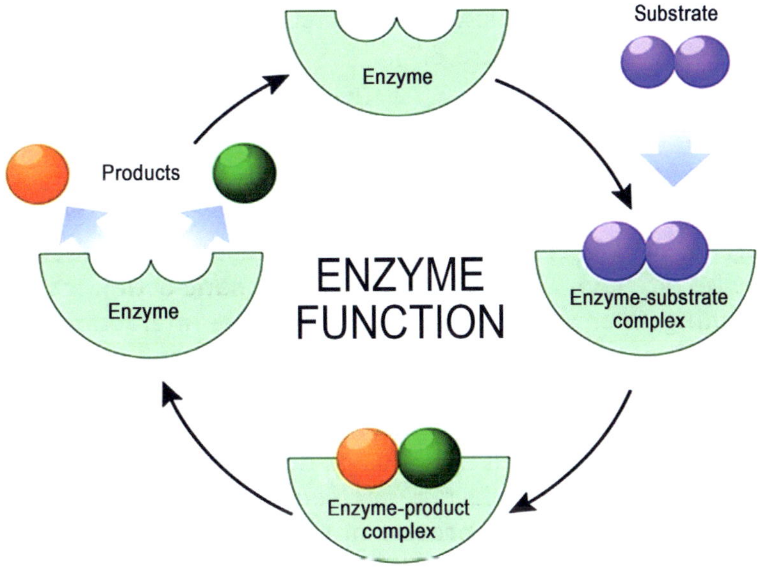

Figure 6.7. Structure and function of enzymes.

Source: https://d2jx2rerrg6sh3.cloudfront.net/image-handler/picture/2018/8/shutterstock_760624435.jpg.

By lowering the activation energy required for reactions, enzymes accelerate the rate of reactions and allow metabolic processes to proceed under physiological conditions, thus sustaining cellular functions and causing energy conversions required by life. Hierarchical organization and modularity of metabolic pathways are critical to achieving the interaction between cellular processes and enzyme activity. Enzymes are more likely to be within metabolic modules, a set of substrates and enzymes performing a particular process within a metabolically larger framework. Not just this, but the modules themselves are also highly conserved across different biological systems, and the modules possess a significant architecture founded on evolutionary solutions. For instance, the presence of co-factors typically at the boundaries of modules illustrates how biochemical pathways are diversified and controlled. The efficacy of enzyme activity hinges on different factors, including temperature, pH, substrate, and product concentration, illustrating the necessity of enzymes for purposes of maintaining homeostasis and optimizing cellular metabolism, hence impacting energy consumption in the organism as a whole (Braakman & Smith, 2013). Enzyme function knowledge is also very critical in fields of nutrition and health, specifically as it pertains to micronutrient metabolism. Enzymes involved in vitamin biosynthesis and metabolism are significant in plant physiology as well as in human health because they play a role in nutrient utilization and absorption. Lack of essential vitamins in the diet will have severe health consequences, illustrating the inter-relationship of enzymic activity and nutritional status. Methods , such as biofortification aim at enhancing the micronutrient content of staple foods while combating deficiencies that affect billions of individuals globally. By leveraging advancements with metabolic engineering, the potential lies to engineer and enhance the activity of key enzymes involved in vitamin biosynthesis and elevate the nutritional content of food sources. This enzyme-micronutrient metabolism interaction demonstrates the tremendous impact of enzymes' structure and function on energy and health processes at the cellular level (Strobbe et al., 2018) (Figure 6.7).

6.2.2. Mechanisms of Enzyme Action

Enzymes, as biological catalysts, play a vital role in accelerating the pace of biochemical reactions, basically changing the face of metabolic processes. These macromolecules act by lowering the activation energy barrier so that reactions can proceed faster and under conditions less harsh than would otherwise be possible. The mechanism of enzymatic action involves substrate binding to their active site, forming an enzyme-substrate complex; the interaction stabilizes the transition state and induces product formation. Molecularly, enzymes are highly specific due to their three-dimensional structure, defined by amino acid arrangement. Thus, the selectivity and efficiency of enzymes are crucial in maintaining the finely tuned balance of metabolic pathways in line with the modular architecture of metabolism, with the focus on their modulator and catalyst roles in cellular networks (Braakman & Smith, 2013). Cofactors and coenzymes significantly contribute to enzyme activity through their participation in the process of reaction. These non-protein molecules, , such as metal ions and organic molecules from vitamins, are key facilitators that support enzyme activity. Metal ions like zinc or magnesium, for instance, stabilize enzyme structures or catalysis directly, while coenzymes like NAD^+ or coenzyme A facilitate electron transfer and acyl group formation, respectively. The enzyme-cofactor interaction forms independent modular units that support enhancing the general efficiency of metabolic pathways. From examining the distinctive roles of these cofactors, the complex hierarchy of the metabolic network is comprehensible as they systematically position themselves on the edge of metabolic modules, regulating connectivity and throughput of biochemical reactions (Braakman & Smith, 2013). Recent studies on enzyme activity under various physiological states have yielded new evidence in relation to the alterations in enzyme activity in pathophysiologic conditions. One such study is the investigation of 2-nitroimidazole treatment with liver cell metabolism, demonstrating significant alteration in enzyme levels and activities due to induced hypoxia. Enhanced carbohydrate-metabolizing enzymes and lysosomal enzymes have been reported, reflecting changes in metabolism patterns under stress. This phenomenon also demonstrates the robustness of enzyme activity, as well as chain reactions that these fluctuations can have in larger-scale metabolic pathways. By analyzing enzyme mechanisms under different conditions, researchers can understand the mechanisms by which enzymatic fluctuations provide for overall metabolic function and disease initiation, underlining their relevance for both basic and applied biosciences (Figure 6.8).

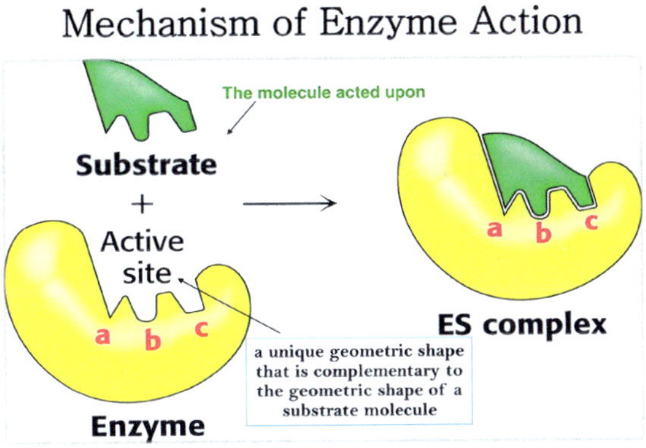

Figure 6.8. Mechanisms of enzyme action.

Source: https://external-content.duckduckgo.com/iu/?u=https%3A%2F%2Fimage3.slideserve.com%2F6886656%2Fmechanism-of-enzyme-action-l.jpg&f=1&nofb=1&ipt=ab973bfe1f1e84cb2e5e6ccb56b9de99404336f1ec105b11619833fa6e772ddf.

6.2.3. Factors Affecting Enzyme Activity

Enzymes, being biocatalysts, are highly affected by some of the factors that increase their activity and total metabolic efficiency. Temperature is among the key factors affecting the kinetic energy of molecules. When temperature increases, the enzyme-substrate interaction will be more efficient, leading to higher reaction rates. But this increased activity only lasts so long; past the enzyme's optimal temperature, it will denature the enzyme and render it inactive. In addition, pH changes can also significantly change an enzyme's shape and functionality, as drastic pH levels can break hydrogen and ionic bonds that keep the enzyme's three-dimensional structure intact. It is crucial in biochemical use, including drug discovery and industrial processes, to understand completely how temperature and pH influence enzyme kinetics. Substrate concentration is also a significant factor that has an impact on enzyme activity, influencing the speed at which enzymatic reactions take place. If substrate concentrations are low, the reaction rate begins rapid as the likelihood of enzymatic binding is more frequent; however, this only persists until the enzyme becomes saturated. On saturation, an entire complement of active sites is occupied, and no further rises in substrate concentration lead to increases in the rate of reaction. This saturation effect brings into relief the importance of the concentration of the enzyme in metabolic reactions. In certain cases, , such as in patients with AKI, the residual enzyme activity can be disturbed, thus affecting drug effectiveness and metabolism (Abosaif et al., 2014). Thus, an understanding of substrate and enzyme concentration interactions becomes critical for the optimization of metabolic efficacies as well as therapeutic treatments. Modulation of enzyme activity by the action of competitive and non-competitive inhibitors is also crucial. Competitive inhibitors bind to the active site of an enzyme and limit access to a substrate, with the effect of decreasing the reaction rate. Conversely, non-competitive inhibitors occupy another site and alter the enzyme's function without interfering with substrate binding directly. Non-competitive inhibitory activity is of considerable value in the study of cancer, where cancer cells often redirect metabolic pathways to bypass standard treatments (Abdel-Aziz et al., 2018). Knowledge of the functioning of such mechanisms enables researchers to develop more efficient pharmacological interventions that target enzyme activity more selectively. These findings can be applied to therapeutic innovations that take into account both enzymatic activity and dynamic environment within which the enzymes operate, ultimately resulting in improved patient outcomes for complex metabolic disease and cancer.

Test Your Knowledge

1. What is the mechanisms of enzyme action?
2. What are the Factors affecting enzyme activity?

6.3. Metabolic Pathways

Learning Outcome

1. Overview of catabolic and anabolic pathways;
2. Key metabolic pathways: glycolysis and Krebs cycle;
3. Regulation of metabolic pathways.

Metabolic pathways are complex sets of biochemical processes required to maintain life. They enable the conversion of nutrients to energy, structural components, and metabolic intermediates essential to cellular function. Pathways can be broadly categorized into catabolic and anabolic processes. Catabolism breaks down molecules to free energy, and anabolism builds up complicated molecules with the utilization of energy. The complex connectivity of such pathways facilitates cells to dynamically respond to changes in their environment, so as to maximize utilization of resources, as well as energy. Understanding the topology of metabolic pathways not only illuminates cellular processes but also aids in comprehending the evolutionary history of living organisms, revealing the molecular processes that set living systems apart from dead matter (Goldford & Segrè, 2018). The value of metabolic pathways extends to basic physiology; they have a significant function in disease and drug development as therapies. The genome-scale metabolic network of Staphylococcus aureus, for example, illustrates the influence of metabolism on virulence and antibiotic resistance. Scientific knowledge is provided to elucidate critical metabolic processes in bacteria by an integrated framework combining genomic data and experimental evidence. Such models enable scientists to predict the physiological characteristics of organisms under various circumstances and hence decide on the potential drug targets. This type of information is particularly significant in the era of novel antibiotic resistance because it guides the identification of novel chemotherapeutic drugs that are resistant to such resistant microorganisms (Heinemann et al., 2005). Furthermore, a comprehension of metabolic pathways is increasingly refined with improvements in systems biology, which provide avenues for quantitative analysis of metabolism. These avenues enable the study of metabolic networks at all scales, from molecular to organism-level behavior. Through the exploration of ancient metabolic pathways, researchers have uncovered crucial innovations that highlight the evolutionary significance of the pathways. By paying attention to collective autocatalysis and bioenergetics coupling, new research describes how the early metabolic networks may have emerged, thus illuminating life's origin. Such synthesis-based research not only makes clear our current conception of metabolic complexity but also provides a more cohesive view of biological systems that could prove useful for guiding future study in biotechnology and medicine (Goldford & Segrè, 2018) (Figure 6.9).

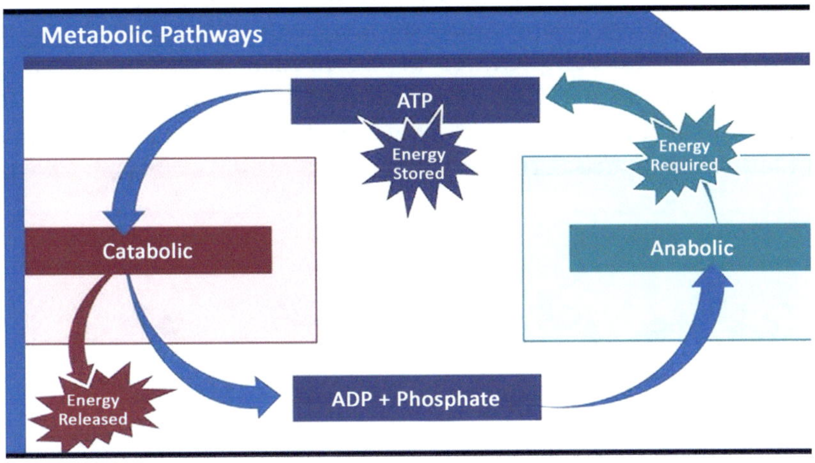

Figure 6.9. Metabolic pathways.

Source: https://i.ytimg.com/vi/m61bQYio9ys/hq720.jpg?sqp=-oaymwEhCK4FEIIDSFryq4qpAxMIARUAAAAAGAElAADIQj0AgKJD&rs=AOn4CLBN18ehEDx8-Jqm7hueyu2sIY6dVg.

6.3.1. Overview of Catabolic and Anabolic Pathways

The metabolism of living things is a complex series of catabolic and anabolic processes, which are essential to the viability of cellular functions and energy balance. Catabolic processes exist to break down organic compounds, , such as carbohydrates, fats, and proteins, into less complex products. The process is characterized by the release of energy, which is used in the form of ATP. Catabolism, in this instance, is facilitated by biochemical processes that oxidize glucose through a series of enzymatically catalyzed processes , such as glycolysis and the citric acid cycle. Catabolic reactions do not only yield ATP but also other energy-rich compounds , such as NADH and $FADH_2$, which serve vital functions in cellular respiration. Catabolic processes thus show how the energy of nutrients can be efficiently tapped and used to drive various cellular activities. Conversely, anabolic processes function to assemble complex biomolecules from their simpler precursors, and in this process, they need a supply of energy. Anabolism plays critical roles in cellular growth, repair, and reproduction because these processes construct macromolecules like proteins, nucleic acids, and polysaccharides. Anabolism is under strict control and literally uses energy gained by catabolic processes to fuel these endergonic processes. Key anabolic processes include the synthesis of proteins, through which amino acids are linked to form proteins, and photosynthesis, a significant metabolic process in plants in which light energy is stored as chemical energy. In cellular metabolism, the balance between the two rival processes allows organisms to adapt to the environment and control homeostasis, showing the dynamic process of metabolic processes. Recent work has unveiled the importance of metabolic modules for catabolic as well as anabolic processes, illustrating that such frameworks control significant evolutionary innovations. Metabolic systems are modular in that certain sets of reactions and enzymes are recyclable within different biological contexts to conserve energy and resources. For example, cofactors play a critical role in such metabolic modules and act as first-order regulators that enable the efficacy of biochemical reactions in both pathways. Metabolic architectures, based on the study, are not only reflective of the current physiological states of organisms but also influence long-term evolutionary plans for resource exploitation and adaptation to environmental change (Braakman & Smith, 2013). In addition, the energetic balance of spending, as has been witnessed in the instance of rumen microbes, is an indication of the complexities of metabolism whereby organisms are unable to maintain energy efficiency while meeting other metabolic requirements (Andries et al., 2015). Such complex understanding of metabolic pathways forms a stage for further studies in fields like biotechnology and medicine (Figure 6.10).

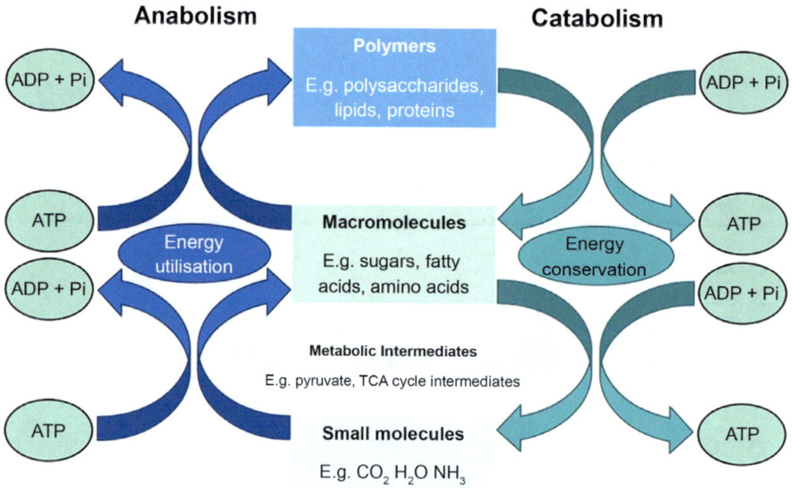

Figure 6.10. Catabolic and anabolic pathways.

Source: https://www.researchgate.net/publication/343835412/figure/fig1/AS:934063879766034@1599709591914/Coupling-of-anabolic-and-catabolic-pathways-Anabolism-utilises-energy-to-make.ppm.

6.3.2. Key Metabolic Pathways: Glycolysis and Krebs Cycle

The Krebs cycle (or citric acid cycle) and glycolysis are two critical phases of cellular respiration, and both are responsible for the production of energy in aerobic organisms. Glycolysis is a process that occurs in the cytoplasm and initiates the breakdown of glucose by degrading glucose to pyruvate, resulting in a net gain of two molecules of ATP and two molecules of NADH. This anaerobic and aerobic process serves a primary role in supplying energetic substrates to the Krebs cycle. Under the presence of oxygen, pyruvate is transferred into the mitochondria, where it is oxidized to yield acetyl-CoA that enters the Krebs cycle. This aerobic process is vital for maximizing ATP yields through oxidative phosphorylation, a testament to its role in cellular energy generation. The Krebs cycle adds another ATP production as part of a series of enzyme-catalyzed reactions within the mitochondrial matrix. During the cycle, acetyl-CoA flows into the series of reactions where it ends up being paired with oxaloacetate to become citrate. During the continuation of the cycle, citrate becomes modified through a series of alterations, eventually reforming oxaloacetate while releasing essential electron carriers NADH and $FADH_2$. These electron carriers are brought into the ETC, contributing to an electrochemical gradient that energizes ATP synthesis. The cycle also produces CO_2 as a waste product that is released as waste, illustrating the synergy between energy production and waste elimination in cell metabolism. This lovely interplay of reactions is important because disarrangements lead to metabolic disorders, illustrating the need for specificity in these pathways. Research has indicated that alterations in metabolism pathways, notably those involving glycolysis and the Krebs cycle, could be connected with a wide range of pathological conditions, including cancer. For transformed cells, as in Richters syndrome, there will often be an aberrant control of energy metabolism with a shift away from traditional pathways towards more glycolytic and glutaminolytic pathways (Micillo et al., 2023). Such metabolic reprogramming defines the way in which cancer cells reconstruct to support their augmented anabolic needs while sustaining a high rate of proliferation. Rather than relying solely on oxidative phosphorylation, these cells exhibit a reliance on aerobic glycolysis, also referred to as the Warburg effect, for the production of intermediates necessary for biosynthesis. Detecting such metabolic derangements, therefore, not only sheds light on energy production mechanisms but also indicates potential therapeutic targets in the realm of metabolic reprogramming, as evidenced by selective inhibitors of the pathways of glycolysis and the Krebs cycle (Periasamy, 2019) (Figure 6.11).

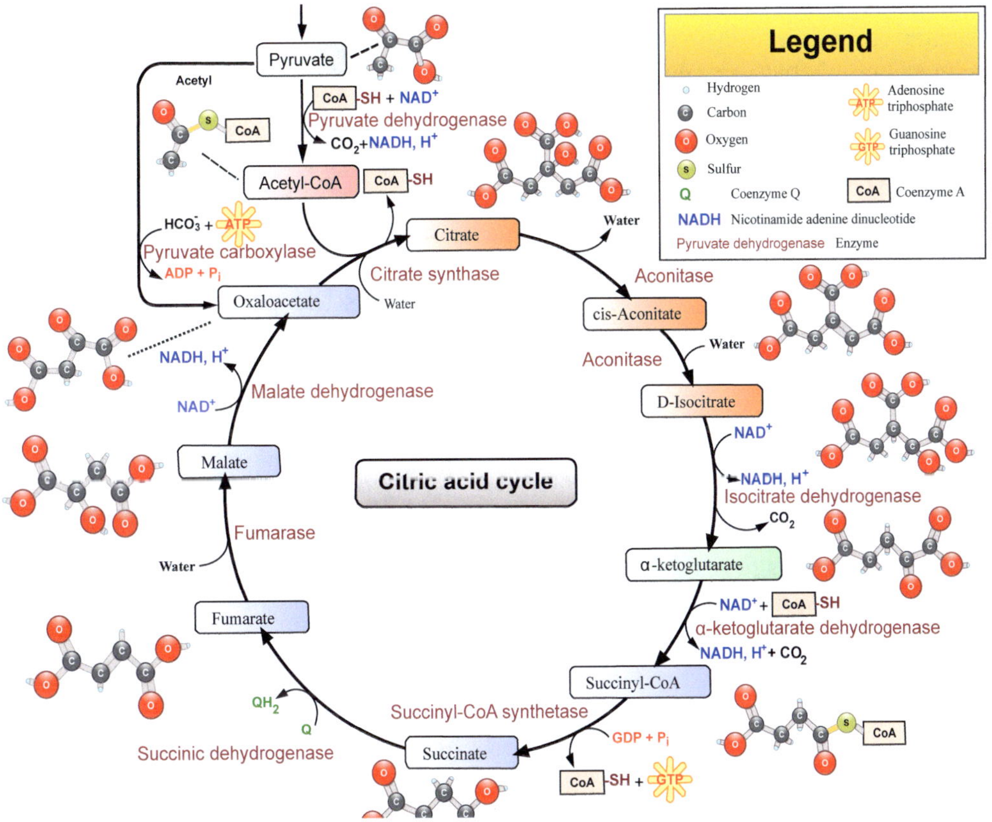

Figure 6.11. Glycolysis and the Krebs cycle.

Source: By Narayanese, WikiUserPedia, YassineMrabet, TotoBaggins - http://biocyc.org/META/NEW-IMAGE?type=PATHWAY&object=TCA. Image adapted from Image:Citric acid cycle noi.svg, CC BY-SA 3.0, https://commons.wikimedia.org/w/index.php?curid=6217701

6.3.3. Regulation of Metabolic Pathways

Regulation of metabolic pathways is a key aspect of cellular homeostasis and control of energy. Enzymes are the primary regulators of metabolism, catalyzing biochemical transformations that are necessary to maintain life. The activity and efficacy of enzymes are not constant but rather controlled through various mechanisms to render metabolic activities commensurate with the cells' energy needs and environmental conditions. For instance, allosteric regulation, covalent modification, and signal cascades that activate or inhibit prominent enzymatic activities may regulate metabolic pathways. These processes are crucial in that they achieve a balance in metabolic outputs in order for organisms to adapt to shifting nutrient availability or physiological state. Recent work has indicated the critical role of certain proteins in metabolic control, of which the 14-3-3 proteins are significant, acting in many signaling pathways that direct energy and nutrient homeostasis. The proteins regulate enzyme functions by binding to target proteins, thus affecting whole metabolic networks. The cell's interaction with 14-3-3 proteins and key signaling pathways, , such as insulin signaling and AMP-dependent kinase pathways, demonstrates their pivotal role in nutrient sensing and metabolism. Such regulation is especially critical in terms of disease, including obesity and diabetes, where metabolic pathway dysregulation leads to deleterious health outcomes. From a comprehension of these paradigms of regulation, scientists can orient therapeutic strategies that target metabolic dysfunctions. Besides, the role of humic substances (HS) in metabolic regulation has been established based on their relevance in plant physiology and nutrient nutrition. It has been established that HS can elicit activation of enzymes in the glycolytic pathway and induce protein synthesis, thus enhancing overall energy metabolism. This indicates the intimate relationship between external and intrinsic metabolic pathways. HS's ability to govern redox homeostasis and the enzymatic function of metabolic pathways gives a clue to more ecological significance, with prospective implications for plant development and ecosystem well-being. Through delineation of these interactions, researchers have a clearer notion of the mechanisms involved in metabolism as well as the potential of natural products to be utilized for enhancing farm productivity and ecological sustainability (Døskeland et al., 2011; Masi et al., 2018).

Test Your Knowledge

1. What are catabolic and anabolic pathways?
2. What are the key metabolic pathways?

6.4. Energy Transfer in Cells

Learning Outcome

1. Processes of cellular respiration;
2. Role of electron transport chain;
3. Importance of oxidative phosphorylation.

Energy transfer from cells is a dynamic and complex process that is pivotal to the support of life. Supporting energy transfer processes are metabolic processes that facilitate the conversion of energy sources into usable forms. This process is largely characterized by two general categories of metabolism: catabolism and anabolism. Catabolic processes break down larger molecules, releasing stores of energy, which are commonly obtained as ATP. Conversely, anabolic reactions employ energy gained from ATP to construct complex molecules necessary for cell growth and function. The balance between these two metabolic processes is regulated in a controlled manner and demonstrates the hierarchical and modular organization of metabolism, highlighting its central role in cellular functionality and responding to environmental changes (Braakman & Smith, 2013). In addition, some enzymes catalyze such metabolic processes, accelerating reactions that would occur at unconceivably slow velocities otherwise. Enzymes lower the activation energy required for biochemical reactions so that cells can respond quickly to energy requirements changes. The catalytic process is strictly controlled by such variables as substrate concentration, temperature, and cofactor availability. Environmental conditions, in the case of photosynthetic cells, influence electron transport and ATP synthesis profoundly. The balance between photosynthesis electron production and utilization in subsequent metabolic pathways is crucial; disrupted, it will lead to damage through the accumulation of lethal ROS and to cells (Saroussi et al., 2019). It is thus crucial to understand enzyme kinetics and regulative processes managing metabolic pathways in an effort to appreciate how energy transfer is optimized for cellular efficiency. The efficiency of energy transfer in cells relies not just on the complex design of metabolic pathways and enzymes but on cooperative groups of biochemical interactions that react to stresses from the external environment. Cells exhibit widespread versatility in their capability to regulate and control energy flow based on the circumstances that prevail at present. For instance, under diminished chloroplast activity, for instance, in Chlamydomonas reinhardtii, organisms employ alternative electron exit points to maintain energy balance. Electron diversion to routes involving flavodiiron proteins and starch biosynthesis are some of the examples of the capacity of the cells to control metabolism. This adaptability is pivotal to the existence of organisms that inhabit diverse environments where energy sources may vary extensively. Thus, the study of energy transfer mechanisms gives insight into broader ecological interactions while informing potential biotechnological applications to boost photosynthesis and energy utilization in cropping systems (Saroussi et al., 2019) (Figure 6.12).

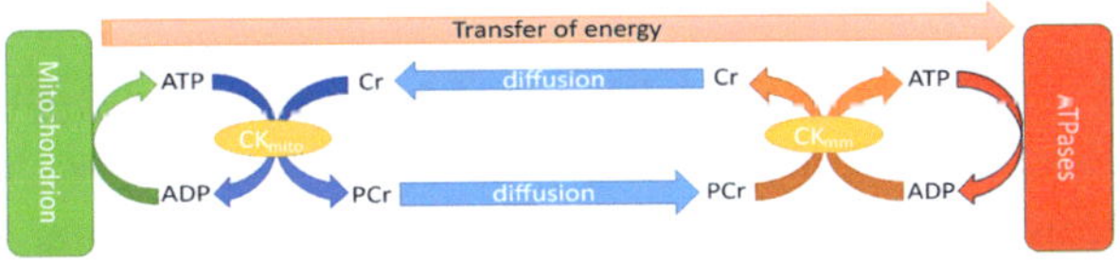

Figure 6.12. Energy transfer in cells.

Source: https://www.researchgate.net/profile/Jack-Miller-14/publication/342496236/figure/fig3/AS:916500017860611@1595522040784/Schematic-of-energy-transfer-through-the-cell.jpg.

6.4.1. Processes of Cellular Respiration

In an effort to perceive the complex processes

of cellular respiration, the fact that the biological process is vital in generating energy in aerobic organisms is worth mentioning. The whole process involves the conversion of glucose and oxygen into CO_2, water, and ATP. Cellular respiration is divided into four key steps: glycolysis, citric acid cycle, ETC, and oxidative phosphorylation. Glycolysis occurs in the cytoplasm, breaking down glucose to pyruvate with minimal side product of ATP and reducing equivalents. Pyruvate is further oxidized in the mitochondria through the citric acid cycle, liberating high-energy electrons that are essential for driving subsequent ATP synthesis.

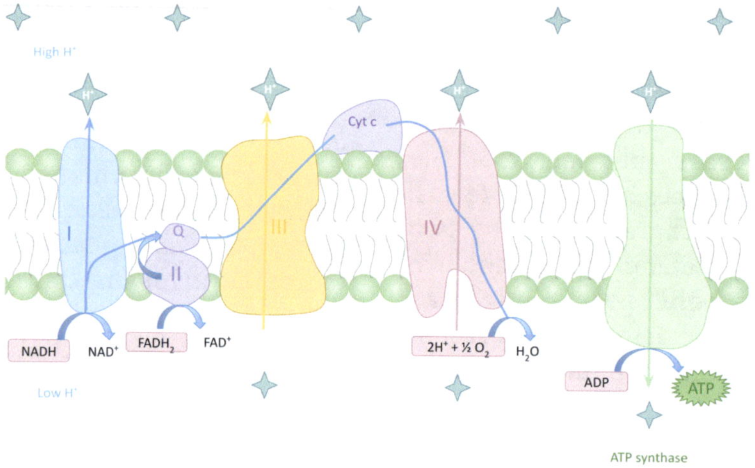

Figure 6.13. Cellular respiration process.

Source: By Irinapickle - Own work, CC BY-SA 4.0, https://commons.wikimedia.org/w/index.php?curid=130463092

This initial breakdown of glucose is the groundwork for the more complex reactions that occur within the mitochondria, illustrating the multi-step cellular respiration. An ETC is the key element in cellular respiration, as it allows electrons to be passed through a series of protein complexes in the mitochondrial membrane. As electrons travel through such complexes, their energy is utilized to drive protons (H^+) across the membrane to establish a gradient of protons. ATP synthase then draws on this gradient, where adenosine diphosphate (ADP) and inorganic phosphate are converted into ATP. Notably, alternative respiratory enzymes , such as alternative NADH dehydrogenases and alternative oxidases are an evolutionary mechanism to enhance metabolic flexibility, particularly under environmental transitions that may impair the standard respiratory route. The presence of these alternative enzymes shows how the organism possesses the ability to maintain bioenergetic efficiency during stressful conditions to gain prolonged ATP production even when the normal pathway is compromised (Gospodaryov et al., 2019). In addition, oxygen delivery and energy metabolism interaction is paramount in cellular respiration. Under hypoxia, when oxygen levels are reduced, the cells must adapt to remain alive and prevent tissue injury. NO plays a vital role herein, as it is a signaling molecule that facilitates hypoxic vasodilation and therefore increases oxygen delivery and blood flow. NO is also involved in the modulation of metabolic processes through the inhibition of oxygen consumption and activation of the utilization of alternative metabolic pathways. Such an adjustment reflects not only the importance of oxygen in cellular respiration but also the dynamic nature of energy management in cells. Awareness of these adjustments provides an illumination of the intricacy of metabolic processes and the fine balances that support cellular functioning under varied conditions of the environment (Umbrello et al., 2013) (Figure 6.13).

6.4.2. Role of Electron Transport Chain (ETC)

The ETC is the core process of aerobic respiration and takes place in the inner mitochondrial membrane of eukaryotic cells. The process is a series of complex protein complexes and electron carriers that enable electron passage from NADH and $FADH_2$ obtained, leading to the production of ATP. As electrons pass along the chain in various complexes, they pass on energy that is used to transport protons across the mitochondrial membrane, generating an electrochemical gradient. The gradient is used to drive ATP synthesis by ATP synthase, a protein enzyme that phosphorylates ADP into ATP.

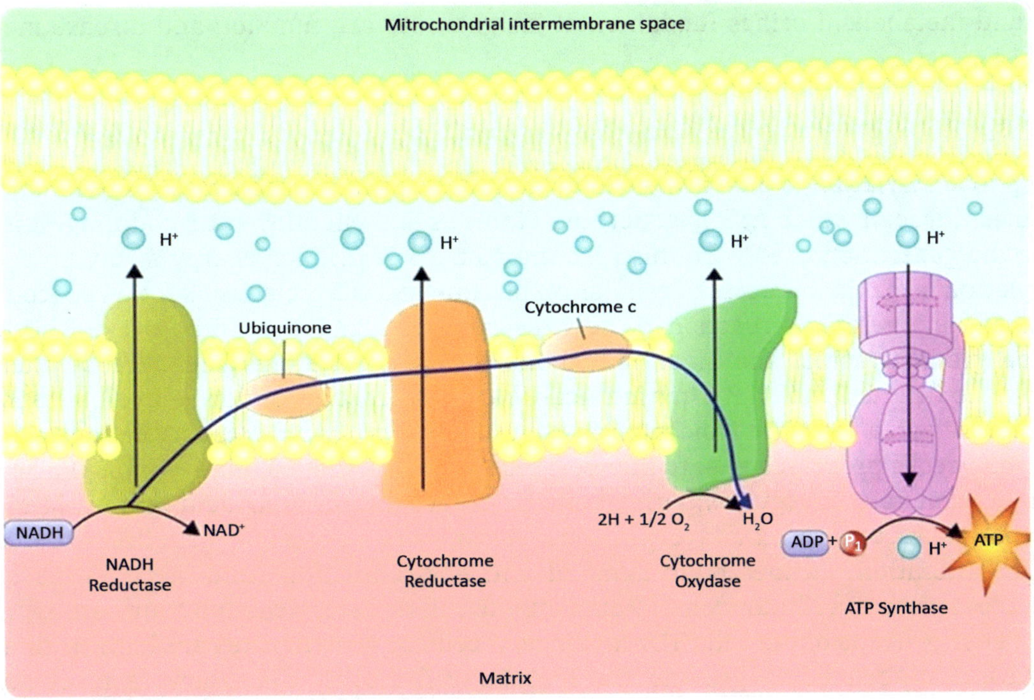

Figure 6.14. Role of the electron transport chain.

Source: https://d2jx2rerrg6sh3.cloudfront.net/image-handler/picture/2018/7/shutterstock_277344065.jpg.

Therefore, the ETC not only facilitates the generation of energy for various cellular processes but also demonstrates the fundamental relationship between metabolism, enzymes, and energy. Further, the ETC is also responsible for the understanding of metabolic regulation. The flow of electrons through the ETC is strictly regulated based on cellular energy requirements. For instance, in conditions of high ATP requirement, high levels of ADP stimulate the ETC, accelerating ATP production. In states of high ATP supply, however, activity of the chains is minimized, showing a feedback mechanism crucial to cellular energy homeostasis maintenance. This regulating mechanism illustrates how the ETC functions as a critical metabolic sensor that responds to fluctuations in energy status to optimize ATP production. Thus, the role of the ETC is not simply restricted to ATP generation but also has a leading role in the complete scenario of metabolic pathways and cellular energy regulation. Disturbances in the ETC can also lead to gross disturbances in metabolism, which have been experienced in various pathological conditions. Imbalanced ETC may cause deranged ATP production and increased oxidative stress and be implicated in the pathogenesis of diseases , such as diabetes, neurodegenerative disease, and cancer. For instance, it has been found that mutations in genes of ETC components give rise to mitochondrial diseases due to energy insufficiency and abnormal metabolic conditions. In addition, altered metabolic regimes linked with the ETC have been implicated in antibiotic resistance in pathogens like Staphylococcus aureus, as has been emphasized in current genome-scale metabolic models (Heinemann et al., 2005). This underscores the significance of the ETC in physiological processes as well as in clinical practice in new therapeutic strategies (Bode et al., 1987). Thus, understanding the role of the ETC in energy, enzymes, and metabolism brings fundamental insights into cell function and disease mechanisms (Figure 6.14).

6.4.3. Importance of Oxidative Phosphorylation

Oxidative phosphorylation is a critical cellular metabolic process, occurring predominantly in the mitochondria, through which ATP is generated via ETCs and chemiosmotic coupling. It is not only crucial for the generation of ATP but also for the balance of cellular energy as a whole. Oxidative phosphorylation is the final process of cellular respiration and traps electrons from reduced cofactors formed in earlier metabolic reactions, , such as glycolysis and the Krebs cycle. The formation of a proton gradient across the inner mitochondrial membrane energizes ATP synthase, which phosphorylates ADP to form ATP. The significance of this process cannot be overstressed because ATP has been referred to as the molecular currency of energy transfer during biological processes, powering many cells processes vital to life. Besides ATP production, oxidative phosphorylation also plays a significant role in regulating metabolic pathways and cellular adaptations. Mitochondria, the principal characters in oxidative phosphorylation, are dynamic organelles with the ability to adjust to alterations in metabolic demand and environmental conditions. These organelles communicate to other cellular components to modulate biological function, adjusting to the fluctuating energy needs of the cells. For instance, if cellular growth rates are high, as in neoplastic transformation, mitochondria provide the metabolic flexibility necessary to meet augmented anabolic requirements. This flexibility ensures the capacity of cells to survive even under stressful conditions, , such as variable oxygen and nutrient supply, thereby underscoring the pivotal role of oxidative phosphorylation in cellular homeostasis regulation and survival enhancement under stress (Cannino et al., 2018). Besides, the efficiency of oxidative phosphorylation is not only critical for normal cellular function but also for more general physiological effects. Dysruption of this process can lead to a range of states of metabolic disease, including mitochondrial myopathies or other aging diseases, further affirming the importance of unimpaired mitochondrial function. The ability to generate ATP through oxidative phosphorylation is the foundation for most biochemical pathways necessary for the preservation of life, and it emphasizes the interconnectedness of energy generation and metabolic well-being. Secondly, the understanding of this process is of critical importance in applications , such as cancer research, where metabolic reprogramming of tumor cells entails altering oxidative phosphorylation. The knowledge gained from this investigation of oxidative phosphorylation forms the backbone of the development of therapeutic interventions aimed at restoring metabolic regulation and optimizing cellular function (Fernandez-de-Cossio-Diaz & Vazquez, 2018) (Figure 6.15).

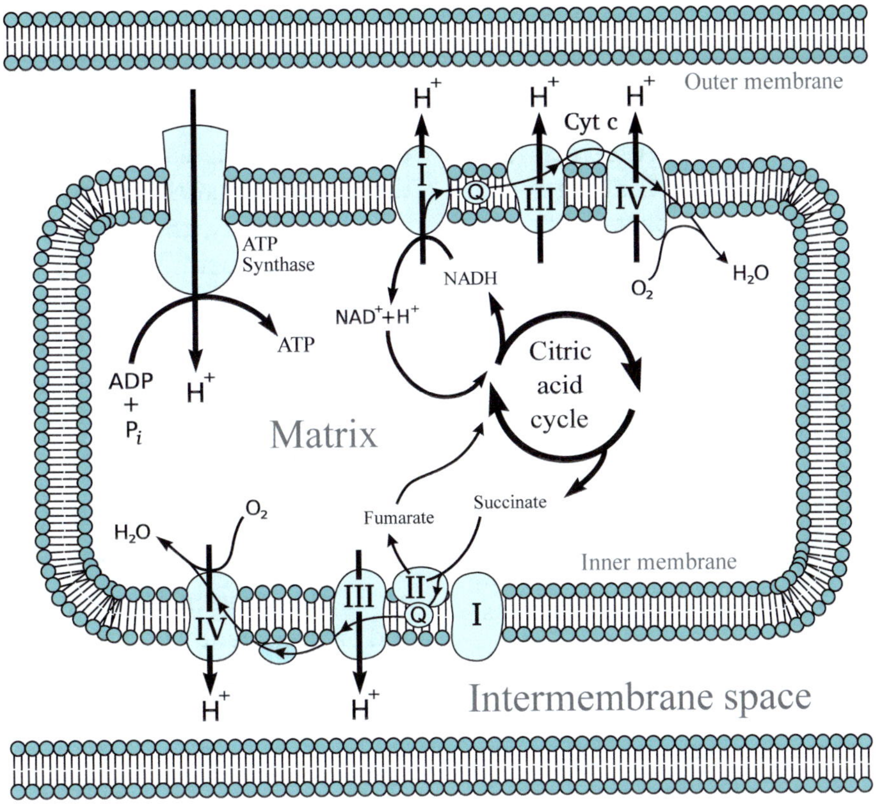

Figure 6.15. Oxidative phosphorylation.

Source: By Fvasconcellos 22:35, 9 September 2007 (UTC) - Vector version of w:Image:Etc4.png by TimVickers, content unchanged.A source (reference) has not been provided for the data in this self-made work., Public Domain, https://commons.wikimedia.org/w/index.php?curid=2716553

Test Your Knowledge

1. What are the processes of cellular respiration?
2. What is the importance of oxidative phosphorylation?

6.5. Enzyme Regulation and Inhibition

Learning Outcome

1. Types of enzyme inhibitors;
2. Allosteric regulation of enzymes;
3. Feedback inhibition in metabolic pathways.

Enzyme inhibition and regulation are fundamental mechanisms that govern metabolic pathways so that organisms are capable of responding accordingly to various environmental conditions and requirements for energy. There are mechanisms that regulate enzyme activity, including allosteric regulation, covalent modification, and feedback inhibition. Allosteric regulation refers to binding of effector molecules to sites other than the active site, and this leads to conformation changes that increase or reduce enzyme activity. Such regulation allows the organism to respond rapidly to changes in metabolite concentrations, thereby allowing for metabolic flux fine-tuning. Upon the change in metabolic needs, enzymatic regulation is crucial for the survival of the organisms through the dynamic coupling between energy status and enzyme activity. Additionally, metabolic inhibition rules illustrate the balance between energy conservation and metabolic efficiency. Inhibition can be by competitive, non-competitive, or uncompetitive means, whereby each inhibits enzyme action differently. Competitive inhibitors, for instance, bind to the active site, being in effect substrate-like molecules, and clog metabolic pathways. Non-competitive inhibitors bind elsewhere on the enzyme, altering its function regardless of substrate presence. This degree of variability in inhibition mechanisms illustrates the importance of enzymes as the nodes of regulation in metabolic networks. Enzyme inhibition is not a means of limiting processes but instead maximizes resource allocation and keeps cellular energy by preventing wasteful metabolic expenses, echoing economics of flux analysis principles in which enzyme costs versus metabolic benefits are most significant (Liebermeister et al., 2014). An intriguing context in which to place value on understanding enzyme regulation is to place it within the context of environmental conditions, particularly levels of nutrients.

For example, iron bioavailability can fundamentally alter enzymatic activities in phytoplankton, as documented in work on the diatom *Thalassiosira pseudonana*. At low iron availability, there is a cascading series of metabolic adjustments involving the induction of proteins involved in intracellular recycling and the pentose phosphate pathway to maintain organism survival in the face of nutrient deficiency. This accommodation emphasizes the importance of enzyme regulation not only in maintaining individual metabolic processes, but also in promoting greater biogeochemical cycles, , such as carbon sequestration. The interactive relationship between enzyme activity and environmental stimuli indicates the broader implications of enzyme regulation on metabolic processes, making its importance robust in the fields of ecology and environmental science (Boyd et al., 2015) (Figure 6.16).

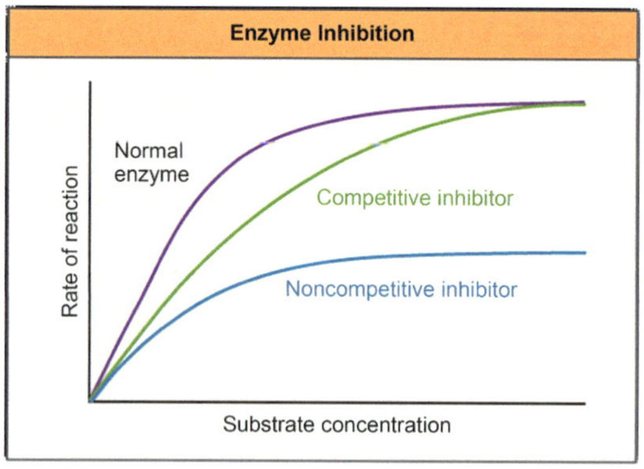

Figure 6.16. Enzyme regulation and inhibition.

Source: https://cdn.kastatic.org/ka-perseus-images/2dc844a85af00f941da8641751c8f312b07a742c.png.

6.5.1. Types of Enzyme Inhibitors

Enzyme inhibitors are key players in the regulation of metabolic processes by controlling enzyme activity with a subsequent effect on systemic energy balance and cellular metabolism. In general classification, the inhibitors are classified into reversible and irreversible inhibitors. The reversible inhibitors bind to enzymes with non-covalent bonds such that the enzyme activity may return to normal upon the removal of the inhibitor. These include competitive inhibitors, which compete with the substrate for the active site, and non-competitive inhibitors, which bind to an allosteric site and modify enzyme conformation. The reversible character of these categories means fine-tuning of metabolic processes, which could prove to be essential in maintaining homeostasis in biological systems. Unraveling such dynamics is key, especially in drug design contexts wherein modulation of enzyme activity can lead to therapeutic benefits for metabolic disease. Irreversible inhibitors, however, bind to enzymes with a covalent bond to irreversibly shut down their catalytic function. Such a form of inhibition is typically linked to an irreversible enzyme inactivation and is often a critical mechanism of action for many drugs. For instance, certain anti-cancer drugs exhibit their effect through irreversible inhibition of some nucleotide biosynthesis enzymes crucial for highly dividing cancer cells. Orotic acid is a pyrimidine metabolic intermediate that is a nice example in which competing inhibitors can disrupt its synthesis by blocking the enzymes. Such interference can cause orotic aciduria, and hence, it is the clinical relevance of knowing different classes of enzyme inhibitors. As noted in enzyme studies like that of dihydroorotate dehydrogenase, enzyme inhibition mechanism hints at the intricacy of metabolic control, further buttressing the need for ongoing research to ascertain their therapeutic potential (Fonteh et al., 2018). Further still, enzyme inhibitor research goes as far as new systems biology applications, making use of model organisms like African trypanosomes. These unicellular eukaryotes present unique opportunities for investigating metabolic pathways since they possess relatively simple cellular architectures and thoroughly characterized genomics. Dynamic modeling enables one to simulate the effects of enzyme inhibitors on metabolic networks and gain understanding of the mechanisms underlying the control of energy production and utilization in cells. As was found in trypanosome studies, kinetic data for glycolytic enzymes are used to direct model-building that can appropriately predict the outcome to a variety of inhibitors. This kind of an approach has promise in discovering new drug development targets, particularly in parasitic diseases, making enzyme inhibition study important in the development of therapeutic strategy (Barbara et al., 2010) (Figure 6.17).

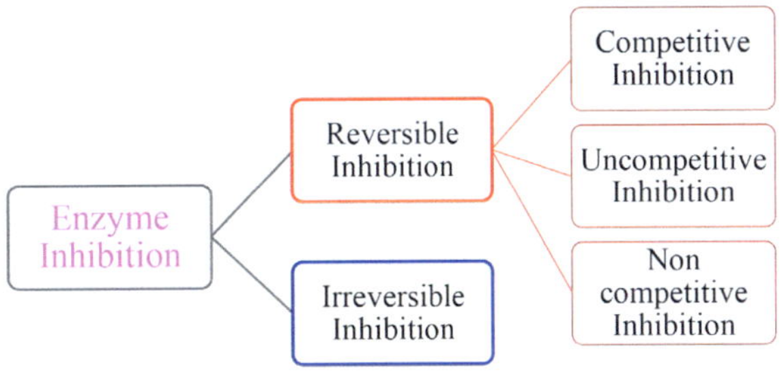

Figure 6.17. Enzyme inhibitors and classification of enzyme inhibition.

Source: https://i.imgur.com/k2IlqYc.png.

6.5.2. Allosteric Regulation of Enzymes

The complex enzymic mechanisms are primarily regulated by allosteric regulation, a significant mechanism that modulates the activity of enzymes in metabolic pathways. Allosteric regulation involves the binding of metabolites to areas outside the active site of an enzyme, causing conformational changes that modulate enzymatic activity. This advanced mechanism allows for strict control of metabolic flux, so that cellular response is sensitively calibrated to changing energetic requirements. Regulation is particularly important in the central carbon metabolism (CCM), where flux balance is accountable for significant processes , such as ATP generation and biosynthetic precursor synthesis. Understanding how allosteric effectors work with these pathways is significant because it uncovers potential therapeutic and metabolic engineering targets, which will ultimately result in the optimization of metabolic processes in natural and constructed systems. Experimental research has uncovered the significance of allosteric regulation in modulating enzyme activity in various organisms and revealed evolutionary significance and conservation of allosteric regulation. More recent studies, for instance, of the yeast *Saccharomyces cerevisiae*, have shown how changes in allosteric regulation correlate with shifts in glycolytic flow as a function of growth conditions. At higher yeast specific growth rates, there is a sharp diminution in the role of metabolite levels, e.g., allosteric effectors. This implies a complex relationship where increased glycolytic flux may lead to decreased allosteric control, and therefore an equilibrium dynamic in metabolic regulation. The findings above indicate the need for further research on the interaction of posttranslational modification and allosteric interaction in managing the flux along key metabolic pathways (Chen et al., 2023). Moreover, recent advances in metabolic control analysis have shed new light on the regulatory small molecule networks (SMRN) behind enzyme regulation. By reconstructing SMRNs for organisms like Escherichia coli, researchers have been able to uncover how small molecule interaction with enzyme activity operates, emphasizing the built-in trade-offs present with regulation and production metabolically. The allosteric regulation of enzymes can profoundly impact metabolic homeostasis, revealing the complex balance that organisms possess in energy production versus resource allocation. It not only enhances our understanding of metabolic pathways but also illustrates the evolutionary conservation of such mechanisms among a broad spectrum of species. Collective outcomes from such studies justify the critical function of allosteric control in promoting metabolic effectiveness and responsiveness to a rapidly fluctuating cellular environment (Briars et al., 2017) (Figure 6.18).

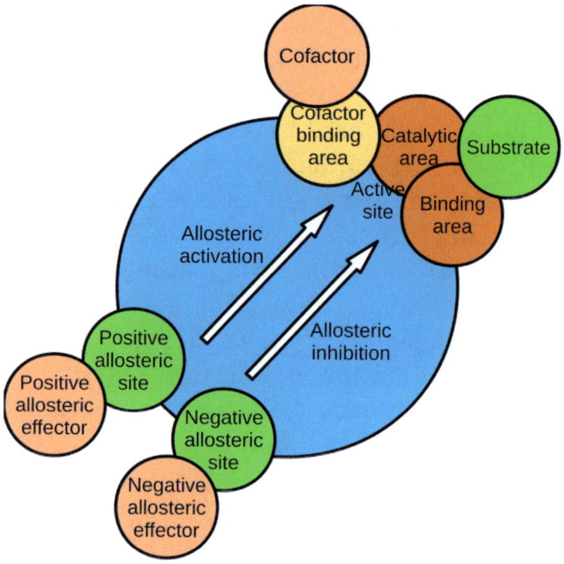

Figure 6.18. Allosteric regulation of enzymes.

Source: By Tenthkrige (talk).Tenthkrige - Own work(Original text: I (Tenthkrige (talk)) created this work entirely by myself.), CC0, https://commons.wikimedia.org/w/index.php?curid=85816837

6.5.3. Feedback Inhibition in Metabolic Pathways

Feedback inhibition is a critical mechanism for maintaining metabolic homeostasis in cells since it acts as a feedback mechanism for the regulation of enzyme activity across various pathways. Feedback inhibition helps to prevent cells from the overaccumulation of metabolites, hence preventing energy wastage. Feedback inhibition regulates the activity of the enzymes involved in producing the major metabolites by inhibiting them once their concentration is sufficient enough, thus regulating metabolic flow. For instance, when a terminal compound in a biosynthetic pathway is accumulated, it may bind to an allosteric site of a related enzyme, triggering a conformational change that decreases the activity of the enzyme. The regulatory strategy is essential for the adaptive behavior of metabolic networks under fluctuating cellular demands to prevent waste and ensure optimal conditions for cellular function. The significance of feedback inhibition also extends beyond mere simple metabolic pathways to influence the larger scheme of cellular signaling and interaction. The regulatory influence of small molecules over enzyme activity has profound implications in various biological scenarios. For instance, disruption of feedback inhibition is the cause of metabolic diseases or even cancer, where metabolic pathways are re-routed towards facilitating rapid growth and cell division. In the context of the tumor metabolism, disrupted enzymatic function may arise from mutational activation or loss of oncogene or tumor suppressor gene function, creating uncontrolled throughput through mechanisms , such as aerobic glycolysis. By defining these complex regulatory networks, researchers are establishing the evolutionary conservation of metabolic control processes across species and are gaining deep insight into the role feedback inhibition plays to ensure homeostasis in differing biological systems (Briars et al., 2017). Moreover, the study of feedback inhibition has gained relevance in clinical environments, particularly in cancer treatment. Cancer cells possess a distinct metabolic signature with enhanced glucose metabolism to lactate under aerobic conditions, known as the Warburg effect. The metabolic shift supports rapid cell division and viability in stressful environments, and thus investigation of the feedback processes is essential in the formulation of targeted therapies. Recent research has highlighted the potential of microRNAs to fine-tune the metabolic pathways by modulating the expression and activity of glucose and lipid metabolism enzymes. Therapeutic modulation of these miRNAs can provide new avenues to rein in cancer's dysregulated metabolic pathways, drawing attention to the complex relationship between feedback inhibition and potential therapeutic intervention in metabolic disorders (Hatziapostolou et al., 2013).

Test Your Knowledge

1. What are the types of enzyme inhibitors?
2. What is feedback inhibition in metabolic pathways?

6.6. Metabolism and Homeostasis

Learning Outcome

1. Relationship between metabolism and homeostasis;
2. Impact of metabolic disorders on health;
3. Role of metabolism in energy balance.

Homeostasis and metabolism are basic interdependent processes that ensure the survival and functions of living organisms. Metabolism is all the biochemical reactions occurring in a single cell, , such as catabolic reactions that break down molecules to supply energy and anabolic reactions that build complex molecules from simple ones. This fine tuning is necessary for homeostasis, the constancy within the internal environment necessary for maximum cellular function in the presence of variable external environments. Disruption of metabolic pathways leads to metabolic disorders, and as such profoundly impact human homeostasis. Thus, it is imperative to have full comprehension of mechanisms of metabolic regulation for the purpose of therapeutic intervention against diseases as metabolic homeostasis is essential for well-being. At the level of metabolic pathways, fatty acids are an essential part, particularly their role in energy generation and cell signaling. Recent research demonstrates that saturated fatty acids (SFAs) have both harmful and beneficial effects on cellular metabolism based on their concentration. For instance, while high levels of palmitate cause lipotoxicity and adverse health effects, low concentrations are found to be advantageous in enhancing mitochondrial function. The modulation of mitochondrial homeostasis by low levels of palmitate means that dietary sources of dietary fats must be approached with sensitivity in the context of homeostasis and metabolism because they can modulate oxidative stress, energy production, and cell health at large (Candas-Green et al., 2020). This sensitivity towards fatty acid metabolism suggests the complexity of achieving homeostatic balance even with variations in dietary habits. Moreover, disruptions in metabolic processes can have catastrophic effects on neurological health. Rett syndrome (RTT), a neurodevelopmental disease caused by mutations within the MECP2 gene, is used to illustrate here the critical interaction between metabolism and homeostasis in the brain. Research has confirmed that lipid metabolism is severely impaired in the brains and livers of MECP2-null models and that this accurately represents an extreme disruption in cholesterol homeostasis. This is evident in biochemical alterations in protein expressions that regulate cholesterol metabolism, e.g., enzymes and receptors with essential roles in lipid balance (Tunno et al., 2014). This points to the possibility of modulation of lipid metabolism as potential for creating novel therapeutic options for RTT and related disorders yet again underscoring the necessity to find metabolic and homeostatic processes in disease and health.

6.6.1. Relationship Between Metabolism and Homeostasis

The complex interaction of homeostasis and metabolism lies at the heart of understanding how bodies maintain internal stability despite external alterations. Metabolism is a highly complex series of biochemical reactions converting food into energy and construction materials needed for repair, growth, and maintenance. In this dynamic world, homeostasis is a control mechanism that makes physiological processes maintain themselves within certain levels so that organisms can cope with stresses like changing temperatures, availability of nutrients, or requirements for energy. Functional metabolic processes, conditioned by enzymatic activity, enable these regulatory processes to proceed by modulating the synthesis of hormones that control energy balance and the allocation of nutrients. For instance, active enzymes of glycolytic pathways can modulate metabolic rates, thus influencing energy homeostasis and the capacity of an organism to respond to varying physiological requirements at various stages of life or environmental conditions. Feedback mechanisms' function is a central aspect of such interaction, particularly the manner through which hormonal signals regulate glucose metabolism, a significant constituent of homeostasis. Feedback mechanisms function by monitoring blood glucose to maintain it at a restricted range optimal for normal cellular processes. When an increase in glucose is detected, insulin is released to stimulate cell uptake, enhancing metabolic pathways leading to energy generation. With the drop in glucose, glucagon is released to activate glycogenolysis, thus releasing glucose stores back into the blood. This complex interaction illustrates the inherent nature of metabolic regulation and its reliance on advanced feedback processes. Employing the conceptual framework of ideas , such as an organicist view of closure of constraints, emphasizing the reliance of metabolic processes and regulatory responses, a more robust understanding of metabolic homeostasis enters the picture beyond simple explanations for feedback, revealing integrating levels between various physiological processes (Bich et al., 2020). Recent studies, , such as the investigation of HS in plant metabolism, demonstrate the significant contribution of environmental factors to metabolic processes and homeostasis. The application of HS revealed alterations in protein expression associated with energy metabolism, indicating that external stimulation can significantly influence internal metabolic conditions. Most notably, glycolytic enzyme activation and ribosomal protein induction have been linked with enhanced energy metabolism and protein synthesis, showing how environmental interaction determines metabolic responses. This interaction emphasizes the necessity of understanding metabolic pathways as isolated biochemical processes as well as intact systems that serve to maintain homeostasis upon being faced with environmental stimuli. The regulation of these enzymatic processes highlights ROS as pivotal mediators and signals of homeostatic adaptation in response to extrinsic dynamics (Masi et al., 2018) (Figure 6.19).

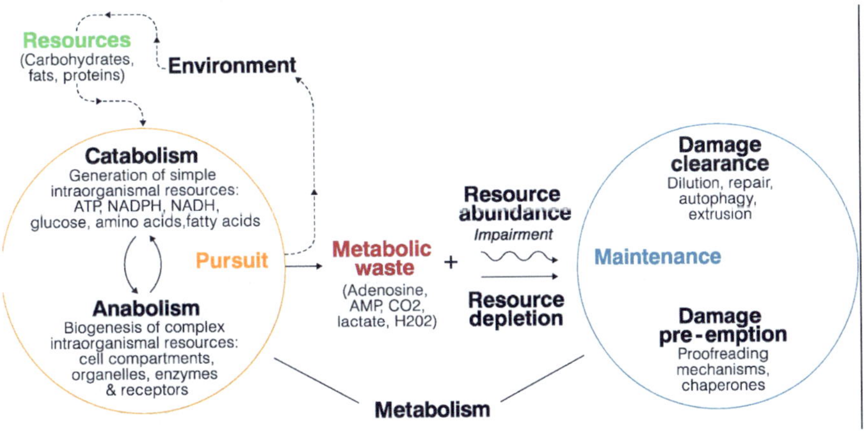

Figure 6.19. Relationship between metabolism and homeostasis.

Source: https://www.cell.com/cms/10.1016/j.tem.2023.01.003/asset/7567493b-dd29-415f-a064-f0365b063ded/main.assets/gr1_lrg.jpg.

6.6.2. Impact of Metabolic Disorders on Health

Metabolic disorders present the world with a significant health problem, impacting not only the health of individuals but also healthcare systems as a whole. Metabolic disorders disrupt nutrient processing as it is supposed to be done, leading to obesity, diabetes, and cardiovascular disease. The integration of metabolic components, diet, and the environment leads to the development of these disorders, which are further hastened through lifestyles characterized by no exercise. When metabolic pathways are disrupted, the body's ability to guarantee homeostasis breaks down, creating a domino effect of health issues. A cure for metabolic disorders has to be based on a deep understanding of biochemical processes, which will lead to the best prevention and cure methods. Through the knowledge of the way such disorders affect the expenditure of energy and the utilization of nutrients, researchers can open doors for future innovation in medicine and public policy to curb their incidence and impact on society (Sarmidi et al., 2012). Metabolic disorders closely relate to nutritional deficiencies in cases of 'hidden hunger' when individuals fail to get adequate micronutrients despite their appropriate caloric intake. This crisis affects over two billion individuals across the globe, including those living in industrialized nations. Low levels of essential vitamins and minerals can have severe health effects, impair immunity, and exacerbate metabolic diseases. Such is the case with the lack of essential B-vitamins , such as B1, B6, and B9, which play significant roles in energy metabolism and cell processes. These deficiencies can result in compromised glycemic control and increased fat deposition, making the management of metabolic disease more challenging. Thus, the inclusion of micronutrient augmentation approaches , such as biofortification, in dietary recommendations is vital. This is in the interest of optimizing overall health outcomes and arresting the effects of metabolic disorders via better nutritional care and interventions (Strobbe et al., 2018). Besides, metabolic diseases have far-reaching consequences going beyond individual health, extending to societal and economic structures. Rising rates of disease , such as obesity and diabetes impose a significant strain on health expenditure, burdening public health systems. With chronic diseases on the rise, medical services and treatment are in greater demand, necessitating sound prevention strategies. This includes increasing physical activity promotion, improving access to healthy food, and informing people of the importance of healthy eating in maintaining metabolic health. By creating environments that make healthy choices easier, society as a whole can attempt to reduce the burden of the metabolic disorders. Collaboration between policymakers, healthcare practitioners, and food producers needs to be initiated to promote an integrated health strategy whose core principles are metabolism and energy balance applied in public health promotion (Sarmidi et al., 2012).

6.6.3. Role of Metabolism in Energy Balance

Metabolism plays a crucial role in the energy balance within organisms as a complex system transforming food into energy and, simultaneously, regulating other biological processes. At its core, metabolism is composed of both catabolic and anabolic processes, which respectively break down nutrients to release their energy and synthesize new materials required for growth and repair. A disturbance in these metabolic activities can lead to conditions , such as obesity, diabetes, and other metabolic diseases, and therefore it is important to understand the interaction between energy intake and expenditure. With the extreme significance of this balance, recent research has focused on the complex roles of energy metabolism in cellular homeostasis, emphasizing the fact that alterations in metabolic control would have far-reaching influences on general well-being and disease states. Among the most important parts of metabolic processes is the mitochondria, commonly referred to as the powerhouses of the cell due to their function in aerobic respiration and energy generation by ATP production. The dysfunction of mitochondria has been linked to various diseases, which underscores the relationship between metabolism and energy homeostasis. It has been elucidated in recent studies that mitochondrial bioenergetic changes can not only impact energy generation but also cellular signaling pathways governing cell survival and growth. Of special interest is the identification of the 5′-AMP activated protein kinase (AMPK) that highlights how metabolic changes can impart crucial signals in cancer metabolism so that the integrity of mitochondria is vital in maintaining energy homeostasis. These findings exemplify how metabolism pathways are interconnected to cellular destiny that constitutes the basis of how energy regulation functions both in disease and healthy states (Iommarini et al., 2017). Further, some enzymes play a fundamental role in offering proper metabolic function and, by extension, energy equilibrium. For example, hypoxanthine guanine phosphoribosyl transferase (HPRT) enzyme is involved in purine metabolism, and abnormality of this enzyme may lead to metabolic dysfunction and deranged mitochondrial energy metabolism. Research suggests that HPRT deficiency may inhibit mitochondrial respiration, producing accumulations of metabolic intermediates, which may disrupt neuronal integrity, particularly under conditions like Lesch-Nyhan syndrome. This suggests how complicated the interactions are among enzyme function, energy metabolism, and overall metabolic status. Uncovering these connections further illuminates how enzymatic regulation of metabolism can influence energy homeostasis and even yield therapeutic strategies for metabolic disease (Vinokurov et al., 2023).

Test Your Knowledge

1. What are the impacts of metabolic disorders on health?
2. What is the role of metabolism in energy balance?

Summary

In comprehending the complex interplay of energy, enzymes, and metabolism, it is clear that enzymes are critical catalysts of metabolic pathways that bring about efficiency and regulation in living systems. These proteins catalyze basic biochemical reactions, allowing organisms to convert energy into useful forms while simultaneously regulating physiological processes. Elucidating the role of significant enzymes, , such as plant hexokinases, provides significant insight into how organisms react to their environmental energy source. As is evident in research on *Physcomitrella patens*, hexokinases are not only involved in the metabolism of glucose but also function as sugar sensors, thus bridging energy regulation with overall metabolic response. These observations underscore the importance of enzymatic activity in energy conversion and metabolic homeostasis, providing a platform to study life at the cellular level (Olsson et al., 2005). Further, research on energy balance, especially in high-yielding dairy cows, indicates the complexity of metabolic regulation. The dependency of cow reproductive performance on their status is the fine balance to be maintained across metabolic networks. Cows with severe negative energy balance (SNEB) have altered gene expression that affects reproductive performance and health adversely. Through finding hepatic gene expression with fluctuating energy balances, it was revealed that these changes could lead to downregulation of such critical biological pathways that they affect hormone levels and levels of inflammation. These results highlight how lactational energy needs can compromise cellular function and metabolic acclimations that are vital to reproduction as well as welfare in cattle (Diskin et al., 2007). Taken collectively, the topic of energy, enzymes, and metabolism highlights the interconnectedness of life processes that regulate life. Enzymes are crucial components of metabolic pathways responsible for catalyzing the conversion and regulation of energy in living organisms. The outcome of energy imbalance, particularly in the agricultural setting, is a timely reminder of the sensitive balance needed for ideal metabolic operation. As continuing research sheds lighter on the intricacies of the interactions, the need for systems thinking in energy management and metabolic health is made increasingly clear by the day. Follow-up research that delves more thoroughly into these subjects, particularly in mixed environments, will be critical to advancing knowledge and improved outcomes in human health and in ag systems. In this regard, the work on metabolism not only advances our scientific knowledge but also informs practical implications for sustainability and productivity in various biological systems.

Metabolism is a complex set of biochemical processes sustaining life, characterized by its modularity and hierarchical structure. Not only are metabolic architectures critical to isolated biochemical reactions, but larger biological networks are also built on them. This modularity allows for an easier grasp of how metabolic pathways function and evolve over time. Various modules made up of specific sets of chemicals and enzymes in combination comprise the metabolic network, reproducing similar patterns to enable the generation and utilization of energy. The origin of these modules dates back to the earliest evolutionary steps, and thus, they form an early building block of life's biochemical processes. These patterns in metabolism inform our understanding of energy circulation in ecosystems and the concepts that govern biospheric organization (Braakman & Smith, 2013). Also, metabolic cofactors have an essential role to play as regulatory scaffolds, which promote enzymatic activity and regulate metabolic flux. Such cofactors are as simple as they are complex, with the more complex forms usually located at interfaces of metabolic modules. This strategic placement allows them to regulate substrate and enzyme interaction effectively. The specific functions and structural diversity of cofactors demonstrate their importance in the biochemical setting. By functioning as keys to place different organic reactions in terms of metabolism, cofactors are the key to maintaining the harmony and efficiency of metabolic pathways. Their response to enzymes demonstrates how biochemical processes have adapted to ensure changes in the environment, educating us about the evolution of life and optimization of energy use (Ebenhoeh et al., 2013). The combination of these fundamental ideas of energy, enzymes, and metabolism provides a framework of comprehension of biochemical networks at various hierarchical levels. By establishing connections between small-molecule substrates, enzyme activities, and metabolic modules, researchers reveal dynamic interactions that sustain cellular processes. Additionally, the conserved nature of core metabolic pathways sheds light on evolutionary histories, illustrating how organisms adapted to different ecological niches. Analysis of such metabolic architectures reveals significant correlations between energy transfer, enzyme-mediated regulation, and the generalized relevance of metabolism to life maintenance. Therefore, a deeper understanding of the metabolism is attained, as its central role in the power machinery and evolutionary biology is revealed (Braakman & Smith, 2013).

KEY TERMS FROM THIS CHAPTER

Activation energy	Active site	Anabolism
ATP (adenosine triphosphate)	Catabolism	Catalyst
Coenzyme	Cofactor	Competitive inhibition
Endergonic reaction	Energy coupling	Enzyme
Enzyme-substrate complex	Exergonic reaction	Feedback inhibition
Inhibitor	Metabolism	Noncompetitive inhibition

CRITICAL THINKING QUESTIONS

1. **What is the main molecule that cells use for energy?**
   a. DNA
   b. Glucose
   c. ATP
   d. Enzyme
2. **Enzymes speed up chemical reactions by:**
   a. Increasing temperature
   b. Increasing the amount of product
   c. Lowering activation energy
   d. Adding more reactants
3. **What term describes all the chemical reactions occurring in a cell?**
   a. Anabolism
   b. Catabolism
   c. Digestion
   d. Metabolism
4. **Which of the following can affect enzyme activity?**
   a. Temperature
   b. pH
   c. Substrate concentration
   d. All of the above
5. **In metabolism, catabolic reactions are those that:**
   a. Build larger molecules from smaller ones
   b. Store energy
   c. Release energy by breaking down molecules
   d. Do not involve enzymes

Short Answer Questions

1. What is metabolism, and what are its two main types?
2. What is the first law of thermodynamics and how does it apply to living organisms?
3. What is ATP and why is it important in cells?
4. Describe how energy is released from ATP. What happens to the molecule?
5. What are enzymes and what role do they play in metabolic reactions?
6. What is activation energy and how do enzymes affect it?

Answers to Critical Thinking Questions

1. (c); 2. (c); 3. (d); 4. (d); 5. (c)

References

1. Abdel-Aziz, H., Ahmadzadeh, M., Al-Zhoughbi, W., Albina, A., Alkan, A., Allard, J., Allen, S., et al. (2018). Metabolism within the tumor microenvironment and its implication on cancer progression: An ongoing therapeutic target. https://core.ac.uk/download/214833072.pdf (accessed on 13th June 2025).
2. Abosaif, N., Andres-Hernando, A., Philips, B., Boucher, R., Bougle, D., Brown, M., et al. (2014). Xenobiotic metabolism: The effect of acute kidney injury on non-renal drug clearance and hepatic drug metabolism. https://core.ac.uk/download/19937736.pdf (accessed on 13th June 2025).
3. Vinokurov, A. Y., Soldatov, V. O., Seregina, E. S., Dolgikh, A. I., Tagunov, P. A., Dunaev, A. V., Skorkina, M. Y., Deykin, A. V., & Abramov, A. Y. (2023). HPRT1 Deficiency Induces Alteration of Mitochondrial Energy Metabolism in the Brain. *Molecular neurobiology*, *60*(6), 3147–3157. https://doi.org/10.1007/s12035-023-03266-2 (accessed on 13th June 2025).
4. Braakman, R., & Smith, E. (2013). The compositional and evolutionary logic of metabolism. *Physical biology*, *10*(1), 011001. https://doi.org/10.1088/1478-3975/10/1/011001 (accessed on 13th June 2025).
5. Amigo, J. M., Baldazzi, V., Beard, D. A., Benyamini, T., Blank, L. M., Boggy, G., Burgard, A. P., et al. (2011). Systems approach to modelling pathways and networks. https://core.ac.uk/download/9634232.pdf (accessed on 13th June 2025).
6. Andries, J., Argyle, J., Armstead, D., Asanuma, N., Atasoglu, C., et al. (2015). Maximizing efficiency of rumen microbial protein production. https://core.ac.uk/download/323070152.pdf (accessed on 13th June 2025).
7. Arrigo, K. R., Comeau, A. M., Comiso, J. C., Devos, C., DiTullio, G. R., Ferrara, S., Gabric, A. J., et al. (2014). Polar microalgae: New approaches towards understanding adaptations to an extreme and changing environment. https://core.ac.uk/download/41989130.pdf (accessed on 13th June 2025).
8. Bakker, B. M., Krauth-Siegel, R. L., Clayton, C., Matthews, K., Girolami, M., Westerhoff, H. V., Michels, P. A., Breitling, R., & Barrett, M. P. (2010). The silicon trypanosome. *Parasitology*, *137*(9), 1333–1341. https://doi.org/10.1017/S0031182010000466Bar-Even, A., Davidi, D., Flamholz, A., Liebermeister, W., et al. (2016). The protein cost of metabolic fluxes: Prediction from enzymatic rate laws and cost minimization. http://arxiv.org/abs/1604.00167 (accessed on 13th June 2025).
9. Bich, L., Mossio, M., & Soto, A. M. (2020). Glycemia Regulation: From Feedback Loops to Organizational Closure. *Frontiers in physiology*, *11*, 69. https://doi.org/10.3389/fphys.2020.00069 (accessed on 13th June 2025).
10. Saroussi, S., Karns, D. A. J., Thomas, D. C., Bloszies, C., Fiehn, O., Posewitz, M. C., & Grossman, A. R. (2019). Alternative outlets for sustaining photosynthetic electron transport during dark-to-light transitions. *Proceedings of the National Academy of Sciences of the United States of America*, *116*(23), 11518–11527. https://doi.org/10.1073/pnas.1903185116(accessed on 13th June 2025).
11. Bode, W., Fischer, R., Huber, R., Rüdiger, W., et al. (1987). ZZE-configuration of chromophore

ß-153 in *C-phycocyanin* from *Mastigocladus laminosus*. https://core.ac.uk/download/12163146.pdf (accessed on 13th June 2025).

12. Bowater, R. P. (2005). Nucleotides: Structure and properties. https://core.ac.uk/download/2776256.pdf (accessed on 13th June 2025).
13. Boyd, P. W., Faux, J. F., Goodlett, D. R., Harvey, E., et al. (2015). Diatom proteomics reveals unique acclimation strategies to mitigate Fe limitation. https://core.ac.uk/download/156628750.pdf (accessed on 13th June 2025).
14. Briars, E., Christodoulou, D., Goldford, J. E., Noor, E., et al. (2017). Genome-scale architecture of small molecule regulatory networks and the fundamental trade-off between regulation and enzymatic activity. https://core.ac.uk/download/142082717.pdf (accessed on 13th June 2025).
15. Candas-Green, D., Chen, H. W., Fan, M., Hu, Y., et al. (2020). Low-level saturated fatty acid palmitate benefits liver cells by boosting mitochondrial metabolism via CDK1-SIRT3-CPT2 cascade. https://core.ac.uk/download/287622302.pdf (accessed on 13th June 2025).
16. Chen, M., Li, H., Nielsen, J. B., Xia, Y., et al. (2023). Yeast increases glycolytic flux to support higher growth rates accompanied by decreased metabolite regulation and lower protein phosphorylation. https://core.ac.uk/download/571662807.pdf (accessed on 13th June 2025).
17. Chen, X., Fredrickson, J. K., Li, M., Liu, C., et al. (2017). Regulation-structured dynamic metabolic model provides a potential mechanism for delayed enzyme response in denitrification process. https://core.ac.uk/download/286729904.pdf (accessed on 13th June 2025).
18. Di Tunno, I., Hayek, J., Pallottini, V., Segatto, M., et al. (2014). Cholesterol metabolism is altered in Rett syndrome: A study on plasma and primary cultured fibroblasts derived from patients. https://core.ac.uk/download/98345941.pdf (accessed on 13th June 2025).
19. Diskin, M. G., Fenwick, M., Fitzpatrick, R., Kenny, D., et al. (2007). Effects of negative energy balance on liver gene and protein expression during the early postpartum period and its impacts on dairy cow fertility. https://core.ac.uk/download/45656678.pdf (accessed on 13th June 2025).
20. Døskeland, S. O., Haavik, J., Kleppe, R., Martinez, A., et al. (2011). The 14-3-3 proteins in regulation of cellular metabolism. https://core.ac.uk/download/pdf/82604421.pdf (accessed on 13th June 2025).
21. Umbrello, M., Dyson, A., Feelisch, M., & Singer, M. (2013). The key role of nitric oxide in hypoxia: hypoxic vasodilation and energy supply-demand matching. *Antioxidants & redox signaling*, *19*(14), 1690–1710. https://doi.org/10.1089/ars.2012.4979Ebenhoeh, O., Flis, A., Pokhilko, A., Stitt, M., et al. (2013). Adjustment of carbon fluxes to light conditions regulates the daily turnover of starch in plants: A computational model. https://core.ac.uk/download/29151105.pdf (accessed on 13th June 2025).
22. Fernandez-de-Cossio-Diaz, J., Vazquez, A. A physical model of cell metabolism. *Sci Rep* **8**, 8349 (2018). https://doi.org/10.1038/s41598-018-26724-7(accessed on 13th June 2025).
23. Fonteh, A. L. (2018). Erotic aciduria. https://core.ac.uk/download/213462592.pdf (accessed on 13th June 2025).
24. G., Nagarjuna. (2004). Towards a model of life and cognition. https://core.ac.uk/download/pdf/86194.pdf (accessed on 13th June 2025).
25. Iommarini, L., Ghelli, A., Gasparre, G., & Porcelli, A. M. (2017). Mitochondrial metabolism and

energy sensing in tumor progression. *Biochimica et biophysica acta. Bioenergetics, 1858*(8), 582–590. https://doi.org/10.1016/j.bbabio.2017.02.006 (accessed on 13th June 2025).

26. Goldford, J. E., & Segrè, D. (2018). Modern views of ancient metabolic networks. https://core.ac.uk/download/322972862.pdf (accessed on 13th June 2025).
27. Goldford, J. E., Hartman, H., Segrè, D., Smith, T. F., et al. (2017). Remnants of an ancient metabolism without phosphate. https://open.bu.edu/bitstream/2144/24313/1/Goldford_etal_Cell.pdf (accessed on 13th June 2025).
28. Gospodaryov, D. V., & McDonald, A. (2019). Alternative NAD(P)H dehydrogenase and alternative oxidase: Proposed physiological roles in animals. https://core.ac.uk/download/232454231.pdf (accessed on 13th June 2025).
29. Grimes, K. (2023). The role and regulation of folate metabolism during bacterial infection of *Drosophila melanogaster*. https://core.ac.uk/download/587834787.pdf (accessed on 13th June 2025).
30. Hatziapostolou, M., Iliopoulos, D., & Polytarchou, C. (2013). miRNAs link metabolic reprogramming to oncogenesis. https://core.ac.uk/download/30650764.pdf (accessed on 13th June 2025).
31. Heinemann, M., Kümmel, A., Panke, S., et al. (2005). *In silico* genome-scale reconstruction and validation of the *Staphylococcus aureus* metabolic network. https://core.ac.uk/download/pdf/12944401.pdf (accessed on 13th June 2025).
32. Liebermeister, W. (2014). How enzyme economy shapes metabolic fluxes. http://arxiv.org/abs/1404.5072 (accessed on 13th June 2025).
33. Masi, A., Arrigoni, G., Conselvan, G. B., Pivato, M., Carletti, P., Trevisan, S., et al. (2018). Protein profiling of *Arabidopsis* roots treated with humic substances: Insights into the metabolic and interactome networks. https://core.ac.uk/download/168406849.pdf (accessed on 13th June 2025).
34. Micillo, M., Brandimarte, L., Omezzolli, G., Bertola, N., Vallone, F. E., Iannello, A., Lavarello, C., et al. (2023). The characterization of metabolic profile and substrates dependencies of Richter's syndrome cells open for translational opportunities. *Blood.* https://www.semanticscholar.org/paper/8db7be341cbf856e1255727d8986185258bedbfc (accessed on 13th June 2025).
35. Cannino, G., Ciscato, F., Masgras, I., Sánchez-Martín, C., & Rasola, A. (2018). Metabolic Plasticity of Tumor Cell Mitochondria. *Frontiers in oncology*, *8*, 333. https://doi.org/10.3389/fonc.2018.00333
36. Olsson, T. (2005). Functional characterization of hexokinases in the moss *Physcomitrella patens*. https://core.ac.uk/download/11694926.pdf (accessed on 13th June 2025).
37. Periasamy, A. (2019). Special issue on metabolism *Cytometry A, 95. Cytometry Part A.* https://www.semanticscholar.org/paper/bbc8eb5c59472e04e670b30d8bc1ab4cc13e7886 (accessed on 13th June 2025).
38. Peyton, J., Planes, F. J., Prada Jiménez de Cisneros, J., Tobalina, L., et al. (2013). A network-based approach for predicting key enzymes explaining metabolite abundance alterations in a disease phenotype. https://core.ac.uk/download/42349597.pdf (accessed on 13th June 2025).
39. Sarmidi, M. R. (2012). Food is our medicine. https://core.ac.uk/download/11801185.pdf (accessed on 13th June 2025).

40. Sharma, R. (2008). The hepatocellular hypoxia criteria: 2′-Nitroimidazole effect on hepatocyte carbohydrate metabolizing enzymes. https://core.ac.uk/download/pdf/288247.pdf (accessed on 13th June 2025).

41. Strobbe, S., & Van Der Straeten, D. (2018). Toward eradication of B-vitamin deficiencies: Considerations for crop biofortification. https://core.ac.uk/download/158346323.pdf (accessed on 13th June 2025).

42. Zou, M. L. (2007). Variation in the apparent fecal digestibility of macronutrients and urinary energy excretion for three diets varying in fat and fiber content—Assessment of the Atwater factors and related energy conversion factors: A thesis presented in partial fulfillment of the requirements for the degree of Master of Science in Nutritional Science at Massey University, Palmerston North, New Zealand. https://core.ac.uk/download/162617248.pdf (accessed on 13th June 2025).

CELLULAR RESPIRATION

7

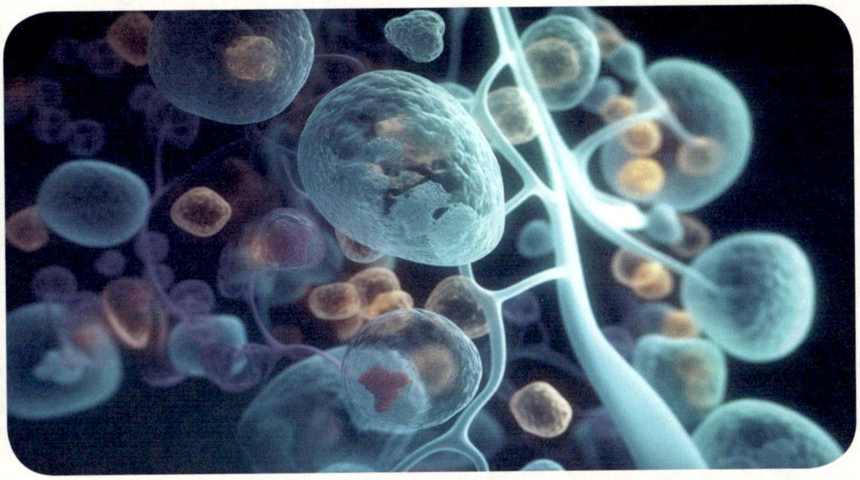

LEARNING OBJECTIVES

After studying this chapter, you will be able to:

1. Explain the process of cellular respiration;
2. Define glycolysis;
3. Describe the Krebs cycle;
4. Explore oxidative phosphorylation;
5. Discuss anaerobic respiration;
6. Focus on the regulation of cellular respiration.

Introduction

Cell respiration is a fundamental biochemical process present in all life forms and serves as the process by which cells obtain energy from food materials for sustaining life. It is a chain of metabolic processes that is accountable for transforming biochemical energy present in organic compounds, primarily glucose, into ATP, the currency of energy of a cell. The significance of cellular respiration extends beyond mere energy production, supporting cellular activities , such as growth, repair, and maintenance of homeostasis. As organisms advance, the complex details of their respiratory process have developed to ensure maximum energy yield from various substrates, reflecting a dynamic balance between metabolic demands and environmental availability. One of the main aspects of cellular respiration knowledge is microbial carbon use efficiency (CUE), which informs us about how efficient microorganisms are in fixing carbon substrates into biomass. CUE is important for soil biogeochemical cycling and ecosystem functioning overall. Remarkably, CUE changes can dramatically regulate soil organic matter dynamics and terrestrial carbon flux, implying a greater ecological role. The variation within CUE—from population-scale carbon use efficiency (CUEP) to community-scale efficiency (CUEC), to ecosystem-scale efficiency (CUEE)—is an integrated model of examining microbial respiration and energy dynamics contributions in varied environmental contexts (Frey et al., 2016). It is imperative to know about these to appreciate the complexity of respiration not just at the cellular but at larger ecological levels. Furthermore, scientific research findings are increasingly uncovering the complex mechanisms of cellular respiration. The role of ethylene as a promoter of physiological processes in phenomena , such as fruit ripening is an exemplar of the interfacing of respiratory processes with other metabolic pathways (Bouzayen et al., 2006). As the genetic and biochemical underpinnings of respiration are pursued through research, emerging genomics and metabolomics techniques open up exciting possibilities for furthering our understanding of how respiration controls growth and quality traits in plants. This convergence of information underscores the importance of cellular respiration in fundamental biology and its applied significance in agriculture. The reconciliation of these observations relates to the reconciliation of an extensive vision on respiration, bridging gaps between cell functions and environmental health (Figure 7.1).

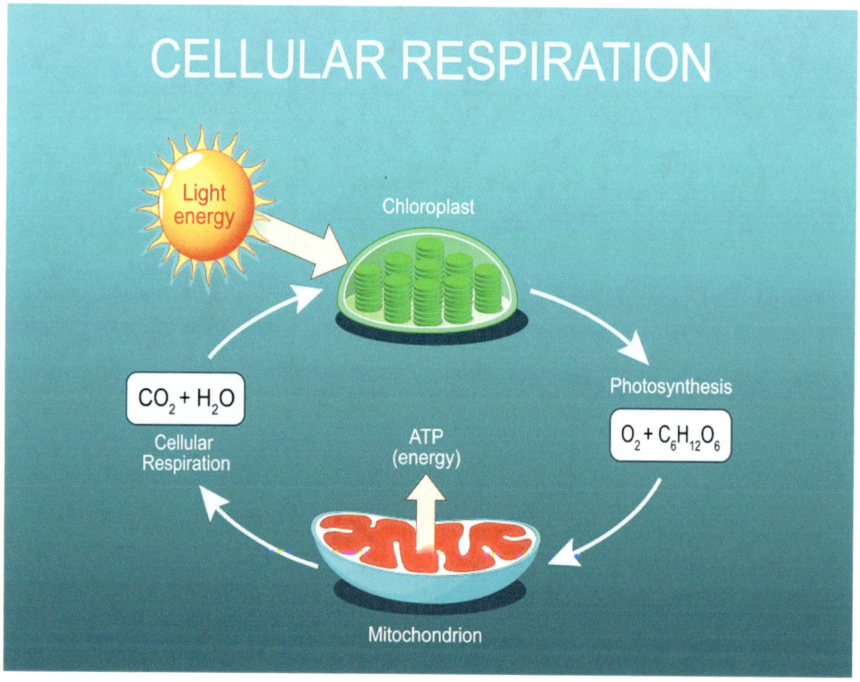

Figure 7.1. Cellular respiration.

Source: https://cdn.vectorstock.com/i/1000v/76/23/cellular-respiration-pocesses-of-aerobic-vector-50337623.jpg.

Definition of Cellular Respiration

Cellular respiration is a critical biochemical process in which cells convert nutrients to ATP, the cell's primary energy currency. The metabolic process is a series of complex reactions that occur in eukaryotic cells' mitochondria, utilizing oxygen to produce energy from glucose and other organic molecules. The net equation is a shorthand definition of the process: glucose and oxygen yield CO_2, water, and energy. This energy production is vital to sustain cellular activity, including growth, repair, and homeostasis. Cellular respiration, in a strictly technical sense, is not simply an energy-producing process; it is really more the convergence of pathways of metabolism which allow organisms to extract energy from the environment, which in turn sustains the multiple biological processes and the complex network of life itself. There are three major stages of cellular respiration: glycolysis, the citric acid cycle, and oxidative phosphorylation. Glycolysis is the initial stage, which is accomplished in the cytoplasm where glucose is fermented into pyruvate, yielding a small amount of ATP and NADH. This is followed by the citric acid cycle, or Krebs cycle, located in the mitochondria, which breaks down the pyruvate and generates further electron carriers, including NADH and $FADH_2$. Finally, there is oxidative phosphorylation in the inner membrane of the mitochondria through the utilization of the ETC and chemiosmosis to generate ATP efficiently through the phosphorylation of ADP. All of these steps are crucial in generating cellular energy, and any disruption at any level will have harmful effects on cellular metabolism and organismal health, highlighting the sophistication of this elementary biological process. Cellular respiration is even more crucial than the mere creation of ATP. Its implications are wide-ranging on ecological systems, influencing carbon cycling and microbial community processes in ecosystems. The process is also associated with microbial CUE that regulates how microorganisms break down organic carbon into biomass. Understanding these correlations widens the perspective of cellular respirations role in ecosystems as it informs models of soil biogeochemistry and provides solutions to the complexities of carbon cycling. Therefore, the expression cellular respiration comes to include ecological contexts, in which it constitutes a central process that aids organismal metabolism and ecosystem function, thereby furthering our understanding of terrestrial biogeochemical cycling and the continuous carbon input/output balance in the environment (Frey et al., 2016; Kumar et al., 2004).

Importance in Biological Systems

The role of cellular respiration extends beyond energy production; it's a ubiquitous feature of biological systems that sustains life via complex biochemical processes. Aerobic respiration, used primarily in the process of converting the energy, is performed in the mitochondria, where glucose is oxidized to produce ATP. ATP is the cells' universal currency of energy, stimulating required processes ranging from the contraction of muscles to conduction of nerve impulses. An understanding of cellular respiration also illuminates how organisms adapt to varying environmental conditions. As a case in point, anaerobic organisms are extremely versatile in the production of energy under oxygen-deprived conditions, as a testament to the evolutionary significance of multiple metabolic pathways. Metabolic versatility in such a manner underscores the importance of cellular respiration in maintaining homeostasis and supporting life in various ecological niches. Moreover, cellular respiration has an intrinsic association with aging and overall organism well-being. Cellular respiration disorders can lead to the accumulation of ROS, which are said to cause cell aging and dysfunction. Recent evidence suggests that specific mutations, like those governing copper metabolism in the filamentous fungus *Podospora anserina*, can prolong lifespan within the framework of alternative respiratory pathways with reduced ROS production. This information is indicative of the significance of understanding the interaction of metabolic processes and aging in complex biological processes. Moreover, the plasticity of the respiratory pathways can direct treatment measures to avoid age disorders and metabolic disorders. As such, the fundamental role of cellular respiration in controlling lifespan and health further underlines its importance in biological processes as a whole (Servos et al., 2012). Lastly, the complex mutual interactions involved in cellular respiration demonstrate the cooperative nature of biological systems. Biological cooperation may manifest in its several forms, and metabolic pathways can be interdependent to optimize energy production and resource allocation. For example, hierarchical organization of macromolecules and cell structure demonstrates how cooperative interactions create resilience and evolutionary success.

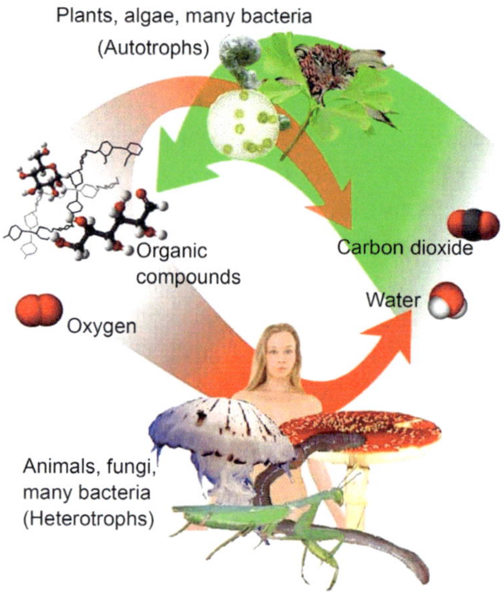

Figure 7.2. Cellular respiration is important in biological systems.

Source: https://www.bioexplorer.net/images/aerobic-respiration.jpg.

The development of hierarchical complexity in these networks has been associated with cooperation behaviors, implying that the evolution of cellular respiration has been influenced heavily by such interactions. Since cooperation is responsible for building resilience against environmental stress, recognizing such dynamics within cellular respiration is essential to recognize how biological systems evolve and adapt over time (Braeckman et al., 2009). The metabolic pathway interdependence serves to highlight the unity of life in supporting the contention that cellular respiration is not merely a mechanical phenomenon, but a part of life's complex weave (Figure 7.2).

7.1. The Process of Cellular Respiration

Learning Outcome

1. Overview of the stages;
2. Role of ATP in energy transfer;
3. Comparison with photosynthesis.

Cellular respiration is a complex biochemical process that occurs within the mitochondria of cells, where it serves to convert biochemical energy from nutrients into ATP, a vital energy currency. The process consists of several stages, including glycolysis, the citric acid cycle (Krebs cycle), and oxidative phosphorylation. In the cytoplasm, glucose is degraded to pyruvate in glycolysis, with a low output of ATP and reducing equivalents as NADH. This is the first step because it supplies the requisite substrates to other oxidative reactions that occur in the mitochondria. The pyruvate is further oxidized in the citric acid cycle to extract the electrons that are important in the ETC. This complicated dance of reactions highlights the crucial importance of cellular respiration in providing cellular functions and balancing energy. The ETC is where the majority of ATP is produced, representing a critical stage in cellular respiration. The ETC is located in the inner mitochondrial membrane and entails a chain of protein complexes that successively pass electrons of NADH and $FADH_2$ produced in the glycosylation and the citric acid pathway. The transfer of electrons along the chain results in the release of energy which is used to transport protons across the membrane establishing an electrochemical gradient. These gradient drives proton back into the mitochondrial matrix through ATP synthase, facilitating the phosphorylation of ADP to form ATP. Thus, ETC does not only maximize the ATP production but also maximizes the significance of the mitochondrial functioning, showing the interdependence of the cellular processes involved in the energy production. Control of cellular respiration is a highly regulated process and makes the process efficient and sensitive to the metabolic needs of the cells.

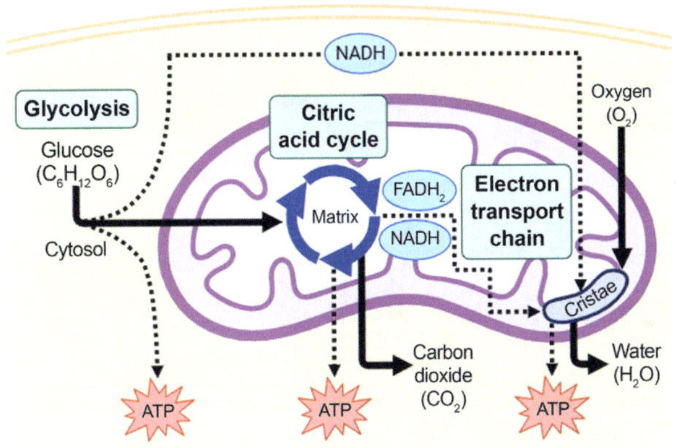

Figure 7.3. The process of cellular respiration.

Source: https://cdn.kastatic.org/ka-content-images/8c9fcb84d41e42ef085cd2f745f6a865359fa275.png.

The availability of substrates, enzyme catalysis, and feedback regulation of metabolism are critical factors that regulate the various respiration phases. Moreover, cellular quality-control mechanisms , including mitophagy have recently been in the limelight due to their roles in the maintenance of mitochondrial health and functioning. The fine control of mitophagy underlines the tight rope walk between energy production and cellular homeostasis. It has reported new mitophagy activators that do not depend on respiration collapse, giving important insights into the complex web of cellular metabolism. These developments highlight the ongoing research efforts directed towards elucidating the connections between mitochondrial function, cellular respiration, and overall cellular health, thereby improving our understanding of various metabolic disorders and potential therapeutic interventions (Manganas et al., 2016; Brunet et al., 2017) (Figure 7.3).

7.1.1. Overview of the Stages

Cellular respiration is a complex process that is carried out over four distinct stages: glycolysis, the citric acid cycle, oxidative phosphorylation, and the ETC. Each is crucial in the biochemical energy from food being transferred to ATP, the currency of the cell's energy. Glycolysis is the initial step, wherein the glucose is degraded to form pyruvate, while releasing a small quantity of ATP and reducing equivalents in the form of NADH. This process occurs in the cytoplasm and is not an oxygen-dependent process, hence it is an anaerobic process. Following glycolysis, pyruvate is transported into the mitochondria, where it continues to be converted and fed into the citric acid cycle, yielding enhanced energy extraction and repeated regeneration of metabolic intermediates. The Krebs cycle, or citric acid cycle, is crucial to energy production and occurs in thc mitochondrial matrix. Here, acetyl-CoA is combined with oxaloacetate to form citrate, which undergoes a series of enzyme-catalyzed transformations that enable the complete oxidation of carbohydrates, fats, and proteins into CO_2. Along this circuitous path, high-energy electrons are harvested in the NADH and $FADH_2$ forms, which are precious to subsequent processes of respiration. These coenzymes not only participate in redox reactions but also play a crucial role in connecting metabolic pathways. The significance of the steps as a middle step cannot be overstated, as it directly influences the provision of reducing equivalents that drive the ETC, the final stage in cellular respiration. The ETC is the terminal phase of cellular respiration that takes place on the inner mitochondrial membrane. Here, the energy-rich electrons from NADH and $FADH_2$ are transferred along a chain of complexes, leading to the generation of a proton gradient across the membrane. This electrochemical gradient is then harnessed by ATP synthase to power ATP synthesis via the process referred to as chemiosmosis. As protons move back to the mitochondrial matrix, ATP is formed from ADP and inorganic phosphate. The efficiency of this stage is critical, as it produces the majority of the ATP during cellular respiration. Understanding these stages provides insight not only into energy production but also into how the disruptions in these mechanisms result in diseases , such as cancer, where cell control is lost, and there is a complex interaction of metabolic dynamics (Costa et al., 2009; Merks et al., 2017).

7.1.2. Role of ATP in Energy Transfer

The complex role of ATP in intracellular energy transfer is key to understanding cellular respiration processes. ATP is utilized as the primary energy currency of the cell to enable multiple biochemical reactions necessary in sustaining life. ATP is generated in cellular respiration through various pathways, primarily glycolysis, the citric acid cycle, and oxidative phosphorylation.

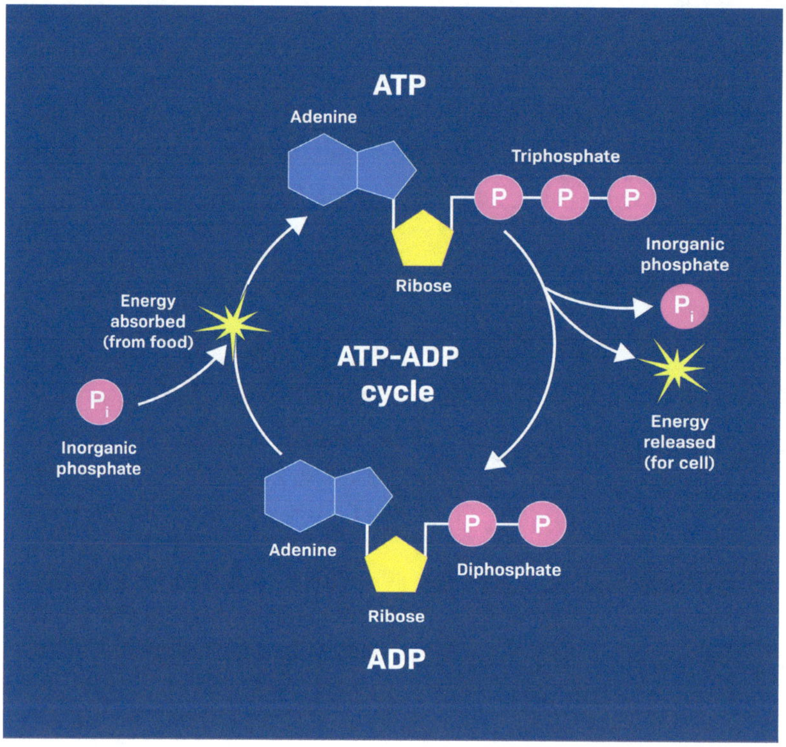

Figure 7.4. Role of ATP in energy transfer.

Source: https://media.labxchange.org/xblocks/lb-LabXchange-8ef89bac-html-1/lxc_ap1_p6_a4_adenosine-triphosphate-atp-and-its-role-in-cellular-energetics_figure-3150x17108455424449-a7fbdb50167c4449f6765c5af40efe35.png.

This complex choreography allows cells to convert chemical energy that is stored in nutrients into a usable form, illustrating the role of ATP to facilitate energy transfer. The continued synthesis of

ATP through these metabolic pathways ensures cells with the capacity to perform functions that are required, from muscle contraction to the transport of molecules across membranes by active transport, thus illustrating the necessity of ATP in bioenergetics. Besides its function as a direct energy source, ATP is also crucial in managing metabolic processes within the cell. When hydrolyzed, it provides energy that powers numerous enzymic reactions, literally coupling exergonic and endergonic reactions within the metabolic process. This coupling is critical in macromolecule assembly, proteins, and nucleic acids, and in cellular homeostasis. The dynamic, high-turnover behavior of ATP guarantees that cellular processes are kept precisely in step with the requirements of the organism. Additionally, new evidence points to an association between ATP synthesis and mitochondrial integrity, such that mitochondrial well-being directly affects ATP levels and thus cellular energy metabolism. This relationship also underscores the significance of ATP as a carrier of energy as well as a controller of cellular functions. The relationship between ATP and cellular metabolism is also observable in pathological conditions, e.g., pulmonary hypertension (PH), wherein mitochondrial dysfunction could lead to change in energy production from oxidative phosphorylation to glycolysis. Under such circumstances, the impaired ATP production can have deep implications for cellular function and overall well-being. The pH was found to comprise reduced mitochondrial respiratory complex activities, which are accountable for reduced ATP production, and thus compensatory use of less efficient glycolytic pathways for energy generation. The metabolic reprogramming not only limits ATP availability but also severely skews the cellular energy landscape, influencing cell viability and function. Thus, exploring ATP's role in energy transfer proves its indispensability in normal physiology, and in disease, further solidifying its status as a required molecular player in cellular respiration (Figure 7.4).

7.1.3. Comparison with Photosynthesis

The relationship between cellular respiration and photosynthesis is the foundation upon which the flow of energy in biological systems occurs. Photosynthesis occurs in plants, algae, and some bacteria, but it is the process by which chemical energy, primarily in the form of glucose, is created from sunlight. Photosynthesis also produces oxygen as a by-product that aerobic organisms utilizing oxygen for survival require. Conversely, cellular respiration is the mechanism through which organisms, including plants and animals, generate ATP, the energy currency of the cell, from glucose. Thus, photosynthesis and cellular respiration may be viewed as inverse processes with products serving as substrates for the other, forming a cyclical route to life on earth. At the center of photosynthesis are the chloroplasts, in which light energy is used to drive the synthesis of glucose in the light-dependent and light-independent reactions. Cellular respiration, however, takes place primarily in the cell's mitochondria, in which glucose is oxidized to produce energy. This type of energy is stored as ATP through glycolysis, citric acid cycle, and oxidative phosphorylation. While photosynthesis is energy capture, cellular respiration is energy release. Interestingly, the two processes contrast starkly in their dependency on the environment; photosynthesis needs sunlight, which restricts its occurrence to daytime and certain conditions, while cellular respiration is a constant process that can happen in either light or dark conditions, allowing organisms to sustain metabolic processes regardless of light availability. The evolutionary consequence of such interdependent processes is deep. Evidence shows that the first photosynthetic organisms played a significant role in the Earth's atmosphere in the context of oxygen levels, paving the way for the development of aerobic organisms and thus the multicellular organism with the ability to carry out cellular respiration. Further, studies involving various organisms, , such as *Halobacterium salinarum*, show the variation in energy-harvesting strategies, the mechanisms whereby some organisms can employ alternative metabolic pathways, including fermentation, alongside respiration and photosynthesis (Oren et al., 2009). This versatility indicates the adaptive edge that organisms enjoy in diverse environments. Lastly, the disparity between photosynthesis and cellular respiration not only accounts for their divergent roles in energy conversion but also reflects the complex interdependence of ecosystems, thus enhancing biological processes' understanding.

Test Your Knowledge

1. What is the role of ATP in energy transfer?
2. How to comparison with photosynthesis?

7.2. Glycolysis

Learning Outcome

1. Location and significance;
2. Steps involved in glycolysis;
3. Products generated from glycolysis.

Glycolysis is one of the most important metabolic pathways in the wider perspective of cellular respiration as it represents the first step of glucose catabolism. This is an anaerobic pathway which takes place in the cytoplasm and degrades the glucose molecules into two molecules of pyruvate, with a reasonable amount of energy being produced as ATP and NADH. The glycolytic pathway consists of 10 enzymatic reactions that can be grouped into two phases, energy investment phase and energy payoff phase. The investment phase requires two ATP molecules to phosphorylate glucose and its derivatives, whilst the payoff phase produces four ATP molecules and two NADH, with the overall result being a net two ATP gain. Such a balance in the utilization and production of energy is an example of how glycolysis serves as a quick source of energy to cells especially in situations when oxygen supply is scarce. Besides, the connection between glycolysis and other physiological states also evidences its importance. For example, in states like PH, where mitochondrial dysfunction occurs, cells may shift their energy production strategies from oxidative phosphorylation to an increased reliance on glycolysis. The feature of this metabolic reprogramming is high glycolytic activity, which represents a compensatory response to meet cellular energy demands even in the presence of perturbed mitochondrial activity. The shift to glycolysis does not only emphasize the flexibility of pathways but also its possible role in diseases with compromised oxidative phosphorylation, including cardiovascular diseases. The appreciation of these changes helps to improve our knowledge of the basic role of glycolysis in normal physiology and pathology, which strengthens its significance in cellular respiration. Lastly, glycolysis regulation is essential to cellular homeostasis and adaptations to metabolic needs. Key regulatory enzymes, , such as hexokinase and phosphofructokinase (PFK), play vital roles in controlling the pathways flux and are influenced by various biochemical signals, including ATP, ADP, and fructose-2,6-bisphosphate levels. These control systems maintain a balance between glycolytic activity and the energy demands of the cell, especially in tissues with variable metabolic demands, e.g., muscles during vigorous exercise. Moreover, the glycolysis dysregulation is frequently linked with several diseases, , such as cancer, where the so-called Warburg effect is witnessed, resulting in the increased glycolytic flux even under oxygen conditions. This adaptability and regulatory complexity render glycolysis not only a fundamental component of cellular energy metabolism but also a potential target for therapeutic interventions in metabolic disorders, thus contributing to our understanding of cellular respiration (Black et al., 2015; Shah et al., 2017) (Figure 7.5).

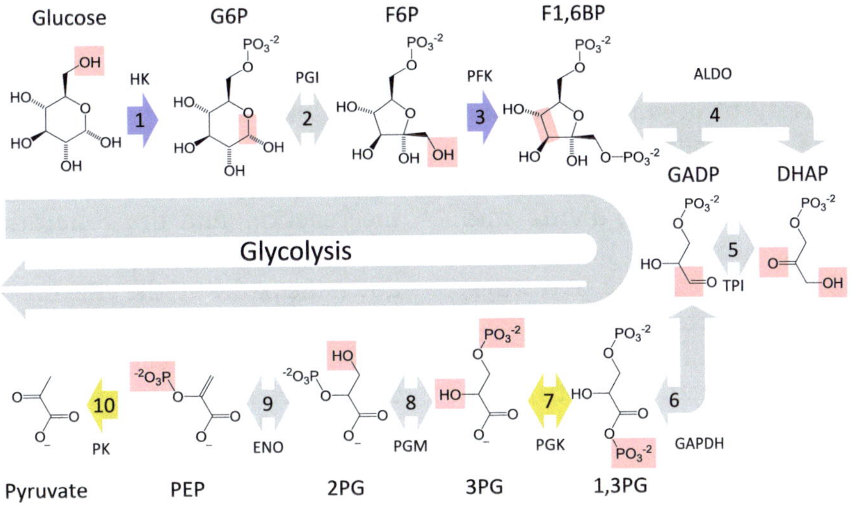

Figure 7.5. Glycolysis.

Source: https://en.wikipedia.org/wiki/File:Glycolysis_metabolic_pathway_3_annotated.svg

7.2.1. Location and Significance

The location of cellular respiration is key in understanding its place in cellular metabolism and the production of energy. Cellular respiration occurs primarily in the mitochondria of eukaryotic cells, which are generally referred to as powerhouses of the cell. The major components of the ETC and ATP synthase are found in the inner mitochondrial membrane, which is responsible for generating ATP. Besides, the spatial organization of the enzymes of the Krebs cycle within the matrix of the mitochondria represents a highly evolved level of integration for conducive energy transfer. The compartmentalization of such metabolic processes ensures that reactive intermediates are quickly channeled into successive processes, ensuring that maximum energy yield is obtained and the potential loss is minimized. By being organized locally, the cells can quickly counteract shifting energy demands and maximize the utilization of resources, emphasizing the utmost biological significance of mitochondria in cellular respiration. Even more significance is placed on localized cellular respiration when the microbial cell and its ATP content are taken into account. Studies reveal that the levels of ATP in such cells can be employed to reflect the entire biomass parameters, displaying how efficient cellular respiration is to cellular well-being as well as metabolic activity (Azam et al., 2025). The intertwined relationship between the levels of ATP and microbial growth reveals that the availability of energy indeed affects the viability and functionality of such organisms. Moreover, understanding the nuances of location-based respiration mechanisms opens up the dynamics of how cells adapt to varied energy substrates and environmental conditions. Such an ability is essential for the preservation of cellular operation but also for balance in the ecosystem due to microbial respiration being a key factor in nutrient cycling and energy transfer among ecosystems. More broadly, the significance of cellular respiration extends beyond mere survival; it is a component of energy metabolism in an enormous variety of organisms ranging from fungi to plants. The energy produced by cellular respiration drives other biosynthetic reactions that allow cells to grow, divide, and maintain homeostasis. Moreover, the process also interacts with other metabolic pathways, reflecting on the interconnectedness of cellular processes. For instance, within photosynthetic organisms, glucose that is generated through photosynthesis can be broken down through cellular respiration, pointing to a cycle of dependency that sustains life processes. This related meaning can be used to describe how cell respiration is crucial not just for the purposes of explaining individual organisms but also in explaining the larger ecological function of energy exchange and metabolic systems (Allan et al., 2005).

7.2.2. Steps Involved in Glycolysis

Glycolysis is a fundamental biochemical reaction involved in cellular respiration, which takes place in the cytoplasm of a cell. It is a multi-step reaction that is started by the phosphorylation of glucose by the enzyme hexokinase. The reaction traps glucose inside the cell and prepares it for further breakdown. Glucose-6-phosphate is then rearranged into fructose-6-phosphate by an isomerization catalyzed by the enzyme phosphoglucose isomerase. The pathway proceeds through a second phosphorylation step, where fructose-6-phosphate is converted to fructose-1,6-bisphosphate by PFK, another regulatory step that commits glucose to glycolysis. The impact of the early steps is evident by their control of the metabolite flow in glycolysis and the overall efficiency of the cell's energy production. As glycolysis continues, the six-carbon molecule fructose-1,6-bisphosphate is split into two three-carbon molecules named glyceraldehyde-3-phosphate (G3P) and dihydroxyacetone phosphate (DHAP), facilitated by aldolase. DHAP catalytically isomerizes readily to G3P, thus ensuring glycolysis continues with two molecules of G3P. At subsequent steps, G3P is phosphorylated and oxidized to yield 1,3-bisphosphoglycerate, with the enzyme G3P dehydrogenase catalyzing the reaction, and the generation of NADH, the required reducing agent for the ETC. Further down the pathway, the enzyme phosphoglycerate kinase removes the phosphate group from 1,3-bisphosphoglycerate and transfers it to ADP to produce ATP. This is a case of substrate-level phosphorylation, in which the energy derived is directly from the breakdown of glucose. Finally, glycolysis concludes in the conversion of phosphoenolpyruvate (PEP) to pyruvate through the enzyme pyruvate kinase, producing a molecule of ATP as well.

This final step serves a purpose because pyruvate can be transported into the mitochondria for aerobic respiration or fed into the process of fermentation when anaerobic conditions exist. The net return in glycolysis is two ATP and two NADH molecules per glucose molecule, and it is a major source of energy and intermediates for various metabolic processes. It is imperative to understand these steps of glycolysis since research indicates that alterations in glycolytic flux may modulate cellular function, , such as in cardiac mesenchymal stem cells (MSCs), where metabolic reprogramming may control cell proliferation and cell death (Jin et al., 2019). Further, the regulation of glycolysis according to cellular demand underscores its significance in homeostasis maintenance (Archiche et al., 2017) (Figure 7.6).

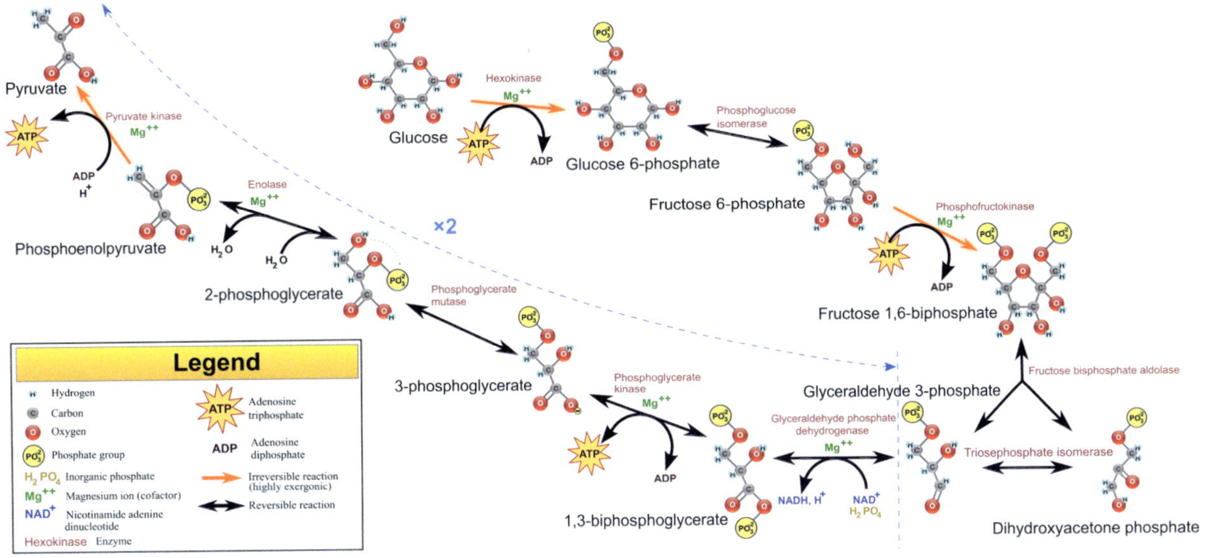

Figure 7.6. Steps involved in glycolysis.

Source: By This W3C-unspecified vector image was created with Inkscape . - Own work, CC BY-SA 3.0, https://commons.wikimedia.org/w/index.php?curid=7724574

7.2.3. Products Generated from Glycolysis

Glycolysis, the initial pathway of cellular respiration, has crucial metabolic functions beyond the breakdown of glucose. One of the predominant products of glycolysis is pyruvate, which can further be subjected to biochemical transformations depending upon the oxygen status of the cells. During aerobic conditions, pyruvate is transported to the mitochondria, where it is transformed into acetyl-CoA, the primary substrate for the citric acid cycle. Alternatively, in an anaerobic process, pyruvate is converted to lactate in animal cells or to ethanol and CO_2 in yeast. This flexibility allows organisms to adjust glycolytic metabolism to various environmental conditions so that energy generation can continue at suboptimal oxygen tensions. Such flexibility is essential for survival, particularly in tissues that will briefly experience hypoxic conditions, allowing cellular physiology's dynamic requirements during diverse activities and stressors. In addition to pyruvate, glycolysis yields a net amount of ATP and NADH, both significant contributors to cellular energy metabolism. The primary energy currency of the cell, ATP, is generated through substrate-level phosphorylation through glycolysis. This occurs in the process of some steps of glycolysis, where high-energy phosphate groups are transferred to ADP to yield ATP. NADH, on the other hand, is produced when NAD^+ receives electrons during oxidation of G3P. These two compounds get significant functions in subsequent metabolic pathways. ATP generated can be used directly for cell processes, while NADH is a reduced cofactor that donates the ETC to multiply energy yield. Thus, double production of glycolytic pathways reflects their inherent importance in cellular homeostasis and energy metabolism. Moreover, the metabolites formed during glycolysis can serve as precursors for many anabolic routes. Conversion via enzymes yields intermediates , such as DHAP and 3-phosphoglycerate, which can be used to synthesize lipids, amino acids, and nucleotides. This versatility is reflective of the central role of glycolysis not only in energy production but also as a provider of precursors for biosynthesis. For instance, 3-phosphoglycerate can enter the biosynthetic pathways to serine and subsequently to other important amino acids. Such convergence of metabolism and synthesis shows the intertwined nature of cellular processes where energy production and biomolecule production are coupled. Thus, even though glycolysis is originally appreciated due to its energy content, its value to cellular building blocks reasserts its broader function in maintaining cellular processes and supporting growth, especially under varied physiological demand and metabolic status.

Test Your Knowledge

1. What are the steps involved in glycolysis?
2. What are the products generated from glycolysis?

7.3. The Krebs Cycle

Learning Outcome

1. Location within the cell;
2. Key reactions and enzymes involved;
3. Importance of electron carriers.

The citric acid cycle or the Krebs Cycle is a critical part of the cellular respiration process that helps in the oxidation of organic fuel molecules. This cycle, found in the matrix of mitochondria, breaks down acetyl-CoA, which is mostly obtained by the breakdown of carbohydrates, fats, and proteins, into energy-containing molecules. The cycle starts with the condensation of acetyl-CoA with oxaloacetate to citrate, which is then acted upon by a sequence of enzyme conversions. In the process of these conversions, there are key intermediates like isocitrate and alpha ketoglutarate that help in the production of NADH and $FADH_2$. These electron carriers prove vital to other phases of cellular respiration, namely the ETC, whereby the stored energy in them eventually results in the formation of ATPs, hence the significance of the Krebs Cycle in energy metabolism. In addition, the Krebs Cycle is closely connected with other metabolic pathways, which makes it a metabolic crossroad of cellular respiration. It does not only play a role in energy production but it also produces precursor molecules which are essential in biosynthetic processes. As an illustration, intermediates generated in the course of the cycle could be diverted to produce amino acids, nucleotides, and additional biomolecules. These two functions of the Krebs Cycle reflect the fact that metabolic pathways are highly interconnected so that the production of energy and the synthesis of cellular components proceed simultaneously. Also, the Krebs Cycle acts as a regulatory checkpoint, and it increases or decreases its activity depending on cellular energy needs. As evidenced by recent research, specific enzymes within the cycle, , such as isocitrate dehydrogenase, are crucial for regulating metabolic responses, demonstrating how the cycle adapts to maintain cellular homeostasis amidst varying physiological conditions (Goetz et al., 2011). The Krebs Cycle is much more than a process of energy production, as it was found to impact aging and related cellular processes. The impairments or disruptions of the cycle may result in the reduced functioning of mitochondria, oxidative stress, and a series of metabolic disorders that lead to rapid aging of cells. Studies have shown that longevity genes, many of which impact mitochondrial function, regulate the Krebs Cycle functionality, potentially delaying aging processes (Cristina, 2008). These findings suggest a complex relationship between metabolism and longevity, whereby an optimized Krebs Cycle may limit the generation of ROS that contribute to cellular damage over time. This makes the efficiency not only a pillar in the process of cellular respiration but also a key regulator of cellular health and lifespan, making it a critical process in energy metabolism and aging (Figure 7.7).

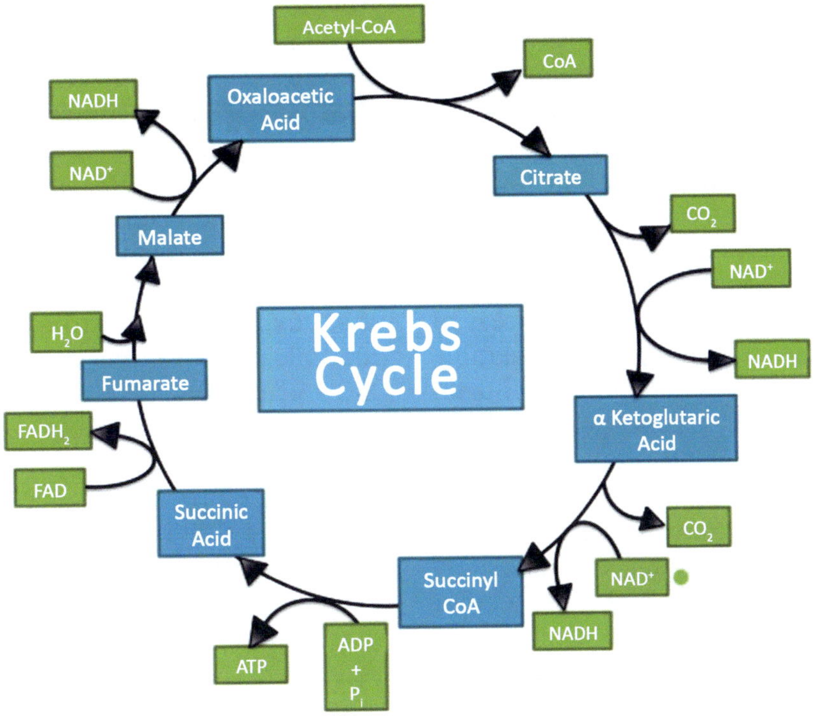

Figure 7.7. The Krebs cycle.

Source: https://cdn.hswstatic.com/gif/krebs-cycle.jpg.

7.3.1. Location Within the Cell

The complex cellular respiration process occurs in different regions of the cell, each with its contribution towards the efficiency and regulation of energy production. The cytoplasm initially offers the space where glycolysis occurs, reducing glucose to pyruvate and producing a low amount of ATP and NADH. The process is significant as it forms the groundwork for the subsequent processes of energy production in the mitochondria. In this regard, positioning glycolytic enzymes in the cytosol emphasizes cellular compartmentalization to facilitate efficient energy conversion of carbohydrates into usable forms. Such subcellular organization optimizes metabolic pathways, which mirrors the complex design of cellular respiration. Finally, understanding these spatial dynamics reveals how changes in these local processes can affect cellular function as a whole, revealing the importance of place in the cell in the determination of energetic homeostasis. Following glycolysis, the pyruvate that is produced is transported into the mitochondria and then further oxidized by the citric acid cycle (Krebs cycle). This occurs in the mitochondrial matrix, again showing the compartmentalization of cellular respiration. The Krebs cycle is making use of acetyl-CoA derived from pyruvate, producing carrier molecules , such as NADH and $FADH_2$ that are necessary for the next phase of cellular respiration—oxidative phosphorylation. This phase is important in having a specialized environment since the inner mitochondrial membrane is where ETC proteins reside which can be utilized to synthesize ATP. The fine architecture of the mitochondria, including its double membrane, facilitates efficient electron transfer and proton pumping, leading ultimately to ATP production by chemiosmosis. The distinctive locations of metabolic pathways thereby accentuate the importance of cell architecture in optimizing energy production. Finally, the final process of cellular respiration, oxidative phosphorylation, takes place along the inner mitochondrial membrane, where the ETC is situated. This is a vital step in the sense that it captures the energy released from electrons that move through the ETC to pump protons from the mitochondrial matrix to the intermembrane space, establishing a proton gradient. This structural arrangement of constituents is integral to the synthesis of ATP since protons move back to the matrix through ATP synthase, known as chemiosmosis. Moreover, the role of oxygen as the terminal electron acceptor in this process cannot be overstated; in its absence, the entire process of aerobic respiration would be halted, which would lead to anaerobic conditions and less efficient energy production. As can be observed, the unique location of metabolic activity in the cell, , such as that highlighted in cellular respiration, exhibits the complexity and efficiency of biological systems (Barnes et al., 2017; Alessio et al., 2019).

7.3.2. Key Reactions and Enzymes Involved

The key feature of cellular respiration is a series of crucial reactions that are tightly regulated by specific enzymes. Glycolysis is the first step, which occurs in the cytoplasm, and is the enzymatic breakdown of glucose to pyruvate. This pathway involves 10 enzymic reactions, hexokinase and PFK being two crucial regulatory enzymes. Hexokinase phosphorylates glucose to glucose-6-phosphate, while PFK is the main regulatory step in the pathway and is sensitive to the cellular energy state. Phosphorylation of glucose to pyruvate in glycolysis not only produces ATP but also reduces NAD^+ to NADH, which is required for the subsequent oxidative phosphorylation event that occurs in mitochondria. Subsequent to glycolysis, the pyruvate molecules enter the mitochondria and are oxidatively decarboxylated. This very critical reaction is catalyzed by the pyruvate dehydrogenase complex, a multi-enzyme complex, by which pyruvate is converted into acetyl-CoA with the release of CO_2 and formation of NADH. Acetyl-CoA enters the citric acid cycle (Krebs cycle), wherein it is further oxidized to produce more NADH and $FADH_2$, as well as GTP or ATP. The enzymes employed in the cycle, , such as citrate synthase and isocitrate dehydrogenase, play significant roles in regulating the flow of metabolites throughout the cycle. Notably, the cycle generates high-energy electron carriers, the most significant members of the ETC, highlighting the role such enzymatic processes play in efficient energy generation during cellular respiration. The ETC, the final stage of cellular respiration, takes place in the inner mitochondrial membrane. It involves a series of protein complexes that facilitate the transfer of electrons from NADH and $FADH_2$ to molecular oxygen as the terminal electron acceptor. The enzymatic activity of cytochrome c oxidase, and others, is significant in this process because it catalyzes the reduction of oxygen to water while pumping protons (H^+) into the intermembrane space at the same time, creating an electrochemical gradient. The gradient is used to power the synthesis of ATP-by-ATP synthase during oxidative phosphorylation. Moreover, NO, produced as a result of hypoxia, plays an important role in controlling mitochondrial function and blood flow and affects energy metabolism under oxygen deficiency (Umbrello et al., 2013). The synergy of these reactions and enzymes not only maximizes the production of ATP but also illustrates the dynamic condition of cellular respiration with varying physiological states (Cannino et al., 2018) (Figure 7.8).

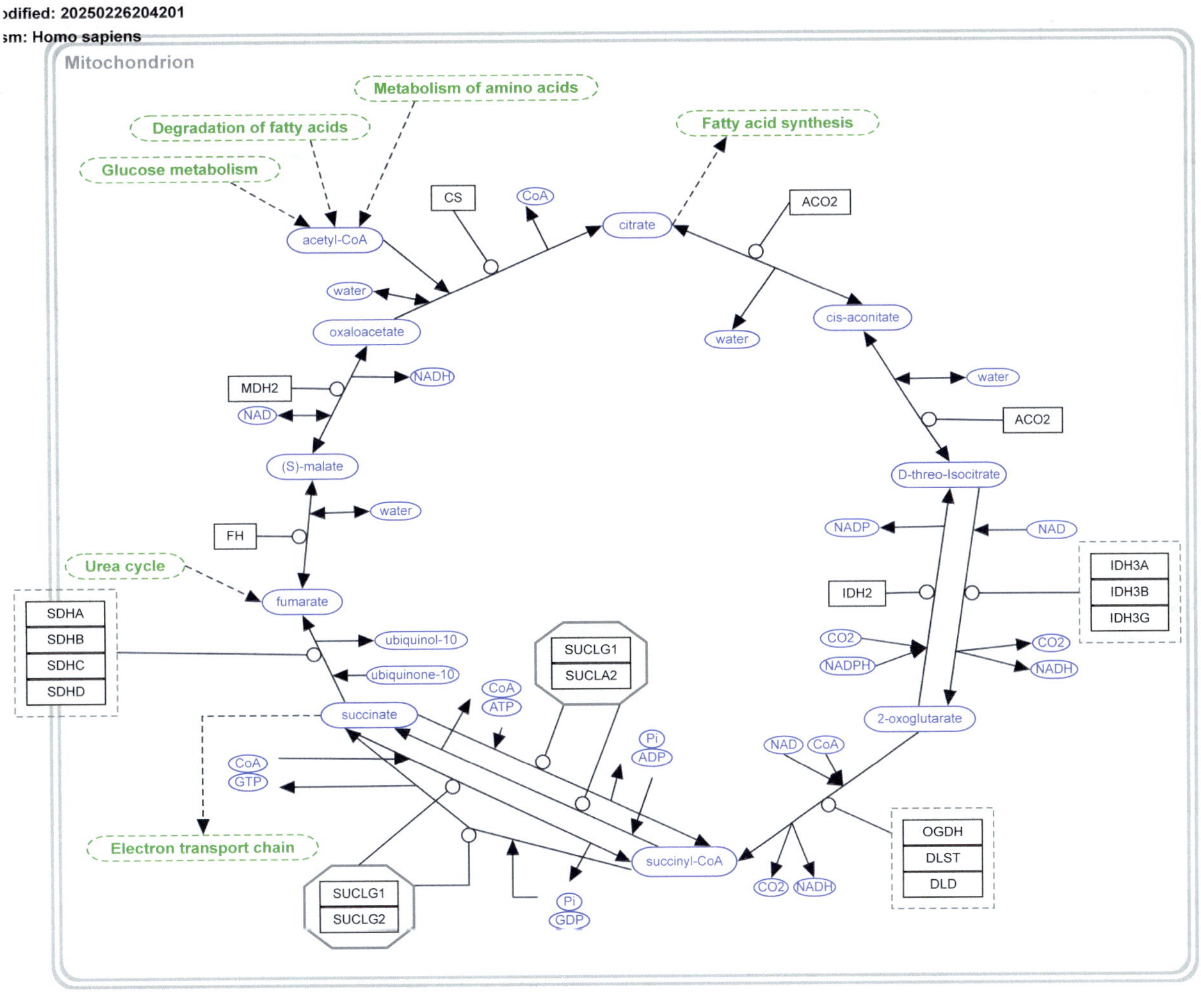

Figure 7.8. The Krebs cycle key reactions and enzymes involved.

Source: https://www.wikipathways.org/pathways/WP78.html

7.3.3. Importance of Electron Carriers

The role of electron carriers in cellular respiration can never be overemphasized since they facilitate the passing of electrons through the ETC in the mitochondria. The carriers , such as NADH and $FADH_2$, are not mere observers; they are crucial in the process of energy conversion from macronutrients into ATP, which is the cell's primary energy currency. By electron shuttling from metabolic substrates, electron carriers enable stepwise energy release without threatening the potential damage ensuing in the event of an unexpected energy release. Furthermore, the function of electron carriers is interconnected with ROS generation, which may serve as signaling molecules or trigger mitochondrial dysfunction. Recognition of the double purpose lends richness to our understanding of cellular metabolism and its control during various physiological conditions, both the intricacy and the necessity of these carriers for respiration. Moreover, structural assembly of the respiratory ETC is not as rigid as once considered. Present research finds that motor protein component structure has a significant influence on mitochondrial efficiency and cell health. Electron carrier interaction with respiratory complexes is necessary for the optimization of ATP production. The discovery of channeling pathways between complexes III and IV demonstrates the way in which the position of these carriers optimizes electron transfer competence while concurrently reducing the possibility of deleterious ROS formation. This phenomenon underlines the process of metabolic plasticity, whereby electron carrier structure is adjusted in line with the metabolic needs of the cell. Consequently, a study of the assembly and regulation of electron carriers is not only teaching biochemistry about respiration but also outlining the wider implications for cellular health and aging (Díaz et al., 2019). In the broader context of cellular respiration, electron carriers have a function more complex than the mere production of energy; they are essential to cellular metabolic homeostasis. The variations in electron transfer that result from varying cellular demands illustrate a sophisticated system of regulation for maintaining metabolic balance. Electron carriers also play central roles in signal transduction cascades that connect respiration to other metabolic activities so that cells may react flexibly to fluctuating nutrient levels or environmental stress. The ability of cytochrome c to become altered after translation further demonstrates such flexibility, such that variations in electron flow can be quickly corrected to accommodate cellular need. This adaptability emphasizes the interconnectedness of metabolic functions and the vital role electron carriers play not only in energy generation but in overall cell metabolic health (Díaz et al., 2019).

Test Your Knowledge

1. What are the key reactions and enzymes involved?
2. What are the importance of electron carriers?

7.4. Oxidative Phosphorylation

Learning Outcome

1. Role of the electron transport chain;
2. Mechanism of ATP synthesis;
3. Importance of oxygen in the process.

Within the complex framework of cellular respiration, oxidative phosphorylation holds a pivotal role by facilitating the production of ATP. This complex process occurs in the mitochondria, where the ETC plays a central function. Electrons are passed on by reduced cofactors (NADH and $FADH_2$) through a sequence of protein complexes and eventually reduce oxygen to water. The energy released during electron transfer is utilized to pump protons (H^+) from the mitochondrial matrix into the intermembrane space, creating a proton gradient. This electrochemical potential is a form of stored energy which can be used to make ATPs through ATP synthase, a process called chemiosmosis. The dependence of these elements depicts that oxidative phosphorylation does not only generate ATP but couples electron transport with proton translocation, which comprises the core of energy metabolism in aerobic life forms. These regulatory characteristics of oxidative phosphorylation are important in the cellular homeostasis especially in adaptation to varying energy requirements. In circumstances of elevated metabolic demand, e.g., during exercise or when physiological stress is elevated, oxidative phosphorylation rate must adjust to provide enough ATP. NO emerges as an influential signaling molecule, mediating this adaptation through its vasodilatory effects and modulation of mitochondrial processes. According to recent insights, NO influences energetic metabolism by enhancing oxygen delivery while concurrently adjusting mitochondrial function to reduce oxygen consumption, thus preventing oxidative stress and potential cellular damage during hypoxic conditions (Umbrello et al., 2013).

Such a dual role of NO represents an example of how different signaling pathways are closely integrated to maximize ATP production and metabolic efficiency according to different physiological conditions. Studies have helped to establish the relationship that exists between body size and mitochondrial function further helping to understand the role of oxidative phosphorylation in various species. For example, studies on frogs reveal that larger animals exhibit increased mitochondrial coupling efficiency, which correlates with higher ATP production compared to their smaller counterparts (Dumet et al., 2015). The phenomenon highlights the adaptive importance of mitochondrial bioenergetics in the regulation of metabolic rates. The findings illustrate how variations in mitochondrial parameters, such as, such as proton leak and ROS production—are intimately linked to organismal size and energetic demands. Consequently, the elucidation of the association between oxidative phosphorylation and body size does not only contribute to the basic biological principles portfolio, but also the general ecological and evolutionary debate on the use of energy among different taxa (Figure 7.9).

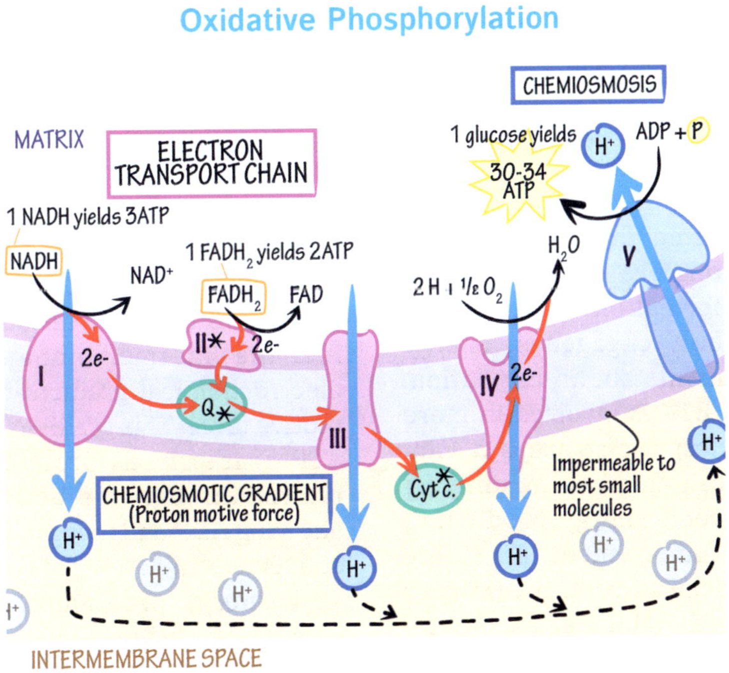

Figure 7.9. Oxidative phosphorylation.

Source: https://d1j63owfs0b5j3.cloudfront.net/term/images/oxidative-phosphorylation-2664.png.

7.4.1. Role of the Electron Transport Chain (ETC)

The ETC is an essential component of cellular respiration, primarily located in the inner mitochondrial membrane of eukaryotic cells. The series of protein complexes and electron carriers conveys electrons from the breakdown of glucose, thus powering the synthesis of ATP, the energy currency of the cell. As electrons move along the chain from one complex to another, they create the energy that is utilized to transport protons from the mitochondrial matrix to the intermembrane space, creating a proton gradient. The gradient not only drives ATP synthesis via ATP synthase but is utilized in maintaining cellular functions as well. Furthermore, unimpeded functioning of the ETC is critical in the integration of various metabolic pathways and emphasizes its central importance in cellular metabolism and energy production. ETC efficiency is also linked to the generation of ROS, which may serve as signaling molecules affecting various metabolic processes. As documented in recent studies, ROS generation is highly correlated with the operation of the respiratory chain, implying an essential interplay between metabolic communication and energy production. Based on electron transport models, it has been demonstrated that variations in succinate concentrations are accountable for variable levels of ROS formation, which implies that the ETC is not only engaged in energy conversion but also in metabolic regulation. This is a sign that the ETC is involved in more than mere bioenergetics; it also plays a vital role as a regulator of cell metabolism, unifying the fine balance between energy production and reactive species management, providing insights into mitochondrial function and broader relevance for cell health (Mark et al., 2012). Moreover, multi-heme cytochromes are an example of the diverse functionality of electron transport proteins across different organisms, particularly prokaryotes. These proteins are not only needed for normal energy conversion but also for widespread biogeochemical cycling of nitrogen, sulfur, and iron, thereby underlining their ecological importance. For instance, organisms , such as *Shewanella* and *Geobacter* employ these cytochromes to facilitate long-distance electron transport, which enables them to both reduce solid substrates , such as iron oxides. This versatility shows the evolutionary significance of cytochromes and their promising utilization in biotechnology, , such as microbial fuel cells and bioremediation processes. As more scientific investigation is conducted on multi-heme cytochromes, computational and experimental approaches will better elucidate their structure-function relationships, setting the stage for subsequent innovative developments harnessing these systems to produce energy and rehabilitate the environment (Brown et al., 2015).

7.4.2. Mechanism of ATP Synthesis

The generation of ATP from cellular respiration depends predominantly on the process that takes place in the mitochondria. The mitochondria are the cell's energy-generating components where the oxidative phosphorylation stage occurs. The production of ATP is essentially catalyzed by the enzyme ATP synthase, which is located in the inner mitochondrial membrane. As electrons are passed along the ETC, protons (H^+) are actively pumped out of the mitochondrial matrix and into the intermembrane space, producing an electrochemical gradient known as the proton motive force. This gradient is crucial in powering ATP synthesis, as protons pass back into the matrix through the action of ATP synthase, allowing ADP and inorganic phosphate (Pi) to be converted into ATP. This process demonstrates how energy production is intertwined with electron transport and proton motion. Other than oxidative phosphorylation, sophistication of ATP production is also modulated by other parameters, including mitochondrial integrity and ion homeostasis. To illustrate, disruptive chemicals , such as phomoxanthone A (PXA) can critically impair mitochondrial function. In a fascinating discovery, PXA was shown to depolarize mitochondria and inhibit electron transport without elevating cellular respiration. This highlights the fine tuning attained by the mitochondria in energy production and buffering of calcium. Such mitochondrial dysfunction can have the potential to induce apoptosis, which is largely due to the release of pro-apoptotic proteins followed by cristae disruption. The delicate coordination of mitochondrial membrane dynamics and efficient synthesis of ATP highlights how maintaining healthy mitochondria is essential for guaranteeing optimal energy metabolism by the cell (Anand et al., 2018). Besides, ATP synthesis is also linked directly with the hypoxic response of the cell. During diminishing oxygen tensions, tissues may be hindered in maintaining adequate energy states, and this evokes a cascade of metabolic adaptations mediated by NO. The messenger molecule facilitates hypoxic vasodilation, thereby maximizing blood flow and augmenting oxygen delivery to tissues. Second, NO modulation can reduce oxygen consumption and stimulate other metabolic pathways that essentially facilitate energy production under low oxygen conditions. The role of NO is not merely vasodilation; it plays an important role in homeostasis management in times of energy supply-demand disturbances. This recognizes ATP's significance not just as a product of cellular respiration but as a critical factor influencing diverse metabolic reactions under conditions of oxygen insufficiency (Umbrello et al., 2013) (Figure 7.10).

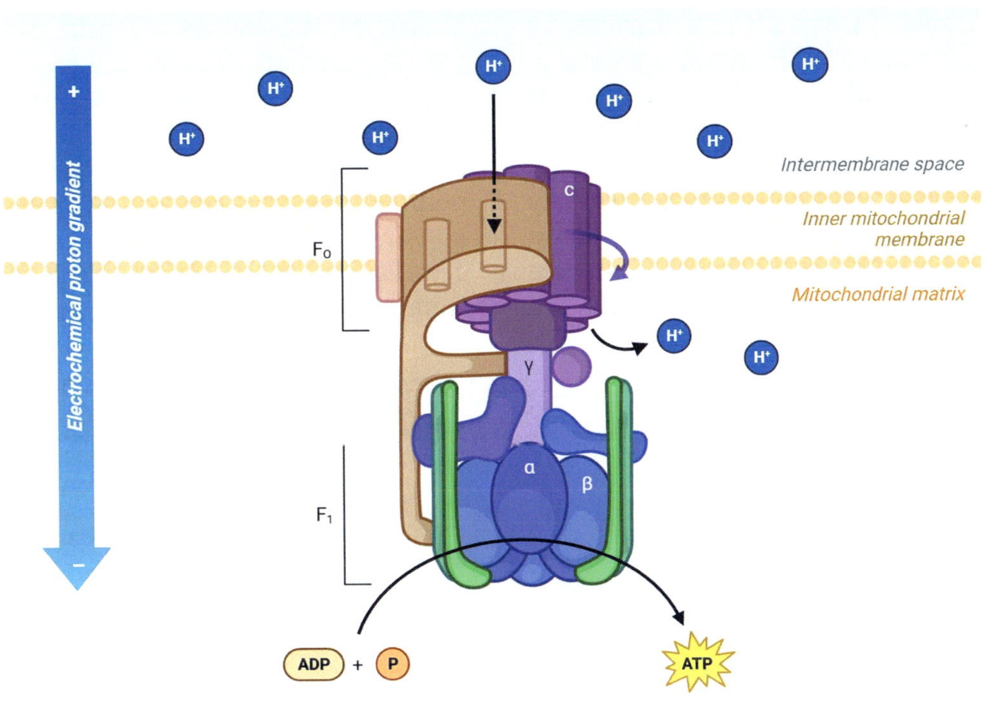

Figure 7.10. Mechanism of ATP synthesis.

Source: https://microbenotes.com/wp-content/uploads/2023/09/Structure-of-ATP-Synthase.jpeg.

7.4.3. Importance of Oxygen in the Process

The function of oxygen in cellular respiration is critical since it is the terminal electron acceptor in the ETC. It is a process that takes place within the inner mitochondrial membrane and is critical to the effective production of ATP. Without oxygen, an element critical to aerobic respiration, cells would rapidly deplete their ability to produce ATP using oxidative phosphorylation. This decrease in energy production severely compromises cell function, leading to adverse effects on cell viability and health. Furthermore, the absence of oxygen provokes a shift towards anaerobic processes, which, although capable of sustaining metabolic operations, yield much lower amounts of ATP molecules per molecule of glucose. Thus, oxygen not only maximizes energy-producing efficacy but also encourages the metabolic health of eukaryotic life, underlining its essential position in homeostasis and the maintenance of life. In addition, the role of oxygen is not restricted to the mere production of ATP; it is also crucial in regulating ROS within the cell. It is important to observe the balance, as higher ROS levels, which are produced during aerobic respiration, could induce oxidative stress, which adversely affects cellular components. Interestingly, there is evidence that certain longevity genes are involved in mitochondrial function and the regulation of ROS. The implication from this is that while oxygen is key to energy metabolism, oxidative stress management is equally essential for cellular integrity and deceleration of aging processes. Thus, a balance of oxygen consumption and ROS production is essential for cell homeostasis, reflecting oxygen's dual role in cell well-being and energy metabolism (Cristina, 2008). Oxygen influence on cellular respiration also manifests in natural environments, e.g., in sewage treatment processes. Degradation of organic matter in sewage largely relies upon oxygen availability, facilitating microbial metabolism. This aerobic breakdown yields CO_2, which has a significant effect on pH in activated sludge systems. Efficient oxygen utilization by microbial communities has a direct correlation with the amount of chemical oxygen demand (COD) that is eliminated from wastewater. Therefore, it becomes essential to understand how oxygen utilization occurs to enhance waste treatment methods and ensure proper metabolic activity. These observations not only attest to the effectiveness of biologic oxidation but also demonstrate the polyfunctionality of oxygen in various biologic processes, validating its significance from cellular to ecosystem levels (Lijklema, 1971).

Test Your Knowledge

1. What is the role of the electron transport chain?
2. What is the mechanism of ATP synthesis?

7.5. Anaerobic Respiration

Learning Outcome

1. Definition and conditions for occurrence;
2. Types of anaerobic pathways;
3. Comparison with aerobic respiration.

Anaerobic respiration is an important biological phenomenon which enables living beings to produce energy in an oxygen-free environment. In contrast to aerobic respiration, where glucose is completely oxidized to CO_2 and water, anaerobic respiration leads to the incomplete breakdown of substrates, thus producing metabolites of lactic acid or ethanol, in addition to ATP. The pathway is critical to the survival of different microorganisms , such as bacteria and yeasts, which grow in low oxygen conditions. In addition to helping cellular functions in hypoxic conditions, anaerobic respiration has been found to be an important process in nutrient recycling in oxygen-limited conditions. As an example, during vigorous exercise, the human muscle tissue temporarily balances the energy requirement, which cannot be satisfied by the availability of oxygen to the rate of ATP production, exemplifying the significance of the process of anaerobic glycolysis in energy metabolism. The overall efficiency of anaerobic respiration can be different among organisms and depends both on the substrate utilized and the particular metabolic routes by which it is processed. As an illustration, in yeast, sugars fermentation produces alcohol and is greatly exploited in brewing and baking sectors. In contrast, other bacteria can use other terminal electron acceptors, , such as nitrates or sulfates, leading to a variety of metabolic end products. These types of strategies point to the versatility of anaerobic metabolism which enables organisms to respond to changing environmental conditions. Remarkably, however, recent research has shown that in environments where aerobic respiration would be expected some microorganisms nonetheless persist in anaerobic metabolism, demonstrating a bioenergetic versatility that is only now being recognized. The presence of simultaneous metabolic pathways can enable organisms to maximize energy extraction, emphasizing the evolutionary significance of anaerobic respiration in various ecological niches (Oren et al., 2009). Besides, the research of anaerobic respiration is not limited to a single organism and also provides insight into ecological and industrial practices. As another example, the end products of anaerobic respiration play a key role in biotechnological applications, like biogas production, waste digestion, or biofuel production. Anaerobic digestion to produce methane is one of the sustainable ways of managing organic waste to produce renewable energy. Also, knowledge of anaerobic pathways helps to solve health-related issues e.g., lactic acidosis in human muscle during oxygen debt conditions. Research into the mechanics of anaerobic respiration will undoubtedly provide valuable insights into the metabolic adaptations of various organisms, their ecological roles, and potential industrial applications, reinforcing the significance of this biological process in both natural and applied contexts (Berkmann et al., 2020) (Figure 7.11).

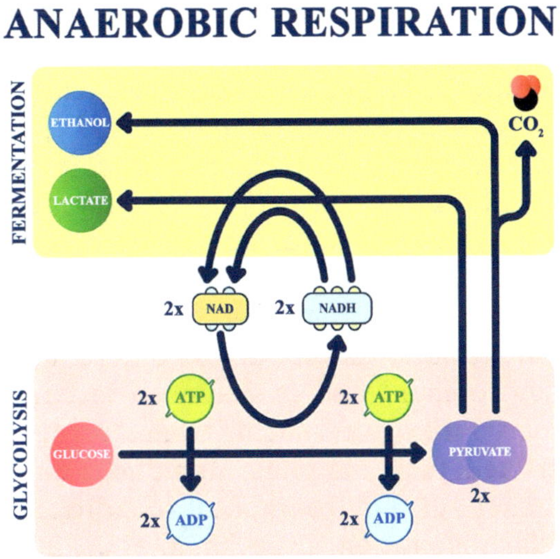

Figure 7.11. Anaerobic respiration.

Source: https://biologydictionary.net/wp-content/uploads/2016/10/Anaerobic_respiration.jpg.

7.5.1. Definition and Conditions for Occurrence

Cellular respiration is a critical biochemical process that converts nutrients into energy to support cellular activities. In the simplest terms, the definition of cellular respiration includes a series of metabolic reactions occurring mostly in cells, to support the metabolism of glucose and produce ATP. It occurs under both aerobic and anaerobic conditions, showing its sensitivity to varied environmental and physiological conditions. Oxygen present aerobic respiration includes the citric acid cycle, glycolysis, and oxidative phosphorylation. Anaerobic respiration without the presence of oxygen typically produces byproducts , such as lactate or ethanol and is reflective of the organisms metabolic flexibility and capacity to survive under different conditions. Its cellular respiration definition thus includes not only its biochemical nature but also its very purpose in the mobilization of energy for different life forms. Cellular respiration is essentially contingent on the supply of substrates and concentration of oxygen for its manifestation. In aerobic conditions, cellular respiration is facilitated by the use of oxygen as the terminal electron acceptor in the electron transport system. This requirement emphasizes the relationship between metabolic processes and environmental variability, as cells must adapt to shifting oxygen and nutrient levels. For example, in Caenorhabditis elegans organisms, metabolic patterns vary across different stages of development and in response to extrinsic stressors like alternating temperatures or restricted food supplies, demonstrating the influence of environmental conditions on cellular respiration (Braeckman et al., 2009).

Conversely, anaerobic respiration occurs when there is a lack of oxygen, and cells are compelled to resort to other metabolic pathways for ATP production, proving the plasticity and variation of forms of life in various conditions. Apart from the fundamental processes of metabolism, cellular respiration has physiological effects, and its impacts are realized on organismal health and lifespan as well. The distinction between aerobic and anaerobic processes affects not only the generation of energy but also the generation of ROS, which are a source of oxidative stress and cellular damage when uncontrolled. As a case in point, with Brugada syndrome, understanding syncope and its mechanism emphasizes the importance of identifying metabolic responses to oxygen availability (Anselmino et al., 2017). Thus, a consideration of the conditions under which anaerobic or aerobic respiration is favored sheds light on the delicate balance between energy production and cell integrity. The definition and conditions governing cellular respiration thus clarify its vital role in sustaining life, adaptation, and wellness across species.

7.5.2. Types of Anaerobic Pathways

Anaerobic processes are crucial mechanisms whereby cells can generate energy in the absence of oxygen, allowing various organisms to survive in oxygen-free habitats. Two significant anaerobic processes are fermentation and anaerobic respiration. Fermentation, predominantly shown by certain bacteria and yeast, entails incomplete hydrolysis of glucose to produce energy with byproducts , such as ethanol or lactic acid. This is essential in situations where oxygen is scarce, ultimately leading to the revival of NAD^+, which is used for the continued glycolysis, therefore energy generation despite the anaerobic state. Anaerobic respiration, however, has an ETC but with alternative electron acceptors , such as sulfate, nitrate, or CO_2 instead of oxygen. This is a normal process in some bacteria and archaea, particularly where organic matter is high, where organic molecules are more easily decomposed than by fermentation. Anaerobic respiration is more efficient than fermentation, the energy-producing process that is required in many various forms of ecosystems. Microbial communities possess complex dynamics under anaerobic respiration, where various species are found to coexist symbiotically to optimize energy conversion processes with tightly interrelated metabolic pathways to stability and community function (Chen et al., 2020). Additional studies on anaerobic pathways demonstrate the role of thermodynamics in anaerobic processes. Present research has provided insight into the level of energy needed for growth-supporting metabolic pathways, which has emphasized interactions in microbial communities. Understanding these anaerobic processes and their accompanying energy dynamics can be applied to predict population dynamics and community resilience within various ecological niches. Of specific interest, the integration of thermodynamics into theoretical frameworks has increased our knowledge of how some bacteria, , such as sulfate reducers and methanogens, utilize low-energy reactions to survive, with the notable impact of anaerobic pathways on global biogeochemical cycling (Chen et al., 2020). Consequently, breaking down such energy-efficient processes not only enlightens our understanding of cellular respiration but also paves the way for potential biotechnological applications, which enhance the sustainability and bioconversion strategies (Figure 7.12).

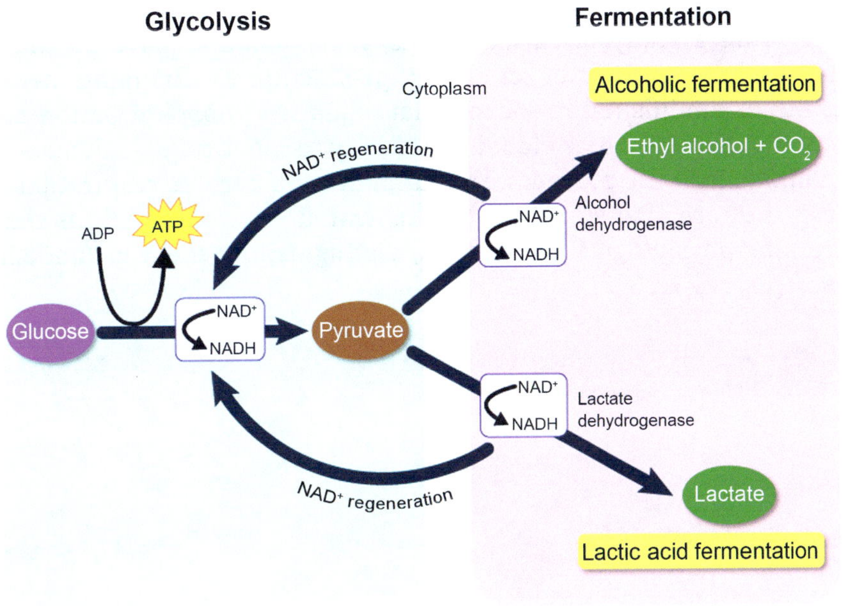

Figure 7.12. Types of anaerobic pathways.

Source: https://www.sciencefacts.net/wp-content/uploads/2021/09/Anaerobic-Respiration-Diagram.jpg.

7.5.3. Comparison with Aerobic Respiration

Cellular respiration is more complicated than merely making energy; an important contrast is between anaerobic and aerobic respiration. Aerobic respiration occurs in the presence of oxygen and involves a complex series of biochemical reactions, namely glycolysis, the Krebs cycle, and oxidative phosphorylation. Up to a certain limit, aerobic respiration leads to the total oxidation of glucose, producing a staggering 36 to 38 ATP molecules per glucose molecule. Contrary to anaerobic respiration under non-oxygen conditions and resulting in incomplete oxidation to produce only 2 ATP molecules through glycolysis, the low yield is a reflection of the efficiency of aerobic processes and dependency on oxygen as a terminal electron acceptor in the ETC. Thus, understanding the primary distinction in energy generation and metabolic by-products, i.e., CO_2 and water, becomes essential in determining the evolutionary advantage of aerobic respiration in high-energy-demanding organisms. Based on the role of mitochondria reveals greater disparity between aerobic and anaerobic metabolism. Mitochondria, more often called the powerhouses of the cell, are the key to the success of aerobic respiration. Experiments indicate that mitochondrial bioenergetics have a profound effect on ATP production efficiency, especially in larger organisms.

For example, studies show that mitochondrial coupling efficiency may rise with body size and thus lead to increased ATP production among larger organisms like the bullfrog compared to their smaller counterparts, the common frog (Dumet et al., 2015). This illustrates that not only is aerobic respiration more effective at a cellular level than anaerobic mechanisms, but it is also based on physiological processes , such as organism size, hence linking metabolic activity to broader biological paradigms. The genetic regulation of metabolic pathways further emphasizes the difference between aerobic and anaerobic respiration. In mammalian cells, the switch between glycolysis and oxidative phosphorylation is regulated by specific genes and enzymes, thus impacting energy metabolism. From a study of cardiac MSCs, it has been discovered that key regulators , such as the Dicer gene have an impact on energy production by changing metabolic tendencies. Dicer deletion results in diminished mitochondrial oxidative phosphorylation and compensatory increased glycolysis, a classic case of metabolic plasticity to meet cellular demand (Jin et al., 2019). These findings illuminate how the biochemical foundations of aerobic respiration not only facilitate production of energy but are also subject to sophisticated controlling inputs that can respond to environmental stress or physiological fluctuation, distinguishing them in fundamental ways from anaerobic processes (Figure 7.13).

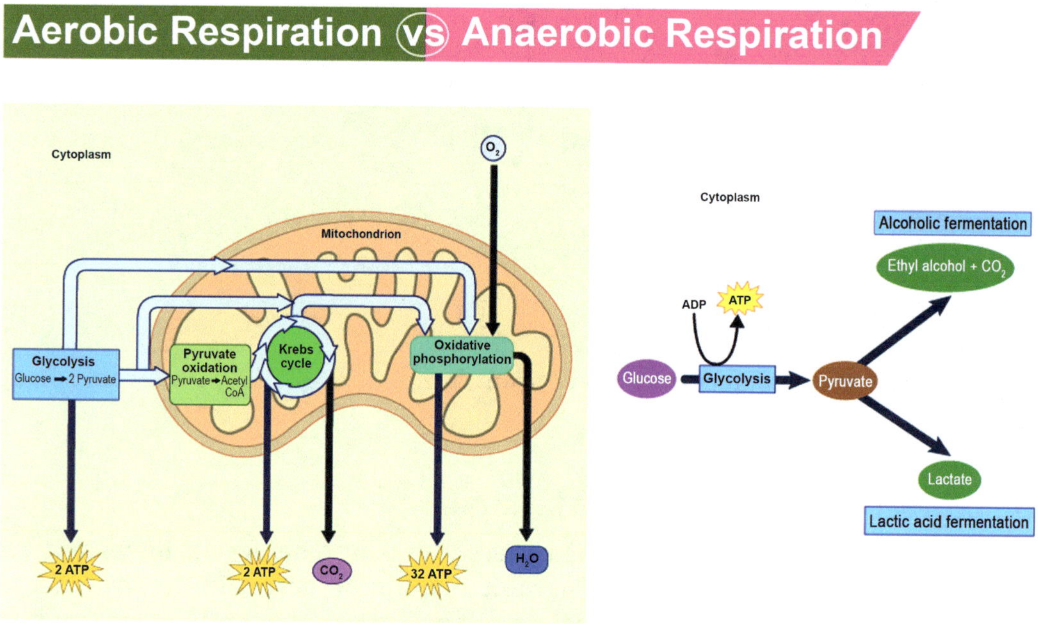

Figure 7.13. Aerobic versus anaerobic respiration.

Source: https://www.sciencefacts.net/wp-content/uploads/2021/09/Aerobic-vs-Anaerobic-Respiration.jpg.

Test Your Knowledge

1. What are the types of anaerobic pathways?
2. How to compare with aerobic respiration?

7.6. Regulation of Cellular Respiration

Learning Outcome

1. Factors affecting the rate of respiration;
2. Role of enzymes in regulation;
3. Feedback mechanisms in metabolic pathways.

Cellular respiration is a complex process that is regulated to match the production of energy with the requirements of the cell. The key point of this regulation is the regulation of metabolic pathways in the mitochondria, where the fine balance of energy substrates and redox state can have a profound effect on rates of respiration. Mitochondria are not only the powerhouse of the cell, but also important in cellular signaling and homeostasis. Recent insights into the redox-controlled pathways within the intermembrane space have shed light on how oxidative stress can modulate protein import and function, ultimately influencing the respiratory processes critical for maintaining cellular health (Manganas et al., 2016). These regulatory processes are critical to understanding how cells sense the need to alter their metabolic processes in reaction to changing physiological states and hence protect against dysfunctions that may jeopardize survival. Another level of complexity in the regulation of cellular respiration is the interactions between respiration and other physiological systems. It has been noted that breathing is closely connected with heart rhythms, blood pressure, and neural activity, which emphasizes a complex system of modulation between breathing and cellular energy metabolism.

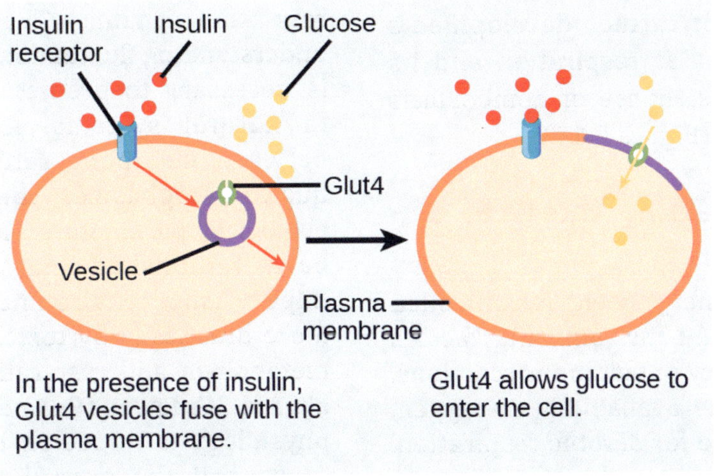

Figure 7.14. Regulation of cellular respiration.

Source: https://s3-us-west-2.amazonaws.com/courses-images/wp-content/uploads/sites/3206/2018/05/03175847/Figure_07_07_01.png.

In particular, the alterations in the intracellular mitochondrial functioning have been observed in the course of the inhalation and exhalation cycle, which allows considering the association between the respiratory patterns and metabolic responses. Such dynamics indicate that the respiratory drive may influence membrane potential alterations, which can further affect cellular activities at various levels (Barnes et al., 2014). This mutual interaction demonstrates the significance of the knowledge of the physiological background, in which the process of cellular respiration is implemented, its regulation in diverse environmental and biological conditions, in particular. Besides, regulation of cellular respiration is not merely an issue of metabolic fine-tuning, but it is also a key to the general cellular wellbeing and disease prevention. The respiratory dysregulation may cause diverse health-related problems, , such as metabolic disorders and vulnerability to oxidative stress. At a higher level, the realization of the interdependence of respiration on other cellular processes can carry far-reaching consequences with regard to medical therapy and intervention protocols. Insights into how ionic changes during respiration can affect overall cellular function open avenues for therapeutic exploration, particularly in addressing conditions like breathing disorders and stress-related ailments (Barnes et al., 2014). With further developments in research, the holistic picture of these regulatory networks will form the basis of further developments in understanding cellular respiration and its importance in the sustenance of homeostasis in biological systems (Figure 7.14).

7.6.1. Factors Affecting the Rate of Respiration

Cellular respiration energy yield is controlled by various biological and environmental factors that affect the efficiency of metabolic reactions. One such factor is the availability of oxygen, which is a prerequisite for aerobic respiration. In low-oxygen conditions, cells will switch to anaerobic respiration, resulting in lower energy yields. Substrate availability also affects the rates of respiration, where adequate supply of glucose and other organic compounds must be provided to sustain metabolic processes. Temperature also affects the kinetic energy of the molecules and hence the enzymatic activity involved in respiration. For instance, a rise in temperature may raise the metabolic rates up to a certain level but then lead to denaturation of enzymes, resulting in lowered efficiency of respiration. Understanding these factors can provide insight into how organisms respond to varied environmental conditions during cellular respiration. Another factor that should be considered is the influence of pH levels on respiration rates. The metabolic activities of cellular respiration are also prone to pH changes since departures from optimum levels can degrade enzymatic functions. For example, Lijklema (1971) suggests that CO_2 produced in the course of cellular metabolism has the potential to affect pH, which in turn continues to affect respiration. In those environments whose pH is not near neutrality, as in certain aquatic habitats, such perturbations have the potential to interfere with cellular activity and affect the rate of respiration among organisms. Sub-lethal toxicity of environmental pollutants also becomes a factor in modulating respiration as the pollutants can induce stress response in organisms, further complicating cellular respiration dynamics. Therefore, monitoring and understanding fluctuations in pH and pollutants is necessary to predict respiration dynamics in natural systems. Additionally, synergy between metabolic rate and environmental quality emphasizes the interplay between ecological parameters and respiration. It has been established that organisms in more impoverished environments may be stressed more drastically by toxicants, so their rates of metabolism and respiration behaviors will be altered. Willows (1994) discusses how feeding physiology is impacted by sub-lethal toxicity to control the overall available energy for respiration. Such non-linear correlation of toxic stress and respiration speaks to the plasticity of the organisms subjected to environmental fluctuation. Successful management of ecosystems involves an understanding of these complex interactions, as they can have ripple effects on population health and biodiversity. Overall, oxygen status, pH, and environmental quality are essential to understanding the intricacies of cellular respiration and the robustness of biological systems (Figure 7.15).

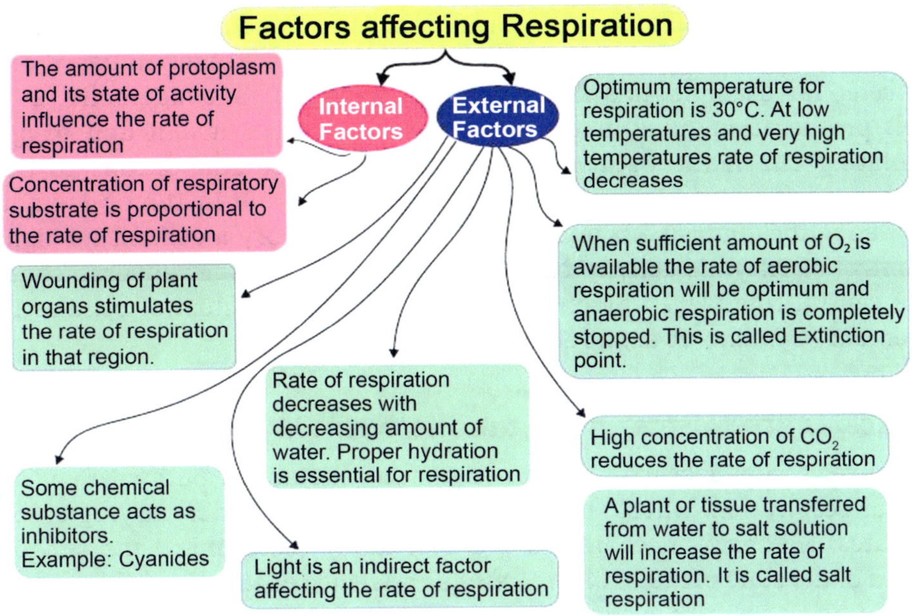

Figure 7.15. Factors affecting the rate of respiration.

Source: https://img.brainkart.com/imagebk34/eBZ46RA.jpg.

7.6.2. Role of Enzymes in Regulation

Enzymes play a critical role in the regulation of biochemical pathways, particularly in metabolic processes like cellular respiration. Through their function as biological catalysts, enzymes accelerate the rate of metabolic reactions without being consumed in the process. Acceleration is crucial to guarantee the process of homeostasis in cells by allowing them to respond quickly to fluctuations in energy demand and metabolic states. As an example, the enzyme PFK is a site of regulation for glycolysis, controlling the entry of glucose via this central metabolic pathway. PFK adjusts the rate of glycolysis to equate energy production with cell needs by monitoring levels of ATP and AMP. This direct control underscores the significance of enzyme activity in ensuring the capacity of cells to carry out efficiently ATP production needed by cellular processes, particularly under fluctuating environmental conditions. Furthermore, enzyme control is directly linked with cellular signaling mechanisms that are conditioned by external situations, , such as oxygen presence. In hypoxia conditions, for example, NO is a critical signaling molecule that controls the activity of various enzymes engaged in energy metabolism. Enzyme activities, including NO-mediated activities, are critical to control metabolic pathways to harness substitute sources of energy during oxygen insufficiency. The elevation of NO under conditions of hypoxia operationally reduces oxygen consumption and promotes the utilization of anaerobic mechanisms, in essence allowing cells to adapt to low-oxygen environments without sacrificing metabolic efficiency. Such regulation is not only required in normal physiology but also in pathology, when abnormal cellular respiration leads to tissue hypoxia and organ failure, highlighting the enzymes central role in providing adequate metabolic response (Umbrello et al., 2013). The capacity of mitochondrial enzymes to respond to altering metabolic demands is also of critical significance in the survival of cells in adverse conditions, e.g., neoplastic transformation. Mitochondria, or dynamic organelles, also exhibit bioenergetic plasticity, enabling them to modulate enzyme function based on cell demands. Within cancer cells, for example, this flexibility allows for rampant metabolic rewiring intended to achieve high proliferation rates in the presence of limited nutrient and oxygen availability. Through modulating enzymatic pathways, mitochondria offer adaptations to blood perfusion and energy supply variability, demonstrating their essential role in tumor metabolism. The co-operation between mitochondrial enzymes and cellular regulatory mechanisms highlights how the coordinated control of enzymic activity can enable cells to live under adverse conditions, further evidence of the central role played by enzymes in the broader context of cellular respiration and energy management (Cannino et al., 2018).

7.6.3. Feedback Mechanisms in Metabolic Pathways

Feedback mechanisms play a crucial role in regulating metabolic pathways, particularly in the context of cellular respiration, where they provide homeostasis under varying environmental and physiological states. Feedback mechanisms are essentially biochemical messengers that inform cellular processes of the state of metabolic reactions and hence enable the cell to react to fluctuations in energy demand and substrate availability.

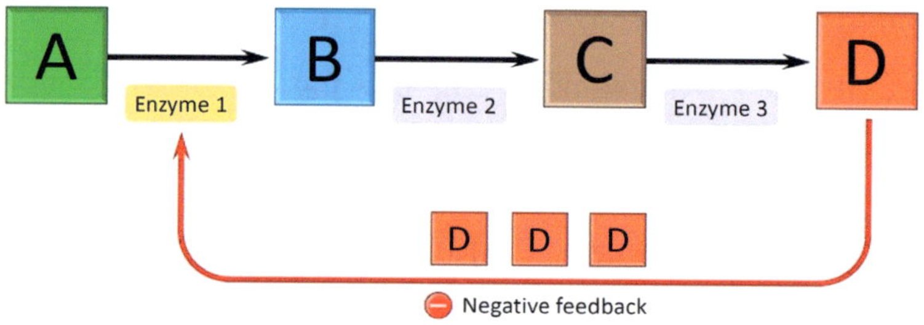

Figure 7.16. Feedback mechanisms in metabolic pathways.

Source: https://old-ib.bioninja.com.au/_Media/feedback-inhibition_med.jpeg.

As an illustration, the activation or inhibition of specific enzymes may be regulated by the concentrations of end products or precursors of metabolism. The system not only maintains maximum cellular activity but also prevents excessive accumulations of metabolites causing cellular stress and malfunction. A significant type of feedback regulation is about the interaction between respiration and cellular signaling. Current studies have shown that dynamics of the respiratory processes influence cellular functions other than energy production, like modulation of the rhythm of the heart and the vascular reactions during the cycles of respiration (Barnes et al., 2014). Because cellular respiration, particularly oxidative phosphorylation, regulates the production of ATP, it also regulates the feedback signals to further refine metabolic processes. Through this mechanism, mitochondrial function changes may impact feedback mechanisms, thus influencing larger-scale physiological responses, including those related to heart function and blood pressure. Understanding these complexities offers more insight into potential therapeutic application for a variety of health conditions, particularly those relating to respiration and metabolic dysregulation. Furthermore, feedback loops extend to pathological conditions, for example, in heart failure and other metabolic disease conditions. Studies indicate that changes in metabolic signaling can aggravate cardiac diseases and there is evidence of an complex interaction between metabolic health and cellular respiration. For instance, treatments for cardiotoxic agents illustrate how altered mitochondrial metabolism and function can lead to severe health conditions , such as heart failure (Ganeshan et al., 2019). This relates to the necessity of careful feedback loops in metabolic pathways so that the risk of cell damage can be minimized. Additionally, more recent research on agents like melatonin shows they can restore balance to disrupted feedback mechanisms caused by stressors and highlight the crucial link between cellular respiration, metabolic regulation, and overall health (Figure 7.16).

Test Your Knowledge

1. What are the factors affecting the rate of respiration?
2. What is the role of enzymes in regulation?

Summary

The complex process of cellular respiration emphasizes the pivotal role of energy generation in the maintenance of life. This biochemical process not only provides the cellular energy currency ATP but also serves as an essential mechanism in the maintenance of cellular processes and homeostasis. The efficiency of cellular respiration relies considerably on various factors, including the availability of oxygen and the integrity of metabolic pathways. Identification of these dynamics allows scientists to study how interruption in respiration can lead to severe health problems, , such as metabolic disorders. Because cellular respiration has repercussions beyond the generation of energy, they spill over into other biological processes and exhibit a close relationship between cellular metabolism and the health of the organism as a whole. In addition, this research into cellular respiration reveals its interconnectedness with other biological systems and emphasizes coordination among cellular components. The observation of hierarchical interactions among macromolecules, cells, and tissues suggests that efficient energy production is contingent upon a perfectly synchronized system of biochemical reactions and regulatory mechanisms. Local and global cooperation in these systems can affect the process of aging and thereby contribute to the overall vitality and adaptability of an organism, recent studies suggest. This holistic view of cellular respiration as an integral part with richness and cooperation colors the larger picture of biological reaction to stresses in the environment, deepening our understanding of life sciences still further. Finally, the implications of cellular respiration extend into realms of research and therapeutic control. Understanding of the nuances of cellular energy metabolism opens up facing the challenges of aging and chronic disease. Mechanism studies governing copper homeostasis, , such as the effects of mutations on respiration, reveal a complex interdependence between energy metabolism and cell integrity. Investigations indicate that derailments of the processes possess the ability to generate conditions influencing not only metabolic health but lifespan (Servos et al., 2012). As science advances, it is crucial to further dissect these associations, explaining how cellular respiration can be used for novel therapies and to advance the quality of life, especially in elderly populations suffering from disorders related to energy.

Cellular respiration, a fundamental biological process, is investigated to discover major mechanisms by which organisms convert biochemical energy present in nutrients into ATP. The process is composed of numerous integrated pathways, including glycolysis, the citric acid cycle, and oxidative phosphorylation. Every step contributes to energy production, suggesting the intricacy of cellular metabolism. Glycolysis occurs in the cytoplasm, transforming glucose to pyruvate and yielding a net gain of ATP and NADH. Pyruvate proceeds to the mitochondria, where the citric acid cycle takes place, oxidizing further metabolites and generating electron carriers for the ETC. The process is carefully balanced, and imbalances can lead to compromised energy availability, illustrating the importance of proper cellular respiration for overall well-being. In addition to energy production, cellular respiratory regulatory mechanisms are essential in keeping cellular and organismal functions in check. Hormonal signals and substrate levels are significant in determining the metabolic processes necessary for dynamic adaptation to the organism's energy requirements. For instance, exercise doubles the requirement for ATP and therefore increases glycolysis and mitochondrial function. Cell respiration is also heavily linked with cooperation between cellular elements, reflecting the general principles of biology. Recent studies emphasize the significance of cooperative networks in maintaining optimal cell functioning and reacting to environmental changes. Quoting Braeckman et al. (2009), local and global cooperation of biological networks contributes to robustness and evolutionary capacity for adaptation of organisms and demonstrates how complex interactions become crucial in efficient energy metabolism, particularly under stress or adaptation. Recent research also indicates that particular molecular regulators, , such as peroxisome proliferator-activated receptor-γ coactivator (PGC)-1β, have deep effects on the muscles' mitochondrial function and endurance capacity. Scientific research verifies that overexpression of PGC-1β enhances the muscle's ability to adapt to endurance training, also enhancing cellular respiration efficiency in energy production. Interestingly, these adjustments suggest training can solidify long-term changes in cell function, as reflected by the preservation of maximal oxygen uptake even following detraining. However, Lee et al. (2017) demonstrate that other physiological responses to detraining were not meaningfully affected, illustrating that the entire context of cellular respiration involves countless variables far beyond easily adaptive responses. All of these observations depict the complex regulation of cellular respiration and its importance in evolution, guaranteeing the complexity of energy metabolism in living organisms.

KEY TERMS FROM THIS CHAPTER

Aerobic respiration	Anaerobic respiration	ATP (adenosine triphosphate)
Cellular respiration	Chemiosmosis	Cristae
Electron transport chain	FAD (flavin adenine dinucleotide)	
FADH2	Glucose	Glycolysis
Oxidative phosphorylation	Lactic acid fermentation	Matrix (mitochondrial matrix)
Mitochondria	NAD^+ (nicotinamide adenine dinucleotide)	
Pyruvate	Pyruvate oxidation	

CRITICAL THINKING QUESTIONS

1. **What is the primary purpose of cellular respiration?**
   a. To produce oxygen
   b. To break down DNA
   c. To convert glucose into usable energy (ATP)
   d. To create proteins
2. **Which organelle is mainly responsible for cellular respiration in eukaryotic cells?**
   a. Nucleus
   b. Ribosome
   c. Mitochondrion
   d. Golgi apparatus
3. **What are the three main stages of aerobic cellular respiration?**
   a. Glycolysis, Krebs cycle, Electron transport chain
   b. Photosynthesis, Fermentation, Respiration
   c. Glycolysis, Calvin cycle, ATP synthesis
   d. Glucose breakdown, Oxidation, Reduction
4. **Where does glycolysis occur in the cell?**
   a. Mitochondria
   b. Cytoplasm
   c. Nucleus
   d. Endoplasmic reticulum
5. **What is the final electron acceptor in the electron transport chain?**
   a. Glucose
   b. NAD^+
   c. Oxygen
   d. Carbon dioxide

Short Answer Questions

1. What is the purpose of cellular respiration in living organisms?
2. What is the overall chemical equation for cellular respiration?
3. In which organelle does cellular respiration primarily take place in eukaryotic cells?
4. What are the three main stages of aerobic cellular respiration?
5. Where in the cell does glycolysis occur, and what are its end products?
6. What molecule is the main electron carrier produced in the Krebs cycle?

Answers to Critical Thinking Questions

1. (c); 2. (c); 3. (a); 4. (b); 5. (c)

References

1. Alessio, E., Buson, L., Chemello, F., Peggion, C., Grespi, F., Martini, P., Massimino, M. L., Pacchioni, B., Millino, C., Romualdi, C., Bertoli, A., Scorrano, L., Lanfranchi, G., & Cagnin, S. (2019). Single cell analysis reveals the involvement of the long non-coding RNA Pvt1 in the modulation of muscle atrophy and mitochondrial network. *Nucleic acids research*, *47*(4), 1653–1670. https://doi.org/10.1093/nar/gkz007 (accessed on 13th June 2025).
2. Allan, I., Gonzalez, C., Greenwood, R., Mills, G., et al. (2005). Water quality monitoring: A 'toolbox' in response to the EU's Water Framework Directive requirements. https://core.ac.uk/download/29586544.pdf (accessed on 13th June 2025).
3. Anand, R., Barbian, A., Berleth, N., Borchardt, R., et al. (2018). The mycotoxin phomoxanthone A disturbs the form and function of the inner mitochondrial membrane. https://core.ac.uk/download/323077269.pdf (accessed on 13th June 2025).
4. Anselmino, M., Barbonaglia, L., Bergamasco, L., Bianchi, A., et al. (2017). Etiological diagnosis, prognostic significance and role of electrophysiological study in patients with Brugada ECG and syncope. https://core.ac.uk/download/323385723.pdf (accessed on 13th June 2025).
5. Archiche, A., Ferreta, A., Kasirer-Friede, A., Lechi, A., Marín-Hernández, A., Schafer, A. I., Verhoeven, A. J. M., et al. (2017). Glycoprotein Ib activation by thrombin stimulates the energy metabolism in human platelets. https://core.ac.uk/download/129069877.pdf (accessed on 13th June 2025).
6. Lee, S., Leone, T. C., Rogosa, L., Rumsey, J., Ayala, J., Coen, P. M., Fitts, R. H., Vega, R. B., & Kelly, D. P. (2017). Skeletal muscle PGC-1β signaling is sufficient to drive an endurance exercise phenotype and to counteract components of detraining in mice. *American journal of physiology. Endocrinology and metabolism*, *312*(5), E394–E406. https://doi.org/10.1152/ajpendo.00380.2016 (accessed on 13th June 2025).
7. Azam, F., Hodson, R., & Holm-Hansen, O. (2025). Applications of adenine nucleotide measurements in oceanography. https://core.ac.uk/download/pdf/42885369.pdf (accessed on 13th June 2025).
8. Barnes, V. A., Cearley, S. M., Junca, S., Ravinder, V., et al. (2017). The dynamic role of breathing and cellular membrane potentials in the experience of consciousness. https://core.ac.uk/download/131210584.pdf (accessed on 13th June 2025).
9. Barnes, V. A., Crawford, M. W., Harden, K., Ravinder, V., et al. (2014). Widespread depolarization during expiration: A source of respiratory drive? https://core.ac.uk/download/131210052.pdf (accessed on 13th June 2025).
10. Berkmann, J. C., Herrera Martin, A. X., Ellinghaus, A., Schlundt, C., Schell, H., Lippens, E., Duda, G. N., Tsitsilonis, S., & Schmidt-Bleek, K. (2020). Early pH Changes in Musculoskeletal Tissues upon Injury-Aerobic Catabolic Pathway Activity Linked to Inter-Individual Differences in Local pH. *International journal of molecular sciences*, *21*(7), 2513. https://doi.org/10.3390/ijms21072513 (accessed on 13th June 2025).
11. Black, S. M., Desai, A. A., Fineman, J. R., Khalpey, Z., et al. (2015). Complex I dysfunction underlies the glycolytic switch in pulmonary hypertensive smooth muscle cells. https://core.ac.uk/download/323064501.pdf (accessed on 13th June 2025).
12. Bouzayen, M., Flores, F. B., Latché, A., Pech, J. C., et al. (2006). Mechanisms of fruit ripening:

Retrospect and prospects. https://core.ac.uk/download/12040271.pdf (accessed on 13th June 2025).

13. Braeckman, B. P., Houthoofd, K., & Vanfleteren, J. R. (2009). Intermediary metabolism. *WormBook : the online review of C. elegans biology*, 1–24. https://doi.org/10.1895/wormbook.1.146.1(accessed on 13th June 2025).

14. Brown, J. P., Collins, M. J., Coursolle, D., Fonseca, B. M., Blumberger, J., Butt, J. N., Rosso, K. M., et al. (2015). Multi-heme cytochromes in *Shewanella oneidensis* MR-1: Structures, functions and opportunities. https://core.ac.uk/download/29111652.pdf (accessed on 13th June 2025).

15. Brunet, A., Cassidy-Stone, A., Hoshino, A., Jain, A., Kazlauskaite, A., Khaminets, A., et al. (2017). The pharmacological regulation of cellular mitophagy. https://core.ac.uk/download/76981017.pdf (accessed on 13th June 2025).

16. Chen, J., Delattre, H., Soyer, O. S., Wade, H., et al. (2020). Thermodynamic modelling of synthetic communities predicts minimum free energy requirements for sulfate reduction and methanogenesis. https://core.ac.uk/download/323058136.pdf (accessed on 13th June 2025).

17. Costa, F., Gianfranceschi, L., Molthoff, J. W., Schouten, H., et al. (2009). Transcription analysis of apple fruit development using cDNA microarrays. https://core.ac.uk/download/pdf/29251611.pdf (accessed on 13th June 2025).

18. David Zeferino d'Azevedo Cristina. (2008). A regulated response to mitochondrial dysfunction modulates longevity and rates of living. Universidade Nova de Lisboa. Instituto de Tecnologia Química e Biológica. https://core.ac.uk/download/157621860.pdf (accessed on 13th June 2025).

19. Díaz Moreno, I., Díaz Quintana, A. J., Guerra Castellano, A., Pérez Mejías, R., et al. (2019). Cytochrome c: Surfing off of the mitochondrial membrane on the tops of complexes III and IV. https://core.ac.uk/download/222572428.pdf (accessed on 13th June 2025).

20. Dumat, C., Pinelli, E., Pourrut, B., Shahid, M., et al. (2011). Lead uptake, toxicity, and detoxification in plants. https://core.ac.uk/download/12043147.pdf (accessed on 13th June 2025).

21. Dumet, A., Rey, B., Romestaing, C., Roussel, D., et al. (2015). Oxidative phosphorylation efficiency, proton conductance and reactive oxygen species production of liver mitochondria correlate with body mass in frogs. https://core.ac.uk/download/42364842.pdf (accessed on 13th June 2025).

22. Umbrello, M., Dyson, A., Feelisch, M., & Singer, M. (2013). The key role of nitric oxide in hypoxia: hypoxic vasodilation and energy supply-demand matching. *Antioxidants & redox signaling*, *19*(14), 1690–1710. https://doi.org/10.1089/ars.2012.4979Frey, S. D., Geyer, K. M., Grandy, A. S., Kyker-Snowman, E., et al. (2016). Microbial carbon use efficiency: Accounting for population, community, and ecosystem-scale controls over the fate of metabolized organic matter. https://core.ac.uk/download/215504071.pdf (accessed on 13th June 2025).

23. Ganeshan, T. K. (2019). Investigating the protective role of the natural hormone melatonin in reducing drug-induced cardiotoxicity in the therapy of chronic diseases. https://core.ac.uk/download/287605215.pdf (accessed on 13th June 2025).

24. Goetz, A., Heckel, K., Schulte-Uentrop, L., Kiefmann, M., Tritt, O., Keller, P., Kiefmann, R., et

al. (2011). A novel alveolar Krebs cycle-triggered CO_2 sensing mechanism regulates regional pulmonary ventilation. https://core.ac.uk/download/pdf/290175.pdf (accessed on 13th June 2025).

25. Servos, J., Hamann, A., Grimm, C., & Osiewacz, H. D. (2012). A differential genome-wide transcriptome analysis: impact of cellular copper on complex biological processes like aging and development. *PloS one*, *7*(11), e49292. https://doi.org/10.1371/journal.pone.0049292 (accessed on 13th June 2025).

26. Jin, Y., Kim, I., Shen, Y., Su, Q., et al. (2019). RNAase III-type enzyme Dicer regulates mitochondrial fatty acid oxidative metabolism in cardiac mesenchymal stem cells. https://core.ac.uk/download/288845921.pdf (accessed on 13th June 2025).

27. Kiss, H. J. M., Mihalik, A., Nánási, T., Ory, B., Spiró, Z., Soti, C., & Csermely, P. (2009). Ageing as a price of cooperation and complexity: Self-organization of complex systems causes the gradual deterioration of constituent networks. *BioEssays, 31*(6), 651–664. https://doi.org/10.1002/bies.200800224 (accessed on 13th June 2025).

28. Kumar, A., & Smith, B. (2004). Towards a proteomics meta-classification. https://core.ac.uk/download/132155256.pdf (accessed on 13th June 2025).

29. Lijklema, L. (1971). Factors affecting pH change in alkaline wastewater treatment–II: Carbon dioxide production. https://core.ac.uk/download/pdf/11471763.pdf (accessed on 13th June 2025).

30. Manganas, P., MacPherson, L., & Tokatlidis, K. (2017). Oxidative protein biogenesis and redox regulation in the mitochondrial intermembrane space. *Cell and tissue research*, *367*(1), 43–57. https://doi.org/10.1007/s00441-016-2488-5 (accessed on 13th June 2025).

31. Merks, R. M. H., & Szabó, A. (2017). Blood vessel tortuosity selects against evolution of aggressive tumor cells in confined tissue environments: A modeling approach. https://core.ac.uk/download/388651915.pdf (accessed on 13th June 2025).

32. Cannino, G., Ciscato, F., Masgras, I., Sánchez-Martín, C., & Rasola, A. (2018). Metabolic Plasticity of Tumor Cell Mitochondria. *Frontiers in oncology*, *8*, 333. https://doi.org/10.3389/fonc.2018.00333Oren, A., Varma, A., Scharf, B., Teusink, B., et al. (2009). Systems analysis of bioenergetics and growth of the extreme halophile *Halobacterium salinarum*. https://core.ac.uk/download/12175290.pdf (accessed on 13th June 2025).

33. Schumaker, M. F., Friedman, M., Cascante, M., Trucco, M., Bourne, P. E., Votyakova, T. V., & Selivanov, V. A. (2012). Multistationary and oscillatory modes of free radicals' generation by the mitochondrial respiratory chain revealed by a bifurcation analysis. https://core.ac.uk/download/12213294.pdf (accessed on 13th June 2025).

34. Shah, H. (2017). PO_2 dependence of oxygen consumption in skeletal muscle of hypertensive and normotensive rats. https://core.ac.uk/download/84613284.pdf (accessed on 13th June 2025).

35. Willows, R. (1994). The ecological impact of different mechanisms of chronic sub-lethal toxicity on feeding and respiratory physiology. Water quality & stress indicators in marine and freshwater systems: linking levels of organization. Freshwater Biological Association. (accessed on 13th June 2025).

PHOTOSYNTHESIS

8

LEARNING OBJECTIVES

After studying this chapter, you will be able to:

1. Explain the process of photosynthesis;
2. Define the role of chlorophyll;
3. List the factors affecting photosynthesis;
4. Explore photosynthesis in different organisms;
5. Focus on the importance of photosynthesis to life on earth;
6. Illustrate the human impact on photosynthesis.

Introduction

Photosynthesis is the cornerstone of life on Earth, playing the crucial role of converting solar radiation into chemical energy stored within glucose. This complex biochemical reaction is carried out primarily by green plants, algae, and certain bacteria, and they possess specialized organelles known as chloroplasts. In these complexes, chlorophyll pigments capture a specific range of wavelengths of light and are also responsible for the conversion of water and CO_2 into glucose and oxygen. Photosynthesis is not only the generation of energy; it organizes whole ecosystems because it is the primary source of energy for nearly all living organisms. Consequently, it is imperative that knowledge of photosynthesis is not only important for biological science but also for addressing fundamental environmental issues, , such as climate change and food security. In current studies, the ability of various plant species to adapt to different environmental conditions has shown the control of external factors on photosynthesis. For instance, research on the growth morphology of *Pennantia corymbosa* reveals that at different stages of development, they react differently towards light intensity and water. Experiments reveal that juvenile stages are best adapted to open conditions with higher rates of photosynthesis under light stress and higher drought tolerance compared to their adult counterparts. These findings are a reflection of the importance of flexibility in photosynthetic processes in determining species existence and distribution, underscoring the need for more research on the physiological process of photosynthesis (Hollows, 1978). Additionally, advancements in genetic research have provided insights into the complexity of photosynthesis under abiotic stress factors, including salinity. The research on the rice nucleolin gene, OsNUC1, shows its effect on photosynthetic efficiency under salt-stress conditions. Transgenic OsNUC1-S-overexpressing plants exhibited increased net photosynthesis and stomatal conductance, particularly during stress conditions, which indicates a molecular adaptation allowing for better energy conversion during unfavorable conditions. The present work presents the way GM can enhance photosynthetic competence and improve food yield on marginal soils. Ongoing findings of these physiological and genetic factors are imperative not only for the conservation of biodiversity but also for encouraging sustainable agriculture amidst global environmental change (Boonchai et al., 2018) (Figure 8.1).

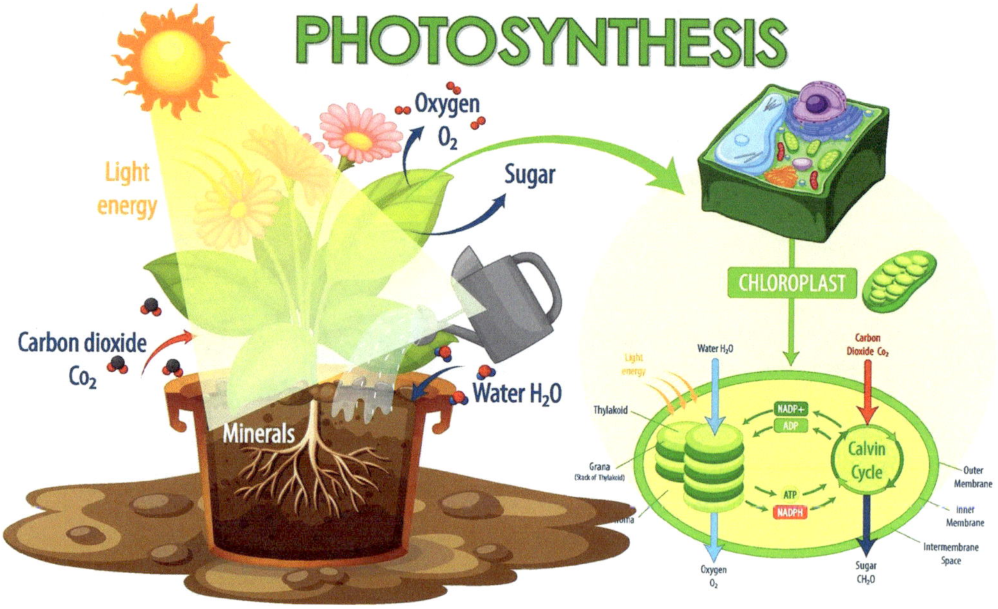

Figure 8.1. Photosynthesis.

Source: https://static.vecteezy.com/system/resources/previews/004/026/220/non_2x/diagram-showing-process-of-photosynthesis-in-plant-free-vector.jpg.

Definition of Photosynthesis

Photosynthesis is the biochemical process whereby green plants, algae, and certain bacteria can utilize light energy and convert it to chemical energy in the form of glucose. Photosynthesis is essential to the persistence of life on Earth because it represents the primary means by which solar energy is captured and converted into a form that living organisms can use. In essence, photosynthesis is a process whereby plants absorb sunlight through chlorophyll, the green pigment found in plant cells. Atmospheric CO_2 and ground water are utilized by plants to produce glucose and, as a byproduct, release oxygen. This remarkable process not only energizes plant growth but also supports the entire food chain, with herbivores relying on these plants as a source of food. The understanding of this complicated process is essential to the understanding of larger-scale ecological processes and the importance of maintaining plant life in combating climate change. In the process of photosynthesis, there are two broad phases that occur: the light-dependent reactions and the light-independent reactions, or more commonly referred to as the Calvin cycle. Light-dependent reactions occur within thylakoid membranes in chloroplasts, where sunlight triggers the breakdown of energy-storing molecules , such as ATP and NADPH. These reactions entail direct sun exposure and result in the production of oxygen gas as a by-product. The Calvin cycle, however, occurs during the stroma of chloroplasts, where energy from the initial reaction drives the fixation of atmospheric CO_2 into glucose. This cooperation of both stages illustrates the complexity of photosynthesis and the effectiveness of converting solar energy. Also, the constructivist nature of science illustrates the challenges teachers face in getting such complex scientific information across to students, which poses the need to reconceptualize the nature of science and the learning processes in science (Barker et al., 1997). Also, photosynthesis has a broader implication than on single organisms; actually, it defines Earth's ecosystems. By trapping solar energy in chemical form, photosynthesis provides energy for sustaining various forms of life. The complex dependency on photosynthetic activity emphasizes interdependence in life, as plants give the nutritional basis for herbivores, who then serve as the source of sustenance for carnivores. In addition, understanding the requirements needed for effective photosynthesis allows for analysis of broader environmental variables. Light intensity, CO_2 concentration, and temperature are found by scientific research to have substantial effects on the efficiency of photosynthesis. These results have not only significance regarding ecological balance but also for agricultural practices, emphasizing the importance of optimizing conditions for plant development. Thus, photosynthesis not only sustains life the way we know it, but it is also a crucial area of study needed to address the dangers posed by climate change and shifts in planetary conditions (Sandora, 2019).

Importance of Photosynthesis in Ecosystems

Photosynthesis is the fundamental mechanism whereby energy enters into ecosystems and establishes the stage for life on Earth. It occurs in plants, algae, and certain bacteria, converting light energy from the sun to chemical energy in the form of glucose. Through the combined reaction of CO_2 and water, these plants produce organic compounds with oxygen as a byproduct. This oxygen not only enables aerobic respiration in heterotrophs but also sustains the many ecological relationships that rely on this essential gas. The effects of photosynthesis reach beyond the direct metabolic requirements of autotrophs to affect entire food webs. Hence, it can be argued that photosynthesis is more than a metabolic process and rather a dominant force in ecosystem processes because it facilitates energy flux and matter cycling within complex ecological systems. Beyond the role of energy conversion, photosynthesis is also essential in carbon cycling, an important process for maintaining the health and stability of ecosystems. Autotrophs absorb atmospheric CO_2 and transform it into organic forms, acting as a significant carbon sink and mitigating human-induced climate change effects. Excessive atmospheric carbon is a critical global concern since it is strongly associated with climate variability that affects biodiversity and species distribution. As freshwater ecosystems are increasingly subjected to human alterations, it becomes important to recognize the contribution of photosynthesis to the management of such ecosystems. For instance, research indicates that the impact of flow cessation can have a detrimental effect on autotrophic organisms in streams and restrict their capacity to perform photosynthesis effectively (Etter & Fischer., 2020). It requires an integrative research framework taking into consideration the junction point of photosynthesis, flow regimes, and ecosystem resilience. Moreover, photosynthetic productivity itself is not uniform across different ecosystems and displays a great range, reflecting its importance. Some of the influences on rates of photosynthesis in terrestrial and aquatic ecosystems include light intensity, temperature, and nutrient availability. Phytoplankton blooms in upwelled water, for example, demonstrate how photosynthesis can be the main driver of primary production in bodies of water that are highly diverse in life forms. Furthermore, the interplay between photosynthesis and other environmental processes , such as nutrient cycling, normally leads to complex dependencies among ecosystems. This information is significant in conservation settings because of amplified pressures of climate change and habitat loss. As dynamic as ecosystems are, it is increasingly necessary to value the importance of photosynthesis and its broader ecological impact on ecological health and biodiversity maintenance (Keulen et al., 2025) (Figure 8.2).

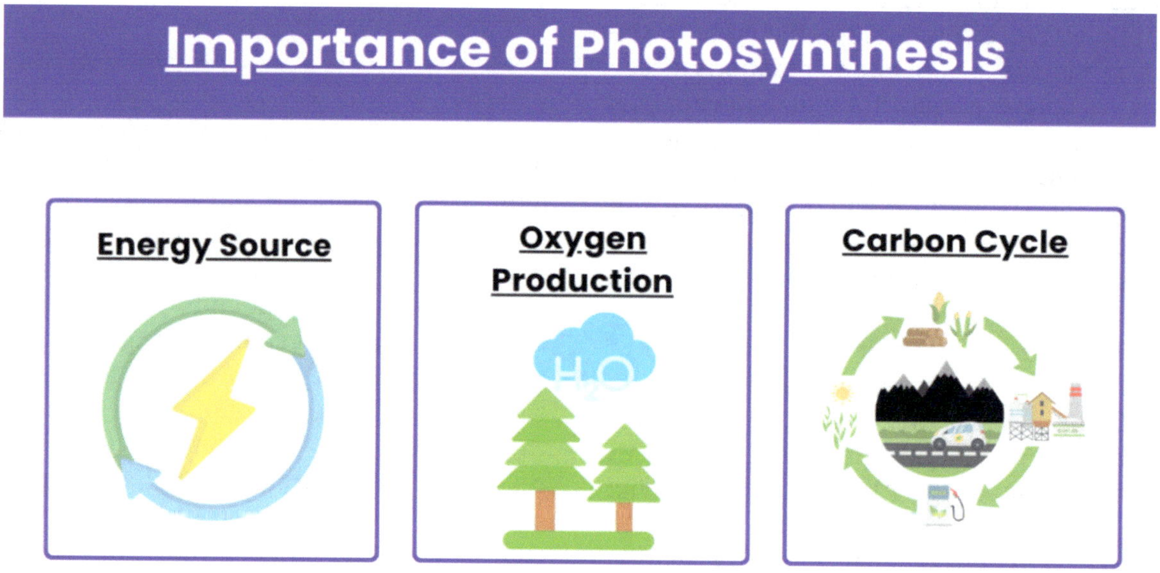

Figure 8.2. Importance of photosynthesis in ecosystems.

Source: https://images.examples.com/wp-content/uploads/2024/07/Importance-of-Photosynthesis.png.

8.1. The Process of Photosynthesis

Learning Outcome

1. Overview of the photosynthetic equation;
2. Stages of photosynthesis: Light-dependent reactions;
3. Stages of photosynthesis: Light-independent reactions (Calvin cycle).

Photosynthesis is essentially crucial to the livelihood here on earth, as it involves the conversion of light energy into chemical energy which is bonded in glucose. The transformation is achieved in plant cells mainly in their chloroplasts in a set of complex reactions. This process can be summarized into two major steps: the light reactions and the Calvin cycle. In the light reactions, sunlight is captured by chlorophyll and a cascade of electron transport reactions occur, leading to the generation of ATP and NADPH. These high-energy molecules are subsequently used in the Calvin cycle, in which CO_2 is incorporated into organic compounds by a subsequent sequence of enzyme-catalyzed reactions. This cyclic process shows the dependence of capture and conversion of energy, which shows the significance of light as a source of energy to all autotrophic organisms in different ecosystems. The physical factors that affect the dynamics of photosynthesis in aquatic ecosystems include the availability of light and time scales of mixing.

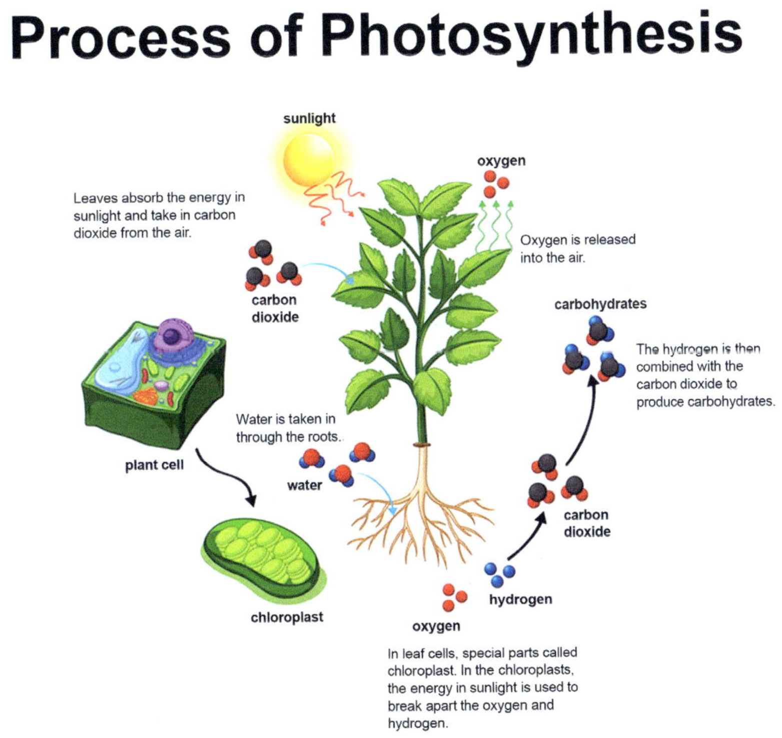

Figure 8.3. The process of photosynthesis.

Source: https://blogcdn.aakash.ac.in/wordpress_media/2022/03/Process-of-photosynthesis.jpg.

The interaction of these factors is important to the photosynthetic organisms (, such as phytoplankton) to grow and be productive. Temporal framework of the physical process influences competition between algal species because various species have dissimilar adjustments to the surrounding. The effects of mixing and transport on algal growth can be dramatic according to recent analyses, fast mixing can cause instant biological activities, i.e., nutrient uptake and photosynthesis whereas slower mixing can cause longer periods of growth. The relative efficiency of these processes can profoundly impact the composition of pelagic ecosystems, underpinning the delicate balance that sustains aquatic life and drives productivity within these systems (Imboden, 1992). Besides, the knowledge of photosynthesis cannot be limited to its mechanical aspects but should include ecological consequences and be affected by external forces like climate change. With the increase in the CO_2 concentration because of anthropogenic activities, photosynthetic organisms are confronted with special problems that may change their contribution to the carbon cycling. Changes in temperature, nutrient status, and light can have large impacts on photosynthetic capacity and growth rates. The health of the whole aquatic ecosystem and its ability to counter the effects of climate change will depend on the resilience of these organisms, as recent research has highlighted. Therefore, enhancing our understanding of photosynthesis is critical not only for ecological studies but also for developing strategies to ensure sustainability in the face of environmental uncertainties and change (Gimbel & Schreiber, 2010) (Figure 8.3).

8.1.1. Overview of the Photosynthetic Equation

he photosynthetic equation is a general equation for the process that plants, algae, and certain bacteria employ to create chemical energy from light energy. In simplest terms, it may be expressed as $6CO_2 + 6H_2O + \text{light energy} \rightarrow C_6H_{12}O_6 + 6O_2$. The stoichiometric equation demonstrates that CO_2 and water are transformed into glucose and oxygen with the energy of sunlight. During this process, chlorophyll and other pigments within plant cells play a crucial role in capturing light energy, predominantly blue and red wavelengths, and bouncing back green light and that's why plants are green. Their ability to successfully transform light energy into chemical molecules is not an evolutionary accident but a fine-tuning for solar energy capture maximization, as seen in recent studies probing photon energy maxima and pigment dynamic absorption (Bebout et al., 2007). Besides, the photosynthetic reaction can be separated into two major phases: the light-dependent reactions and the light-independent reactions, or Calvin cycle. During the light-dependent reactions, occurring in thylakoid membranes of chloroplasts, sunlight's energy drives the generation of ATP and NADPH with oxygen produced as waste. These energy carriers are essential for the subsequent Calvin cycle, in which carbon fixation occurs. Atmospheric aerobic CO_2 is added to organic molecules in a series of enzymatic reactions that end in glucose production. Internal process dynamics of such occurrences represent the sophisticated interdependencies of environmental parameters , such as light intensity and water levels, which are all vital to determining photosynthetic efficiency and overall plant productivity at large (Amthor et al., 2001). The significance of the photosynthetic equation does not lie with mere plant biology. It is the basis of terrestrial land ecosystems since it is the primary process by which atmospheric CO_2 is immobilized into organic matter and hence maintains the entire food chain. The biological process impacts the global carbon cycles and climatic conditions and highlights the need for photosynthesis in maintaining ecologic balance. Understanding the nuances of the photosynthetic equation informs not only agricultural management but also conservation efforts and climate change mitigation measures, since possible interference with photosynthetic efficacy will cascade in effects both locally and globally. While increasing research continues to uncover the complexity of this vital process, photosynthesis' role in environmental sustainability and management of carbon is increasingly obvious (Amthor et al., 2001; Bebout et al., 2007) (Figure 8.4).

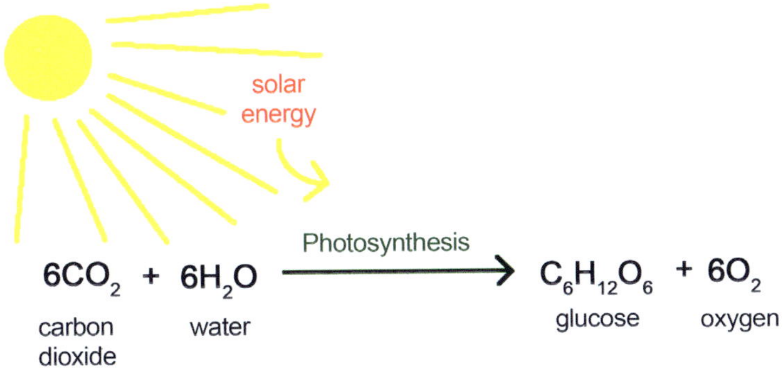

Figure 8.4. Photosynthetic equation.

Source: https://cdn.kastatic.org/ka-perseus-images/b5696ba86426f4fcc8be09e1a910f0033d241d24.png.

8.1.2. Stages of Photosynthesis: Light-Dependent Reactions

The light reactions of photosynthesis are the initial stage of this important biological process, during which chemical energy is formed using the energy of the sun. The light reactions are started in the thylakoid membranes of the chloroplast by the capture of photons by accessory pigments and chlorophyll. The energy from photons stimulates the electrons, which are relayed down through a series of protein complexes of the ETC. This electron movement helps in the photolysis of water, a reaction that yields oxygen as a byproduct and provides protons and electrons for ATP and NADPH formation. This is crucial conversion of solar energy into chemical forms that serves as energy currency for subsequent light-independent reactions that help in the biosynthesis of glucose and other carbohydrates that are crucial for plant metabolism. At the core of the light-dependent reaction are the processes of photophosphorylation, which occur in both non-cyclic and cyclic electron flow. In non-cyclic photophosphorylation, the energized electrons move through the ETC, producing NADPH and ATP, but also allowing for the subsequent photolysis of water. In cyclic photophosphorylation, electrons are recycled to the photosystem, generating yet more ATP without NADPH production. This adaptability allows the plant to manage its energy production according to needs, particularly under fluctuating light regimes. These adaptive processes highlight the complexity of photosynthesis as a metabolic process, underscoring its importance not just in relation to energy conversion but also in support of cellular homeostasis in the physiological system of the plant. Recent studies have emphasized the contribution of environmental stresses towards influencing the photosynthetic function of organisms by means of light-dependent reaction efficiency. As an example, stress caused by limiting nutrients and infection can dramatically influence the gene expression of photosynthesis genes. It has been shown in experiments that plants exhibit downregulated expression of light reaction genes during infected conditions while at the same time remodeling sugar metabolic pathways. Dissociation of the source-sink interactions can critically affect plant growth and development, with poor energy conversion leading to reduced yields and compromised plant vigor, as agriculture continues to struggle with these challenges, the complexities of light-dependent reaction continue to be of importance. Greater insights will drive the development of new strategies that enhance resource use efficiency and robustness in photosynthesis (Pascale & Stanghellini, 2011; Martinelli et al., 2013) (Figure 8.5).

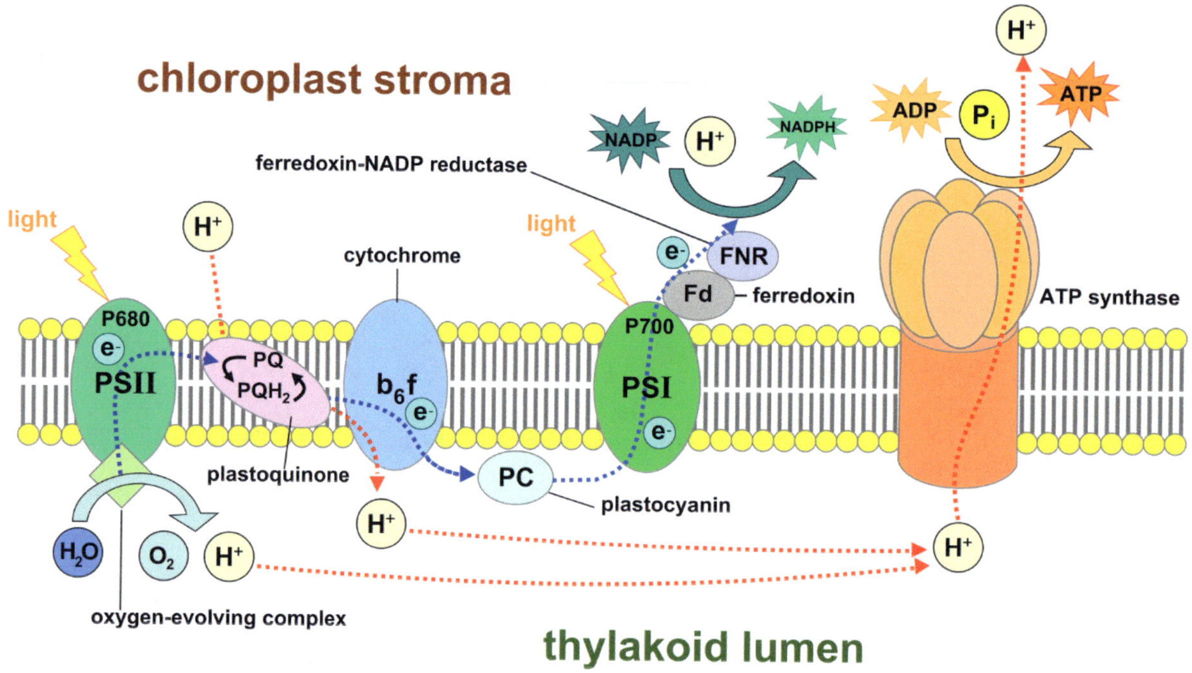

Figure 8.5. Light-dependent reactions.

Source: By Somepics - Own work, CC BY-SA 4.0, https://commons.wikimedia.org/w/index.php?curid=38088695

8.1.3. Stages of Photosynthesis: Light-Independent Reactions (Calvin Cycle)

The light-independent reactions, or the Calvin Cycle, are a crucial stage of photosynthesis carried out in the stroma of chloroplasts. Unlike the light-dependent reactions, where sunlight is used as the energy source, the Calvin Cycle uses atmospheric CO_2 to synthesize glucose and is therefore a crucial process in carbon fixation. The Calvin Cycle has three primary stages: carbon fixation, reduction, and regeneration. During carbon fixation, a five-carbon sugar ribulose bisphosphate (RuBP) is combined with CO_2, an enzyme catalyzed by the activity of ribulose bisphosphate carboxylase/oxygenase (RuBisCO). The initial step is significant in the creation of a six-carbon intermediate that quickly splits into two three-carbon compounds, 3-phosphoglycerate (3-PGA), which become modified through the subsequent steps of the cycle. In the reduction step of the Calvin Cycle, the 3-PGA molecules become phosphorylated with ATP and reduced by NADPH, both of which are light reaction products. This transformation results in the synthesis of G3P, which is a three-carbon sugar and one of the structural components of glucose and other carbohydrates. The utilization of ATP and NADPH not only indicates the dependence of the light-dependent and light-independent reactions on each other but also indicates the efficiency of energy transfer in the process of photosynthesis. Following this reduction, some of the G3P is transported onward to the production of glucose and the remainder of the molecules are recycled in a bid to regenerate RuBP so that the cycle continues. The finely tuned balance represents the principal interplay among the metabolic processes within the chloroplast. The final stage, the regeneration of RuBP, ensures the continuity of the Calvin Cycle and its capability for continuous carbon fixation. It is in this stage that G3P molecules are rearranged by a series of enzyme-catalyzed reactions until they eventually restore RuBP, the condition required for CO_2 to re-enter the cycle once more. The completion of the Calvin Cycle is regulated by a variety of factors, , such as light intensity, concentration of CO_2, and the specific metabolic needs of plants. Moreover, plastid protein import mechanisms studies indicate regulation of the cycle is critical in adapting to varying physiological conditions since most proteins in the process rely on effective import systems , such as Toc and Tic complexes (Alba et al., 2012; Stengel, 2009). This precise control emphasizes the significance of the Calvin Cycle in not only producing energy but also in maintaining plant growth and health (Figure 8.6).

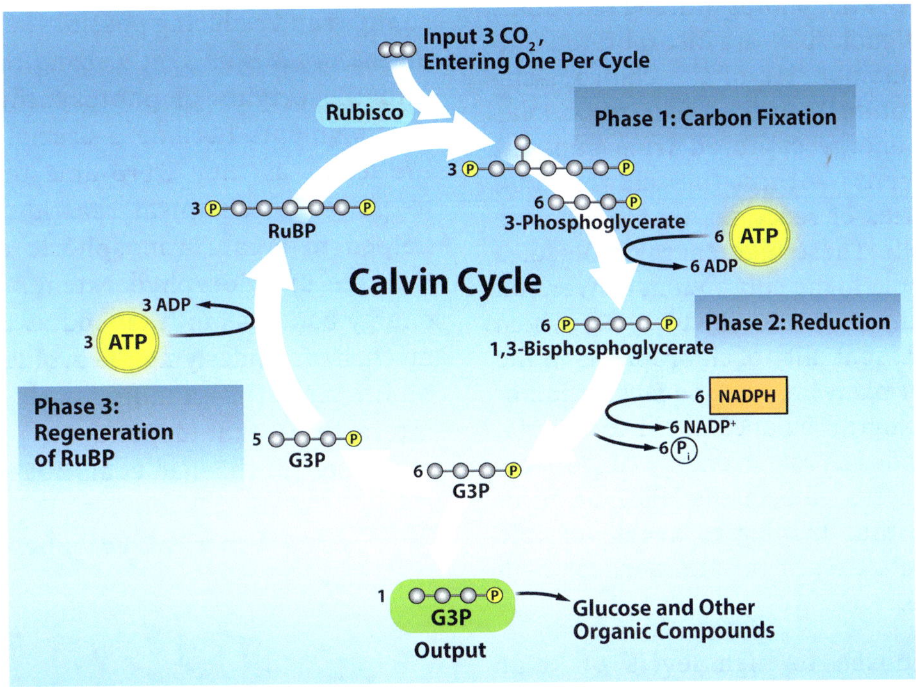

Figure 8.6. The light-independent reactions of photosynthesis.

Source: https://openintrobiology.pressbooks.tru.ca/wp-content/uploads/sites/59/2022/06/4-12-Calvin-cycle.png.

Test Your Knowledge

1. What is the photosynthetic equation?
2. What are the stages of photosynthesis?

8.2. The Role of Chlorophyll

Learning Outcome

1. Structure and function of chlorophyll;
2. Absorption of light energy;
3. Importance of chlorophyll in photosynthesis.

The green pigment in plants, chlorophyll is essential in photosynthesis, the overall conversion of solar energy into chemical energy. This pigment can absorb light best at blue and red regions of the spectrum, and it plays roles in the initial steps of the photosynthetic reactions. The chlorophyll molecules are incorporated into the thylakoid membranes of the chloroplasts, and they are optimally located to harvest solar photons. The energy captured from sunlight excites electrons within the chlorophyll, initiating a series of reactions that synthesize ATP and NADPH. These high-energy molecules play a critical role in the subsequent conversion of CO_2 into glucose in the Calvin cycle, which highlights the essentiality of chlorophylls in the continuation of plant life and the facilitation of the overall ecological balance in the terrestrial ecosystem. Besides its role in energy conversion, chlorophyll plays a role in the dissipation of excess energy, thus helping to avoid possible damage of plant tissues in situations of high light incidence.

Plants exposed to high levels of solar radiation may experience an overflow of energy which may cause the generation of harmful ROS. The risk is reduced by chlorophyll in ways that allow the energy to be harmlessly dissipated as heat or to be passed on to accessory pigments, e.g., carotenoids. These carotenoids do not just help in the light absorption but also increase the overall photosynthetic performance by increasing the range of light which can be utilized in photosynthesis. By optimizing energy capture and ensuring that excess energy is managed effectively, chlorophyll plays a dual role in the maximal efficiency and protection of plants, thereby reinforcing its centrality in photosynthesis (Alastair et al., 2018). The evolutionary meaning of chlorophyll and the related pigments could be interpreted in terms of thermodynamic and emergent complexities of life on Earth. Chlorophyll development is assumed to be an autocatalytic process which follows thermodynamic imperatives to enhance the overall rate of photon dissipation in the world. This evolutionary trend suggests that organic pigments , such as chlorophyll have evolved to optimize their usefulness in harvesting sunlight and reducing possible harmful impacts on the organisms synthesizing them. Research into the origins of photosynthesis indicates that pigments became a crucial role in early life forms as they were able to use sunlight to power biochemical reactions, which also helped to create atmospheric oxygen. Thus, the role of chlorophyll extends beyond mere energy conversion; it embodies a foundational mechanism underlying the evolutionary history of life and the complex web of ecological interactions that depend on photosynthetic processes (Michaelian et al., 2014) (Figure 8.7).

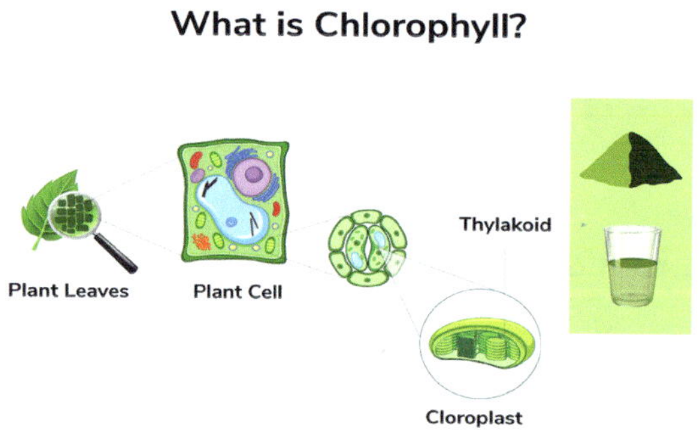

Figure 8.7. The role of chlorophyll.

Source: https://media.geeksforgeeks.org/wp-content/uploads/20240604175644/What-is-Chlorophyll.png.

8.2.1. Structure and Function of Chlorophyll

The chlorophyll molecule, one of the crucial pigments of photosynthesis, has a molecular structure characterized by a porphyrin ring containing a central magnesium atom. This specific arrangement allows chlorophyll to easily absorb light energy, primarily in blue and red regions of the visible spectrum. The unique composition of chlorophyll molecules renders them essential in the conversion of light energy to chemical energy that initiates photosynthesis. By the process of absorbing photons, chlorophyll excites electrons within the molecule, moving them to a state of increased energy. This is significant to subsequent reactions in the process of photosynthesis and demonstrates how the molecular shape of chlorophyll directly relates to its function in energy conversion and storage in plant cells for the benefit of life on Earth.

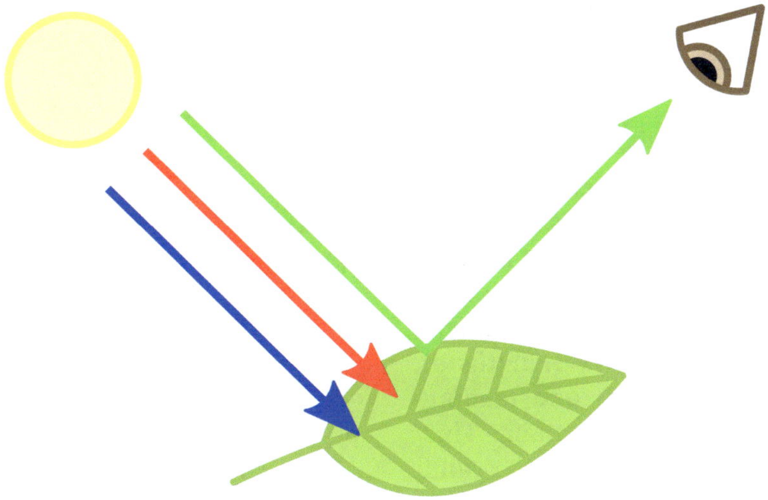

Figure 8.8. Structure and function of chlorophyll.

Source: By Nefronus - Own work, CC BY-SA 4.0, https://commons.wikimedia.org/w/index.php?curid=75964559

Functionally, there are mainly two forms of chlorophyll, chlorophyll a and chlorophyll b, each with its own complementary but distinct functions during photosynthesis. Chlorophyll a is the major pigment for light-dependent reactions, whereas chlorophyll b helps absorb other wavelengths of light and transfers the energy to chlorophyll a. This synergy enhances the photosynthetic efficiency of plants to survive in different light environments. Research, as in the example of the effect of fertilization on the content of chlorophyll in plants like *Capsicum annum*, highlights the ways in which external factors affect the production of chlorophyll and, consequently, the efficiency of photosynthesis. The findings suggest an increase in chlorophyll content by optimal nutritional levels, pointing to a related relationship between nutrient availability, the structure of chlorophyll, and plant general well-being, as seen through the example of controlled cultivation conditions (Suharja & Sutarno, 2009). Further, the current findings of the knowledge of chlorophyll derivatives , such as methyl pheophorbides, provide an insight into the adaptability and resilience mechanisms of this pigment. The research on cyanobacteria, *Spirulina geitleri*, proves the adaptability of altered chlorophyll structures to suit varying conditions of the environment. The results from research of the derivatives indicate that how molecular arrangements of chlorophyll vary makes its absorption properties unique, thus enabling photosynthesis even under suboptimum conditions. Such findings underscore the importance of chlorophyll not only in its traditional role in photosynthesis but also in its capacity to evolve and maximize efficiency in different forms of ecosystems. Such findings underscore the fine balance between structure and functionality in chlorophyll, solidifying its place as a crucial component in the photosynthesis machinery of life (Budzikiewicz et al., 1986) (Figure 8.8).

8.2.2. Absorption of Light Energy

The capture of light energy is the focus of photosynthesis, in effect energizing the biochemical processes that sustain plant life and thus the ecosystems that they in turn sustain. Chlorophyll molecules within chloroplasts capture light energy largely in wavelengths of blue and red, exciting electrons and leading to a series of photochemical reactions. This is necessary not only to generate ATP and NADPH but also to facilitate carbon fixation, the conversion of CO_2 in the atmosphere into organic compounds. Additionally, processes to compensate for photoinhibition, whereby excessive light kills the photosynthetic machinery, have evolved in plants. These adaptations enable efficient capture of energy without damage, emphasizing the delicate balance plants tread between trapping light energy and the integrity of cells in serving healthy ecosystems and climate change resilience in changing conditions (Ana et al., 2008). The regulation of light energy capture is complex and multi-layered. As an example, energy-dependent quenching (qE) is a critical process in which the plant manages light harvesting under varying light. Mechanism of qE depends upon the pH gradient established across thylakoid membranes, and this affects photosystem II (PSII) energy transfer processes. Based on variations in the intensity of the light, the plant can manage its energy uptake appropriately, varying between quenched and unquenched states to prevent over-excitation of chlorophyll and consequent cellular stress. This adaptability plays a pivotal role in the optimization of photosynthetic efficiency and plant survival under varying environmental conditions. The complex relation between light absorption and physiological behavior demonstrates that plants have evolved to optimize energy capture at the expense of functional integrity (Amarnath et al., 2013). In addition, the understanding of light absorption is supplemented by the investigation of morphological adaptation in plants, particularly C_4 plants.

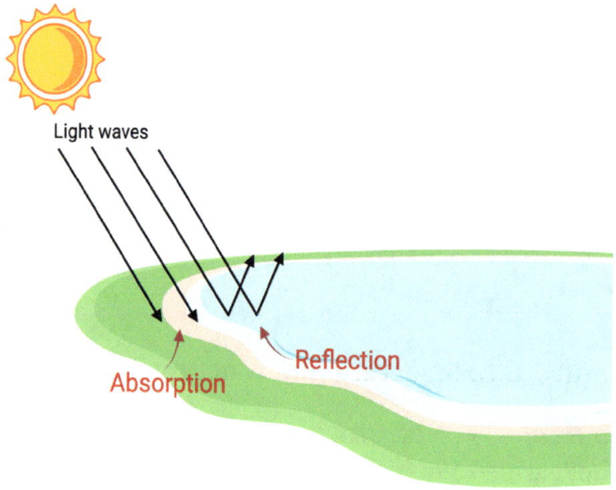

Figure 8.9. Absorption of light energy.

Source: https://cdn.kastatic.org/ka-perseus-images/2cb350967fe4c41ee8f601e1e998601df188603a.png.

These plants possess specialized leaf structures that allow them to conduct photosynthesis independently on each leaf surface, and they are responsive to the variations in light and water. For example, during water stress, the leaves of C_4 plants curl, influencing the area of the leaf surfaces exposed to sunlight. This remarkable flexibility enables stomata on the abaxial side to be open for gas exchange but remain closed on the adaxial side, thereby regulating photosynthetic metabolism effectively. Such a mechanism provides significant survival advantage over C_3 plants, which are incapable of controlling their dorsoventral functions in respect of light conditions as effectively. Thus, an understanding of the uptake of light energy through physiological and morphological views subjects us to the complex mechanisms plants employ to enhance their survival and performance in diverse ecosystems (Ana et al., 2008) (Figure 8.9).

8.2.3. Importance of Chlorophyll in Photosynthesis

Chlorophyll is the focal point of photosynthesis and works predominantly as a green pigment that captures the energy of sunlight. Chlorophyll resides in thylakoid membranes of chloroplasts, where light energy conversion to chemical energy takes place, sustaining photosynthetic organisms' life. Chlorophyll picks up light blue and red wavelengths and bounces green wavelengths, giving plants their green color. The energy from absorbed light causes the electrons in the chlorophyll molecule to become excited, initiating a chain of reactions that result in the synthesis of ATP and NADPH, the energy carriers needed for the subsequent Calvin cycle that synthesizes glucose. This process not only highlights the basic function of chlorophyll in the acquisition of solar energy but also its central place in the broader picture of ecosystem processes, where it is a fundamental component of food chains and communities. Furthermore, the efficiency with which chlorophyll absorbs light energy largely controls the rate of photosynthesis as well as plant growth. Variations in chlorophyll content directly influence the capacity of a plant to photosynthesize, and, therefore, agriculture and ecosystem health are affected. For instance, under stressed conditions , such as drought or excessive salinity, photorespiration is more pronounced, leading to inefficiencies in carbon fixation. However, chlorophyll plays a function in regulating such an effect because it inhibits the over-reduction of photosynthetic ETCs. It has been proven, through research, that chlorophyll deficiencies will lead to wasted energy that threatens the metabolic efficacy of the plants, thus demonstrating the importance of maintaining the optimum concentration of chlorophyll for promoting plant health, vigor, and productivity in various environmental conditions (Anderson et al., 2000). Besides its role in harvesting energy, chlorophyll is involved in the core processes of metabolism of the plant. Since an integral component of photosynthesis, not only is it held accountable for the formation of glucose but for that of by-products critical to various metabolic activities. For example, the sugar metabolites formed by photosynthetic pathways are accountable for amino acid, lipid, and vitamin biosynthesis, which are critical to plant structure and function. Chlorophyll is also implicated in stress response through the development of chemicals that are protective of oxidative damage. The general realization of the multifunctional aspect of chlorophyll, aside from its function in complex metabolic processes and interaction with other plant physiology constituents (, such as coenzymes, prosthetic groups, and pigments), emphasizes its general importance to the efficiency and adaptability of photosynthetic organisms under various and stressful environments (Lea-Smith et al., 2020).

Test Your Knowledge

1. Explain the absorption of light energy.
2. What is the importance of chlorophyll in photosynthesis?

8.3. Factors Affecting Photosynthesis

Learning Outcome

1. Light intensity and its effects;
2. Carbon dioxide concentration;
3. Temperature and its impact on the rate of photosynthesis.

Photosynthesis is a complicated biochemical process, which is strongly affected by environmental factors, which define its efficiency. One of the main factors is light intensity; the photosynthetic apparatus cannot work at its best in the conditions of insufficient light. The intensity of light also affects the rate of photosynthesis whereby, as the intensity of light rises, the rate of photosynthesis rises only to a certain level which is the saturation point after which further increase of light does not increase the rate of photosynthesis. This interaction highlights the importance of the availability of light in different ecosystems, especially with regard to climate change and farming activities. Furthermore, the concept of photosynthesis use efficiency, representing the ability of plants to convert light energy into chemical energy, indicates how improvements in this area could substantially affect food security and agricultural sustainability (Berger et al., 2019). In this way, the control of lighting situations in farming situations can result in considerable improvements in harvest volumes and sustainability measures. CO_2 concentration is also another important parameter in photosynthesis, serving as a substrate to the photosynthetic reactions. Availability of CO_2 is directly related to the carbon fixation rate in photosynthesis. In most cases, the photosynthetic rate rises with a rise in CO_2 levels enhancing plant growth and productivity. This effect may however differ across species and environmental conditions. As an example, some plants, the C_3 species, react more to elevated CO_2 than the C_4 species because of the variations in their metabolic pathways. Photosynthesis dynamics are further complicated by the interaction of CO_2 concentration with other parameters , including light intensity and temperature. Understanding these relationships is critically important for researchers and agricultural scientists who aim to optimize photosynthetic efficiency and improve food production systems (Ai et al., 2013). Moreover, temperature has a great impact on enzyme processes of photosynthesis. Every enzyme also possesses an optimum temperature range of functioning, and a departure of this range may change the overall photosynthetic efficiency.

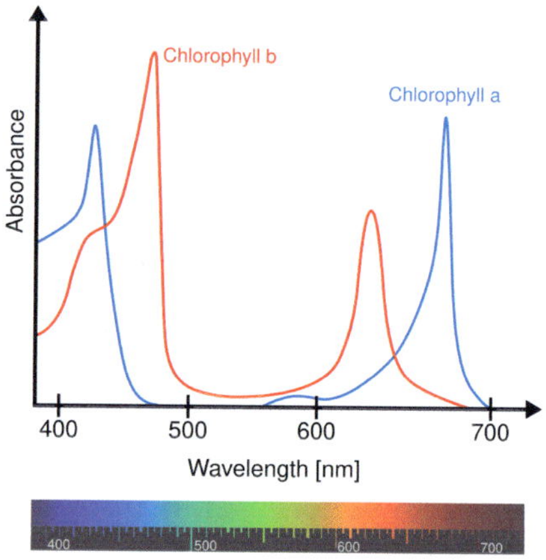

Figure 8.10. Factors affecting photosynthesis.

Source: By Original: Daniele Pugliesi Vector: M0tty - This file was derived from: Chlorophyll ab spectra2.PNG:, CC BY-SA 3.0, https://commons.wikimedia.org/w/index.php?curid=20509583

Rising temperatures can stimulate photosynthetic rates to an optimum level, beyond which enzyme denaturation could take place resulting in low rates and stunted plant health. Further, temperature affects the stomatal control of gas exchange, which has a direct effect on CO_2 uptake. With a trend in the rise of global temperatures caused by climate change, the study of the impact of temperature variation on photosynthesis in different ecosystems is of more critical importance. This knowledge can inform breeding programs aimed at developing thermally resilient crop varieties that maintain productivity despite climatic stressors, ultimately contributing to food security in a changing world (Berger et al., 2019) (Figure 8.10).

8.3.1. Light Intensity and Its Effects

Light intensity is also an important stimulator of the process of photosynthesis, which is an intrinsic biological process that enables plants to convert light energy into chemical energy. The greater the light intensity, the higher the photosynthesis rate, but only up to a certain point because it is dependent on the light saturation point. Above this peak point, additional light does not significantly enhance photosynthetic rates, and may even induce photoinhibition, in which too much light modifies chlorophyll and other cell structures.

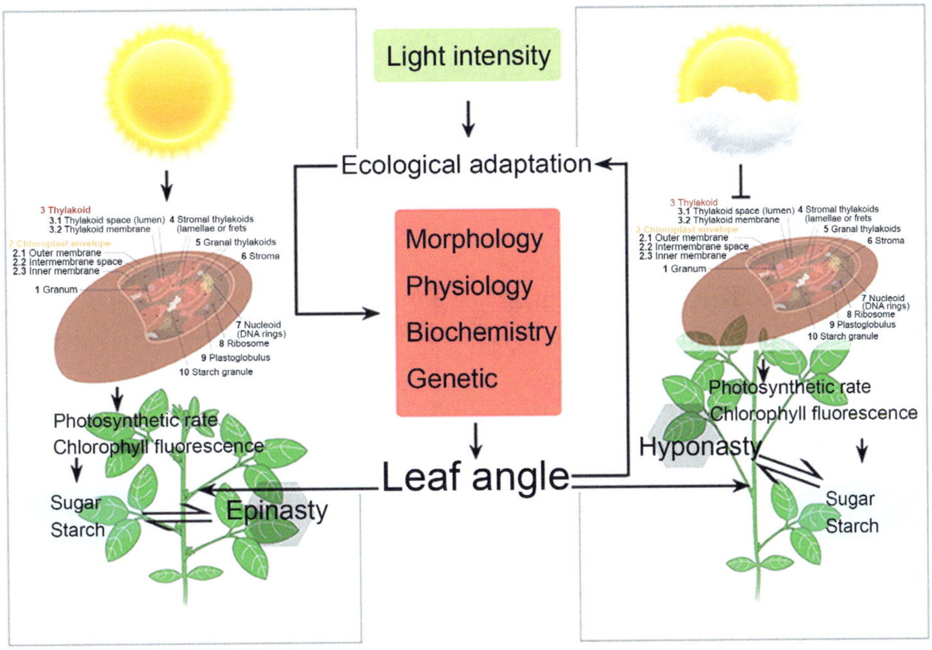

Figure 8.11. Light intensity and its effects.

Source: Feng L, Raza MA, Li Z, Chen Y, Khalid MHB, Du J, Liu W, Wu X, Song C, Yu L, Zhang Z, Yuan S, Yang W and Yang F (2019) The Influence of Light Intensity and Leaf Movement on Photosynthesis Characteristics and Carbon Balance of Soybean. Front. Plant Sci. 9:1952. doi: 10.3389/fpls.2018.01952

This complex response indicates the delicate balance by which plants survive to use light energy for growth at no damage to cells. It is crucial to know the effect of light intensity on photosynthesis to optimize farming methods, particularly in indoor environments, where artificial light can be modified for optimal crop yield and efficiency. LED technology has changed the horticultural lighting industry with an efficient alternative to conventional high-intensity discharge lighting. LEDs not just offer enhanced energy efficiency but also allow for accuracy in light spectrum beneficial for plant growth, particularly for photosynthesis. LEDs have shown higher luminous efficiency compared to conventional lighting in studies, leading to significant improvement in plant biomass and growth rate. For instance, studies have demonstrated that lettuce plants cultivated using a combination of red, green, and blue LED light exhibited greater growth and reduced nitrate concentrations, suggesting the value of tailored lighting methods to meet specific plant physiological needs. Thus, controlled conditions use of LEDs is capable of maximizing light intensity and enhancing photosynthetic performance, which has implications for yields and long-term sustainability (Grundy et al., 2017). In addition to light intensity, other conditions in the environment , such as soil moisture, interact with photosynthesis and demonstrate the complexity of plant response to shifting conditions. For instance, research on the perennial grass *Leymus chinensis* demonstrated that drought stress can initially enhance photosynthetic potential but recovery following drought is typically variable. The study showed that even if the drought-stressed plants had higher relative growth rates and enhanced photosynthetic efficiency, recovery was contingent on the level of water deficit the plants experienced. The capacity of stomatal conductance, being a key regulator of photosynthetic efficiency, to recover to pre-drought levels was impaired. This indicates that while light intensity may stimulate photosynthesis, other environmental factors like soil moisture also must be considered to more effectively comprehend the complex dynamics influencing plant growth and productivity (Shimizu et al., 2009) (Figure 8.11).

8.3.2. Carbon Dioxide (CO_2) Concentration

Atmospheric CO_2 concentration is a critical process of photosynthesis, which is crucial for plant growth and the carbon cycle in the world. As plants utilize CO_2 in producing organic matter through the process of photosynthesis, elevated concentrations of CO_2 can enhance the process as well as biomass accumulation. Terrestrial ecosystems benefit highly through this process, as an increased vegetation density has the capacity to sequester more carbon. However, it is also necessary to consider consequent ecological problems; elevated CO_2 levels can induce growth in vegetation, but they can alter species distribution and disrupt present ecological balance. Understanding these interactions lies at the heart of implementing photosynthesis in carbon management practices under rising atmospheric CO_2 levels, as illustrated in the carbon cycle, which regulates the interaction between ecosystems and the atmosphere, requiring keen observation of human activities influencing CO_2 emissions and concentrations (Borate et al., 2023). Research has established that elevated atmospheric CO_2 concentration possesses the ability to stimulate photosynthesis, particularly in species found in grassland. For example, a perennial ryegrass long-term study revealed increased light-saturated leaf photosynthesis and daily CO_2 uptake rates under higher levels of CO_2 after 10 years. In particular, scientists observed an increase of 43% in photosynthesis and a 36% rise in CO_2 uptake integral, which was indicative of a strong photosynthetic reaction to increased CO_2. But acclimation of physiological parameters, , such as maximum carboxylation velocity and electron transport rates, indicates that chronic elevations in CO_2 could lead to variable long-term consequences on photosynthetic efficiency. These results show the fineness of the photosynthetic response to increasing CO_2 levels and that, while enhancement is observed in the short term, the elevated CO_2 conditions may lead to altered growth patterns and ecosystem functions (Ainsworth et al., 2003). Moreover, the impacts of elevated CO_2 levels are not only confined to some plant species but also propagate to wider ecological processes. As CO_2 concentrations continue to increase in the air, the threat of saturation for the carbon sinks necessitates careful consideration of ecosystem management practices. Such saturation could restrict natural systems' ability to effectively sequester carbon, thus hastening climate change impacts. Combating these risks involves global efforts to manage carbon emissions, preserve forests, and apply sustainable agriculture for enhanced photosynthesis with carbon balance. Collective efforts can enhance the resilience of ecosystems against shocks and enhance the effectiveness of photosynthesis in minimizing atmospheric CO_2 impacts, calling for a comprehensive understanding of the interaction among carbon cycling, plant physiology, and climate change (Borate et al., 2023).

8.3.3. Temperature and Its Impact on the Rate of Photosynthesis

The temperature-rate of photosynthesis relationship is complex and significant, defining the physiological response of plants to their changing environments. Photosynthesis relies on enzymatic reactions that are sensitive to temperature. For the majority of instances, the rate of photosynthesis will be enhanced with rising temperatures up to an optimum point; beyond which the efficiency of photosynthesis begins being disrupted by protein denaturation and other metabolic disruptions. Experiments have also demonstrated that high temperature episodes can affect photosynthetic performance negatively, particularly under changing salinity regimes as a result of environmental stresses like climate change (Piedras et al., 2019).

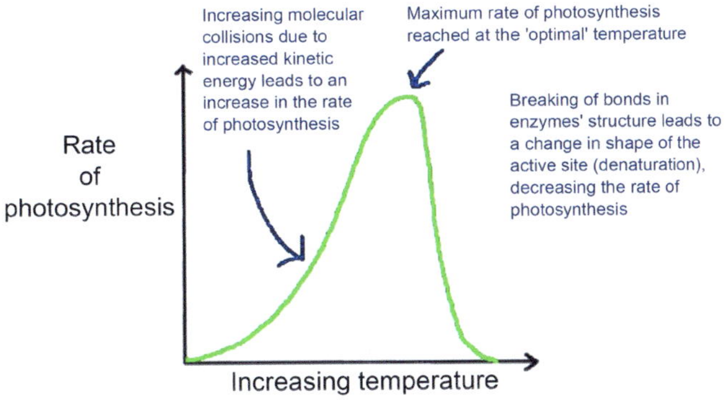

Figure 8.12. Temperature and its impact on the rate of photosynthesis.

Source: https://i.ytimg.com/vi/PQ1zZ9G5d7M/maxresdefault.jpg?sqp=-oaymwEmCIAKENAF8quKqQMa8AEB-AH-CYAC0AWKAgwIABABGGUgZShlMA8=&rs=AOn4CLDzHezOvYhVQcLYp-DaRpKewSW_8Q.

This interaction reflects how delicate plants are in getting a balance of their metabolic processes, where temperature and other abiotic factors come into play. The adverse effect of increasing temperatures on photosynthesis is also dependent on changes in nutrient availability. As observed in recent studies, climate change impacts essential nutrients required for photosynthesis , such as nitrogen (N) and phosphorus (P), which are crucial for plant growth and metabolic function. Rising temperatures have been projected to enhance soil weathering, altering the cycling of these nutrients and thus impacting their availability for plant nutrient uptake (Lagomarsino et al., 2010). This interaction creates a feedback loop whereby not only does temperature directly affect the process of photosynthesis but also changes the nutrient dynamics of ecosystems and thus impacts plant health and production. An understanding of these interactions is central to making predictions about how increasing global temperatures could impact photosynthesis as well as overall ecosystem function. Moreover, the variations in the response of photosynthesis to shifts in temperature contain important information regarding the adaptive processes of different plant species. For instance, some species possess greater thermotolerance, whereby they retain photosynthetic efficiency even in stressful conditions. Such flexibility can be explained by enhanced biochemical processes that allow for photosynthesis under increased temperatures, whereby such plants can dissipate excess energy and resist oxidative stress more effectively (Piedras et al., 2019). Conversely, species lacking such adaptations may be severely affected in the form of lowered photosynthetic capabilities and hence lower biomass and poor ecological recovery. Therefore, detection of these thermotolerance traits gains enhanced importance, as is the case with today's global warming scenario, as it directs conservation efforts and selection of plants to grow for agricultural and reforestation intents. To ensure biodiversity and food security in a changing climate, it is crucial to be aware of the complex relationship between temperature and photosynthetic efficiency (Figure 8.12).

Test Your Knowledge

1. What is light intensity and its effects?
2. What is carbon dioxide concentration?

8.4. Photosynthesis in Different Organisms

Learning Outcome

1. Photosynthesis in plants;
2. Photosynthesis in algae;
3. Photosynthesis in cyanobacteria.

Photosynthesis is considerably different in various organisms, which enables them to adapt to many ecological niches. Photosynthesis in green plants is mainly found in chloroplasts, and it consists of two phases, the light-dependent reactions, and the Calvin cycle. This is an efficient process because it involves the presence of chlorophyll that helps to trap sunlight and turn CO_2 and water into glucose and oxygen. Moreover, similar processes are observed in some algae and cyanobacteria, which also highlights the evolutionary importance of photosynthesis in water. These organisms have adapted to maximize light capture using specialized pigments and can grow under different light conditions, representing the flexibility of photosynthetic approaches among taxa. These adaptations play an important role in the continuation of the energy flow in the ecosystem and the balance of the atmosphere which again proves the main importance of photosynthesis in sustaining life on the earth. Conversely, there are other organisms which employ different carbon fixation methodologies that do not follow the conventional photosynthesis process. A good example occurs in the purple and green sulfur bacteria, where anoxygenic photosynthesis occurs. It is a process which utilizes light energy, however, uses other substances instead of water to reduce CO_2, namely hydrogen sulfide. Consequently, these bacteria have a role in unusual biogeochemical cycles, especially those operating in anaerobic conditions, at deep-sea hydrothermal vents and in salt marshes. Their success in low-oxygen environment demonstrates the flexibility of life and highlights the evolutionary routes that photosynthetic entities have undertaken to utilize the available resources efficiently. As research progresses, understanding these diverse mechanisms will be fundamental in exploring sustainable practices, especially in contexts like artificial photosynthesis aimed at carbon sequestration and renewable energy production, as highlighted by Nesamma et al. (2020). The recent developments in the area of intermarriage between biological and synthetic systems further demonstrate the promise of photosynthetic systems in solving some of the modern-day challenges like carbon emission. Through the natural properties of CO_2 fixation by biological systems and novel nanomaterials, scientists have made a nanomaterial-biological interface that improves the photosynthetic process. As an example, by introducing cadmium sulfide nanoparticles into certain microorganisms, these beings can convert solar energy to carbon conversion in a more efficient manner. This new method not only promises to surpass the drawbacks that exist in the traditional catalysts but also demonstrates a bright direction in developing renewable energy technologies. With additional research focused on improving biocompatibility and understanding charge transfer mechanisms, the intersection of biology and technology could pave the way for groundbreaking developments in photosynthesis, particularly in harnessing solar energy for producing value-added carbon products, as emphasized by Cestellos-Blanco et al. (2019) (Figure 8.13).

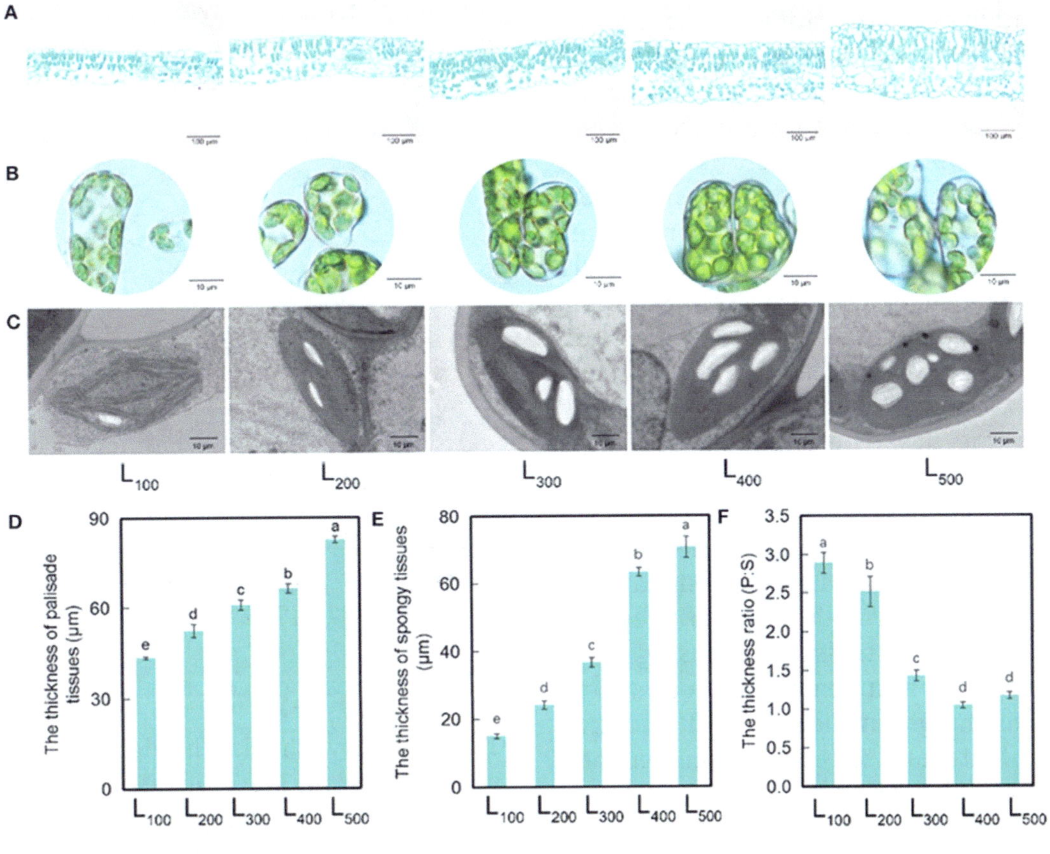

Figure 8.13. Photosynthesis in different organisms.

Source: Feng L, Raza MA, Li Z, Chen Y, Khalid MHB, Du J, Liu W, Wu X, Song C, Yu L, Zhang Z, Yuan S, Yang W and Yang F (2019) The Influence of Light Intensity and Leaf Movement on Photosynthesis Characteristics and Carbon Balance of Soybean. Front. Plant Sci. 9:1952. doi: 10.3389/fpls.2018.01952

8.4.1. Photosynthesis in Plants

Photosynthesis is a critical process of life on Earth, enabling plants to convert light energy into chemical energy in the form of glucose. This complex process not only sustains plants but is also the foundation of the majority of food webs in various ecosystems. In light-dependent and light-independent stages, plants employ sunlight to split water molecules and oxygen as byproducts in the formation of ATP and NADPH, which are required for oxygenation in the Calvin cycle where CO_2 gets converted into organic molecules that subsequently result in glucose. This duality of purpose, to yield oxygen and organic substance, demonstrates the significance of photosynthesis in maintaining equilibrium in the atmosphere and supporting diverse forms of life. Regulation of photosynthesis is not a mere biochemical reaction; it is deeply connected with environmental conditions outside, particularly light. Photosynthesis efficiency and rate depend upon light intensity and quality and induce plants to vary their physiological makeup according to different conditions. Interestingly, the plant circadian clock plays the lead role in managing these processes. Scientific research shows that photosynthesis and related energy cycles are under the control of a clock, which allows plants to optimize their energy use during the day. The circadian oscillator succeeds in synchronizing these processes with environmental cues, thereby allowing the plants to utilize the light energy efficiently. This association of internal and external signals not only enables plant growth but also prepares plants to deal with stressors, imposing their vigor to resist environmental stressors, , such as variations in light availability and oxidative stress. As of late, due to recent advancements, the relationship between photosynthesis and ROS has been an area of particular interest within the field of plant physiology. Although ROS are generally considered to be deleterious byproducts of metabolism, recent studies have established their roles as crucial signaling molecules with the potential to impact numerous different cellular processes. Notably, both the generation and clearance of ROS exhibit circadian regulation under the control of the plant circadian clock, demonstrating systematic regulation. For instance, CCA1, a critical regulator in the clock system, was identified to regulate ROS homeostasis and gene expression related to photosynthetic activity. This emphasizes a sophisticated feedback mechanism where the initiation of ROS production overlaps with plants metabolism processes, finally influencing photosynthesis efficiency. Hence, insights into these interactions lead to windows to enhance plant resilience and productivity under challenges like climate change and environmental stresses, and it is a reflection of the complex dance between photosynthesis and broader physiological frameworks (Lai, 2012; Fan et al., 2013) (Figure 8.14).

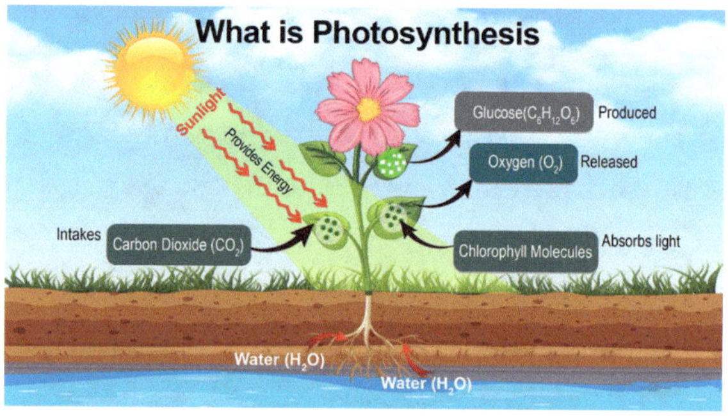

Figure 8.14. Photosynthesis in plants.

Source: https://cdn.educba.com/academy/wp-content/uploads/2024/04/What-is-Photosynthesis.png.

8.4.2. Photosynthesis in Algae

The photosynthesis in algae is of highest significance in the world's ecosystems as it is part of the primary producers in aquatic ecosystems. Algae produce glucose using sunlight, CO_2, and water, a process that occurs predominantly in their chloroplasts, which contain chlorophyll pigments that absorb light energy. It not only yields organic matter that provides a nutritional foundation for diverse marine life but is also significant in the supply of atmospheric oxygen. Algae might present a structurally uncomplicated appearance, but they have an incredible degree of metabolic heterogeneity, whereby they can thrive in a variety of habitats from high-nutrient waters to extreme polar environments. Studies indicate that polar microalgae's adaptation mechanisms enable them to manage critical environmental stresses, thus their ecological relevance even under harsh conditions (Arrigo et al., 2014). Photosynthesis by tropical reef system benthic primary producers like algae has critical impacts on biogeochemical cycling and microbial activity. Fresh research points out that benthic algae, in addition to photosynthesizing, release dissolved organic carbon (DOC) into the water, a significant portion of their daily carbon fixation. The release of DOC has an effect on microbial community structures and function and causes differential metabolic responses in planktonic and benthic microbes. In particular, studies have determined that the algae and coral exudates have been found to shift microbial community composition, with severe consequences for the oxygen content of the water column. These interactions attest to the complex feedback mechanisms between photosynthetic algae and microbial communities within reef environments, and they underscore the sensitive relationships keeping these biodiverse ecosystems intact (Carlson et al., 2013). Moreover, the photosynthetic efficiency of algae provides the foundation for high growth rates as well as the capture of solar energy required in mitigating climate change. Algal blooms supported by nutrients in the form of runoff from coastal waters demonstrate the ability of these organisms to respond to changing environmental conditions and enhance photosynthetic efficiency. Adaptability is particularly crucial with the commencement of global temperatures in different aquatic ecosystems. Moreover, genetic and physiological insight drawn from studies of algal photosynthesis opens up opportunities for biotechnological utilization in bioenergy and carbon sequestration. Thus, an understanding of the simple mechanisms of photosynthesis in algae not only gives us valuable knowledge of ecological interactions but also addresses fundamental challenges in sustainability and environmental conservation endeavors on the global stage (Carlson et al., 2013; Arrigo et al., 2014) (Figure 8.15).

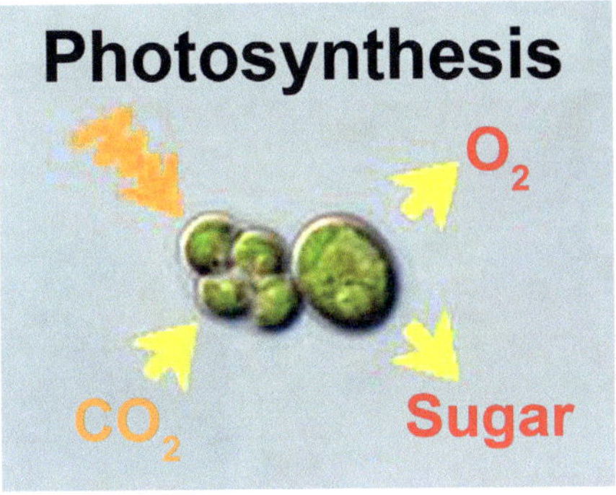

Figure 8.15. Photosynthesis in algae.

Source: https://www.frontiersin.org/image/researchtopic/25172.

8.4.3. Photosynthesis in Cyanobacteria

The cyanobacterial photosynthesis is a key process in global biogeochemical cycles, primarily because the microorganisms occupy a specific ecological niche. As they are phototrophs, cyanobacteria utilize sunlight as the principal source of energy to metabolize CO_2 and water into glucose and oxygen. The process is aided by chlorophyll, a pigment that captures light energy, particularly at wavelengths of blue and red. The ability of cyanobacteria to endure nearly all environmental conditions, , such as extreme climate conditions, enhances their photosynthetic efficiency. For instance, some organisms exhibit capabilities to endure conditions in polar climates, where light and temperature fluctuations complicate metabolic activities. By modifying their photosynthetic machinery to such extreme conditions, cyanobacteria contribute significant primary productivity to areas where other forms of life may be harmed, eventually contributing to their position as integral components of aquatic food webs. Most notably, cyanobacteria have developed advanced mechanisms for protecting their photosynthetic machinery against environmental stressors, , such as oxidative damage and nutrient stress. In addition to producing oxygen from photosynthesis, cyanobacteria can also fix nitrogen (N_2) from the atmosphere. The ability is significant in environments where nitrogen is a limiting factor. The oxygen produced in photosynthesis is, however, toxic to the nitrogenase enzyme responsible for fixing nitrogen since nitrogenase is highly oxygen-sensitive.

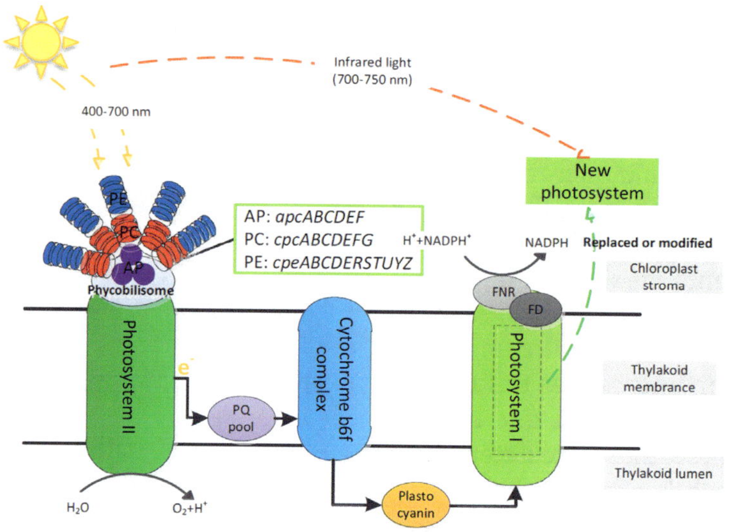

Figure 8.16. Photosynthesis in cyanobacteria.

Source: https://www.researchgate.net/profile/Tao-Sun-72/publication/294920136/figure/fig1/AS:941182846763085@1601406885800/A-summary-of-photosynthesis-in-cyanobacteria-Phycobilisome-related-genes-AP.gif.

To avoid this issue, most cyanobacteria exhibit spatial or temporal separation of these two metabolic activities. For instance, hopanoid-rich membranes of non-heterocyst-forming cyanobacterial diazotrophs may serve as an oxygen diffusion barrier, thus nitrogen fixation may be maintained even in oxic conditions (Cornejo-Castillo et al., 2019). Such complex balance portrays the remarkable evolutionary proficiency of cyanobacteria in tolerating their environment without compromising vital biogeochemical processes. Aside from that, the environmental significance of cyanobacteria is that they are primary producers in a wide array of aquatic habitats, especially in extreme environments found in high-latitude habitats. The ability of these microorganisms to photosynthesize best under varying light and temperature regimes is very crucial in regulating global carbon balance and climate. Faced with perpetual darkness during winter and continuous light in summer, cyanobacterial populations have adapted through molecular changes, making them more efficient photosynthetic and capable of survival. Genomic, transcriptomic, and proteomic studies of polar cyanobacteria give insights into the specific genes and processes involved in these adaptations, with possible implications for biotechnological use (Arrigo et al., 2014). Because these regions are highly climate-sensitive, the knowledge about the resiliency of cyanobacteria ensures more certain knowledge about the future changes in ecosystem processes and strengthens their important role in the Earth's biosphere (Figure 8.16).

Test Your Knowledge

1. Explore photosynthesis in plants.
2. Explain photosynthesis in algae.

8.5. The Importance of Photosynthesis to Life on Earth

Learning Outcome

1. Oxygen production and its significance;
2. Role in the food chain;
3. Contribution to carbon cycling.

Photosynthesis is one of the most important biological processes that have enabled the continuity of life on Earth as it presents a vital window of opportunity through which solar energy is converted to chemical energy. Photosynthetic organisms, , such as plants, algae, and a few bacteria, use sunlight to convert CO_2 and water into glucose, with oxygen being a byproduct. This complex biochemical reaction not only provides the energy to grow and develop these organisms but also forms the basis of the entire food web. Photosynthesis is even more important when it comes to the regulation of the atmosphere in terms of the amount of CO_2, climatic systems and the way energy is transferred in ecosystems. As highlighted in various studies, the efficiency of this process is often much lower than thermodynamic predictions would suggest, underscoring the complex balance between energy capture and the material exchanges necessary for life, suggesting that understanding this process requires a comprehensive evaluation of the interconnectedness of various environmental factors (Kleidon, 2020). The physical and ecological significance of photosynthesis is profound, as demonstrated by the interrelationship between photosynthesis and transpiration. Transpiration, the process through which plants lose water vapor to the atmosphere, is important to the overall climatic processes as well as to nutrient transport through the plant. The process is a contributor to entropy production on various organizational levels: of the plant, of the ecosystem, and of the biosphere. When we recognize that transpiration is a thermodynamic necessity, and not just a physiological requirement, then we can realize that plants have adapted to maximize water loss, without sacrificing photosynthetic rate. Such adaptations are critical not only for individual plant survival but also for influencing local and regional climates, ultimately feeding back into the broader dynamics of planetary life (Candia et al., 2010). Moreover, photosynthesis performance and inconsistency are influenced by numerous environmental conditions, and it is critical to support different ecosystems. On land, studies show that the average photosynthetic conversion efficiency is approximately 0.78%, which demonstrates the concrete effect of the material exchange on this process. In particular, parameters like evaporation rates, which are linked to photosynthesis, can be used to explain how sunlight can also lead to transport of not only energy but also essential resources , such as water and CO_2. This association highlights the thermodynamic limitations that form the basis of the biosphere and points to the importance of taking into account these material flows when analyzing the processes of life. As photosynthesis continues to be a focal point of ecological and biological research, its importance can be seen not just in energy production but in maintaining the delicate balance that sustains life on Earth (Kleidon, 2020) (Figure 8.17).

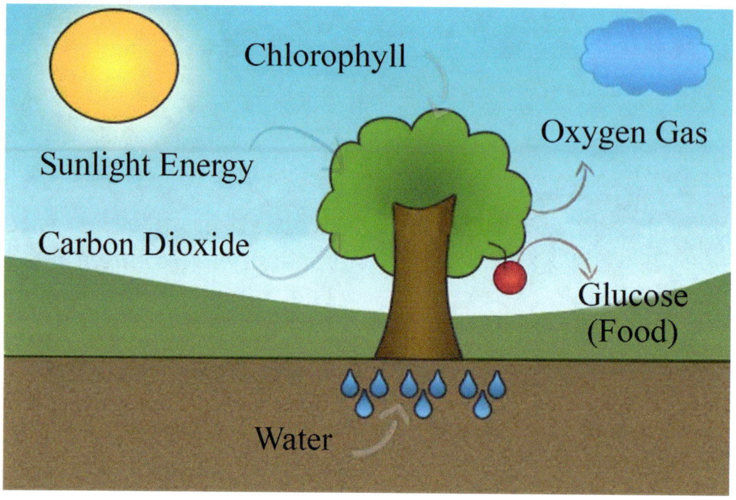

Figure 8.17. The importance of photosynthesis to life on Earth.

Source: https://www.smartsciencepro.com/wp-content/uploads/2014/09/Photosynthesis-Process-Flow.jpg.

8.5.1. Oxygen Production and Its Significance

Photosynthesis is not only important to plant life's survival but also to overall ecosystems due to its role in oxygen production. During photosynthesis, green plants and certain algae capture sunlight and transform it into chemical energy, releasing oxygen as a byproduct. This oxygen is consumed by aerobic life forms, including human beings, as it's required for cellular respiration—the mechanism by which energy is produced within cells. Without this continuous supply of oxygen as a result of photosynthesis, life as we know it would not exist. The fine balance produced by the process emphasizes the interdependence among various life forms and emphasizes the ecological significance of photosynthetic organisms such that they become indispensable for the maintenance of biodiversity and ecological balance. Secondly, the function of oxygen production is not restricted to the terrestrial ecosystem alone; it is equally crucial in aquatic ecosystems. Photosynthetic phytoplankton and algae are oxygen-producing factories, and particularly in oceanic ecosystems. These microorganisms form the foundation element of marine food webs that support a broad variety of life forms ranging from small zooplankton to large marine mammals. The synergy between oxygen production and nutrient cycling within these ecosystems illustrates the fundamental role played by photosynthetic microorganisms. Apart from sustaining the marine life's food, they also influence global biogeochemical cycles, influencing atmospheric composition and climate regulation. In high latitudes, where environmental conditions are harsh, such photosynthetic autotrophs exhibit remarkable adaptations, thereby optimizing their productivity and, in turn, oxygen yield (Arrigo et al., 2014). Industrialization in the context of oxygen production through photosynthetic processes has far-reaching consequences. Through advancements in biotechnology, it is conceivable to genetically optimize the algae strains so as to enhance their photosynthetic efficiency and, consequently, produce higher biomass yields for the generation of biofuels as well as other purposes. By developing math models that characterize the drivers of alga productivity, scientists are now beginning to harness the ability of these organisms in a systematic manner. Such advances not only serve justice in addressing energy challenges but also make carbon capture programs, thus mitigating climate change. With our understanding of the complex interaction between environment and genetics in algal systems, the implications for large-scale production of oxygen and its application in different areas are more obvious (Perin et al., 2017).

8.5.2. Role in the Food Chain

The role of photosynthesis in the food web is a fundamental keystone to life on this planet, since it provides the primary energy that powers nearly all biomes. Photosynthesis, the process of transforming sunlight into chemical state, is carried out by photosynthetic living things, including green plants, algae, and some bacteria, by fixing glucose from water and CO_2. This stored glucose is not just a source of fuel for the plant itself but also a vital energy source for the herbivores, the primary consumers of the food chain. As these herbivores eat the plant material, they metabolize the stored energy, thereby playing a significant role in transmitting the energy to the higher trophic levels, i.e., the omnivores and the carnivores. The dependence of these interactions shows the basic requirement of photosynthesis in the food chain, demonstrating the process of energy transfer between producers and consumers, sustaining life within different ecosystems. Based on the complex interactions of the food chain, the efficiency of the energy transfer between trophic levels largely relies on the nature of photosynthetic organisms. Carotenoids, for instance, enhance photosynthesis by extending the accessibility of light to be utilized by chlorophyll—a process at the heart of plant productivity. Light energy absorbed by carotenoids is passed on maximally to chlorophyll to optimize energy conversion into the biomass growth and upkeep of the plant, and thus herbivores and other species can consume them. These adaptations reflect the complex interrelations between various organisms, where photosynthetic ability not only influences the individual plants but also influences the overall architecture of the community by determining population growth and diversity across the system, highlighting the interconnected nature of life forms sustained by photosynthesis for sustenance and survival (Alastair et al., 2018). Besides, the coherence of energy transfer in photosynthesis has implications for the balance within food chains. Current research indicates that energy transfer from chlorophyll to the reaction center where glucose synthesis occurs is coherent even for normal cellular conditions, which is different from what was previously assumed about energy loss. This coherent energy transfer depicts an efficient process that highlights the efficiency of photosynthesis as a process supporting the food web. The implications of such efficiency are far-reaching, as it maintains the viability of higher trophic levels and, therefore, the productivity of ecosystems in general. Through these mechanisms, photosynthesis not only serves its immediate role in generating food for primary consumers but also supports a life organization that is stable and crucial for the health and stability of ecosystems, further augmenting the central role of photosynthesis in the food web (Patel et al., 2011).

8.5.3. Contribution to Carbon Cycling

Photosynthesis is a central process of the carbon cycle, locking atmospheric CO_2 into organic carbon compounds, primarily through the work of green plants, algae, and certain bacteria. It is an important method of sequestration for carbon in that it actually lowers atmospheric concentrations of CO_2 while also being the foundation of terrestrial food webs. The complex relationship between nitrogen availability and carbon capture also determines net primary production (NPP), reflecting the sensitivity of ecosystems to changes in these nutrient flows. Studies have demonstrated that increased CO_2 can stimulate global NPP, with 4.1 Pg C yr^{-1} increase reported when atmospheric CO_2 is doubled, emphasizing the key role of photosynthesis in carbon sequestration and overall terrestrial ecosystem health (McGuire et al., 1997).

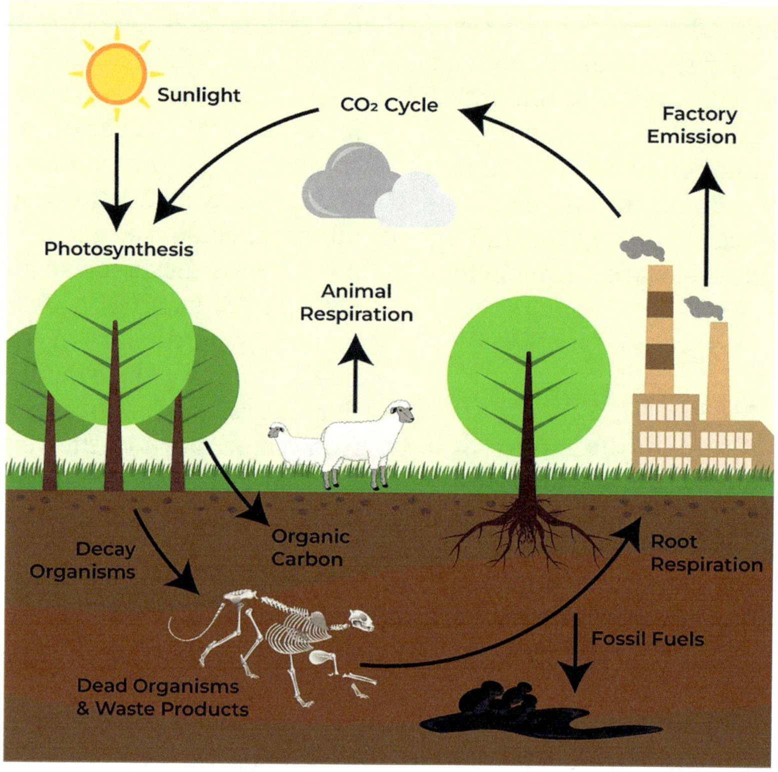

Figure 8.18. Photosynthesis contribution to carbon cycling.

Source: https://static.wixstatic.com/media/31c1cf_acea5959608841e585cebff1c9f03ef6~mv2.png/v1/fill/w_980,h_980,al_c,q_90,usm_0.66_1.00_0.01,enc_avif,quality_auto/31c1cf_acea5959608841e585cebff1c9f03ef6~mv2.png.

Also, the contribution of peatland ecosystems, , such as that of bryophytes , such as Sphagnum, emphasizes the role of photosynthesis in the carbon cycle. Peatlands are known to store more than one-third of land carbon, primarily due to the unique properties of Sphagnum mosses that cause the accumulation of organic matter in waterlogged situations. These microbes accumulate carbon not only with their photosynthetic capability but also through regulation of carbon cycles in the soil by their associated microbiomes. The present studies on the ecological role of Sphagnum under climate change conditions provide an understanding of how Sphagnum can be integrated into broader carbon cycling concerns. Information regarding interactions between photosynthesis, microbial communities, and nutrient cycling in peatlands is relevant to developing an effective system for managing climate change (Weston et al., 2014). Also, the relationship among high atmospheric CO_2, plant nitrogen level, and processes in the soil carbon defines a complex landscape in which photosynthesis can strongly influence carbon cycling. For instance, reductions in vegetation nitrogen level can induce dramatic alterations in NPP and ensuing soil carbon storage. As seen in all simulations, with each stepwise decrease in nitrogen, it is associated with higher soil carbon, highlighting the way changes in nutrient availabilities can extend carbon sequestration limits. This interdependence reveals how a proper interpretation of ecosystem responses to both biotic and abiotic change is crucial to predict future carbon cycling trends. With scientists exploring these linkages, the findings will play a crucial role in informing conservation efforts on how to optimize carbon sequestration in a rapidly changing world (McGuire et al., 1997) (Figure 8.18).

Test Your Knowledge

1. What is oxygen production and its significance?
2. What is the role in the food chain?

8.6. Human Impact on Photosynthesis

Learning Outcome

1. Deforestation and its effects;
2. Climate change and its influence on photosynthesis;
3. Agricultural practices and their implications.

Photosynthesis is also greatly influenced by human activities, and this is mainly caused by the change in atmospheric composition and use of land. The increase in CO_2 levels, driven largely by fossil fuel combustion and deforestation, has immediate implications for photosynthetic processes. Increased levels of CO_2 can boost photosynthesis because plants need CO_2 as an important raw material. The impact of this is however not consistent among the various species and ecosystems. Some plants will benefit with higher CO_2, others will suffer because they will lack some necessary nutrients, as one of the effects of the climate change will be the disturbance of the nutrient cycling and availability. This complexity underscores the necessity for comprehensive research examining not only CO_2 increases but also how these changes interact with other environmental variables, , such as nutrient availability altered by elevated temperature and shifting precipitation patterns (Lagomarsino et al., 2010). In addition, anthropogenic climate change worsens the conditions plants have to photosynthesize by affecting the availability of water and temperature. Ecosystems are also getting predisposed to severe weather conditions, which result in extended droughts or heavy floods, either of which can restrict photosynthetic capacity. Especially drought stress can lead to stomatal closure to limit water loss, which eventually limits CO_2 uptake. Rising temperatures on the globe may also cause heat stress, which further damages photosynthetic activity. When plants cope with these abiotic stresses, their total productivity and growth are impaired, which influence food security and stability of the ecosystem. These interrelated factors highlight how human impacts extend beyond mere CO_2 emissions, showcasing the urgent need for integrative climate models that assess the cumulative effects on photosynthesis across various environments (Keulen et al., 2025). Besides change in the atmosphere, land use practices have a central role in changing photosynthesis. The conversion of forests to other uses , such as deforestation, urbanization, and agricultural activities tip the balance of ecosystems, which in most cases lead to fragmentation of habitats and loss of biodiversity. These changes have the potential to cause the loss of native plant species which could be better suited to local conditions than introduced varieties. Replacement of agricultural systems with monocultures, though useful in the short-term production, tends to decrease the total plant diversity and resilience, affecting the photosynthetic capacity of the ecosystems. Moreover, soil degradation and pollution, which are caused by intensive agricultural methods, may disturb the nutrient dynamics that are essential to be optimized to photosynthesis. To effectively combat these issues and promote sustainable management of ecosystems, it is essential to understand the complex relationships between human actions, nutrient cycling, and the fundamental processes of photosynthesis (Lagomarsino et al., 2010).

8.6.1. Deforestation and Its Effects

Forest clearing turns into a significant threat to the process of photosynthesis, changing the carbon exchange dynamics of ecosystems in a very basic manner. Forests serve as important carbon sinks, absorbing CO_2 during the process of photosynthesis and sequestering it as biomass, which is essential in maintaining global carbon balance. Loss of tree cover significantly reduces the capacity of such ecosystems to sequester carbon, leading to increased atmospheric CO_2 concentrations. As per the reports, "stocks of carbon as biomass in tropical forests and woodlands amount to 271 ± 16 Pg," highlighting the enormous reservoir of carbon to be released due to deforestation. The resulting imbalance supports a vicious cycle in which amplified carbon releases hasten climate change, further weakening the photosynthetic activity of current vegetation due to more harsh environmental conditions. Thus, the impacts of deforestation extend far beyond localized environmental degradation to affect the climatic patterns and photosynthesis in general on a world scale. The interrelatedness between deforestation and climate change is a portentous feedback mechanism. As observed, "higher global temperature will promote a rise of greenhouse gas levels," making known that changes in the climate further destabilize ecosystems based upon photosynthesis. Increased temperatures result in altered precipitation patterns, and the consequence is most often drought that disrupts plant growth and photosynthetic output. In degraded ecosystems where forests have been lost or diminished, not only is sequestration of carbon destroyed, but susceptibility of these systems to climate fluctuation and storms is enhanced. This negative feedback indicates an unsafe process in which less forest cover accelerates climate change, which in turn causes further reductions in forest health. The solution therefore is to put an end to deforestation in a bid to maintain the balance that enables effective photosynthesis and global world health. Additionally, the socio-economic impacts of deforestation cannot be overlooked in analyzing its environmental impacts. Livelihood-based societies that are reliant on forests are being confronted with devastating consequences as a result of deforestation, as it not only disrupts natural systems and leads to loss of biodiversity but also affects the availability of resources and food security, particularly in regions that are highly dependent on forest systems. With further degradation of forest lands, not only are the essential photosynthetic activities disrupted, but regional communities also have problems with climate adaptation and resilience. The interplay between deforestation, climatic processes, and human activities underscores the importance of sustainable forest management. Lastly, integration of conservation activities with an emphasis on the health of ecosystems and the well-being of communities is essential in reducing the adverse effects of deforestation on photosynthesis and global climatic stability (Figure 8.19).

Figure 8.19. Deforestation affect photosynthesis.

Source: https://i.ytimg.com/vi/vJY3DTaE0sI/maxresdefault.jpg.

8.6.2. Climate Change and Its Influence on Photosynthesis

The complex relationship between climate change and photosynthesis is becoming increasingly evident with the current global warming. Photosynthesis, by which plants synthesize chemical energy from light energy, is both affected by CO_2 oscillations and temperature. Elevated atmospheric CO_2 concentrations can enhance photosynthetic rates for the majority of plant taxa; however, this is not even among taxa. For instance, research indicates that certain peat vegetation, for instance, Sphagnum species, can have their photosynthetic activity altered as a result of altered temperature and hydrology changes. When such processes unravel, it is essential to understand how plants' exchange with the environment might be impacted by temperature as well as hydrological variability, particularly with reference to carbon sequestration and nutrient cycling, which are key processes of the photosynthetic pathway (Weston et al., 2014). Moreover, the nutrient cycles necessary for photosynthesis are also at risk with climate change. Such key nutrients, nitrogen (N) and phosphorus (P), are integral parts of the plant's photosynthetic equipment, but their supply is made vulnerable by alteration of climate conditions. Current evidence is that with rising temperatures and changed precipitation regimes, the mechanisms governing the cycling of the nutrients will be disrupted. For example, soil water fluctuations may have an impact on nutrient mobilization and tree nutritional status and, in turn, their photosynthetic potential. Future modeling involves a shift from an analysis of the effects of single climatic variables to an understanding of the multi-variable interaction between variables , such as rising CO_2, temperature increase, and changed water availability (Lagomarsino et al., 2010). This holistic view will be vital to the development of effective strategies for mitigating the effects of climatic change-induced nutrient limitation. Finally, the effects of climatic change extend beyond species to the ecosystem processes underlying photosynthesis. Ecosystems operating on specialized plant communities, , such as peatlands, are particularly vulnerable to climatic fluctuations due to their unique biogeochemical cycling processes. Sphagnum moss, known for its massive contribution to carbon capture, is a good example of this susceptibility. Its sensitivity to environmental fluctuations makes it a good model organism for studying adaptation to changing climatic conditions. As climate change unfolds, photosynthetic responses of keystone species , such as Sphagnum will not only give insights into the biology of plants but also contribute to forecasting the implications on carbon and nutrient cycling at a broader level. Finally, a proper understanding of these relationships is crucial for evaluating the future of photosynthesis under a changing climate and hence affecting climate mitigation and ecosystem management policies (Weston et al., 2014) (Figure 8.20).

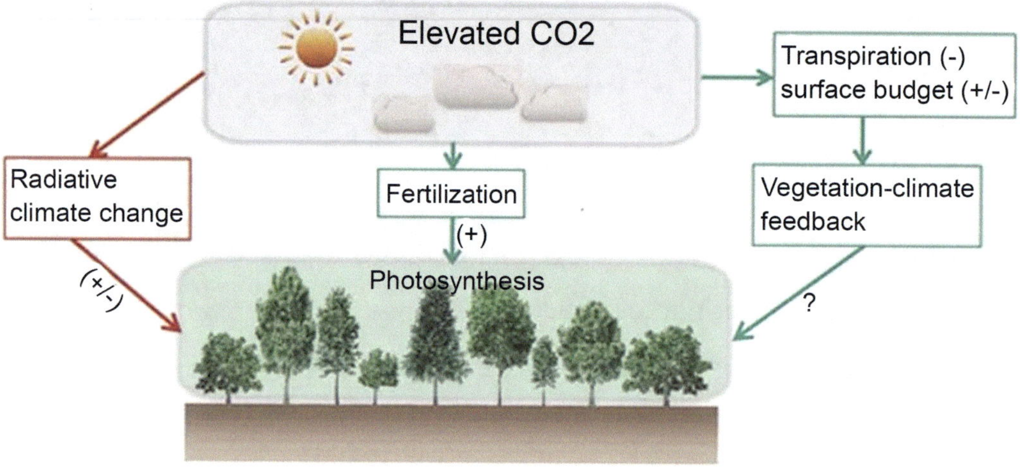

Figure 8.20. Climate change affecting photosynthesis.

Source: https://cdn.cid-inc.com/app/uploads/2022/11/29093404/Picture1-300x137.jpg.webp.

8.6.3. Agricultural Practices and Their Implications

Farming techniques are also of paramount significance in calculating food security and sustainability of the environment, with photosynthesis context being particularly important. Staple foods , such as rice, for instance, rely on optimal photosynthetic activities for maximum production of yields. These methods, however, are now increasingly threatened by the impacts of climate change , such as higher temperatures and water scarcity. As rice constitutes the majority of the world's caloric supply, optimizing productive photosynthesis by way of adaptative agriculture has become a necessity. Unfortunately, limitations on the expansion of land for production and the competing uses of necessary resources render it hard to increase yields in a sustainable manner. The integration of these challenges calls for creative solutions, for instance, the application of simulation modeling to improve handling of spatial and temporal factors in agricultural development for climate-resilient cropping systems for resistance of weather to environmental variability (Bastiaens et al., 2009). The consequences of agricultural practices are not just for current crop production since they significantly contribute to climate change and ecosystem health. Conventional agriculture methods can lead to soil erosion, lowered biodiversity, and increased greenhouse gas emissions, all of which degrade the ecosystem balance necessary to facilitate efficient photosynthesis. Shifting to sustainable agricultural practices, i.e., organic cultivation, crop rotation, and integrated pest management, improves soil fertility and enhances biodiversity, thereby enhancing plant photosynthetic efficiency. Moreover, these practices align with world climatic change mitigation measures. The research commissioned by organizations , such as the UKs Fairtrade Foundation highlights the importance of adapting farming procedures to account for climate exposure along value chains, particularly for vulnerable farmers (Burt et al., 2010). Adaptations not only benefit individual farmers but also make entire communities more resilient in the face of climatic limitations. Finally, the incorporation of enhanced agricultural practice has critical implications on world food security and socio-economic security. New breeding schemes strive toward the development of crop varieties enhanced to address changing climatic trends, thereby enhancing photosynthetic ability and productivity. Agriculturalists and policymakers must join efforts to develop policies that enable the communication of such enhanced practices to farmers across the globe. This approach must take note of local ecological conditions to permit suitably adapted strategies that retain respect for traditional knowledge while taking advantage of scientific progress. By actively addressing the multi-faceted challenges of climate change and resource scarcity, the agricultural industry can make genuine inroads toward sustainable development. In doing so, we not only enhance food security but also provide environmentally responsible stewardship, ultimately leading to agricultural resilience in the presence of an evolving global environment (Bastiaens et al., 2009).

Test Your Knowledge

1. What is deforestation and its effects?
2. What is climate change and its influence on photosynthesis?

Summary

The process of photosynthesis is a linchpin of life on Earth, extending far beyond its simple role of converting solar energy into chemical form. During the conversion of water and CO_2 into glucose and oxygen, photosynthesis not only sustains the very fundamental energy needs of plants but is also crucial in the Earth's carbon cycle and climate regulation. The productivity of this process is central to agricultural biotechnology, where enhanced knowledge of plant genetics can result in better crops that more effectively cope with environmental stresses. The research profiled by Dunwell et al. (2010) evinces the significance of photo-biochemical efficiency improvement in elevating food output to combat food security issues in the context of an increasing world population. Moreover, the uses of amplified photosynthesis extend from agriculture to bioengineering and green energy. In optimizing and harnessing the mechanisms of photosynthesis, scientists aim for alternatives to fossil fuels in the production of biofuels from plant biomass. Not only can this breakthrough method control greenhouse gases, but also a circular economy where by-products from processing plants can be reconstituted into usable products. As explained by ongoing research, the understanding of photosynthesis will drive future agroparadigms and technological advancements in the generation of energy, which marks a profound nexus between environmental understanding and human innovation (Bernacchi et al., 2006). For the future, the climate change and limited resource concerns highlight the imperative for investment in research that widens the understanding of photosynthesis. Emphasizing genetic progress and biotechnological advances, there is a need to promote cross-disciplinary research initiatives that combine ecological inquiry and technological innovation. The advancements in crop biotechnology discussed by Dunwell et al. (2010) constitute a promising direction to not only heighten food security but also contribute to meeting the goals of sustainable agriculture. To summarize, the complex process of photosynthesis is not only a scientific oddity but a vital process that holds the secrets of a sustainable future, linking ecological well-being with human progress.

To appreciate the complexity of photosynthesis is to understand its various forms, in this instance, the divergence of C_3 and C_4 pathways. C_3 photosynthesis is the most ubiquitous form, found primarily in temperate plants, in which CO_2 is directly fixed into a molecule made up of three carbon atoms. C_4 photosynthesis has evolved in several plant lineages, testifying to the evolutionary plasticity of photosynthetic processes. Research indicates C_4 photosynthesis emerged independently at least 60 times in plant lineages, with adaptive evolutionary paths having multiple primary directions. This polyphyletic trend supports that specific environmental stress and phenotypic diversification lead to adaptations towards more efficient photosynthetic operations, illustrating the complex interdependence between metabolic optimization and environmental adaptation during plant evolution (Williams et al., 2013). C_4 plants , such as maize, have a unique adaptation that enhances the efficiency of CO_2 assimilation. With CO_2 concentration in specialized cells, C_4 plants circumvent the limitation posed by invasive oxygen concentrations, which have a tendency to reduce the efficiency of C_3 photosynthesis. Through this adaptation, C_4 plants can thrive in environments where the availability of CO_2 may not be forthcoming. Contemporary studies utilize nonlinear optimization models to characterize the dynamic behavior of these plants in response to environmental fluctuations. Through the inclusion of reaction rate bounds, researchers have successfully predicted the metabolic activity of C_4 systems and demonstrate the intricacies of plant physiology and environmental interaction (Bogart et al., 2015). These findings highlight the value of C_4 photosynthesis in the overall framework of plant adaptability and metabolic engineering. Also, evolutionary photosynthetic pathway adaptation shows optimization of plant adaptation to various ecosystems. Evolution of C_4 photosynthesis from previous C_3 mechanisms illustrates how non-photosynthetic components affect the evolution of these complex traits. Trait flexibility emphasizes that adaptations may arise because of more than one selective pressure, unconnected with photosynthetic efficiency. As more and more adaptations are progressively revealed by science at the genetic and metabolic levels, knowledge of the complex interplay of photosynthesis with the environment and evolutionary biology will be advanced. Such integrated knowledge is required for regulating food security and climate resilience in the scenario of global environmental change.

KEY TERMS FROM THIS CHAPTER

ATP (adenosine triphosphate)	ATP synthase	Calvin cycle
Carbon fixation	Chlorophyll	Chloroplast
Electron transport chain (ETC)	Glucose	Granum
Light-dependent reactions	Light-independent reactions	
NADP+	NADPH	Rubisco
Oxygen (O2)	Photolysis	Thylakoid
Photon	Photosynthesis	Stroma

CRITICAL THINKING QUESTIONS

1. **What is the main purpose of photosynthesis?**
   a. To break down glucose
   b. To produce carbon dioxide
   c. To convert solar energy into chemical energy
   d. To release oxygen from water
2. **In which organelle does photosynthesis take place?**
   a. Mitochondrion
   b. Chloroplast
   c. Nucleus
   d. Ribosome
3. **What are the two main stages of photosynthesis?**
   a. Calvin cycle and Glycolysis
   b. Light reactions and Krebs cycle
   c. Light reactions and Calvin cycle
   d. Glycolysis and Fermentation
4. **What gas is used during photosynthesis?**
   a. Oxygen
   b. Carbon dioxide
   c. Nitrogen
   d. Hydrogen
5. **What is the main pigment involved in capturing light energy for photosynthesis?**
   a. Carotene
   b. Xanthophyll
   c. Chlorophyll
   d. Hemoglobin

Short Answer Questions

1. What is the purpose of photosynthesis in plants and other autotrophs?
2. What is the overall chemical equation for photosynthesis?
3. In which organelle does photosynthesis take place?
4. What are the two main stages of photosynthesis, and where in the chloroplast do each occur?
5. What is the role of chlorophyll in photosynthesis?
6. What are the light-dependent reactions, and what are their main products?
7. What is the Calvin cycle and what does it produce?

Answers to Critical Thinking Questions

1. (c); 2. (b); 3. (c); 4. (b); 5. (c)

References

1. Williams, B. P., Johnston, I. G., Covshoff, S., & Hibberd, J. M. (2013). Phenotypic landscape inference reveals multiple evolutionary paths to C4 photosynthesis. *eLife*, *2*, e00961. https://doi.org/10.7554/eLife.00961 (accessed on 13th June 2025).
2. Achard, F., Achten, W. M. J., Alongi, D. M., Amundson, R., Aragão, L. E. O. C., Asner, G. P., et al. (2014). Perturbations in the carbon budget of the tropics. https://core.ac.uk/download/28977996.pdf (accessed on 13th June 2025).
3. Ai, Q., Cheng, Y. C., Jin, B. Y., & Yen, H. (2013). Clustered geometries exploiting quantum coherence effects for efficient energy transfer in light harvesting. http://arxiv.org/abs/1307.5590 (accessed on 13th June 2025).
4. AINSWORTH, E.A., DAVEY, P.A., HYMUS, G.J., OSBORNE, C.P., ROGERS, A., BLUM, H., NÖSBERGER, J. and LONG, S.P. (2003), Is stimulation of leaf photosynthesis by elevated carbon dioxide concentration maintained in the long term? A test with *Lolium perenne* grown for 10 years at two nitrogen fertilization levels under *F*ree *A*ir *CO_2E*nrichment (FACE). Plant, Cell & Environment, 26: 705-714. https://doi.org/10.1046/j.1365-3040.2003.01007.x (accessed on 13th June 2025).
5. Alba, R., Aldridge, S., Altschul, S. F., Andersson, B., Andon, N. L., Armbruster, W. S., Asano, T., et al. (2012). Proteomic analysis of chloroplast-to-chromoplast transition in tomato reveals metabolic shifts coupled with disrupted thylakoid biogenesis machinery and elevated energy-production components. https://core.ac.uk/download/12043889.pdf (accessed on 13th June 2025).
6. Martinelli, F., Reagan, R. L., Uratsu, S. L., Phu, M. L., Albrecht, U., Zhao, W., Davis, C. E., Bowman, K. D., & Dandekar, A. M. (2013). Gene regulatory networks elucidating huanglongbing disease mechanisms. *PloS one*, *8*(9), e74256. https://doi.org/10.1371/journal.pone.0074256 (accessed on 13th June 2025).
7. Amarnath, K., Fleming, G. R., Sylak-Glassman, E., Zaks, J., et al. (2013). Models and measurements of energy-dependent quenching. https://core.ac.uk/download/323080709.pdf (accessed on 13th June 2025).
8. Amthor, J. S., Chen, J. M., Clein, J. S., Frolking, S. E., et al. (2001). Comparison of boreal ecosystem model sensitivity to variability in climate and forest site parameters. https://core.ac.uk/download/72054858.pdf (accessed on 13th June 2025).
9. Anderson, L. E., Wingler, A., Osmond, C. B., Foyer, C. H., Downton, W. J. S., Bock, G., Kacser, H., et al. (2000). Photorespiration: Metabolic pathways and their role in stress protection. https://core.ac.uk/download/59898.pdf (accessed on 13th June 2025).
10. Arrigo, K. R., Comeau, L. A., Comiso, J. C., Devos, D. P., DiTullio, G. R., Ferrara, M., Gabric, A., et al. (2014). Polar microalgae: New approaches towards understanding adaptations to an extreme and changing environment. https://core.ac.uk/download/41989130.pdf (accessed on 13th June 2025).
11. Barcia Piedras, J. M., Mateos Naranjo, E., Pérez Romero, J. A., Redondo Gómez, S., et al. (2019). Impact of short-term extreme temperature events on physiological performance of *Salicornia ramosissima* J. Woods under optimal and sub-optimal saline conditions. https://core.ac.uk/download/187420495.pdf (accessed on 13th June 2025).

12. Barker, M., Bell, B., Biddulph, F., Carr, M., et al. (1997). The constructivist paradigm and some implications for science content and pedagogy. https://core.ac.uk/download/29200649.pdf (accessed on 13th June 2025).

13. Bastiaens, L., Bouman, B. A. M., Dingkuhn, M., Gaydon, D. S., et al. (2009). An international collaborative research network helps to design climate robust rice systems. http://agritrop.cirad.fr/556000/1/document_556000.pdf (accessed on 13th June 2025).

14. Bebout, B. M., Des Marais, D. J., Ehrenreich, A., Govindjee, Siefert, J., Joliot, P., et al. (2007). Spectral signatures of photosynthesis I: Review of Earth organisms. http://arxiv.org/abs/astro-ph/0701382 (accessed on 13th June 2025).

15. Berger, F. W., Chen, W., DeFries, R., Dehghan, M., Deng, X., Denison, R., Dong, Z., et al. (2019). Food chain inefficiency (FCI): Accounting conversion efficiencies across entire food supply chains to re-define food loss and waste. https://core.ac.uk/download/323057602.pdf (accessed on 13th June 2025).

16. Bernacchi, C. J., Heady, L. E., Leakey, A. D. B., Long, S. P., et al. (2006). Hourly and seasonal variation in photosynthesis and stomatal conductance of soybean grown at future CO_2 and ozone concentrations for three years under fully open-air conditions. https://core.ac.uk/download/4818896.pdf (accessed on 13th June 2025).

17. Bogart, E., & Myers, C. R. (2015). Multiscale metabolic modeling of C_4 plants: Connecting nonlinear genome-scale models to leaf-scale metabolism in developing maize leaves. http://arxiv.org/abs/1502.07969 (accessed on 13th June 2025).

18. Boonchai, C., Buaboocha, T., Chadchawan, S., Comai, L., et al. (2018). Rice overexpressing OsNUC1-S reveals differential gene expression leading to yield loss reduction after salt stress at the booting stage. https://core.ac.uk/download/323071559.pdf (accessed on 13th June 2025).

19. Borate, R. B., Dayal, P., Gairola, A., Kumar, V., Prajapati, S. K., & Srivastava, R. (2023). The role of carbon in life's blueprint and carbon cycle understanding Earth's essential cycling system: Benefits and harms to our planet. https://core.ac.uk/download/587418662.pdf (accessed on 13th June 2025).

20. Budzikiewicz, H., Dismukes, G. C., Dornemann, D., Feher, G., Fischer, H., et al. (1986). Structure of methyl pheophorbide-RCI. https://core.ac.uk/download/12163099.pdf (accessed on 13th June 2025).

21. Burt, P., Chancellor, T., Morton, J. F., Nelson, R., et al. (2010). Climate change, agriculture and Fairtrade: Identifying the challenges and opportunities. https://core.ac.uk/download/316279.pdf (accessed on 13th June 2025).

22. Carlson, C. A., Haas, A. F., Hatay, M., Leichter, J. J., et al. (2013). Influence of coral and algal exudates on microbially mediated reef metabolism. https://core.ac.uk/download/323066328.pdf (accessed on 13th June 2025).

23. Cestellos-Blanco, S., Yang, P., & Zhang, H. (2019). Solar-driven carbon dioxide fixation using photosynthetic semiconductor bio-hybrids. https://core.ac.uk/download/287623651.pdf (accessed on 13th June 2025).

24. Cornejo-Castillo, F. M., & Zehr, J. P. (2019). Hopanoid lipids may facilitate aerobic nitrogen fixation in the ocean. https://core.ac.uk/download/323063928.pdf (accessed on 13th June 2025).

25. Drégelyi-Kiss, G., Drégelyi-Kiss, Á., & Hufnagel, L. (2008). Ecosystems as climate controllers–biotic feedbacks (a review). https://core.ac.uk/download/18426065.pdf (accessed on 13th June 2025).

26. Dunwell, J. M. (2010). Crop biotechnology: Prospects and opportunities. https://core.ac.uk/download/9050.pdf (accessed on 13th June 2025).

27. Etter, A., & Fischer, S. (2020). Stream biofilm response to an increasing number of non-flow periods. ZHAW Zürcher Hochschule für Angewandte Wissenschafte. (accessed on 13th June 2025).

28. Fan, H., Hao, Y., Thung, L., Wu, Y., et al. (2013). Proteomic analysis of glucohexaose induced resistance to downy mildew in *Cucumis sativus*. https://core.ac.uk/download/18447730.pdf (accessed on 13th June 2025).

29. Gardiner, A. T., Britton, G., Uragami, C., Frank, H. A., Hashimoto, H., et al. (2018). Understanding/unravelling carotenoid excited singlet states. https://core.ac.uk/download/157582964.pdf (accessed on 13th June 2025).

30. Gimbel, S., & Schreiber, A. (2010). Evolution and the Second Law of Thermodynamics: Effectively communicating to non-technicians. https://core.ac.uk/download/pdf/81820422.pdf (accessed on 13th June 2025).

31. Grundy, S., Jiang, N., & Lu, C. (2017). Uncovering LED light effects on plant growth: New angles and perspectives. https://core.ac.uk/download/145637219.pdf (accessed on 13th June 2025).

32. WESTON, D.J., TIMM, C.M., WALKER, A.P., GU, L., MUCHERO, W., SCHMUTZ, J., SHAW, A.J., TUSKAN, G.A., WARREN, J.M. and WULLSCHLEGER, S.D. (2015), *Sphagnum* physiology and changing climate. Plant Cell Environ, 38: 1737-1751. https://doi.org/10.1111/pce.12458 (accessed on 13th June 2025).

33. Hernandez Candia, C. N., & Michaelian, K. (2010). Transpiration in plants: A thermodynamic imperative. https://core.ac.uk/download/pdf/289733.pdf (accessed on 13th June 2025).

34. Hollows, D. G. (1978). A physiological investigation of the adaptive significance of juvenility in *Pennantia corymbosa* Forst: A thesis presented in partial fulfilment of the requirements for the degree of Master of Science in Botany at Massey University. https://core.ac.uk/download/159449301.pdf (accessed on 13th June 2025).

35. Imboden, D. M. (1992). The impact of physical processes on algal growth. https://core.ac.uk/download/11020721.pdf (accessed on 13th June 2025).

36. Keulen, H. van, Laar, H. H. van, & Rabbinge, R., et al. (2025). *40 Years Theory and Model at Wageningen UR*. https://core.ac.uk/download/pdf/29259614.pdf (accessed on 13th June 2025).

37. Kleidon, A. (2020). What limits photosynthesis? Identifying the thermodynamic constraints of the biosphere within the Earth system. http://arxiv.org/abs/2005.06529 (accessed on 13th June 2025).

38. Lagomarsino, A., Ainsworth, E. A., Atkin, O. K., Beier, C., Bradford, M. A., Calfapietra, C., et al. (2010). Global climate change and tree nutrition: Effects of elevated CO_2 and temperature. https://core.ac.uk/download/9323.pdf (accessed on 13th June 2025).

39. Lai, A. G. (2012). Coordinated transcriptional regulation between a reactive oxygen species-responsive gene network and the circadian clock in *Arabidopsis thaliana*: A thesis presented in

partial fulfillment of the requirements for the degree of Doctor of Philosophy in Plant Biology at Massey University, Plant Biology, New Zealand. https://core.ac.uk/download/148638571.pdf (accessed on 13th June 2025).

40. Lea-Smith, D., McCormick, A. J., & Mills, L. (2020). Current knowledge and recent advances in understanding metabolism of the model cyanobacterium *Synechocystis* sp. PCC 6803. https://core.ac.uk/download/288394480.pdf (accessed on 13th June 2025).

41. McGuire, A. D., Aber, J., Ågren, G. I., Schloss, A. L., Auchmoody, L., Bartholomew, D., Moore, B., et al. (1997). Equilibrium responses of global net primary production and carbon storage to doubled atmospheric carbon dioxide: Sensitivity to changes in vegetation nitrogen concentration. https://core.ac.uk/download/86427184.pdf (accessed on 13th June 2025).

42. Michaelian, K., & Simeonov, A. (2014). Fundamental molecules of life are pigments which arose and evolved to dissipate the solar spectrum. http://arxiv.org/abs/1405.4059 (accessed on 13th June 2025).

43. Perin, G., Bernardi, A., Bellan, A., Bezzo, F., & Morosinotto, T. (2017). A mathematical model to guide genetic engineering of photosynthetic metabolism. *Metabolic engineering*, *44*, 337–347. https://doi.org/10.1016/j.ymben.2017.11.002 (accessed on 13th June 2025).

44. Nesamma, A. A., Mariam, I., Kareya, M. S., & Jutur, P. P. (2020). CO_2 sequestration by hybrid integrative photosynthesis (CO_2-SHIP): A green initiative for multi-product biorefineries. https://core.ac.uk/download/572846205.pdf (accessed on 13th June 2025).

45. De Pascale, S. and Stanghellini, C. (2011). HIGH TEMPERATURE CONTROL IN MEDITERRANEAN GREENHOUSE PRODUCTION: THE CONSTRAINTS AND THE OPTIONS. Acta Hortic. 893, 103-116 DOI: 10.17660/ActaHortic.2011.893.6 https://doi.org/10.17660/ActaHortic.2011.893.6(accessed on 13th June 2025).

46. Patel, A. D. (2011). Efficient energy transport in photosynthesis: Roles of coherence and entanglement. http://arxiv.org/abs/1104.1253 (accessed on 13th June 2025).

47. Sandora, M. (2019). Multiverse predictions for habitability: Fraction of planets that develop life. http://arxiv.org/abs/1903.06283 (accessed on 13th June 2025).

48. Shimizu, H., Xu, Z., & Zhou, G. (2009). Are plant growth and photosynthesis limited by pre-drought following rewatering in grass? https://core.ac.uk/download/pdf/8324918.pdf (accessed on 13th June 2025).

49. Soares, A. S., Foyer, C., Olmos, E., Arrabaç, M., Driscoll, S. P., & Pellny, T. K. (2008). Incident light orientation lets C_4 monocotyledonous leaves make light work differently. https://core.ac.uk/download/pdf/287717.pdf (accessed on 13th June 2025).

50. Stengel, A. (2009). Characterization of components and mechanisms involved in redox-regulation of protein import into chloroplasts. https://core.ac.uk/download/11030856.pdf (accessed on 13th June 2025).

51. Suharja, S., & Sutarno. (2009). Biomass, chlorophyll and nitrogen content of leaves of two chili pepper varieties (*Capsicum annuum*) in different fertilization treatments. https://core.ac.uk/download/12345928.pdf (accessed on 13th June 2025).

CELL DIVISION

9

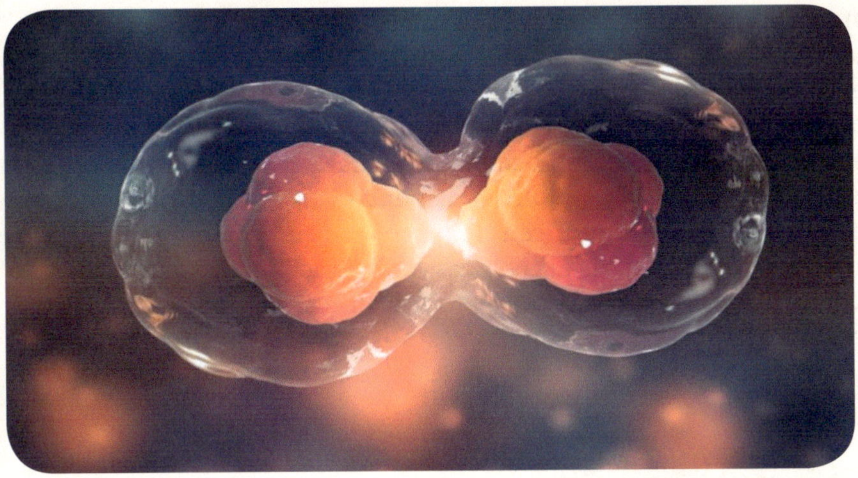

LEARNING OBJECTIVES

After studying this chapter, you will be able to:

1. List the types of cell division;
2. Discuss the cell cycle;
3. Explain mitosis in detail;
4. Explore meiosis in detail;
5. List the factors affecting cell division;
6. Describe the disorders related to cell division.

Introduction

Cell division is fundamental to the science of biology since it is a critical process for growth, repair, and reproduction in organisms. Cell division entails several mechanisms, the most familiar of which are mitosis and meiosis. Mitosis yields two genetically identical daughter cells, and it is critical to the proliferation of somatic cells, while meiosis yields gametes with half the number of chromosomes and plays a role in sexual reproduction organism genetic diversity. An understanding of these processes can inform a model of how cellular organization works at the microscopic level, reflecting on broader physiological principles that define multicellular life. Such a question illuminates the importance of cellular communication, control, and structural stability, thereby providing a comprehensive framework for understanding not only development and growth but also diseases that may arise from cellular failure, e.g., cancer. The intricacy of cell division raises challenging issues concerning functional organization of multicellular systems. Biological organisms are not mere accumulations of cells; they possess a sophisticated interrelation that allows them to develop complex functionalities. Research indicates that structuring intercellular space plays a basic role of maintaining an integrated physiological structure and hence facilitating effective communication and sharing resources among cells. ECM, for instance, plays a significant function in organizing the space by determining cellular fate and behavior. As cells become specialized and differentiated in a multicellular organization, these interactions with their environment are regulated by this structural system. Therefore, comprehension of the principles that control cell division not only reveals the processes themselves but also underscores the relational dynamics that are critical to the survival and normal functioning of multicellular organisms, demonstrating the importance of a structured approach to biological understanding (Bich et al., 2019). Also, the exploration of the concept of biological dispositions extends our knowledge about cell division and differentiation. Dispositions refer to the inherent potentialities of biological systems that determine their behavior under given conditions. The manifestation of cellular properties during division is contingent on a variety of factors that comprise genetic, environmental, and biochemical signals. By studying these dispositions, researchers can establish the different conditions that control how cells respond to their context and alter state. The knowledge has significant implications in developmental biology and regenerative medicine, where cellular potentialities are subject to manipulation to create new therapies for a range of conditions. As a result, the research on cell division relates to bigger biological issues, giving a consistent background to discuss the complexities of life at the organismal and cell levels, which will give access to future life science findings (Hüttemann & Kaiser, 2018) (Figure 9.1).

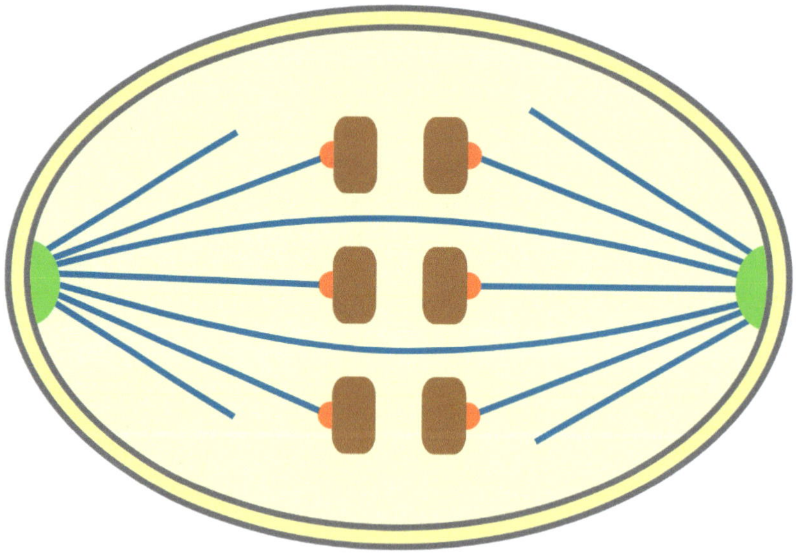

Figure 9.1. Fluorescence microscopy image of a dividing cell, showing the cytoskeleton (red), DNA (yellow), and other cellular structures (blue) during mitosis.

Source: By Baba66 - Own work, CC BY 4.0, https://commons.wikimedia.org/w/index.php?curid=42035755

Definition of Cell Division

Cell division is a basic biological process whereby a single cell divides to form two daughter cells, thereby providing us with growth, development, and tissue repair in multicellular organisms. The process can be divided into broad categories, primarily into two types: mitosis and meiosis. Mitosis is an important process involved in normal growth and cell maintenance, forming daughter cells which are genetically identical to the parent cell. In contrast, meiosis introduces genetic diversity and is necessary for sexual reproduction to produce gametes consisting of half the amount of genetic material as the parent cell. Both processes are tightly controlled by a cascade of checkpoints and regulatory proteins that ensure proper functioning and accuracy during division, indicative of the precise control systems that are integral to cellular biology. The complexity of cell division is also explained through advances in live microscopy and computational modeling, which have significantly enhanced our understanding of embryonic development. A recent emphasis has been on the quantitative nature of morphogenetic processes, allowing researchers to explore the division and differentiation of cells in real-time. With the use of high-content imaging techniques, scientists can analyze hundreds or even thousands of cells simultaneously and reveal the complex dynamics of cell division within living organisms. Not only has this technological breakthrough led to the identification of basic challenges in developmental biology, but it also made systems analysis of embryonic morphogenesis possible. As these technological advances continue, they reveal an unprecedented window on cellular mechanisms of division and development and enable more advanced experimental approaches in cell biology research (Truong & Supatto, 2011). Moreover, the regulatory networks governing cell division are complex and interdependent and mirror intrinsic cellular signals as well as extrinsic environmental stimuli. As discussed by more recent studies, the interaction between the cellular, tissue, and organismal levels of growth is primarily responsible for cell proliferation and growth. This can be most glaringly observed in model species like *Arabidopsis thaliana*, where the molecular process of growth processes is applied to increase agricultural productivity. As the study continues, the integration of knowledge of cellular division into general developmental paradigms will be crucial in resolving global problems in agriculture and sustainability. By investigating growth regulation at various levels, researchers can develop novel crop improvement techniques that bridge the gap between laboratory success and field application, ultimately transforming the discipline of agriculture to meet future food security needs (Wuyts et al., 2015).

Importance of Cell Division in Living Organisms

Cell division is an elementary biological process that allows living organisms to grow, repair, and reproduce. From unicellular bacteria to multicellular organisms, cell division allows cells to replace injured cells as well as tissue growth. In multicellular life, growth and development from embryonic development require it, as cells keep dividing to form the tissues and organs needed to exist. It is strictly controlled to the extent that cells will only divide when needed. Mechanisms for controlling cell cycle are essential; uncontrolled cell division can lead to pathological conditions , such as tumors and cancers. The ability of living organisms to be homeostatic and respond to environmental change relies primarily upon its ability to perform structured cell division, emphasizing its significant role to life's continuity and adaptability. Cell division is not all about growth and repair; it is also necessary for genetic evolution and stability. Cell division makes a copy of genetic information with accuracy and delivers it to daughter cells, hence ensuring the genome integrity of an organism. It is necessary for proper cell functioning since replication mistakes lead to mutations whose effects are damaging. Furthermore, the role of cell division through meiosis and mitosis helps to distinguish sexual from asexual reproduction, respectively. By way of the creation of genetic diversity through recombination during meiosis, cell division is promoting evolutionary processes whereby populations can modify and adapt to new environmental stressors. Cell division therefore not only assists in individual organism survival but also contributes to determining the evolutionary trajectory of species, suggesting its important position within biology. Investigations into the processes of cell division have provided significant insights into its health and disease implications. As cytogerontology uncovers the interface between cell division and aging, the restricted division capacity of somatic cells, introduced by August Weismann, stands out to have significant implications for deciphering cellular aging and longevity (Carrel et al., 1982). Regulation of cell division is not so much a growth process but is very much interlinked with patterns of aging and maintaining cell function over time. Similarly, cellular communication in the ECM illustrates how spatial structure enables coordinated physiological function in multicellular systems (Bich et al., 2019). It is necessary to understand such complex interactions so that therapeutic interventions targeted towards cell division-related illnesses like cancer can be developed, thus the urgency to study this biological process further.

Overview of the Types of Cell Division

Cell division is a biological process of reproduction and growth of cells, which basically takes place in two different ways namely mitosis and meiosis. Mitosis is the way somatic cells separate, and two genetically identical daughter cells are obtained, which is essential to growth, tissue repair, and asexual reproduction in multicellular organisms. It is divided into various stages, namely prophase, metaphase, anaphase, and telophase, with each having events that occur to make sure that the chromosomes are distributed equally. On the other hand, a specialized type of cell division called meiosis, takes place in germ cells and results in the production of gametes or sperm and eggs in sexually reproducing organisms. The process reduces the number of chromosomes by half and provides genetic variation by recombination and independent assortment, which play a central role in evolution and adaptation of populations, as is covered in any textbook of biology. The cell division apparatus is a highly complex molecular machine. It is mainly dependent on the cytoskeleton, which is made up of microtubules, intermediate filaments, as well as actin filaments, all of which are crucial in cell shaping and movement. In mitosis, a spindle apparatus made of microtubules drives the process of separating sister chromatids to different poles of the cell. Likewise, in meiosis, the pattern and association of homologous chromosomes are imperative in the process of making certain that every gamete gets a correct collection of chromosomes. As described by Risler et al. (2009), the polymerization-depolymerization dynamics of these cytoskeletal components are integral to creating the forces required for cell division, influencing cellular motility and division as well. Recognition of these molecular contributions gives clues to the coordination of the proper cell cycle progression and offers potential targets of therapeutic intervention in diseases associated with cell division. The role of a number of biochemical factors that can regulate the process of cell division has also been pointed out by the research, stressing the possibility of extrinsic control. Particular phytochemicals, which are plentiful in herbs and plants, have been shown to affect cell division, stimulating it or inhibiting it depending upon the biological situation. For instance, certain compounds can stimulate protective enzyme synthesis while counteracting processes involved in uncontrolled cell division, a hallmark of cancer; as noted by Amle et al. (2022). The relevance of such findings is growing in the area of medical research, as the knowledge of how natural compounds influence the process of cell division may result in the innovative methods of cancer prevention and treatment. Taken together, these discoveries confirm the importance of the mechanical and biochemical nature of cell division and guarantee a multi-faceted interpretation of the mechanisms that drive cell reproduction (Figure 9.2).

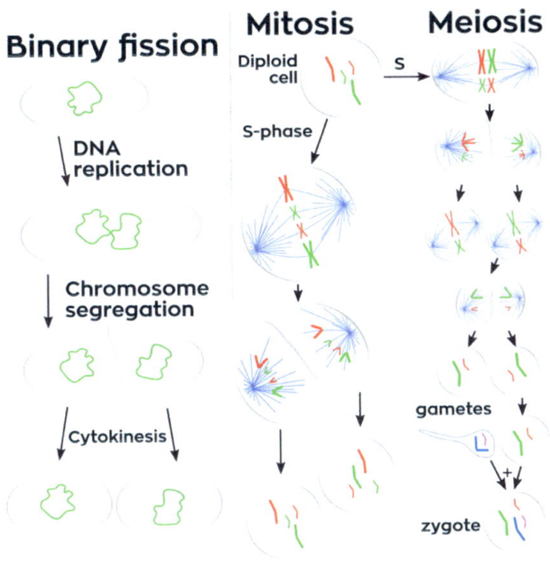

Figure 9.2. Types of cell division.

Source: By domdomegg - Own work, CC BY-SA 4.0, https://commons.wikimedia.org/w/index.php?curid=46512209

9.1. Types of Cell Division

Learning Outcome

1. Mitosis;
2. Meiosis;
3. Binary fission.

Cell division is broadly classified into two types: mitosis and meiosis, each with distinct biological functions. Mitosis refers to the method through which somatic cells, or body cells, duplicate themselves, forming two daughter cells that are genetically similar. This is significant to growth, repair of tissue, and asexual reproduction in certain animals. During mitosis, specific stages, prophase, metaphase, anaphase, and telophase, are responsible for ensuring chromosomes are accurately duplicated and distributed. The significance of this form of division extends beyond mere cellular reproduction; it is also critical to the potential for maintaining genetic stability from generation to generation of cells. The fidelity of mitotic division is critical to organism development and function, affirming its infrastructural role in multicellular life systems. On the other hand, meiosis is a type of specialized cell division that occurs in germ cells to produce gametes—sperm and eggs—that are needed in sexual reproduction. Unlike mitosis, however, meiosis consists of two consecutive sets of division: meiosis I and meiosis II, which give rise to four non-identical daughter cells, each containing half of the original number of chromosomes. This halving is significant, as it guarantees that when gametes combine during fertilization, the resulting zygote will possess the appropriate diploid number of chromosomes. Meiosis also introduces genetic diversity via such mechanisms as crossing over and independent assortment, thus markedly increasing genetic diversity in offspring. Such mechanisms not only introducc cvolutionary flexibility but also lay the ground for the biological complexity characteristic of advanced organisms. Recent synthetic biology breakthroughs have opened up new understanding of cell division, particularly in the regulation of cellular processes for desired purposes. Researchers are exploring the potential of engineered microbial consortia to optimize cell functions through specialized division of labor among different cell types. This new approach separates the metabolic burden from individual cells and allows higher system stability. For instance, cells apply quorum sensing to talk to each other efficiently to perform complex biological circuits and in the process, emulate some of the characteristics of multicellular organisms without full cellular differentiation. Identification of such synthetic processes not only broadens the vision of cell division but also enables potential applications in biotechnology, medicine, and ecological engineering, which shows the continuous evolution of cell division concepts amidst scientific advancement (Kaya et al., 2019; Donoghue et al., 2021) (Figure 9.3).

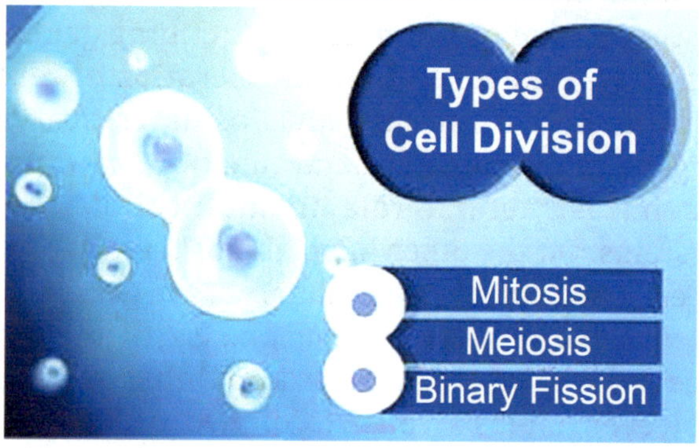

Figure 9.3. Types of cell division.

Source: https://encrypted-tbn0.gstatic.com/images?q=tbn:ANd9GcSww94GgCY5MQ9AWCu9PU0S3rB8SaWfKEAvgw&s.

9.1.1. Mitosis

Mitosis is a crucial phase of the cell cycle in which a parent cell separates to produce two daughter cells that are genetically similar. It is a critical process for the growth, development, and repair of multicellular tissues. Initiated by the separation of chromosomes, mitosis is demarcated phases: prophase, metaphase, anaphase, and telophase, each sequenced to accomplish exact genetic material distribution. By prophase, the chromatin is condensed into visible chromosomes that are later translocated by the mitotic spindle to their respective poles at anaphase. The precision of such a segregation is of enormous importance since loss of precision can lead to aneuploidy and is implicated in many diseases, including cancer. Knowing about the intricacy of mitosis is thus crucial to understand how the integrity of cells is maintained throughout the life of an organism. Research still explores the effect of other external forces on mitosis, the effect of low-frequency magnetic fields in particular. Exposure to these fields has been shown in some research to affect cell behavior, potentially making them accountable for increased numbers of certain cancers, , such as leukemia.

In gamma-ray and very low-frequency (ELF) magnetic field exposed HeLa cell experiments, results indicated that the exposed cells proceeded to mitosis slightly sooner than those that were not exposed to a magnetic field, suggesting the interference with the regular cycle of mitosis. The consequences are of serious concern with the environmental determinants of cell division processes and the possible implications to cell cycle control and overall health of the cells. Investigation of these interactions underscores the complexity of mitosis and susceptibility to intrinsic and extrinsic forces (Harris et al., 2002).

Advances in our understanding of the molecular mechanisms that dictate mitosis have also been instrumental in breakthroughs in cancer chemotherapy. For instance, recognition of HSET, a microtubule motor protein, has shed light on its role in adapting to supernumerary centrosomes in cancer cells. The clustering of these centrosomes in two poles is crucial to prevent multipolar mitosis and ensuing cell death. Such inhibitors as CW069 have been designed that disrupt HSET, being effective in inducing multipolar mitosis selectively in supernumerary centrosome-containing cancer cells. Such selectivity not only reflects the sophistication of the process of mitosis but also implicates targeted therapeutic interventions that can attack cancer cell proliferation selectively. Discovery of such molecular intricacies brings researchers to the doorstep of developing more effective therapies that exploit the distinct vulnerabilities of cancer cells (Bender et al., 2013) (Figure 9.4).

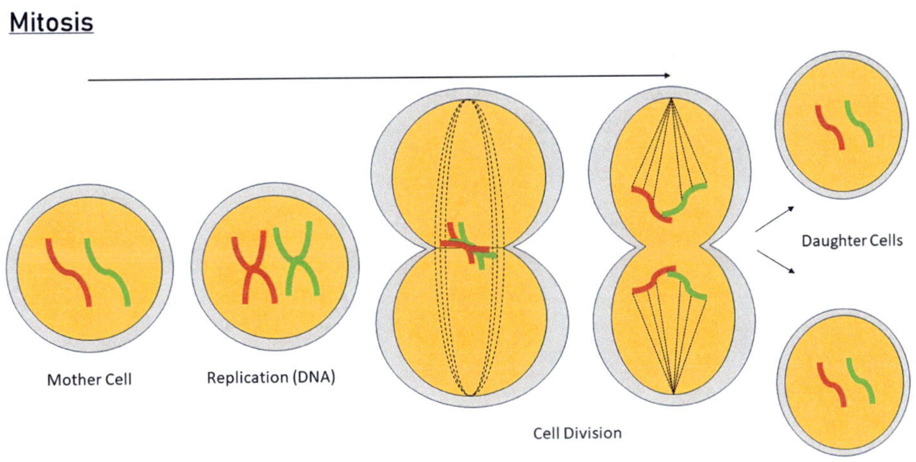

Figure 9.4. Mitosis.

Source: https://www.biotechreality.com/wp-content/uploads/2023/12/Mitosis_diagram.png.

9.1.2. Meiosis

Of all processes of cell division, meiosis is a process that is fundamentally involved in sexual reproduction and genetic variation. While mitosis divides into two genetically identical cells, meiosis is a process that takes place in two consecutive divisions which ultimately lead to four non-identical cells with half the number of chromosomes the original cells had. This reduction in chromosome number is required for maintaining genetic integrity at the time of fertilization, as the gamete union ensures the resulting offspring contain the correct diploid chromosome complement. The unique stages of meiosis—meiosis I and meiosis II—allow homologous chromosomes to be subjected to recombination, a process of jumbling genes and contributing to genetic diversity in populations. This heterogeneity is crucial for the adaptive process of organisms to variant environments, proving the significance of this division process from an evolutionary point of view. The evolutionary function of meiosis also involves something more than just cell division, showing a more complex integration into reproductive mechanisms.

The sexual reproduction cycle (SRC), which Bai and Xu have speculated in their 2013 review, illustrates how meiosis, sex differentiation, and fertilization are interdependent to enhance genetic variation. This model reinterprets meiosis not only as a detached process, but as one component of a broad strategy of reproduction that was developed from unicellular eukaryotes to the vast range of multicellular organisms that currently exist. By placing meiosis in such a perspective, scientists can more fruitfully understand the adaptive mechanisms that have evolved over the course of millions of years, emphasizing the action of genetic diversity as a driver of species survival and evolution. Besides, understanding the meiotic mechanics offers an insight into essential biological processes that regulate cell behavior. Zac Candes' research on cell division and cytoskeleton dynamics emphasizes the roles of these structures during meiosis. Proper regulation of meiotic divisions relies on the cytoskeleton, which facilitates the transport and alignment of chromosomes. This complex interaction between cellular components and meiotic processes is an ideal demonstration of the significance of spatial and temporal controls towards the successful gamete formation. Knowledge of such processes not only reveals the biological significance of meiosis but confirms the holistic nature of cellular activities, paving the way for potential applications in genetics, reproductive biology, and evolutionary studies (Figure 9.5).

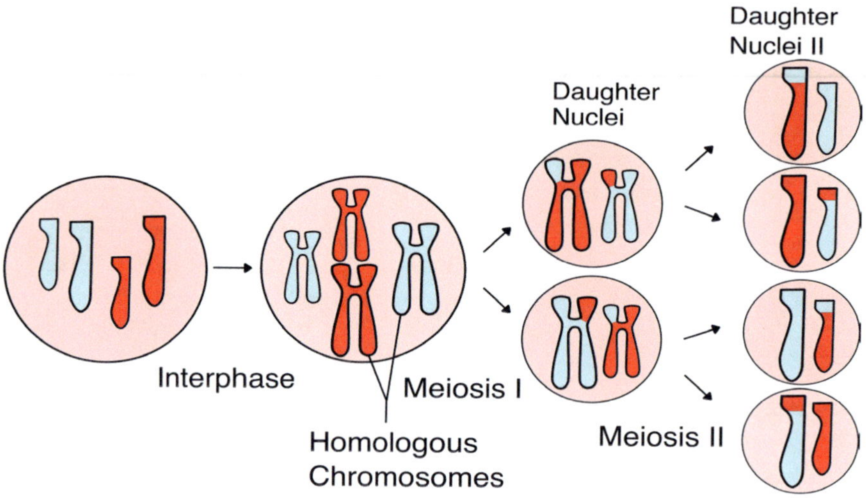

Figure 9.5. Meiosis.

Source: https://upload.wikimedia.org/wikipedia/commons/thumb/9/96/Meiosis_Overview_new.svg/1200px-Meiosis_Overview_new.svg.png.

9.1.3. Binary Fission

Binary fission is a basic form of cell division used by prokaryotic organisms that allows them to grow and reproduce exponentially. The first step in this process involves the duplication of the circular DNA of the organism, which is important to guarantee genetic continuity. After DNA replication has been initiated, the cell lengthens, and the two DNA molecules are pushed to the opposite poles. This extension not only acts as a physical means of separation but also as an important means of functional integration in the cell. As identified in the literature, understanding this integration sheds light on the collective regulatory mechanisms that underpin cellular organization and individuality, revealing the evolutionary significance of binary fission in fostering complex life systems (Militello, 2021).

Therefore, binary fission represents an effective reproduction strategy, which enables prokaryotes to inhabit a wide range of environments, and to do so, genomic integrity is preserved. Formation of the septum, a partition that will eventually divide the mother cell into two daughter cells, is the next step in binary fission. The formation of this septum involves the activity of several proteins which become localized in the middle of the long cell, organizing the cytoskeleton network and making sure that it divides correctly. This phenomenon shows how prokaryotic cells use functional integration to divide cells accurately. Moreover, control of septum formation is also important to control the size and shape of the resultant cells, and so, spatial constraint is important in the organization of biology. The insights gained from these processes have broader implications, as they tie into evolutionary transitions in cell structure, , such as the development of eukaryotic cells from their prokaryotic ancestors, making binary fission a subject of great interest in both evolutionary biology and the philosophy of biology (Militello, 2021). When considering the implications of binary fission, it must be considered how it applies to the development of biological individuality and autonomy among organisms.

High rates of division facilitated by this process permit populations of prokaryotes to respond rapidly to shifting environmental stresses and thus provide an example of how functional integration may impact survival and evolutionary pathways. Through evaluation of the dependencies between cellular processes in the course of binary fission, scientists can shed light on how these processes figure into the overall dynamic stability of prokaryotic existence and its propagation. This exploration ultimately enriches our understanding of major transitions in evolutionary theory, underscoring the significance of binary fission as not merely a reproductive strategy but as a critical factor in the complexity of living systems (Militello, 2021) (Figure 9.6).

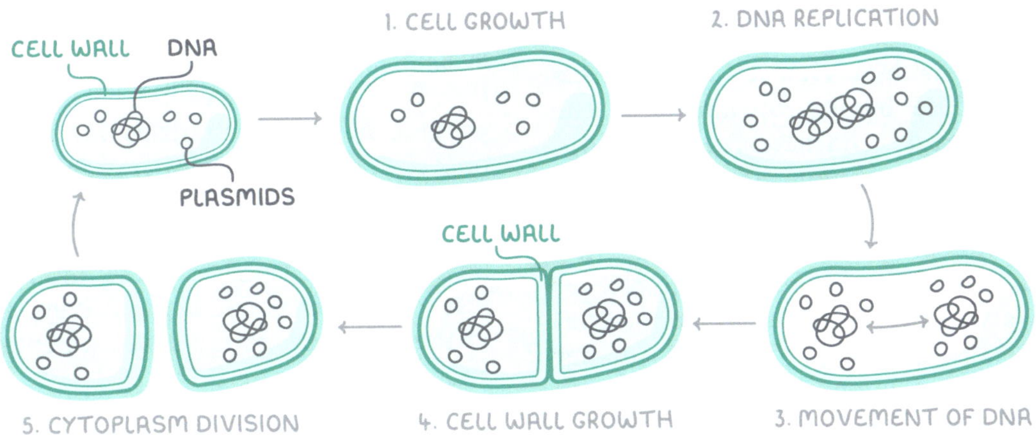

Figure 9.6. Binary fission.

Source: https://s3.eu-west-2.amazonaws.com/elements.cognitoedu.org/18004d09-57b5-41dc-819f-ea80be8faec8/binary-fission-prokaryotic-cells.png.

Test Your Knowledge

1. What is Mitosis?
2. What is Meiosis?

9.2. The Cell Cycle

Learning Outcome

1. Phases of the cell cycle;
2. Regulation of the cell cycle;
3. Importance of checkpoints in the cell cycle.

The cell cycle is a vital and well-controlled series of events in which a cell increases and divides to give rise to two daughter cells. It consists of distinct phases: G1 (gap 1), S (synthesis), G2 (gap 2), and M (mitosis). Each stage is essential to ensure the fidelity and integrity of cell reproduction. At the G1 stage, the cells increase in number and prepare for DNA replication, assessing environmental circumstances with a view to determining if division would be appropriate. S is the subsequent stage, characterized by DNA replication, when the genetic material is copied to ensure that both daughter cells share an equal number of chromosomes. Following the S phase, the G2 phase readies the cell for mitosis by producing the protein and organelles needed for the division of the cell, further emphasizing the planning involved in the cell cycle. Mitosis, the M phase of the cell cycle, is a complex process that organizes the division of replicated chromosomes into two distinct daughter cells. This phase is subdivided into several stages: prophase, metaphase, anaphase, and telophase, each characterized by specific cellular activities. During prophase, chromatin condenses into visible chromosomes, and the mitotic spindle begins to form. As the cell progresses to metaphase, chromosomes align along the metaphase plate, ensuring proper segregation. Anaphase follows, with the sister chromatids being split apart to opposite poles of the cell so that the two daughter cells that are formed end up having the same set of chromosomes. Telophase is the reformation of the nuclear membrane across the split chromatids, culminating in cytokinesis, in which the cytoplasm splits to create two genetically identical daughter cells. Such meticulous choreography of mitosis is evidence of the significance of the cell cycle in preserving genetic continuity across generations. Investigations of the specific cellular mechanisms regulating the cell cycle of specific organisms have uncovered numerous new adaptations. For instance, studies on Trypanosoma brucei indicate anomalies in its cell cycle that differ markedly from those of standard model organisms. The very process of investigating the cell cycle of such parasites is testifying to their life cycle, which necessitates high adaptability to environmental signals. The methodologies employed, , such as cell-cycle arrest with hydroxyurea, indicate the capabilities and limitations of research techniques currently available. Furthermore, new techniques involving flow cytometry and selective staining methods provide new opportunities for studying the phases of the cell cycle of Trypanosoma brucei, , such as indicated by experiments that made possible selective enrichment of single phases, including the G1 phase. These findings add not just to the knowledge of the cell cycle, but also to the implications for developmental biology and disease (Figure 9.7).

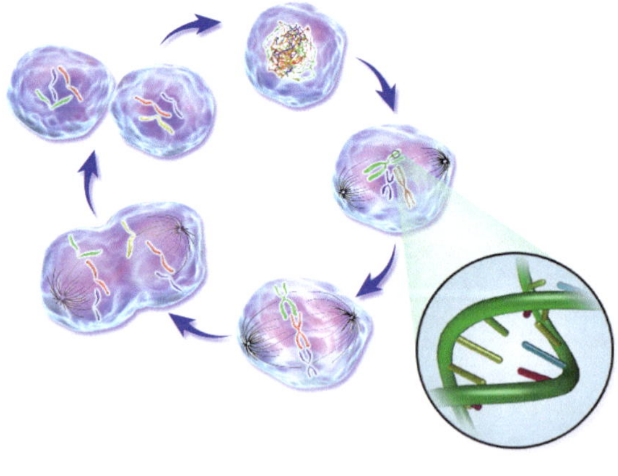

Figure 9.7. The cell cycle.

Source: https://en.wikipedia.org/wiki/File:Normal_Cell_Life_Cycle.png

9.2.1. Phases of the Cell Cycle

The cell cycle is a fundamental process through which a cell prepares for division and subsequently divides into two daughter cells. It consists of several distinct stages: G1, S, G2, and M, and each of these is crucial to ensure that cell division occurs without any problems. The G1 phase is primarily responsible for cell growth and the synthesis of proteins needed for DNA replication. This phase is where the cells look around to determine whether they possess the resources required to proceed with division.

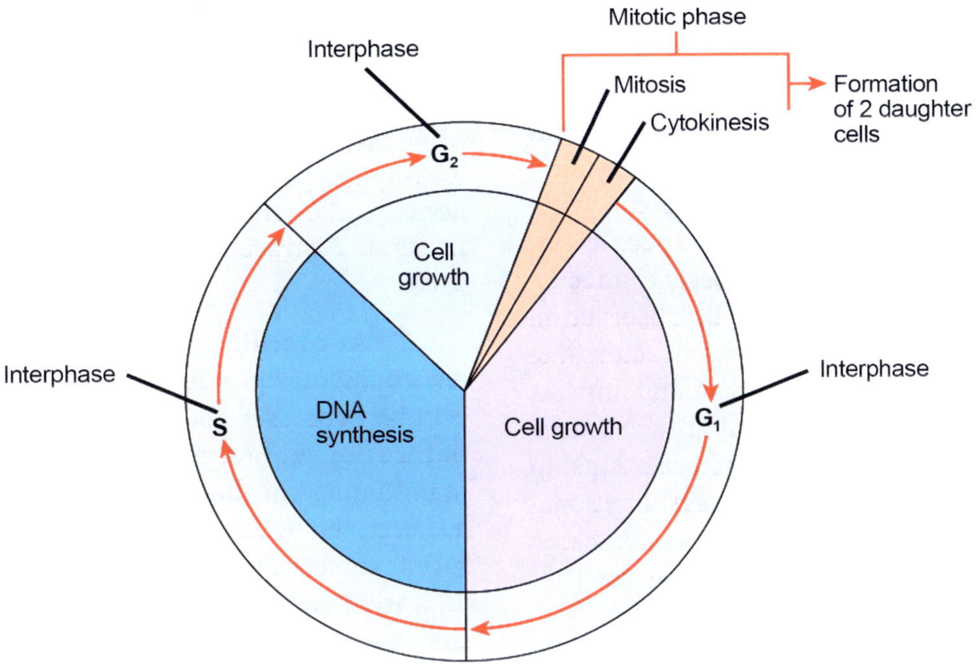

Figure 9.8. Phases of the cell cycle.

Source: https://cdn.kastatic.org/ka-perseus-images/6afac0fce62ec8a30c86f3e02c457224339bb697.png.

This is a critical time because if there is a mistake here, the cell will proceed and divide incorrectly, leading to dysregulation of cellular functions. The functionality of the cell cycle's control mechanisms, for example, the checkpoints between such phases, plays a significant role in maintaining the health of the cells and, consequently, the organisms as a whole. Following the G1 phase, cells enter the S phase where DNA replication occurs. During this phase, each chromosome is replicated to produce two identical sister chromatids so that both daughter cells will have an equal complement of genetic information. The precision of regulation of the S phase is highly critical since any mistake in DNA replication could lead to a mutation that could bring about cancerous events. Following the S phase, the cell transitions into the G2 phase during which it undergoes additional growth and prepares for the terminal phases of division. The G2 phase is a quality control checkpoint, confirming that DNA has been replicated and that the cell is adequately prepared for mitosis. This is a stage of increased protein and organelle synthesis that will be required during cell division. The M phase, incorporating both the mitosis and the cytokinesis, is where the division occurs. Mitosis is divided into a series of stages, prophase, metaphase, anaphase, and telophase, each contributing to the segregation of the sister chromatids into two distinct nuclei. Most importantly, mitosis is designed to be a rapid process, independent of modifications during the prior stages, as supported in recent studies confirming its unchanging duration despite the different lengths of the cell cycle (Araujo et al., 2016). Moreover, specific cell functions like calcification are closely connected with stages like G1, which can be observed in coccolithophore experiments where factors like the presence of nutrients had profound impacts on cell cycle processes and cell growth (Antia et al., 2008). Such a sophisticated relationship between various cellular processes, nutritional states, and specific phases merely underscores the advanced regulatory mechanisms that govern cell division (Figure 9.8).

9.2.2. Regulation of the Cell Cycle

Control of the cell cycle is a complex interplay of various molecular mechanisms, essential for ensuring proper cellular division and function. The central to this control are the cyclins and the cyclin-dependent kinases (CDKs), which work together to induce and control the progression of the cell cycle through its typical phases. Cyclins bind to CDKs, activating their kinase activity and enabling the phosphorylation of target proteins essential for cell cycle transitions. This communication ensures that each stage, G1 to S and subsequently to G2, is executed in a timely and controlled manner. Secondly, the involvement of checkpoint proteins serves as a quality-control process to stop the cell from progressing into the next phase unless certain conditions, , such as the integrity of DNA and proper alignment of chromosomes, are properly satisfied, ensuring genomic stability. Ubiquitin-proteolysis is another major regulatory mechanism in cell cycle regulation. It involves the tagging of specific proteins with ubiquitin, directing them towards proteasomal degradation, a process required for timely cyclin loss and other regulatory units. In the yeast *Saccharomyces cerevisiae*, for example, the SCF(Met30) ubiquitin ligase provides essential regulation of the transcription factor Met4, linking environmental cues to cell cycle progression. The Cdc34/SCF(Met30)-mediated K48-linked polyubiquitin chain controls Met4 activity, allowing for reversible toggling that integrates intracellular conditions into the cell cycle.

This coordination demonstrates how environmental pressures might be applied to cellular processes and indicates the ubiquitous nature of ubiquitination as a means of maintaining the fidelity of cell cycle control, as described by Flick et al. (2006). Furthermore, recent studies have revealed that cellular energy allocation and nutrient availability are some of the non-classical regulators of the cell cycle. As an illustration, plant phloem development shows the interplay between cellular energy allocation and cell division. Sieve elements, as required for the transport of nutrients, undergo a stringently regulated cell division and differentiation pattern. Emerging evidence indicates the involvement of the regulatory networks of translational events, including RNA G-quadruplex structures and specific proteins, , such as JULGI, in coordinating these events. This observation further highlights the complexity of cell cycle regulation, yielding an implication that cellular response to environmental signals and intracellular metabolic conditions are strongly interrelated and determine developmental outcomes during the life cycle of an organism, as explicated by Cho et al. (2018) (Figure 9.9).

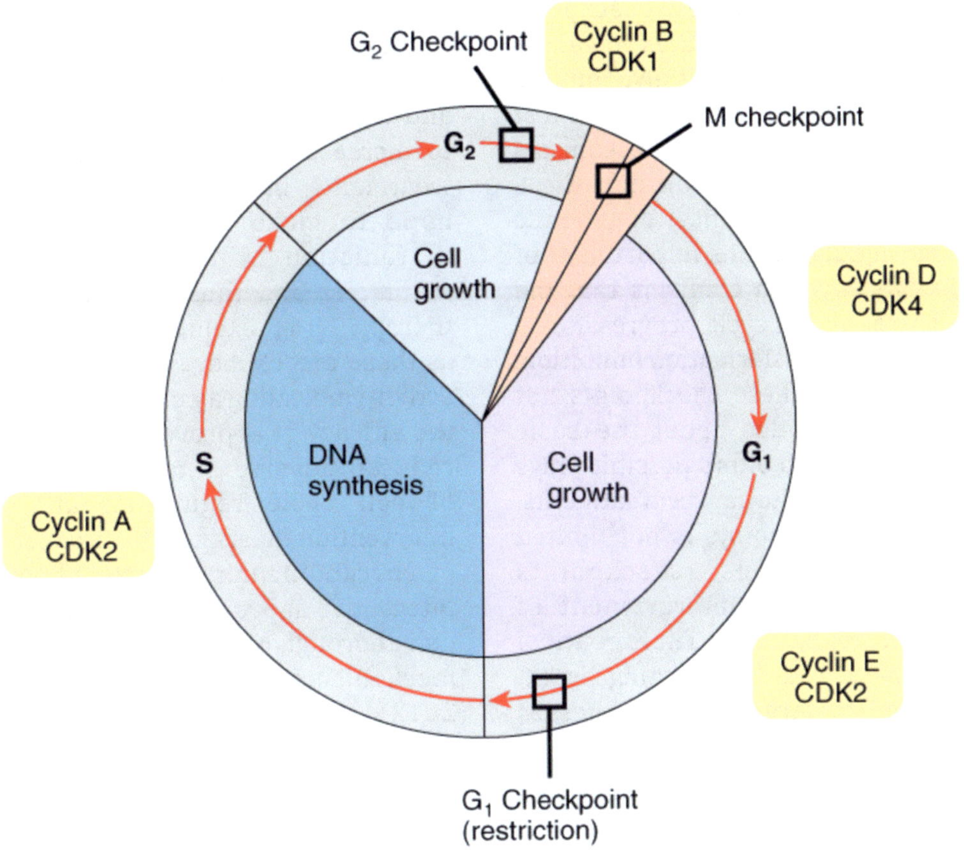

Figure 9.9. Regulation of the cell cycle.

Source: By OpenStax - https://cnx.org/contents/FPtK1zmh@8.25:fEI3C8Ot@10/Preface, CC BY 4.0, https://commons.wikimedia.org/w/index.php?curid=30131218

9.2.3. Importance of Checkpoints in the Cell Cycle

One of the most fundamental processes in cellular biology is the cell cycle; this process is highly regulated to ensure that cells divide with accuracy and efficiency. The key to this regulation is the checkpoints, which are the special control systems that observe the integrity of cell DNA and the correct course of the cellular events. Every checkpoint is a decision checkpoint in which the cell decides whether the environment is favorable for division. As an illustration, the G1 checkpoint assesses the size of the cells, their nutrient status, as well as the integrity of the DNA before committing to DNA replication. Should any abnormalities be present, the cell can pause the cycle to fix the damage, thus averting the transmission of genomic mistakes, a process that is critical in the upkeep of cellular wellbeing and the prevention of illnesses , such as cancer. Hence, checkpoints cannot be considered only as regulatory; they play a central role in the maintenance of genetic information between cell generations, which adds to the overall stability of organisms. To mediate these checkpoint pathways, specialized proteins have evolved that can function as sensors and effectors and can cause cell cycle arrest in response to a variety of stresses. The detection of DNA damage is one of the major pathways; this activates a signaling pathway cascade, which inhibits the cell cycle. It has recently been discovered that these checkpoint processes are frequently deregulated in cancer cells, permitting unrestrained cell division and playing a role in tumorigenesis. Indeed, most cancer cells have broken regulatory mechanisms that are involved in the transition into the S phase, which increases the chances of genomic instability. This highlights the importance of checkpoint activities, which confirms that the appropriate control of cell cycle progression is essential to oncogenic transformation inhibition. Therefore, the study of these checkpoints not only reveals the knowledge about the basic cellular processes but also offers possible ways of therapeutic approaches to cancer treatments. The significance of checkpoints is not limited to physiological functions; checkpoints have a major role in the improvement of cancer treatment modalities. Therapeutical approaches , such as radio sensitizing agent, have been devised to target the weaknesses in the checkpoint functions of cancer cells. Research has also found key elements of the DNA damage signaling pathway, including ATM and the Mre11-Rad50-NBS1 complex, as ways to increase the effectiveness of DNA damage treatments, including radiotherapy. Clinicians hope to make tumor cells more sensitive to radiation by blocking these checkpoint regulators and thus enhancing the effects of therapy. As an example, new inhibitors, , such as those of ATM have demonstrated preclinical activity, indicating a possible capacity to enhance the efficacy of approved cancer therapies, with reduced adverse outcomes. This intersection of cell cycle regulation and therapeutic intervention illustrates the broader implications of checkpoint importance, advocating for an integrative approach to cancer care that hinges on a thorough understanding of cellular division mechanisms (Tvegård, 2008; Fujiwara et al., 2012).

Test Your Knowledge

1. What are the phases of the cell cycle?
2. What is the importance of checkpoints in the cell cycle?

9.3. Mitosis in Detail

Learning Outcome

1. Stages of mitosis;
2. Role of mitotic spindle;
3. Significance of mitosis in growth and repair.

Mitosis is a critical phase of cell division that creates two genetically equivalent daughter cells from a single parent cell. Mitosis is complex and tightly regulated, consisting of several phases: prophase, metaphase, anaphase, and telophase. Each phase is critical, beginning with chromatin condensation into visible chromosomes in prophase. Following this, metaphase aligns the chromosomes on the mitotic spindle to ensure equitable dispensation of the genetic material. Anaphase begins the separation of sister chromatids, which move toward opposite poles of the cell, a defining moment in the cell reproduction cycle. Finally, in telophase, chromosomes de-condense, and nuclear membranes reform, setting the stage for cytokinesis, or true cytoplasmic division resulting in two distinct cells. Regulation of mitosis is required for genomic integrity, a function that is linked closely with post-translational protein modifications in the cell cycle. Specifically, ubiquitylation has been identified as a key process involved in regulating various proteins necessary for advancing during mitosis. The AAA-ATPase CDC-48 has been identified as crucial in this regulatory process, especially in organisms , such as Caenorhabditis elegans. This protein has an important role in DNA replication and mitosis, impacting turnover of factors , such as CDT-1, which is needed to trigger DNA synthesis. Disturbing CDC-48 leads to abnormal stabilization of CDT-1 on chromatin, which can lead to erroneous genomic replication and instability. The complex dynamics between substrates and CDC-48 underlie stable regulatory networks being critical in controlling the fidelity of cell division. Besides, the balance of licensing and elongation functions during replication is tightly controlled by the action of the CDC-48/UFD-1/NPL-4 complex. The control is not just degradation, but a general coordination of protein-protein interactions required for orderly mitosis. For instance, proteins , such as CDC-45 and the GINS complex that are involved in chromatin association must be controlled under stringent control in order for the replicative functions to be efficient. In embryos lacking CDC-48 complex, these proteins remain attached to chromatin longer than necessary, indicating potential deficiencies in regulation that are capable of disrupting cell integrity. Through the studies of such molecular movements, complexity in mitosis is better appreciated with a focus on the essential interplay between structure, function, and regulation in maintaining cellular health and preventing malfunction capable of leading to oncogenesis or other disease (Franz, 2012; Fletcher et al., 2013) (Figure 9.10).

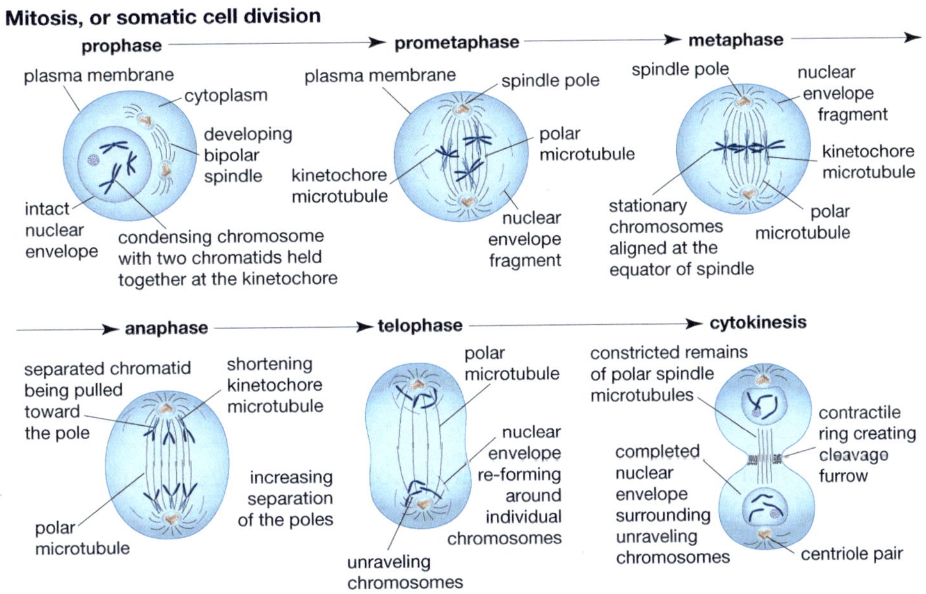

Figure 9.10. Mitosis in detail.

Source: https://cdn.britannica.com/60/53060-050-E3A48FD7/process-cell-division-mitosis.jpg.

9.3.1. Stages of Mitosis

Mitosis is the key cell division mechanism that allows for the equal distribution of genetic material into daughter cells. Mitosis can be subdivided into several distinct phases: prophase, metaphase, anaphase, and telophase, with distinct events permitting orderly segregation of the chromosomes. Prophase is the phase during which chromatin condenses to visible chromosomes that consist of two sister chromatids. Aside from that, the nuclear envelope begins to disintegrate, allowing spindle fibers to be freed from centrosomes, which migrate to the opposite sides of the cell. This initial phase paces the rest of the stages by ordering the chromosomes and opening the door for the cell to divide, showing the elegant choreography that goes into effective cell duplication and the vital role that the nucleus plays in preserving genetic fidelity, as exhibited by the complexity of nuclear form and processes (Amon et al., 2013). As mitosis progresses into metaphase, the position of chromosomes on the metaphase plate is the focal point for the cell. In this alignment, spindle fibers extend from centrosomes and become connected with kinetochores of chromosomes so that each chromatid is ready to be separated.

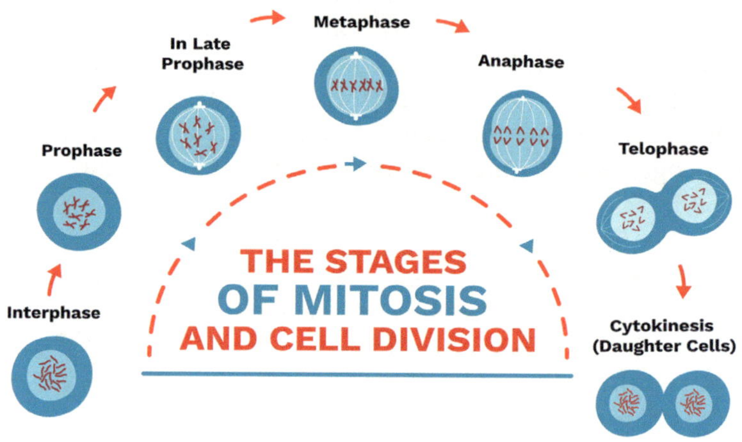

Figure 9.11. The stages of mitosis and cell division.

Source: https://www.thoughtco.com/thmb/sbB2WrLksPIsFgGLvrv5UgVdjkI=/1500x0/filters:no_upscale():max_bytes(150000):strip_icc()/stages-of-mitosis-373534-V5-5b84992cc9e77c00574f03d3.png.

This is a very critical alignment because it is a checkpoint where each daughter cell will receive an equal complement of chromosomes. The metaphase stage is evidence of the cellular structural evolution processes, particularly in eukaryotes, where mitotic apparatus complexity facilitates effective maintenance and segregation of the genome. The level of organization in this state not only underscores the importance of mitosis during cell division but also the elegance of the eukaryotic organisms dependent on stringently controlled mechanisms to ensure survival (Amon et al., 2013). Following metaphase, the cell moves into anaphase, where the sister chromatids are split apart from one another by severing the cohesion, relocating them to the opposite ends of the cell. This magnificent separation is accomplished by the shrinking of spindle fibers, with a refined mechanism of cytoskeletal dynamics and motor proteins. Anaphase is a critical moment in the integrity of having a full set of chromosomes in every daughter cell, representing the precision that occurs in cell division. Finally, mitosis concludes with telophase, where the single chromatid decondenses into chromatin and the nuclear envelope reforms around every complement of chromosomes. This entire process of mitosis, with these pivotal stages, not only enables the duplication of cells but also contributes to growth, development, and healing of tissues in multicellular organisms, demonstrating the positional function of mitosis in biological systems (McCulloch & Navarro, 2016) (Figure 9.11).

9.3.2. Role of Mitotic Spindle

During the complex process of cell division, the mitotic spindle plays a crucial role in the segregation of chromosomes into daughter cells. The microtubule spindle device is engaged in binding to chromosomes by structures referred to as kinetochores in the centromeric regions of sister chromatids. Effective microtubule attachment on these kinetochores is fundamental for genetic material equal segregation. Without a doubt, the spindle assembly checkpoint (SAC) is essential as it verifies these attachments with meticulous care.

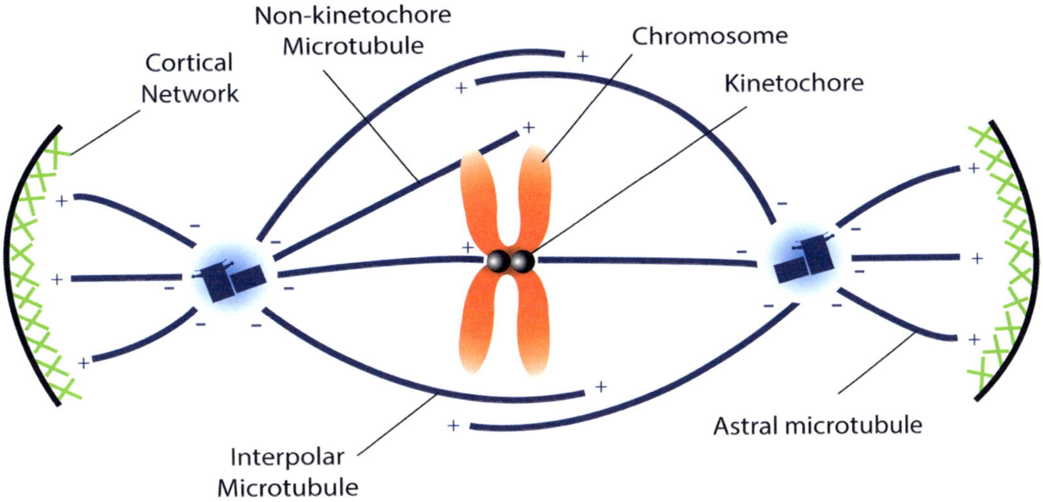

Figure 9.12. Role of the mitotic spindle.

Source: https://www.researchgate.net/profile/Gregory-Rogers/publication/354329850/figure/fig3/AS:1063817454837779@1630645254188/The-mitotic-spindle-is-composed-of-different-populations-of-microtubules-Many-but-not.png.

On improper attachment, SAC halts progression so that missegregation of chromosomes occurring can induce aneuploidy and can promote cancer growth, thereby highlighting the function of the mitotic spindle to protect genomic integrity and cell viability in the course of division (Kukkonen-Macchi et al., 2010). Furthermore, component interaction within the mitotic spindle, including various proteins that are inherent to its functioning, points to its complexity. Proteins , such as Shugoshin 1 (Sgo1), INCENP, and p38 MAP kinase play key roles in kinetochore stability regulation and structure of the entire spindle apparatus. Sgo1, for instance, is a centromeric cohesion protector crucial in tension-sensing SAC. Though INCENP is a crucial part of the chromosomal passenger complex that has a variety of functions to play during mitosis, including the regulation of chromosome movements and spindle dynamics. Studies of these proteins as well as their interactions give useful information regarding coordination of mitotic spindle function and thus pointing to the spindles vital role in guaranteeing successful cell division (Kukkonen-Macchi et al., 2010). Finally, studies have shown that even when cell function is compromised, the mitotic spindle may remain functional, attesting to its robustness and importance. For instance, experiments on PtK1 cells showed that even after cell lysis, the mitotic apparatus remains functional, allowing chromosomes to be segregated to the poles. This suggests that microtubule dynamics, and not actomyosin activity, are responsible for chromosome segregation during anaphase. These findings are in support of the theory that microtubule depolymerization is a rate-limiting factor of chromosome motion. These findings not only cast new insight into the spindles mechanics but also uncover possible avenues for therapeutic interventions in cell division defects disease (Figure 9.12).

9.3.3. Significance of Mitosis in Growth and Repair

Mitosis is central to growth and repair, and because of it, organisms can grow and be able to sustain themselves in terms of cellular structure in their life cycle. This is a sequence of ordered and controlled stages, prophase, metaphase, anaphase, and telophase, and it enables the precise separation of the duplicated chromosomes to two daughter cells. The importance of mitotic divisions may be best seen in situations of rapid growth, e.g., in embryonic development or tissue regeneration, where cells have to divide to allow the size and functioning of the organism. Moreover, the accuracy of mitosis reduces the occurrence of genetic instability, which might result in developmental abnormalities or disorders, especially cancer, in which the unregulated cell division is the major issue of concern. Besides growth, mitosis is also important in the process of cell repair. It enables the replacement of lost or dysfunctional cells after an injury or damage occurred, thus preserving homeostasis and tissue architecture. This repair capability is especially critical in tissues with high turnover rates, including the skin and the lining of the gastrointestinal tract. Mitosis helps in the repair of worn-out cells by genetically identical cells, hence, maintaining the particular functions of the tissue. Moreover, research has highlighted the importance of regulatory proteins, , such as those involved in the DNA repair processes, which are intricately linked to the successful completion of mitosis and thus vital for maintaining cellular and genomic stability (Bhattacharjee et al., 2013).

The specifics of mitosis also have a complicated interaction with the processes of DNA repair. As an illustration, it has been demonstrated that some proteins, including the RPA2a subunit of Replication Protein A, play a critical role in coupling the process of cell division to DNA repair processes. This relationship shows that mitosis is not only a tool of growth but also a very important part of the cell defense mechanism against possible genomic danger. In plants, CDK-mediated phosphorylation of RPA2a has been demonstrated to influence both cell cycle progression and DNA repair pathways, highlighting an evolutionary conservation that underscores the importance of mitosis across different life forms (Dong et al., 2019). These discoveries highlight the complexity of the relationships between growth, repair and fidelity of cells and point to mitosis as a pillar of life at the cellular scale.

Test Your Knowledge

1. What are the stages of mitosis?
2. What is the role of the mitotic spindle?

9.4. Meiosis in Detail

Learning Outcome

1. Stages of meiosis;
2. Genetic variation through meiosis;
3. Importance of meiosis in sexual reproduction.

In meiosis, a form of specialized cell division, the genetic material is meticulously halved to produce gametes. Reducing the number of chromosomes is crucial to maintaining genetic integrity from one generation to the next, since it gives offspring a single set of chromosomes from each parent. Meiosis occurs in two consecutive phases: meiosis I and meiosis II, both comprising different phases like prophase, metaphase, anaphase, and telophase. During prophase I, homologous chromosomes recombine and synapse, and there is an exchange of genetic material. Crossover activities produce genetic diversity that is essential for evolution and adaptation. Products of meiosis not only provide haploid cells but also develop genetic variation through independent assortment and recombination, forming the basis of the organism's ability to adapt to changing environments. The complexity of meiosis is not merely chromosomal segregation; the processes engaged are very sophisticated to ensure genetic fidelity and stability. Homologous recombination (HR), for instance, is key to ensuring the exchange of genetic information in meiosis to ensure accurate segregation of chromosomes. HR is implicated in DNA repair of damage and genomic integrity as a mechanism for protection against mutational events arising during cell division. Proteins like Rad54 are of vital importance in the process. Rad54 acts in conjunction with RAD51 to form homologous pairing and strand invasion, which are vital steps in HR that create opportunities for genetic exchange and repair mechanisms in the course of meiosis.

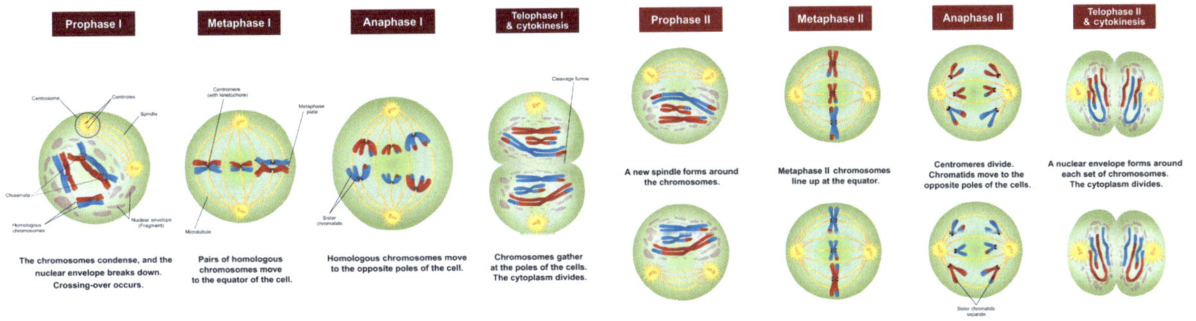

Figure 9.13. Meiosis in detail.

Source: By Ali Zifan - Own work using: information from Campbell Biology (10th Edition) by: Jane B. Reece & Steven A. Wasserman., CC BY-SA 4.0, https://commons.wikimedia.org/w/index.php?curid=49630204

These interactions exhibit a sophisticated molecular ballet that preserves the evolutionary potential while correcting potential causes for genetic instability (Heyer et al., 2006). The importance of meiotic processes is not only confined to individual organisms but also finds itself in broader biological contexts, particularly for organisms , such as Trypanosoma brucei. It has been illustrated through research that the organism that causes sleeping sickness has a remarkable form of genetic exchange through meiosis despite its unique evolutionary background. Genetic mapping due to recent genome projects showed that meiosis, crossing over, and independent assortment operate in *T. brucei*, where they reinstate classical Mendelian principles. Such findings highlight the variability and flexibility of meiotic processes in various organisms. This research not only deepens our understanding of genetic processes in parasites but also highlights the evolution of meiosis as a stable mechanism for facilitating genetic variation in changing environments towards ensuring survival and adaptability of populations (MacLeod et al., 2007) (Figure 9.13).

9.4.1. Stages of Meiosis

Meiosis stages are of crucial importance to cell division in that they ensure genetic diversity and correct chromosome segregation in gametes. Meiosis consists of two major divisions: meiosis I and meiosis II. In meiosis I, the homologous chromosomes synapse to produce tetrads, thus facilitating crossing over of genetic recombination. This takes place in prophase I and is crucial for recombining genetic material, which creates increased variability in offspring. Then, following this, homologous chromosomes are split during anaphase I, and this shows how proper chromosomal separation is crucial to maintaining ploidy stability. Failure at this point can lead to aneuploidy, which is usually deleterious to the resulting gametes and consequently to the survival of the organisms. With meiosis going to meiosis II, it is a regular mitotic division except for haploid cells. Meiosis II includes prophase II, metaphase II, anaphase II, and telophase II, during which the sister chromatids are separated and attracted to the opposite ends of the cell. The nuclear membrane disintegrates again in prophase II, and spindle fibers re-form for the second division. The chromosomes must be properly aligned in metaphase II since misalignment can result in incorrect distribution. Once meiosis II is completed, the cell divides to produce four haploid cells that are different from one another, which eventually develop into gametes. This is a critical phase for the creation of genetic diversity, which is inevitable for the adaptation and evolution of organisms (Valuchova et al., 2020). Molecular processes that are accountable for meiosis, particularly in the segregation of the chromatids, are crucial in understanding how genetic stability is maintained.

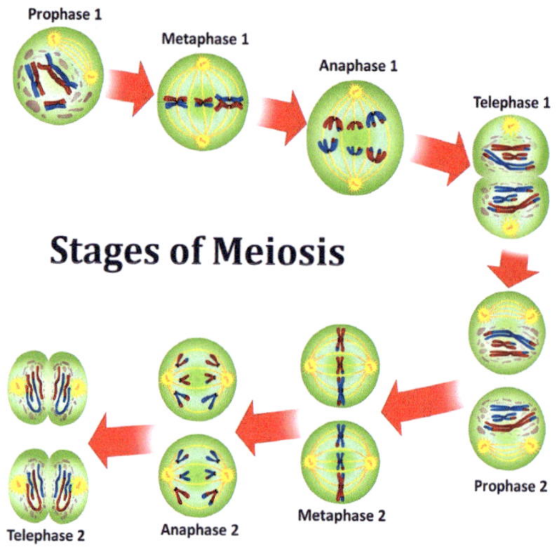

Figure 9.14. Stages of meiosis.

Source: https://www.researchgate.net/profile/Shouvik-Mahanty/publication/332625567/figure/fig3/AS:751263742435331@1556126642575/Stages-of-Meiosis-Photo-Credit-Ali-Zifan-Wikimedia-commons.jpg.

Achritical processes , such as chromosome synapsis, recombination, and checkpoint compliance, ensure faithful chromosomal segregation. Formation of chiasmata during metaphase I, for instance, is a structural basis for the faithful segregation of homologous chromosomes, while ordered release of cohesin during anaphase I is vital for chromatid separation. Furthermore, the surveillance mechanisms incorporated ensure proper attachment of kinetochores to microtubules to prevent failures in segregation. These complex processes underscore the importance of meiosis in maintaining genetic integrity and, in doing so, underscore the merits of investigating meiotic stages against the broader background of cell division and its applicability to plant and animal reproduction (Valuchova et al., 2020) (Figure 9.14).

9.4.2. Genetic Variation Through Meiosis

Meiosis is a fundamental process that achieves genetic diversity, and it is a key force behind evolutionary trend in organisms. A diploid cell is reduced to four haploid gametes due to meiosis. This reduction of chromosome number by half guarantees ploidy stability from one generation to the next. HR and independent assortment during meiosis I are a key process for generating genetic diversity. Homologous chromosomes are juxtaposed and undergo genetic exchange, facilitating recombination of alleles.

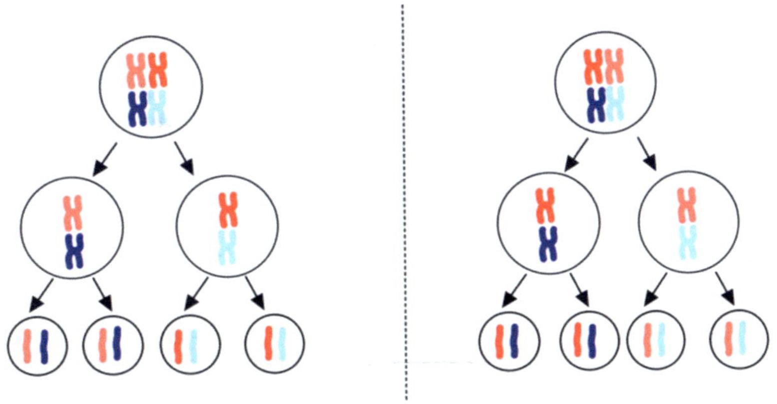

Figure 9.15. Meiosis–genetic variation.

Source: https://i0.wp.com/thescienceandmathszone.com/wp-content/uploads/2021/04/IMG_1122.jpg?resize=688%2C368&ssl=1.

Moreover, the random separation of two pairs of chromosomes into daughter cells ensures that every gamete will possess a unique set of genes. Such mechanisms highlight the importance of meiosis as a generator of genetic diversity, which ultimately enhances adaptability in changing environments. The crossing over during prophase I of meiosis is another important source of genetic variation. This involves the exchange of chromatid segments between non-sister chromatids, thereby creating new allele combinations not found in the parent generation. The resulting genetic recombination is accountable for the creation of new phenotypes among populations. This process makes errors that lead to chromosomal abnormalities, with extreme impacts on reproductive success and offspring survival. For instance, segregation errors at the maternal side during oogenesis have the potential to produce aneuploid conditions , such as trisomy 21, which illustrates the fine balance maintained during meiosis. These errors highlight a fundamental characteristic of meiosis: while it is an apparatus for variance generation, it is also susceptible to disruptions that can compromise genetic integrity and organ health (Dyban et al., 2010). Apart from genetic recombination, accurate segregation of chromosomes during meiosis is also a critical parameter in the upkeep of genetic variation. Accurate chromosome segregation consists of complex regulatory processes that coordinate homologous chromosome pairing, chiasms formation, and biorientation of sister kinetochores. All these are required for the precise transmission of genetic information to gametes. Faults in these mechanisms can lead to the generation of aneuploid or polyploid progeny, contributing to genetic variation within plant and animal kingdoms. In plants, for example, mutations disrupting chromosome segregation can produce unbalanced or unreduced gametes, showing the subtle interaction of genetic and environmental components governing development and fertility (Storme et al., 2014). Understanding these mechanisms is significant not only for the exploration of genetic diversity but also for the treatment of infertility and chromosomal disease in assisted reproduction and internal medicine (Figure 9.15).

9.4.3. Importance of Meiosis in Sexual Reproduction

Meiosis is bound to sexual reproduction and is the mechanism through which genetic variation is acquired. It is a form of specialized cell division that halves the chromosome number, yielding haploid gametes, sperm and eggs, of sexually reproducing organisms. Halving is necessary because it guarantees the zygote formed by fertilization to have the correct diploid number of chromosomes. Also, homologous chromosomes swap genetic material during meiosis, creating genetic diversity among offspring. This is important for populations to be able to adapt to evolutionary stress and environmental pressures. Therefore, meiosis plays not only a fundamental role for preserving chromosome integrity but also gene pool diversification to guarantee healthy populations. Apart from facilitating genetic diversity, meiosis is also significant in correct segregation of chromosomes. In the complex process, various checkpoints and monitors , such as those involving kinases of DNA damage, play to assure and verify chromosomal integrity. For example, ATR (ATM and Rad3-related) kinase plays a vital role during meiotic prophase in facilitating proper chromosome synapsis and in addressing DNA double-strand breaks (DSBs). Evidence from studies has indicated that ATR performs special roles in modulating the timing and efficacy of DNA repair processes through its ability to influence the recruitment of RAD51 and DMC1 to spermatocyte DSBs. These regulatory functions are crucial to successful gamete formation, highlighting the importance of meiosis as a way of evading genetic flaws caused by defective cell division mechanisms, thereby promoting reproductive success and species endurance (Widger, 2018). Meiotic implications extend beyond the organism itself, to larger ecological and evolutionary processes. By producing genetically heterogeneous gametes, meiosis contributes to evolutionary fitness by allowing populations to adapt to changing environments. The process holds broad relevance to agricultural and biotechnological settings, where knowledge of meiotic processes can be employed to guide crop improvement and conservation breeding programs. For instance, yeast organisms, with meiotic systems similar to those of higher eukaryotes, have helped advance the understanding of the basic principles of meiosis and its promise in biotechnology (Lachance & Walker, 2018). *Saccharomyces cerevisiae*, as a model system, has played an important role in helping us understand genetic principles, demonstrating the wider relevance of meiosis beyond sexual reproduction. Thus, the significance of meiosis in the generation of genetic diversity cannot be overemphasized since it predisposes evolution to determine the destiny of species.

Test Your Knowledge

1. What are the stages of meiosis?
2. What is genetic variation through meiosis?

9.5. Factors Affecting Cell Division

Learning Outcome

1. Environmental influences;
2. Genetic factors;
3. Role of hormones in cell division.

Cell division, the growth- and reproduction-necessary process by which a cell divides into two daughter cells, is significantly influenced by a range of internal and external factors. Among these, control of the cell cycle is extremely important since it dictates when and in what order cellular processes take place before division. Cyclins and CDKs are at the forefront of orchestrating these events in such a manner that cells divide only when they are well equipped. In addition, environmental factors , such as levels of nutrients and cell density also control the progression of cell cycle. For instance, in instances of restricted availability of resources or when cell density becomes too high, cells become quiescent and withhold division until environmental conditions prevail once again. This regulatory process emphasizes the contribution of both intrinsic cellular mechanisms and extrinsic agents to imposing accurate control on cell division to maintain cellular function and organism health. Integrity and faithfulness of genetic information are another key cell division modulator. DNA damage induced by environmental factors , such as radiation or chemicals may have the potential to halt the cell cycle to prevent the transmission of defects. It is carried out by various checkpoint processes that check for DNA integrity at various points in the cell cycle. On detection of damage to DNA, p53 and other proteins concerned are activated to initiate repair processes or, if the damage is irreparable, to induce apoptosis. In a similar way, replication errors during DNA replication can compromise genomic stability, leading to outcomes like aneuploidy or other genetic abnormalities. Accurate DNA replication and repair thus become a necessity, enabling the genome to be transmitted to daughter cells consistently, and illustrating how genetic factors exert a strong influence on the cell division process (Gartner et al., 2012). Cell division is also influenced by the extracellular environment through growth factor-responding signaling pathways as well as nutrient supply. These signals initiate intracellular cascades that induce or suppress cell division, with most of this activity being to regulate the expression of genes required for cell cycle progression. , such as, GFs , such as epidermal growth factor (EGF) bind to their receptors and activate protein kinases that direct a cascade toward cell division. Conversely, starvation of GFs can lead to the arrest of cell cycle, a characteristic found in the majority of tissues where cellular growth has to be tightly regulated. External conditions of this sort can not only regulate the rate of the division of the cells but even their capacity to commit to specialized fates, representing a dynamic interaction between external signals and the inner machinery governing cell division (Storme et al., 2014).

9.5.1. Environmental Influences

Environmental factors have a central role in the control of cell division and have a significant effect on organismal growth and development. Knowing how external factors influence cellular processes is important, and especially so within plant biology. For example, plants are involved in complex sign-mediated communication with a range of symbiotic partners both above and below the ground level, which is important for efficient allocation of resources and adaptation to variable environmental conditions. It is this dynamic interaction that allows plants to modify the direction of their growth in response to the cues they receive from the environment. The perception that plants continuously perceive and respond to biotic and abiotic stimuli testifies to the relative importance of environmental factors on cellular activities, particularly those governing cell division and differentiation. By highlighting these interactions, research highlights the complex interplay among environmental parameters and cell processes, specifically the adaptive strategies embraced by organisms in the face of changing conditions (Witzany, 2007).

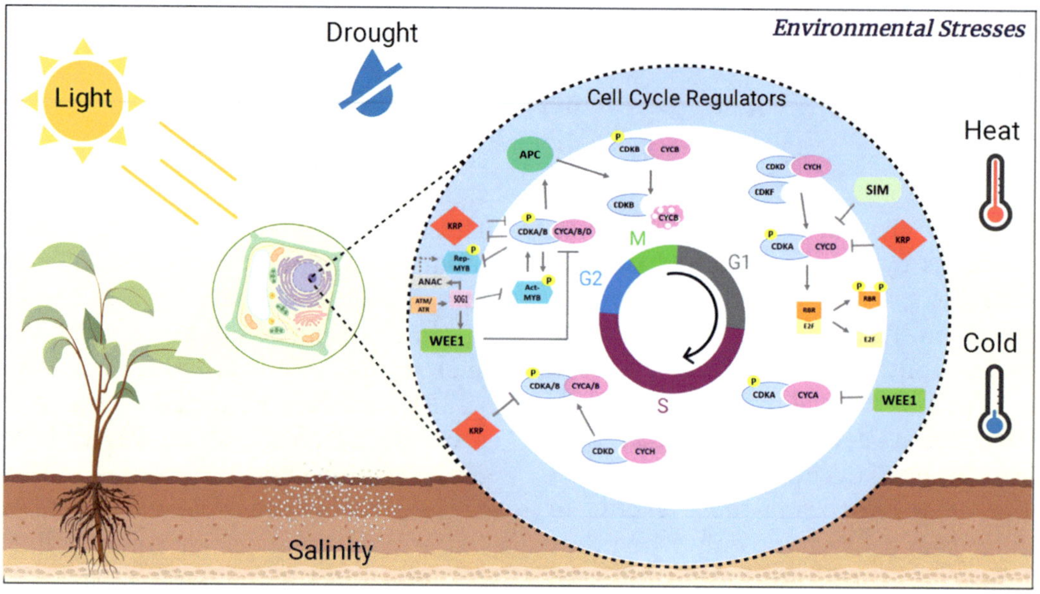

Figure 9.16. Environmental influences of cell division.

Source: https://www.researchgate.net/publication/354242613/figure/fig1/AS:11431281413025039@1745969698635/The-influence-of-the-environment-on-cell-cycle-regulation-The-proliferation-of-cells-in.tif.

Growth in trees, particularly through xylogenesis, provides valuable insights into how environmental parameters influence cell division in the long term. Research has established that cambial timing of activity, taking into account secondary growth in woody plants—is significantly heterogeneous depending on climatic factors. Long-term experiments involving trees , such as European larch and Norway spruce have shown that temperatures can directly influence cell division rates and exhibit a very accurate relationship between environmental and cellular processes. Decadal records illustrate that initiation and cessation of growth operations actually reverse due to external temperature variations. These findings illustrate not only the immediate effect of environmental factors on cell division but also the history behind these responses, highlighting the need for constant monitoring to better understand how trees adapt to climate variation in short- and long-term scenarios (Witzany, 2007). As environmental influences on cell division keep becoming increasingly obvious, it is highly important to consider such influences in the examination of growth patterns within and among taxa. Environmental factors like available light, water, and soil nutrients exert great influence on cellular processes, which consequently impact processes like mitosis and differentiation. For example, in the case of plant cell division, exposure to light will initiate physiological pathways governing growth and developmental timing. Besides, the interactions between other environmental factors pose complex scenarios affecting the rate and efficiency of cell division in plants and other organisms. Scientists are therefore forced to adopt a holistic approach and include a number of environmental variables when modeling cell division dynamics. This general information regarding the impact of external conditions on cellular processes will eventually solidify our comprehension of growth trends and evolutionary adaptations between species (Witzany, 2007) (Figure 9.16).

9.5.2. Genetic Factors

Cell division is not only a physical process but also an complex interplay between genetic factors governing distinct cellular behaviors, e.g., growth, replication, and morphology. In this complex control system, single proteins play a critical role in organisms , such as *Fusobacterium nucleatum*. Genetic factors in biofilm formation have also been demonstrated to regulate cell division in more recent studies, where the cell division proteins EnvC and FtsX have drawn specific interest. Mutations in the genes of these proteins not only cause aberrant cell division but also interfere with the properties of the bacterial cell surface, demonstrating the integration of cellular structure and genetic determinants.

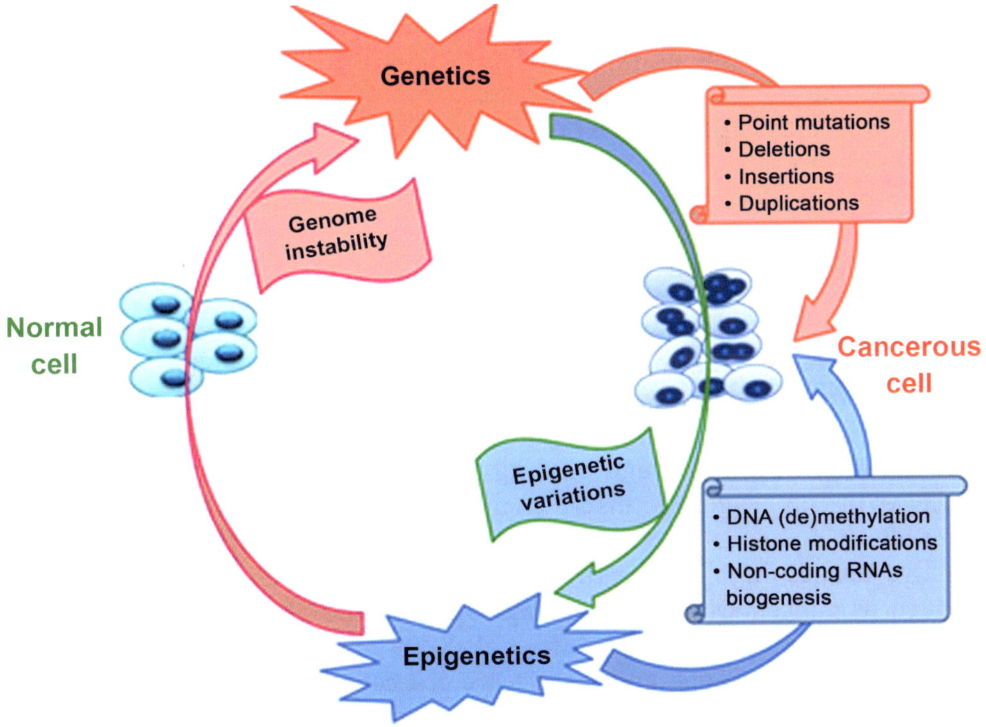

Figure 9.17. Genetic and epigenetic perspectives of cell division.

Source: https://www.researchgate.net/publication/330477164/figure/fig1/AS:716427300454400@1547820987196/Genetic-and-epigenetic-perspectives-of-cell-division-under-the-changing-environmental.png.

This connection of genetics to the physical manifestation of cellular behavior illuminates the extent to which cellular processes are regulated for survival and functioning in various environments, further verifying the importance of genetic foundations in the fundamental process of cell division (Wu et al., 2018). Further investigation into the role of genetic components in cell division reveals contact inhibition, an essential regulatory process that maintains normal tissue architecture and function. This effect, which is typical of epithelial cells, is a fine example of how mechanical restrictions and genetic programming may come together to regulate cell growth. Cells cease dividing at confluence due to contact inhibition, a process induced by a combination of reduced cell motility as well as genetic regulation of the cell cycle. The dynamics of cell area impacting mitotic rates is further evidence for this relationship, where one of the important thresholds of cell size triggers the cessation of proliferation. This sensitive interaction between genes and environmental conditions underscores the need to understand how genetics prescribes cellular response to mechanical stress in phenomena like contact inhibition (Puliafito et al., 2011). Moreover, the extension of the effects of genetic influences from straightforward cell division to understanding pathological states , such as cancer, is a significant aspect. Abnormal control of cell division may lead to uncontrolled proliferation, a feature of tumorigenesis. Understanding the genetic factors that control normal cell cycle regulation has significant ramifications in the study of cancer, particularly in the search for targets of therapeutic intervention. The relationship between mechanisms of cell division and genetic mutations has grave implications, demonstrating how genetic factors not only control the functioning of cells in a healthy condition but also create states of disease. An understanding of such genetic factors is essential to design effective mechanisms to combat cancers, where disruption of cell proliferation assumes a prominent position. This link of genetic determinants to the mechanics of cell division is paramount to the comprehension of complex biological phenomena (Puliafito et al., 2011; Wu et al., 2018) (Figure 9.17).

9.5.3. Role of Hormones in Cell Division

The complex mechanism of cellular division is regulated by several hormones, which assist cellular growth to occur in a structured and chronological manner. Steroid hormones, , such as testosterone and estrogen, are significant in regulating cell cycle advancement. For instance, regulation of gene expression necessary for transition through phases of the cell cycle is established through these hormones. They accomplish this through binding to receptors that then get in contact with the cell's DNA to trigger or inhibit gene transcription required for replication and cell division. This steroid hormone effect is particularly vital in tissues that need growth to happen rapidly, , such as the reproductive tissues, highlighting the pivotal function of steroid hormones in physiology and disease, , such as cancer, where the dysregulation of the above processes can lead to cell division becoming uncontrollable (Maheshwari et al., 2009). Aside from steroid hormones, peptide hormones , such as growth hormone and insulin are critical to the mediation of cell proliferation and division. For instance, insulin facilitates glucose uptake and energy storage, thus providing the substrates needed for cellular proliferation and metabolism. Furthermore, insulin has been found to enhance the cell growth-related signaling pathways, particularly via PI3K/Akt pathway activation, which functions to trigger cell cycle progression. Growth hormone, on the contrary, controls insulin-like growth factors (IGFs) production, which then triggers cellular growth and division via receptor activation triggering distinct signaling cascades important in mitosis. Therefore, these peptide hormones not only maintain the metabolic status of the cell but also directly affect regulation of mechanisms of cell division and growth (Dutta et al., 2014). Coordination between cells and hormones emphasizes the necessity for physiological balance in proper cellular function. Hormonal imbalance may lead to inappropriate cell division, with serious implications on health and disease. For example, overabundance of estrogen has been cited as a cause of some cancers, e.g., breast cancer, because high levels of it might lead to the growth of cells uncontrollably. Moreover, disruptions in growth hormone signaling can lead to growth disorders or acromegaly due to overstimulation of the growth pathways. It is thus important to comprehend the hormonal regulation of cell division since it offers hints towards therapeutic intervention for disease that is characterized by abnormal cell proliferation. Interference with certain pathways of hormones may be beneficial to induce normal control of the cell cycle and avert the impacts of disease based on cell dysregulation (Maheshwari et al., 2009).

Test Your Knowledge

1. What are the genetic factors?
2 What is the role of hormones in cell division?

9.6. Disorders Related to Cell Division

Learning Outcome

1. Cancer and uncontrolled cell division;
2. Genetic disorders linked to meiosis;
3. Implications of cell division disorders on health.

Cell division is a critical process in biology that facilitates growth, development, and general maintenance of the organism. Abnormalities in cell division may lead to severe health disorders, ranging from cancer to developmental disorders. One of the most important elements of cell division is the centrosome, which is the primary microtubule-organizing center in dividing cells. Not only does the centrosome form the mitotic spindle, which is essential in separating chromosomes, but it also regulates numerous cellular functions, , such as adhesion and signaling (Bettencourt-Dias et al., 2013). Centrosome malfunction can lead to defective cell division, which is responsible for ailments like aneuploidy, where cells have an unusual number of chromosomes. Information on the role of centrosomes informs us about the pathogenesis of various disorders of cell division and suggests possible therapeutic interventions. Yet another significant category of cell division-associated disorders arises as a consequence of genetic mutations affecting key enzymes and proteins controlling a cell. For example, mutations of proteins that govern the checkpoints of the cell cycle may lead to uncontrolled cell growth seen in cancer. Furthermore, there are also uncommon genetic diseases caused by mutations in genes , such as *SC4MOL*, associated with cholesterol biosynthesis, and correlated with severe health conditions , such as congenital cataracts and developmental delay (Vallejo et al., 2008). In patients with *SC4MOL* mutations, increased rates of mitosis in cholesterol-deprived conditions demonstrate a complex relationship between lipid metabolism and cell cycle. Thus, understanding these underlying genetic factors brings to light the interface between cellular processes and metabolic pathways that provides an integrative view of how alterations in cell division manifest as clinical phenotypes.

Ultimately, the implications of impaired cell division are not restricted to individual health but to public health issues and therapeutic strategies. Conditions , such as cancer emphasize the critical necessity for targeted therapies to avoid the consequences of abnormal cell division. There has also been further advancement in molecular biology and genetics that has opened new windows of insight into the cellular processes that lead to cancer and other cell division disorders. Moreover, the research on methyl sterols has demonstrated that there is a relationship between cholesterol synthesis and immune response, unveiling new means by which cell division could be controlled (Vallejo et al., 2008). As studies on these conditions and their causative mechanisms go on, there is potential for the development of more efficient and improved treatments, eventually resulting in improved patient outcomes and relief of the effects of these conditions on society (Figure 9.18).

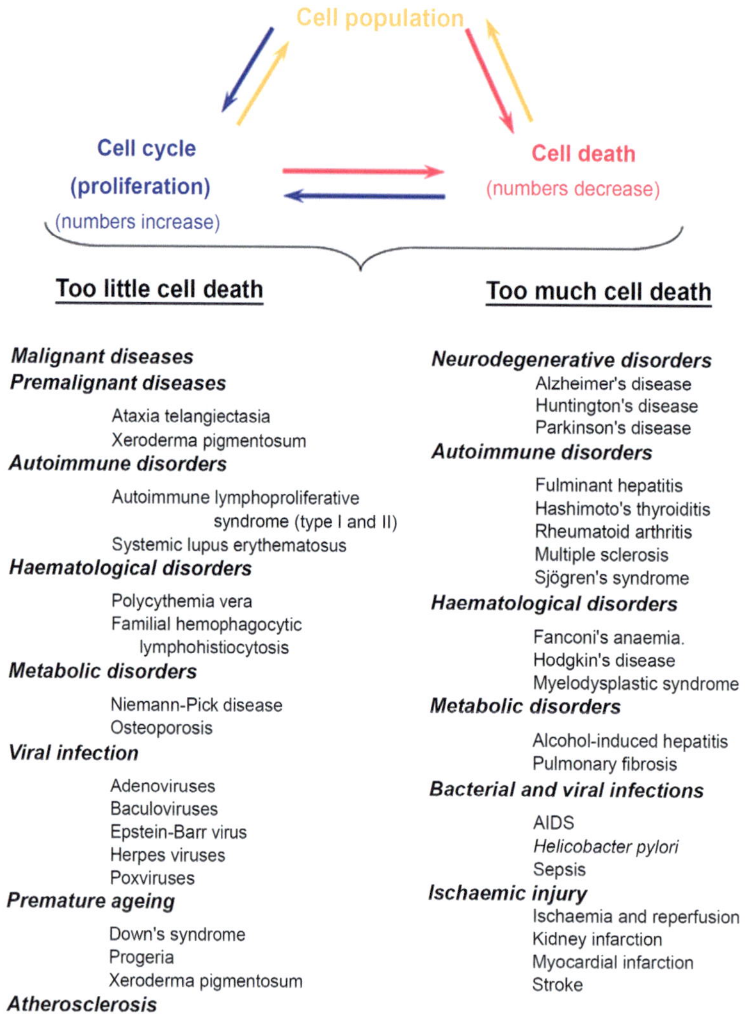

Figure 9.18. Disorders related to cell division.

Source: https://onlinelibrary.wiley.com/cms/asset/cca4b513-e6a1-41c8-a829-b7af48753704/joim_2282_f1.gif.

9.6.1. Cancer and Uncontrolled Cell Division

The cancer process can be essentially summarized as an imbalance of the regular cell division processes. Under normal circumstances, the cell cycle is finely tuned to ensure that it allows for effective growth, development, and apoptosis, yet maintains cellular proliferation under control. But various factors like genetic mutations and external forces like viruses have the potential to override the control processes.

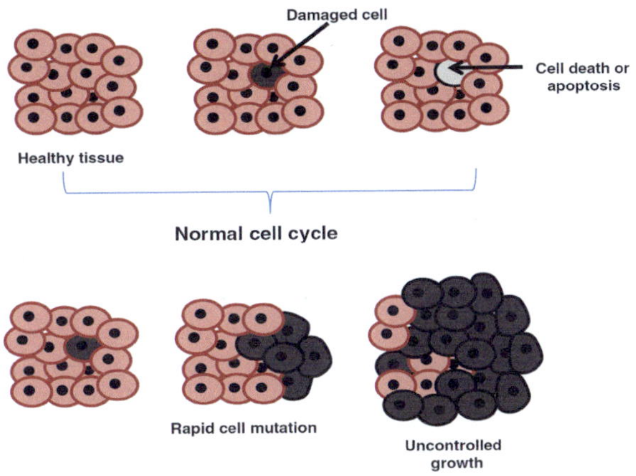

Figure 9.19. Cancer and uncontrolled cell division.

Source: https://www.researchgate.net/profile/Tamalika-Das-2/publication/269098315/figure/fig2/AS:614147964276745@1523435693678/Normal-cell-cycle-and-uncontrolled-division-in-cells-leading-to-formation-of-tumors.png.

For instance, certain oncogenes become active, which leads to overexpression of proteins that promote cell division, whereas tumor suppressor genes, like p53, which play the role of preventing cell growth or inducing apoptosis, are rendered inactive. This disruption creates the platform for cells to divide beyond their normal limit, giving rise to tumors, which, if not controlled sufficiently, will develop into malignancy. Viruses, prominent characters in the cancer drama, hijack the machinery of the host cell to propagate themselves. These intracellular parasites can hijack the hosts cellular machinery to control cell cycle regulation machinery of the cell to propel cell division day in and day out, thereby propagating themselves. For instance, viral proteins can interfere with the signal pathway transducing the GF instructions to the CDKs controlling cell cycle progression. Therefore, the infected cells evade the checks and balances imposed by the regulatory systems of the body, establishing uncontrolled growth. Further, because viruses induce adjacent cells to go through a similar phase of abrogated control, this results in an avalanche effect that can cause the diffuse growth of malignant tumors, establishing an evident connection between viral activity and cancer development (Mitin, 2011). The implications of unregulated cell growth in cancer highlight the critical need for effective therapeutic strategies. Natural products, , such as resveratrol, have been the focus of recent research with potential activity in manipulating cancer cell growth patterns. Results indicate that resveratrol can influence central processes implicated in cell cycle control, triggering apoptosis in cancer cells and inhibiting viability and proliferation. Especially, this compound has proved to be successful in breast and ovarian cancer models, arresting the cell cycle and reducing tumor mass through several mechanisms, including the regulation of cyclins and CDKs (Andreato et al., 2024). With its anti-inflammatory activity and capacity against oxidative stress, resveratrol presents a viable option within the adjunctive cancer treatment that emphasizes the potential for future investigation of dietary interventions that can modulate cell growth and minimize cancer risk effectively (Figure 9.19).

9.6.2. Genetic Disorders Linked to Meiosis

Meiosis is the key process of sexual reproduction that ensures genetic diversity through the re-shuffling of chromosomes. But errors in meiosis can lead to unreserved genetic disease. One of the most well-known consequences of meiotic error is aneuploidy, which produces an abnormal number of chromosomes. Conditions , such as Down syndrome, resulting from an extra copy of chromosome 21, are typically the consequence of errors in chromosome segregation during oogenesis. Most aneuploid fetuses have been discovered to be the consequence of maternal chromosome segregation defects during the critical periods of meiotic chromosome pairing and recombination that take place between early fetal development and ovulation.

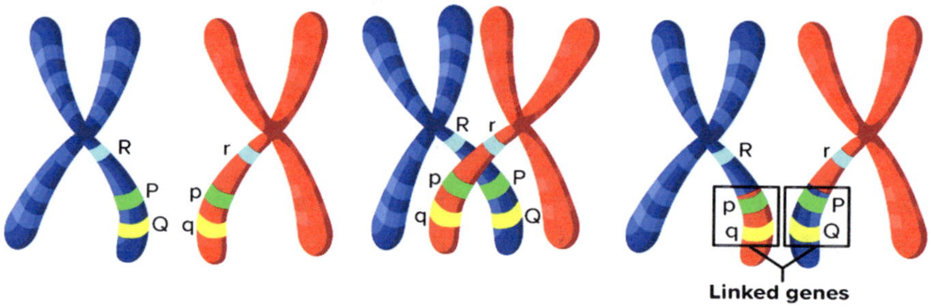

Figure 9.20. Genetic disorders linked to meiosis.

Source: https://dr282zn36sxxg.cloudfront.net/datastreams/f-d%3A50513830ae2ff1bca8b60b4caf3d8abf74731bb2e2676565abb7ac0b%2BIMAGE_TINY%2BIMAGE_TINY.1.

This understanding highlights the need to keep investigating the mechanisms for such errors because increasing our knowledge can help refine reproductive technology and prenatal genetic screening (Dyban et al., 2010). Meiotic errors not only lead to aneuploidies but also account for a number of individual genetic diseases with wide implications in the clinic. Certain metabolic disorders, for instance, have come to be recognized as being linked to meiotic inaccuracies. One such compelling example is the inborn errors of cholesterol biosynthesis, in which *SC4MOL* mutations have been reported to be associated with congenital features , such as cataracts and developmental delay. The relationship between genetic abnormality and meiotic defects reveals a larger scenario wherein enzyme defects disrupt essential metabolic processes, citing the complex interplay between cell division and genetics. Further study of such disorders would uncover further details on how genetic regulation and metabolic health interact, eventually influencing therapeutic strategies (Vallejo et al., 2008). Meiotic mistakes may also have significant post-zygotic consequences that contribute to the initial genetic errors. Following fertilization, the resulting embryo undergoes a series of rapid cell divisions, and it is in these early few mitotic divisions that additional mistakes occur. Consequently, initial stages of development play a crucial role in the determination of offspring health. Empirical data indicate that mitotic errors can interact with meiotic inaccuracies, and it is suspected that both procedures may influence the genetic landscape of an individual. Moreover, unraveling the nature of these interactions will be crucial to practices , such as assisted reproduction, where understanding potential genetic disorders presented by meiotic errors may inform clinical interventions and lead to better outcomes for the infertile patient. By synthesizing the work on meiosis and the disruptions in it, scientists can pool efforts towards novel approaches to genetic disorders (Dyban et al., 2010) (Figure 9.20).

9.6.3. Implications of Cell Division Disorders on Health

Cell division is a fundamental biological process that, when altered, can give rise to many health issues. Disorders of cell division , such as cancer occur as a consequence of aberrations in the cell processes that regulate proliferation. Cancer, characterized by uncontrolled proliferation of cells, can be used as an illustration of the serious health implications of cell division disorders. Because the cells no longer adhere to standard division procedures, the resultant tumors become invasive in adjacent tissues and migratory to distant organs. Not only does this account for increased morbidity and mortality but also presents an extremely heavy load on healthcare systems globally. For that reason, the details of cell division are vital to knowing how to design effective interventions halting dysregulations of the involved mechanisms, which is why it is crucial to keep abreast of research and clinical interest in this area. Moreover, the mutation of genes associated with cell division processes can cause many inherited diseases, which further signifies the significance of this biological process. These are the mutations that result in diseases like Down syndrome and Turner syndrome, where there is abnormal duplication or segregation of the chromosomes during cell division. These diseases can have severe impacts on an individual's physical and intellectual well-being, , such as influencing lifespan, patterns of growth, and overall health. In the past few years, development in high-throughput technologies has been successful in allowing for the investigation of genomic variation related to these diseases and offering insights into their pathobiology and therapeutic potential. Through continued investigation, the journey towards personalized treatments for every patient's genomic makeup is being made increasingly likely, hence enhancing the prospect of improved health outcomes within suffering populations (Oh et al., 2015). The scope of cell division disorders' implications extends beyond genetic mutation and oncogenic alteration to encompass epigenetic changes that affect gene expression and cell behavior. It has been shown by studies that environmental factors and lifestyle choices have the capacity to alter the epigenetic landscape, resulting in alterations that lead to derailing normal patterns of cell division. Such convergence of genetics and the environment indicates the complexity of cell division disorder health impacts. For instance, stress, poisons, and nutritional factors can cause epigenetic changes predisposing an individual to conditions like obesity and diabetes, both of which are increasingly viewed as having cellular proliferation components. Knowledge of how genetic predispositions combine with extrinsic factors allows health experts to maximize prevention planning and intervention to lower the long-term impacts associated with cell division mistakes (Oh et al., 2015).

Test Your Knowledge

1. What is cancer and uncontrolled cell division?
2. What are the implications of cell division disorders on health?

Summary

The complex cell division process is not merely a replication process but is itself a subsidiary column in the establishment and upkeep of biology. Cell division is closely connected with implications on broader fields like developmental biology and oncology. Cell differentiation during division, as indicated by experiments involving *Volvox carteri*, shows how variation in cell size affects various cell fates and therefore development paths. Research highlights that larger cells have a higher tendency to form reproductive cells, and the importance of size in the context of genetic expression and specification. Such findings advance our knowledge of not only cellular function but evolutionary significance of cell division in multicellular organisms and its application in the research of developmental malformations and diseases. Furthermore, the importance of understanding cell division extends far into the realm of cancer research, where dysregulation of such a process is a causative agent of malignancy. The Cancer Genome Atlas project demonstrates the multi-faceted complexity of genomic alterations in cancer and how alteration in cellular processes can lead to uncontrolled cell growth and tumorigenesis. This detailed overview recognizes critical mutations, , such as TP53 and PIK3R1, which are critical to cancer formation. By examining the interactions between genetic mutations and epigenetic alterations, scientists can better tailor treatment plans. This emphasis on cooperative genetic influences during cell division provides potential avenues of therapeutic treatment and highlights how incorrect cell division can establish life-threatening situations, thus supporting the relevance of studying this biological process. Lastly, the convergence of basic knowledge about cell division and its applications within various biological disciplines creates salient questions regarding the future of cellular science and the discipline of medicine. With ongoing scientific exploration of the complexities that facilitate cell division, most importantly its role in carcinogenesis and tissue development, the need for an interdisciplinarity approach becomes increasingly obvious. By combining insights from developmental biology, genetics, and tumor and oncology research, e.g., findings from studies on *Volvox carteri* and the data yielded by TCGA, scientists can achieve a deeper appreciation of the cellular mechanism that is responsible for life. The continued investigation of cell division holds not only the potential for furthering our biological understanding but also for guiding pioneering therapeutic approaches that may drastically enhance health outcomes in diseases that have evolved from cellular malfunction, ultimately leading to a holistic strategy for biological research and medical innovation.

Cell division is a simple biological process that ensures the growth, development, and survival of all living beings. Mechanisms of cell division are complex molecular events with regulatory proteins governing various stages like mitosis and cytokinesis. Satellite cells, the muscle stem cells, have been highlighted in recent studies for their role in tissue repair after injury. The ability of these cells to move is essential in allowing rapid regeneration. Importantly, satellite cells display unique migratory characteristics of defined phases of movement where they initially migrate through the basal lamina and then have an increase in speed on the myofiber surface. Such findings indicate the complexity of cellular motion during division and differentiation and pave the way for future interventions aimed at stimulating tissue regeneration and functional recovery. At the center of our understanding of cell division is the concept of the mitochondrial DNA (mtDNA) bottleneck that controls the passing of mitochondrial diseases. The process of the bottleneck occurs during development of the cell, in which a certain number of mtDNA molecules are distributed randomly during cell divisions. This process affects the population of mtDNA worldwide, leading to a possible loss of mtDNA copy number and eventually influencing cellular function. There has been a new theoretical model proposed in recent times that models mtDNA population dynamics in support of the hypothesis that the bottle-neck mechanism incorporates elements of binomial partitioning and decay. Such a model not only provides statistical support for analyzing mtDNA inheritance but also aids us in refining our predictive capacity with regard to mtDNA expression of disease and diagnosis, pointing out the critical relationship between cell division and genetic stability. The implications of such research are profound beyond the borders of scholarly interest; they are of vital importance to clinical practice and therapeutic protocols. Satellite cell migratory behavior has been determined to be an important factor in devising successful muscle regeneration therapy, particularly in degenerative skeletal conditions. Furthermore, insight obtained through the mtDNA bottleneck process can be applied to inform clinical sampling strategies and interventions for mtDNA disease amelioration. With the marriage of theoretical models and experimental data, researchers have begun to assemble an integrated system that is likely to make future treatments more effective. With our better understanding of the fundamental processes that regulate cell division, the promise of new approaches derived from the individual patient's unique genetic background increases, ushering in new frontiers for medical advances in regenerative medicine and genetic disease.

KEY TERMS FROM THIS CHAPTER

Anaphase	Cell cycle	Centrioles
Centromere	Chromatid	Chromosome
Crossing over	Cytokinesis	G1 phase
G2 phase	Homologous chromosomes	Interphase
M phase (mitotic phase)	Meiosis	Metaphase
Mitosis	Prophase	S phase
Spindle fibers	Telophase	

Critical Thinking Questions

1. **What is the main purpose of mitosis?**
   a. To produce sex cells
   b. To reduce the chromosome number
   c. To create identical daughter cells for growth and repair
   d. To increase genetic variation
2. **During which phase of the cell cycle does DNA replication occur?**
   a. M phase
   b. G1 phase
   c. S phase
   d. G2 phase
3. **What is the correct order of the phases of mitosis?**
   a. Prophase → Metaphase → Anaphase → Telophase
   b. Interphase → Prophase → Telophase → Metaphase
   c. Metaphase → Prophase → Anaphase → Telophase
   d. Prophase → Anaphase → Metaphase → Telophase
4. **Meiosis results in the formation of:**
   a. Two diploid cells
   b. Two identical cells
   c. Four haploid cells
   d. Four diploid cells
5. **Which structure helps separate chromosomes during cell division?**
   a. Nucleus
   b. Cell wall
   c. Spindle fibers
   d. Chloroplast

Short Answer Questions

1. What are the main stages of the cell cycle?
2. What is the difference between mitosis and cytokinesis?
3. List the phases of mitosis in order and describe what happens in each.
4. Why is DNA replication important before a cell divides?
5. How does mitosis ensure that daughter cells are genetically identical to the parent cell?
6. What is the purpose of meiosis, and how is it different from mitosis?

Answers to Critical Thinking Questions

1. (c); 2. (c); 3. (a); 4. (c); 5. (c)

References

1. Aiken, J. M., Apostolova, N., Bacman, S. R., Beaumont, C., Bergstrom, C., Bogenhagen, D. F., Brand, M. D., et al. (2015). Stochastic modelling, Bayesian inference, and new *in vivo* measurements elucidate the debated mtDNA bottleneck mechanism. https://core.ac.uk/download/209204724.pdf (accessed on 13th June 2025).
2. Wu C, Al Mamun AAM, Luong TT, Hu B, Gu J, Lee JH, D'Amore M, Das A, Ton-That H. Forward Genetic Dissection of Biofilm Development by Fusobacterium nucleatum: Novel Functions of Cell Division Proteins FtsX and EnvC. mBio. 2018 Apr 24;9(2). doi: 10.1128/mBio.00360-18. PubMed PMID: 29691334; PubMed Central PMCID: PMC5915739. (accessed on 13th June 2025).
3. Amle, V. S., Dudhate, G. S., Jha, R., Keshamma, E., et al. (2022). Breast cancer treatment relying on herbal bioactive components. https://core.ac.uk/download/597862211.pdf (accessed on 13th June 2025).
4. Amon, A., Bardin, A., Ciliberto, A., Goldbeter, A., Skirycz, A., Brown, A. J. P., Dahmann, C., et al. (2013). From START to FINISH: The influence of osmotic stress on the cell cycle. https://core.ac.uk/download/19888927.pdf (accessed on 13th June 2025).
5. Andreato, C., Fett, C. A., Fett, W. C. R., Kulpe, R., et al. (2024). The use of resveratrol in the fight against cancer in women: A review. https://core.ac.uk/download/620555447.pdf (accessed on 13th June 2025).
6. Antia, A., LaRoche, J., & Müller, M. (2008). Influence of cell cycle phase on calcification in the coccolithophore *Emiliania huxleyi*. https://core.ac.uk/download/11896095.pdf (accessed on 13th June 2025).
7. Araujo, A. R., Gelens, L., Santos, S., Sheriff, R. S., et al. (2016). Positive feedback keeps duration of mitosis temporally insulated from upstream cell-cycle events. https://core.ac.uk/download/77018775.pdf (accessed on 13th June 2025).
8. Bai, S. N. (2015). The concept of the sexual reproduction cycle and its evolutionary significance. https://core.ac.uk/download/pdf/82850883.pdf (accessed on 13th June 2025).
9. Sona Valuchova, Pavlina Mikulkova, Jana Pecinkova, Jana Klimova, Michal Krumnikl, Petr Bainar, Stefan Heckmann, Pavel Tomancak, Karel Riha (2020) Imaging plant germline differentiation within Arabidopsis flowers by light sheet microscopy eLife 9:e52546 (accessed on 13th June 2025).
10. Oh, S., Ji, H., Barzman, D., Lin, P. I., & Hutton, J. (2015). Pediatric asthma and autism-genomic perspectives. *Clinical and translational medicine*, *4*(1), 37. https://doi.org/10.1186/s40169-015-0078-x (accessed on 13th June 2025).
11. Bender, A., Bond, P. J., Gergely, F., Jodrell, D. I., et al. (2013). Design, synthesis, and biological evaluation of an allosteric inhibitor of HSET that targets cancer cells with supernumerary centrosomes. https://core.ac.uk/download/42364306.pdf (accessed on 13th June 2025).
12. Bettencourt-Dias, M. (2013). Q&A: Who needs a centrosome? https://core.ac.uk/download/74283984.pdf (accessed on 13th June 2025).
13. Bhattacharjee, S., Bryer, C., Feeney, L., Lorenz, C., et al. (2013). MHF1–2/CENP-S-X performs distinct roles in centromere metabolism and genetic recombination. https://core.ac.uk/download/16701382.pdf (accessed on 13th June 2025).

14. Bich L, Pradeu T and Moreau J-F (2019) Understanding Multicellularity: The Functional Organization of the Intercellular Space. Front. Physiol. 10:1170. doi: 10.3389/fphys.2019.01170 (accessed on 13th June 2025).

15. Andreas Hüttemann & Marie I. Kaiser (kaiser.m@uni-koeln.de) Published in: Engelhard, K./ Quante, M. (2018) (eds.): Handbook of Potentiality. Dordrecht: Springer, 401-427. (accessed on 13th June 2025).

16. Cancer Genome Atlas Research Network. (2008). Comprehensive genomic characterization defines human glioblastoma genes and core pathways. https://core.ac.uk/download/323062204.pdf (accessed on 13th June 2025).

17. Cande, W. Z. (2007). W. Zacheus Cande: Evolutionary biologist in cell biologist's clothing. Interview by Ruth Williams. https://core.ac.uk/download/323081909.pdf (accessed on 13th June 2025).

18. Cande, W. Z., McDonald, K., & Meeusen, R. (1981). A permeabilized cell model for studying cell division: A comparison of anaphase chromosome movement and cleavage furrow constriction in lysed PtK1 cells. https://core.ac.uk/download/323082060.pdf (accessed on 13th June 2025).

19. Carrel, A., Comfort, A., Götte, A., Weismann, A., et al. (1982). Cytogerontology since 1881: A reappraisal of August Weismann and a review of modern progress. https://core.ac.uk/download/12170010.pdf (accessed on 13th June 2025).

20. Cho, H. S., Cho, H., & Hwang, I. (2018). *In vivo* action of RNA G-quadruplex in phloem development. https://core.ac.uk/download/222576034.pdf (accessed on 13th June 2025).

21. Chowdhury, S., Cooper, J., Das, S., Ellis, L., Forsythe, G., et al. (2010). Cell-cycle synchronization of bloodstream forms of *Trypanosoma brucei* using Vibrant Dye Cycle Violet-based sorting. https://core.ac.uk/download/28964409.pdf (accessed on 13th June 2025).

22. Collins-Hooper, H., Dash, P. R., Otto, A., Patel, K., et al. (2011). Adult skeletal muscle stem cell migration is mediated by a blebbing/amoeboid mechanism. https://core.ac.uk/download/9077.pdf (accessed on 13th June 2025).

23. De Storme, N., Geelen, D., Pawlowski, W., Tiang, C. L., et al. (2014). Chromosome segregation in plant meiosis. https://core.ac.uk/download/55854046.pdf (accessed on 13th June 2025).

24. Wuyts, N., Dhondt, S., & Inzé, D. (2015). Measurement of plant growth in view of an integrative analysis of regulatory networks. *Current opinion in plant biology*, *25*, 90–97. https://doi.org/10.1016/j.pbi.2015.05.002 (accessed on 13th June 2025).

25. Dong, J., Hu, X., Jiang, M., Le, B., et al. (2019). A conserved but plant-specific CDK-mediated regulation of DNA replication protein A2 in the precise control of stomatal terminal division. https://core.ac.uk/download/227040922.pdf (accessed on 13th June 2025).

26. Donoghue, P. C. J., Li, J., Liu, P., Sun, Y., et al. (2021). Ultrastructure and *in-situ* chemical characterization of intracellular granules of embryo-like fossils from the early Ediacaran Weng 'an biota. https://core.ac.uk/download/492594252.pdf (accessed on 13th June 2025).

27. Dutta, S. (2014). Steroid hormone regulation of fetal mouse oocyte development. https://core.ac.uk/download/215700118.pdf (accessed on 13th June 2025).

28. Dyban, A., Kuliev, A., Mantzouratou, A., Pujol, A., Barlow, A. L., Gutierrez-Mateo, C., et al. (2010). Errors in chromosome segregation during oogenesis and early embryogenesis. https://core.ac.uk/download/45628.pdf (accessed on 13th June 2025).

29. Fletcher, A. G., Gavaghan, D. J., Maini, P. K., Osborne, J. M., et al. (2013). Implementing vertex dynamics models of cell populations in biology within a consistent computational framework. https://core.ac.uk/download/pdf/18452024.pdf (accessed on 13th June 2025).

30. Flick, K., Kaiser, P., Ouni, I., Su, N. Y., et al. (2006). The yeast ubiquitin ligase SCF(Met30): Connecting environmental and intracellular conditions to cell division. https://core.ac.uk/download/pdf/4058971.pdf (accessed on 13th June 2025).

31. Franz, A. (2012). Role of the ubiquitin-selective CDC-48/UFD-1/NPL-4 chaperone in DNA replication. https://core.ac.uk/download/12011230.pdf (accessed on 13th June 2025).

32. Fujiwara, T., Kuroda, S., & Urata, Y. (2012). Ataxia-Telangiectasia Mutated and the Mre11-Rad50-NBS1 complex: Promising targets for radio sensitization. https://core.ac.uk/download/12544038.pdf (accessed on 13th June 2025).

33. Gartner, A., Craig, A., Baldinger, T., Blow, J. J., Brauchle, C., Brenner, S., et al. (2012). The dynamics of replication licensing in live *Caenorhabditis elegans* embryos. https://core.ac.uk/download/20452176.pdf (accessed on 13th June 2025).

34. Harris, P. A., Heaton, B., Lamb, J., Wheatley, D. N., et al. (2002). Possible attenuation of the G2 DNA damage cell cycle checkpoint in HeLa cells by extremely low frequency (ELF) electromagnetic fields. https://core.ac.uk/download/60220.pdf (accessed on 13th June 2025).

35. Heyer, W. D., Li, X., Rolfsmeier, M., Zhang, X., et al. (2006). Rad54: The Swiss Army knife of homologous recombination? https://core.ac.uk/download/323070919.pdf (accessed on 13th June 2025).

36. Kaya, M. E. (2019). Engineering cellular communication systems for synthetic biology. https://core.ac.uk/download/224643433.pdf (accessed on 13th June 2025).

37. Kirk, D. L., Kirk, M. M., McRae, S. E., Ransick, A., et al. (1993). The relationship between cell size and cell fate in *Volvox carteri*. https://core.ac.uk/download/4877829.pdf (accessed on 13th June 2025).

38. Kukkonen-Macchi, A. (2010). Functional characterization of proteins required for mitotic progression and the spindle assembly checkpoint. https://core.ac.uk/download/197989714.pdf (accessed on 13th June 2025).

39. Lachance, Marc-André and Walker, Graeme M (October 2018) Yeasts. In: eLS. John Wiley & Sons, Ltd: Chichester,(accessed on 13th June 2025).

40. MacLeod, A., Tait, A., & Turner, C. M. R. (2007). The system of genetic exchange in *Trypanosoma brucei* and other trypanosomatids. https://core.ac.uk/download/1394862.pdf (accessed on 13th June 2025).

41. Maheshwari, A., Baker, T. G., Donovan, P., Gumienny, T., Konishi, I., Fowler, P. A., Pinkerton, J., et al. (2009). Primordial follicular assembly in humans: Revisited. https://core.ac.uk/download/9631230.pdf (accessed on 13th June 2025).

42. Matthews, K. R. (2005). The developmental cell biology of *Trypanosoma brucei*. https://core.ac.uk/download/28964420.pdf (accessed on 13th June 2025).

43. McCulloch, R., & Navarro, M. (2016). The protozoan nucleus. *Molecular and biochemical parasitology*, *209*(1-2), 76–87. https://doi.org/10.1016/j.molbiopara.2016.05.002 (accessed on 13th June 2025).

44. Militello, Guglielmo (2021). *Structural and organisational conditions for the appearance of*

a functionally integrated organisation in the transition from prokaryotic to eukaryotic cell. Dissertation, Universidad Del Pais Vasco (accessed on 13th June 2025).

45. Mitin, V. (2011). Cancer is a side effect of evolution of viruses and bacteria. https://core.ac.uk/download/pdf/290263.pdf (accessed on 13th June 2025).
46. Prendin, A. L., Castagneri, D., von Arx, G., Giai Petit, G., & Carrer, M. (2017). Retrospective analysis of wood anatomical traits reveals a recent extension in tree cambial activity in two high-elevation conifers. https://core.ac.uk/download/84920771.pdf (accessed on 13th June 2025).
47. Puliafito, A., Sigal, A., Abercrombie, M., Shraiman, B. I., Castor, T. P., Chaffer, C. L., et al. (2011). Collective and single cell behavior in epithelial contact inhibition. https://core.ac.uk/download/4893713.pdf (accessed on 13th June 2025).
48. Risler, T. (2009). Cytoskeleton and cell motility. http://arxiv.org/abs/1105.2423 (accessed on 13th June 2025).
49. Truong, T. V., & Supatto, W. (2011). Toward high-content/high-throughput imaging and analysis of embryonic morphogenesis. *Genesis (New York, N.Y. : 2000)*, *49*(7), 555–569. https://doi.org/10.1002/dvg.20760 (accessed on 13th June 2025).
50. Tvegård, T. (2008). Effects of UV light on cell cycle regulation in fission yeast. https://core.ac.uk/download/30832286.pdf (accessed on 13th June 2025).
51. Vallejo, A., Vockley, G., Hoover, J., Michel, J., Ferris, L., Kratz, L., He, M., et al. (2008). Mutations in the *SC4MOL* gene encoding a novel methyl sterol oxidase cause autosomal recessive psoriasis form dermatitis, microcephaly and developmental delay. https://core.ac.uk/download/pdf/287971.pdf (accessed on 13th June 2025).
52. Widger, A. D. (2018). Ablating ATR in mouse meiosis and its consequences for synapsis, recombination and meiotic surveillance mechanisms. https://core.ac.uk/download/154747438.pdf (accessed on 13th June 2025).
53. Witzany, G. (2007). Bio-communication of plants. Publisher: Springer. Editor: Guenther Witzany, Frantisek Baluska. ISBN: 978-3-642-23523-8

MEIOSIS AND INHERITANCE

10

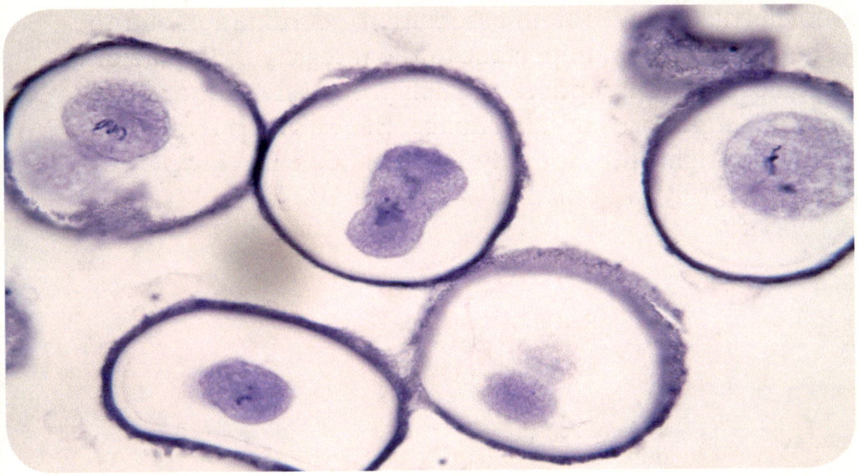

LEARNING OBJECTIVES

After studying this chapter, you will be able to:

1. Explain the process of meiosis;
2. Discuss about genetic variation;
3. Define gametogenesis;
4. Describe mendelian genetics;
5. Explore the non-mendelian inheritance;
6. List the applications of meiosis and inheritance.

Introduction

Knowledge of meiosis is key to the topic of inheritance since it is a unique type of cell division that leads to chromosome reduction during gamete production. Meiosis is crucial in generating genetic diversity through crossing over and independent assortment. These processes not only produce gametes with varying genetic makeups but also ensure that offspring inherit a mix of traits from their parents. This genetic variation is vital to evolution, allowing populations to change in response to changing conditions through generations. Meiosis is, therefore a cornerstone to interpreting patterns of heredity and trait inheritance, emphasizing the beautiful ballet between genetic information from one generation to the next. Moreover, the cellular complexities of meiosis reflect an exquisite choreography of molecular mechanisms governing gamete development. Cellular organization during meiosis involves a cascade of regulatory mechanisms that enable diploid to haploid gamete conversion. For instance, current studies have revealed that during budding yeast meiotic differentiation, mitochondrial dynamics are important as these organelles also undergo morphogenetic changes. The regulation of mitochondrial release from the cell cortex illustrates a strong level of control over organelle structure in response to meiotic signals. This level of coordination highlights not just the importance of segregation of genetic information but also the importance of organelle dynamics to overall meiotic success, highlighting the integration of cellular function and heritance (Berchowitz et al., 2018). In addition, the asymmetric nature of meiotic divisions yields special outcomes that contribute significantly to reproductive plans. In women's meiosis, as an instance, the formation of one large egg and little polar bodies is a result of the unique mechanics of meiotic cell division, distinct from mitosis mechanisms. Asymmetric divisions ensure most of the material in the cytoplasm goes to the egg, enhancing its potential for fertilization. Protein association, coordination and the assembly of the contractile ring during such divisions are important to the integrity of the meiotic spindle and proper segregation of the chromosomes. Disruptions in such processes lead to gamete formation abnormalities and consequently affect patterns of inheritance. Hence, an understanding of the molecular mechanism of meiosis gives insight into not only basic biological principles but also the human fertility implications and health (Flynn et al., 2017) (Figure 10.1).

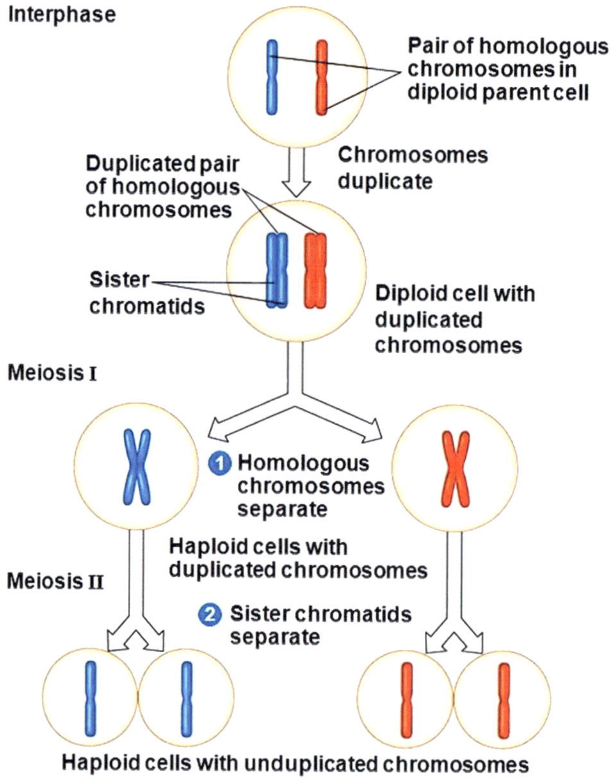

Figure 10.1. Meiosis and inheritance diagram.

Source: https://o.quizlet.com/FD08mWvZFHSg2sGzDC7TiQ_b.jpg.

Definition of Meiosis

The term meiosis refers to a characteristic of cell division employed in sexual reproduction. It reduces the number of chromosomes by half, generating four genetically distinct gametes from one diploid progenitor cell. It begins with a diploid cell dividing in two successive phases: meiosis I and meiosis II. In meiosis I, homologous chromosomes are divided into two different daughter cells, whereas meiosis II is quite similar to mitotic division, dividing sister chromatids. The outcome of meiosis not only explains genetic variation but also ensures species-specific chromosome number integrity from one generation to the next, laying down the basis of understanding inheritance patterns. Such variations are achieved through crossing over and independent assortment, crucial processes that enhance genetic variation in populations. Meiosis's role in inheritance is expressed through its complex complexity. This process enables genetic recombination to occur, with pieces of DNA being exchanged between homologous chromosomes, thereby creating new combinations of alleles. As stated, the instrumental gene theory is a valuable tool in determining how genotypes lead to phenotypic traits, supporting experimental research that connects genotype to observable traits. The relevance of genetic transfer in meiosis has far-reaching implications across fields , such as evolutionary biology and population genetics. Moreover, the complexity of the definition of meiosis extending beyond mere genetic reduction includes the consideration of the organization, maintenance, and succession of genetic information across generations. The complexity of meiosis further points to the importance of cellular mechanisms that regulate gamete formation and genetic fidelity. Research shows that elevated apoptosis levels are present in meiosis prophase I, particularly in oocytes at preceding developmental stages. The findings demonstrate that oocytes are thoroughly screened through meiosis, where unutilizable genetic configurations are eliminated and only the fittest gametes are left to contribute to the next generations. Cleaved PARP-1 and TUNEL staining protocols have reported variation in apoptotic signaling, which underlines that meiotic development and cell survival are closely connected. They illustrate the delicate balance between exporting genetic variety by meiosis and eliminating possibly deleterious genetic differences, thus fortifying the wider designs of meiosis in fertility as well as heredity (Ghafari et al., 2007).

Importance of Meiosis in Sexual Reproduction

The significance of meiosis in sexual reproduction is primarily attributed to its role in generating genetic variation among offspring. During meiosis, recombination of homologous chromosomes and independent assortment occur, processes which are critical in randomizing genetic material. Such variation is not merely critical for populations' adaptability in response to changing environments but also for the very long-term survival of the species. The exchange of genetic material between chromosomes via the mechanism of crossovers generates new combinations of alleles, and by doing so, the fitness of an organism can be enhanced. Thus, such genetic variation is an evolutionary force because it allows populations to occupy new ecological niches as well as respond to selective pressures. From studies on various species, meiotic mechanisms have highlighted the importance of genetic variation in the field of evolutionary biology. Therefore, meiosis knowledge forms the core of understanding the big picture of inheritance and diversity of species. Moreover, the meiotic mechanisms themselves can provide insightful information into hybridization dynamics and their role in inheritance patterns. Hybridization in certain animals, for instance, the cichlids, has already been discovered to facilitate the transfer of significant genes that enable adaptability, as discovered in the meiotic segregation of hybrid research. These tests indicate that hybrids are capable of maintaining the genetic stability of parent species but also exhibiting new traits beneficial to surviving in a diverse environment. The proof of normal Mendelian segregation of hybrids indicates that meiosis is important to ensuring genetic differences can be expressed and selected for. These are indications that the evolutionary role of hybridization is possibly pivotal in diversification mechanisms, further underpinning meiosis' contribution towards sexual reproduction and wider inheritance mechanisms, as suggested by relevant genomic studies (Baroiller et al., 2012). Finally, the understanding of meiosis extends beyond sheer theory to include applied aspects across various sectors like agriculture and conservation. As meiosis produces genetic diversity, it is also crucial for breeding programs designed to increase crop production and resistance against diseases. With applied knowledge of genetic mechanisms, scientists can select the best traits through planned crossbreeding, thus guaranteeing increased productivity in agriculture. An awareness of meiosis also informs conservation practice in the aspect that it demonstrates the importance of conserving genetic diversity among endangered species. Genetic bottlenecks would terribly impair the viability of small populations, and therefore, an understanding of meiotic processes is crucial for ensuring biodiversity. Thus, the application of meiotic principles transcends basic biological research, underscoring its central role in natural and man-created ecosystems (Bekish et al., 2017) (Figure 10.2).

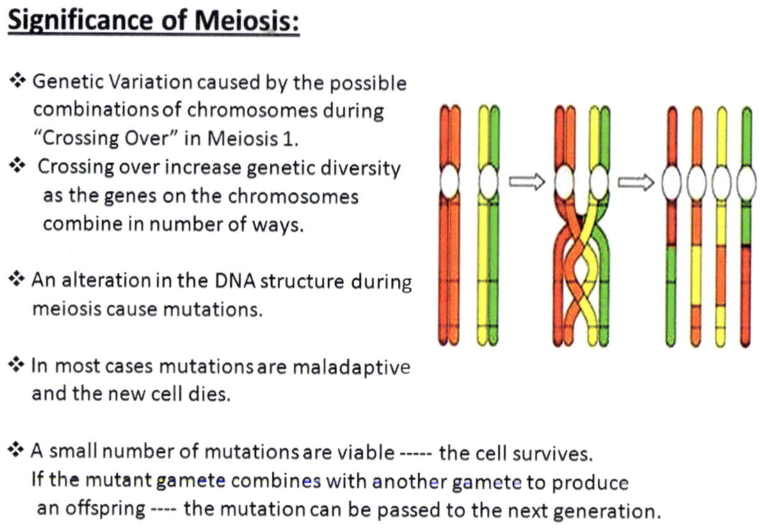

Figure 10.2. Significance of meiosis.

Source: https://slideplayer.com/9535800/30/images/slide_1.jpg.

Overview of Inheritance and Genetic Variation

Genetic inheritance is one of the factors influencing the diversity of life, basically facilitating biological evolution. The process of inheritance results from specialized cell division that reduces the number of chromosomes by half, and it is done via meiosis. During meiosis, homologous chromosome recombination takes place in which there are novel combinations of alleles. Genetic blending is of great importance as it expands genetic variation within populations, providing natural selection with raw material. The functioning of chance events within this process illustrates how genetic consequences are not merely outcomes of deterministic procedures but are influenced by stochasticity. The modern synthesis highlighted these functions of chance in their explanations, underlining their complex co-relations with evolutionary processes, which ultimately define our understanding of adaptation and speciation in diverse ecosystems (Matthews et al., 2016). It is also demonstrated through the study of polyploidy and has profound impacts on adaptation and survival. An example is that *Fennoscandian flora* has a common appearance of polyploid plants, , such as *Cerastium alpinum* L., which have persisted in past climatic change-generated environments. These polyploid species demonstrate greater capacity for colonizing postglacial landscapes, with evidence of greater competitiveness and plasticity in challenging environments, , such as poisonous serpentine soils. Genetic research also indicates varying populations of *C. alpinum* with varying adaptive traits, highlighting the importance of genetic diversity in facilitating colonization and stabilization of species within specific niches. These investigations illuminate the role of polyploidy in guiding evolutionary directions and the persistence of plant taxa in responding to environmental changes, further enriching our general comprehension of mechanisms of heredity (Berglund et al., 2005). Moreover, research on genetic variation extends to not just single species but also ecological processes and evolutionary change at the population level. The relationship between chance events, genetic inheritance, and environmental pressures demands a universal study of adaptive mechanisms across organisms. Scientists may determine the mechanisms through which organisms adapt to changing environments by examining trends in genetic diversity and inheritance. For instance, the discovery of how some genotypes confer a survival advantage in extreme environments deepens our understanding of the evolutionary significance of genetic variation. These searches support the view that genetic inheritance is not a simple process of transmission of alleles but an active and complex force managing life. Synthesis of thought from genetic inheritance and evolutionary theory thus reveals the extensive impacts of genetic variation on biodiversity and survival of species in a changing world (Matthews et al., 2016) (Figure 10.3).

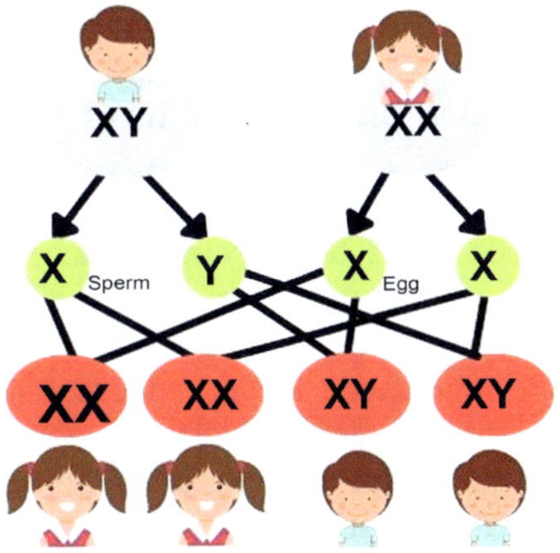

Figure 10.3. Inheritance and genetic variation.

Source: https://media.gosciencegirls.com/wp-content/uploads/2021/05/Heredityevolution-1.jpg.

10.1. The Process of Meiosis

Learning Outcome

1. Stages of meiosis: Meiosis I and Meiosis II;
2. Key events: Crossing over and independent assortment;
3. Comparison with mitosis.

As one of the simpler sexual reproduction mechanisms, meiosis is charged with generating genetic diversity within offspring. This specialized form of cell division occurs in germ cells, dividing one diploid parent cell into four haploid cells. Meiosis can be divided into two general stages: meiosis I and meiosis II. Through meiosis I, homologous chromosomes, two members of a maternal and paternal chromosome pair, pair up and exchange. This exchange of genetic information not only generates variety but also ensures each gamete will carry a different combination of genes. The intricacies of meiotic processes like chromosome alignment and segregation showcase the complex mechanisms underlying inheritance and evolutionary dynamics, the importance of which makes it pivotal to biodiversity as well as adaptation. Meiosis is comprised of a number of important stages, each of which plays a significant role in its overall function of halving the number of chromosomes and increasing genetic variation. During prophase I, the chromosomes condense and the homologous chromosomes form pairs through a process called synapsis. It is significant that this alignment is where crossing over happens—where sections of DNA are exchanged between non-sister chromatids. The resultant genetic recombination is a source of gametic variation, which is the foundation of evolution and ensures that offspring differ from parents and one another. Once through prophase I, the cell transitions to metaphase I, anaphase I, and telophase I, resulting in two daughter cells, each with one set of chromosomes. Moving to meiosis II, essentially a duplicate of mitosis, further shapes the genetic content of the daughter cells to yield a gamete not only genetically distinct but also of the form to generate the next generation. In addition to the role of genetic diversity, meiosis is also precisely regulated for the sake of genome integrity. Such fundamental genomic events as programmed death of cellular entities, i.e., mitochondrial-endoplasmic reticulum-cortex anchors (MECA), demonstrate the complex regulatory mechanisms of meiotic division. Current studies, including those on meiotic transcription factor roles like Ndt80 and kinase roles like Ime2, demonstrate how cellular structure becomes specialized to accommodate the drastic restructuring meiosis entails. This specialization underscores the importance of not only genetic restructuring but of restructuring cellular materials as well, something viable gamete formation requires. Thus, meiosis is a flawless example of the interconnectedness of developmental and genetic mechanisms, emphasizing even more so its significance in examining inheritance and biological diversity (MacLeod et al., 2007; Berchowitz et al., 2018).

10.1.1. Stages of Meiosis: Meiosis I and Meiosis II

Meiosis is divided into two major phases: Meiosis I and Meiosis II, which both consist of distinct cellular processes that are responsible for genetic variation and proper inheritance. During Meiosis I, the halving of the number of chromosomes is the main job. It is achieved by a series of phases consisting of prophase I, metaphase I, anaphase I, and telophase I. During prophase I, homologous chromosomes become synapsed and undergo recombination, which results in genetic exchange and variation in gametes. Such a genetic exchange is supported by organism genetic mapping evidence, for instance, Trypanosoma brucei, demonstrating that meiosis, independent assortment, and crossing over are vital processes in such organisms, in agreement with classical Mendelian genetic expectations, as further evidenced by data on models of genetic exchange from genome projects (MacLeod et al., 2007). Following Meiosis I, the cell advances to Meiosis II, which is a replica of mitosis but without a further doubling of chromosomes. The two resultant haploid cells from Meiosis I continue on to Meiosis II, wherein the sister chromatids are separated through a series of phases: prophase II, metaphase II, anaphase II, and telophase II. Attention is here focused on chromatid separation to ensure that each resulting gamete has its own unique set of alleles. This stage is important in creating genetic diversity in contributing to the variations seen among offspring. Understanding the process is further complemented by research into the significance of mitochondrial inheritance, where nuclear and mitochondrial genomes coordination is vital for gamete maturation. These observations underscore the complexity of inheritance from Mendelian traits to the incorporation of organelle genetics within the broader context of meiosis (Alam et al., 2006). Ultimately, the stages of meiosis not only facilitate the reduction in chromosomal numbers but also propel genetic diversity through recombination and independent assortment. All these processes have far-reaching effects that dictate patterns of inheritance and condition population genetics. With science evolving further, the significance of meiosis extends to applications , such as genetic mapping and the research of mitochondrial diseases, where inheritance processes have significant implications for reproduction. As described in genetic exchange studies, such meiotic phenomena have been instrumental in revealing the complexities of inheritance, hence furthering our knowledge of evolutionary biology and genetics. Having proper meiosis execution, therefore, is of paramount significance in the maintenance of genetic integrity across generations, placing it firmly within the principles of meiosis and inheritance studies (Figure 10.4).

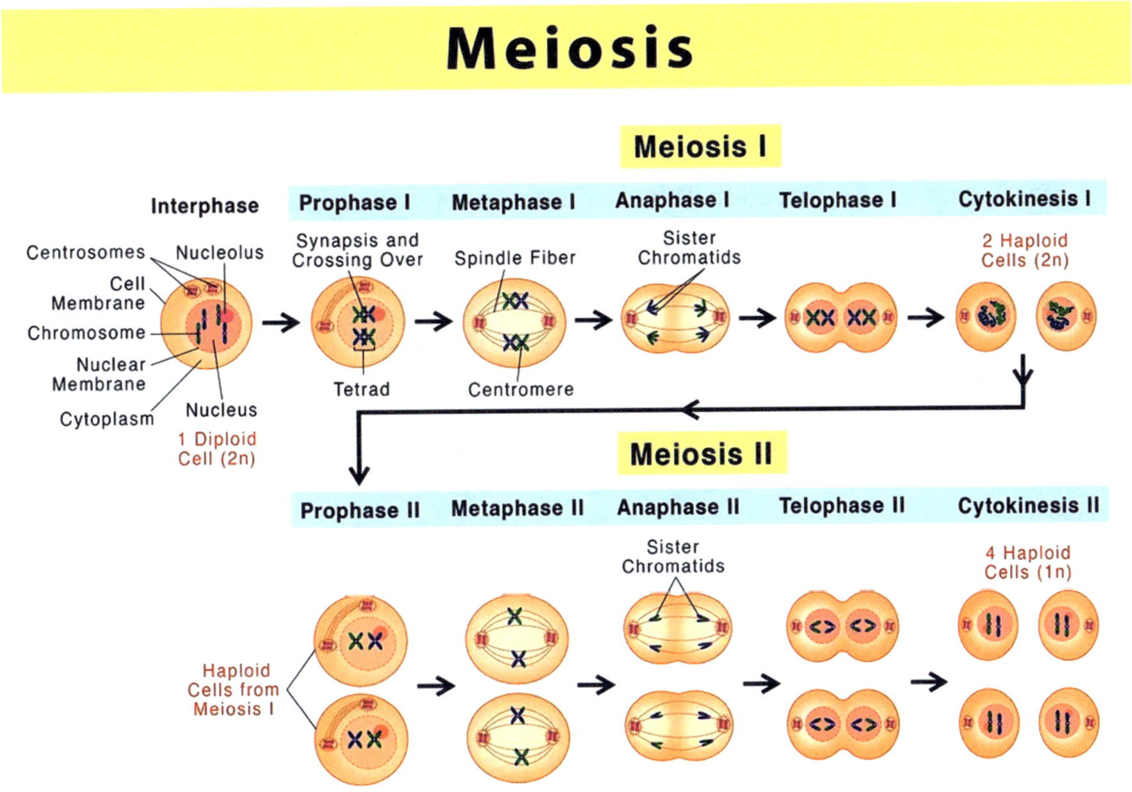

Figure 10.4. The process of meiosis.

Source: https://www.sciencefacts.net/wp-content/uploads/2020/09/Meiosis-Stages-Diagram.jpg.

10.1.2. Key Events: Crossing Over and Independent Assortment

Companion to meiosis and inheritance, crossing over and independent assortment are significant events that contribute significantly to genetic diversity. Crossing over occurs in the course of prophase I of meiosis when homologous chromosomes exchange segments of their genetic material.

The exchange results in novel combinations of alleles on each chromosome, differing from parental forms. Consequently, this shuffling of genes not only contributes to the diversity of the gene pool but also serves as a basis for variation in a population. Such variation is essential to evolution since it conditions organisms with varied traits that may provide survival advantages in varying conditions. In organisms , such as Trypanosoma brucei, the event of crossing over has been confirmed by way of genetic mapping, reinforcing the notion that in non-model species as well, meiosis obeys simple rules , such as those in traditional Mendelian genetics (MacLeod et al., 2007). Independent assortment, occurring at the stage of metaphase I of meiosis, complements the process of crossing over by determining the way different genes segregate independently from one another. Both pairs of homologous chromosomes align randomly on the metaphase plate, and with so many combinations when gametes are formed, the potential is limitless. Random segregation is also a source of genetic variation because it provides for the combining of the maternal and paternal alleles in the offspring gametes. The complexity of the pattern of inheritance is therefore increased to create a state where offspring will express an array of characteristics along a continuum of potential combinations. This is a procedure that conforms to Mendels law of independent assortment, which has been established via experiments in a variety of organisms, illustrating the central role played by this mechanism in producing genetic diversity from one generation to another (MacLeod et al., 2007). The role of crossing over and independent assortment goes beyond simple genetic variation; they are fundamental procedures that determine the remaining principles of inheritance and evolution. When organisms reproduce, genetic diversity generated by these meiotic processes ensures that populations would be capable of adapting to environmental pressures. Apart from that, this diversity becomes critical in fields of conservation biology and agriculture, where understanding the genetic basis of traits is needed for species management and breeding crops. The revelation of these landmark events not only confirms classical genetics but also demands attention to highlighting equitable and culturally sensitive approaches in science instruction, particularly to those underrepresented groups in STEM education. By addressing the means by which these principles apply to real-world concerns, educators can encourage the future generation to gain a greater sense of the complexity of genetics and its implications for society (Figure 10.5).

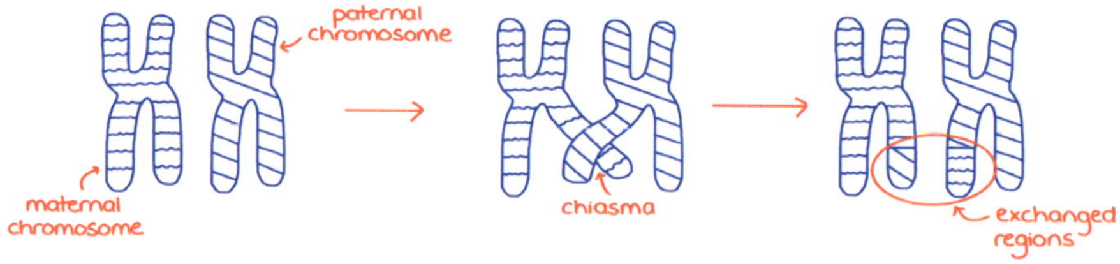

Figure 10.5. Crossing over and independent assortment.

Source: https://i0.wp.com/zhtutorials.com/wp-content/uploads/2024/04/Crossing-over.png?resize=736%2C201&ssl=1.

10.1.3. Comparison with Mitosis

In the domain of cell division, meiosis has to be contrasted with mitosis, as both are pivotal for organismal development and heredity. Mitosis entails the accurate replication and separation of DNA into two daughter cells, ensuring genetic stability and continuity in somatic cells. Each daughter cell in mitosis retains the same diploid number of chromosomes as the parent cell, a necessity to maintain an organism's genetic identity. In contrast, meiosis consists of two distinct cycles of division, meiosis I and meiosis II, yielding four haploid gametes with a distinct genetic makeup.

Such a reduction in chromosomes is essential for sexual reproduction since it prepares gametes for fertilization as well as allows genetic diversity through independent assortment and recombination. Mechanistic differences between these processes are used to indicate the unique roles that mitosis and meiosis play in cellular function and in inheritance. Genetic ramifications of meiosis diverge significantly from mitosis, emphasizing the essential nature of meiosis for inheritance. While mitosis makes exact copies of existing genetic material without alteration, meiosis generates genetic variation through several crucial mechanisms. One of these operations is crossing over, which occurs at prophase I and allows homologous chromosomes to exchange genetic information. Not only does the operation redistribute alleles, but also increases genetic diversity in offspring, a vital parameter allowing populations to evolve to fit into environmental changes. Second, the law of independent assortment during metaphase I ensures that maternal and paternal chromosome segregation into gametes is random.

These processes are all enormously different from mitosis, where the daughter cells inherit exact copies of the parent cell's chromosomes. This intrinsic variation in meiosis is vital to the evolutionary processes and survival of species and thus proves its significance when speaking of inheritance. Comparing meiosis's chromatin movement to that of mitosis reveals even more differences in their purposes and implications for inheritance. During mitosis, chromatin is condensed to observable chromosomes to allow for the accurate segregation of sister chromatids to each daughter cell in support of the postulate of genetic fidelity. Meiosis, on the other hand, involves not only the condensation of chromatin but also the strategic pairing and recombination of the chromosomes. The necessity for chromatin remodeling to ensure these operations occurs is brought into prominence by current studies, showing that chromatin plasticity is significant in adaptations to varied environmental stresses, as discussed in plant research (Pecinka et al., 2019). This adjustability allows meiotic cells to refine and ensure successful gamete formation. Since it becomes evident by comparative study, the unique chromatin organization and activities during meiosis versus mitosis reflect specialist functions necessary for gamete production and ultimately for the inheritance of diverse genetic traits in offspring.

Test Your Knowledge

1. What are the stages of meiosis?
2. What is crossing over and independent assortment?

10.2. Genetic Variation

Learning Outcome

1. Role of crossing over in genetic diversity;
2. Independent assortment and its effects on gametes;
3. Mutations and their contribution to variation.

Genetic variation is a process vital to the evolutionary process and is inseparably related to the processes of meiosis and inheritance. Meiosis, while dissociating the paired chromosomes, possesses the process of HR in which there is an exchange of genetic material between the paired chromosomes. It is not only accountable for a random assortment of alleles but also for introducing new combinations of genes that can manifest themselves as phenotypic variations in the ensuing generations. The uniqueness of such processes is illustrated in *Trypanosoma brucei* experiments, where genetic crosses have demonstrated the working implications of genetic exchange. The achievement of its genome project has brought to light the presence of meiosis, independent assortment, and crossing over in this organism, which led to the identification of polymorphic markers that have a primary function in explicating the inheritance pattern of traits in this parasite and, by virtue, in other organisms as well (MacLeod et al., 2007).

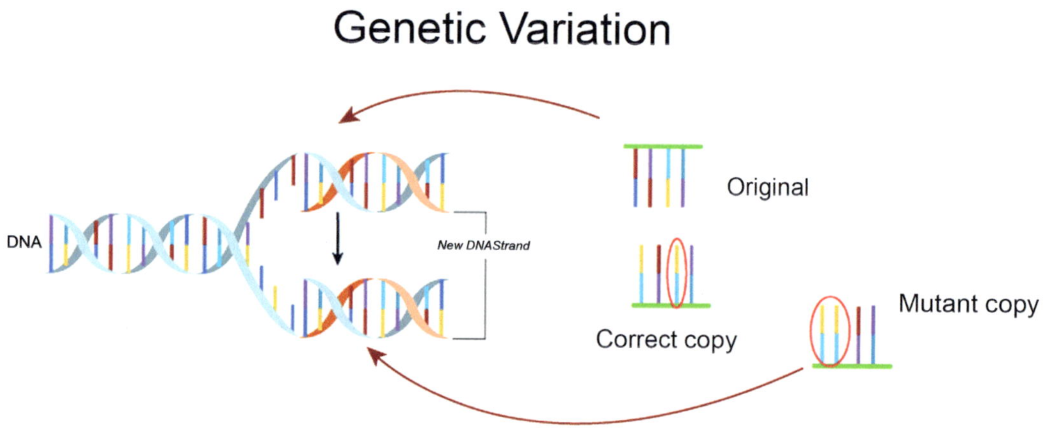

Figure 10.6. Genetic variation.

Source: https://media.geeksforgeeks.org/wp-content/uploads/20240119101553/Genetic-Variation.png.

In addition to such well-documented mechanisms, heritability can be introduced by mutations that arise during the process of sexual reproduction, as in the case of species like *Pseudonitzschia multistriata*. The diatom is a case where genotypic stability is enhanced by the introduction of mutations by a high frequency of both vegetative cell division and sexual reproduction. Researching microsatellite loci among various strains, researchers identified deviations from characteristic Mendelian inheritance, discovering that genetic variation mechanisms are more complex than the segregation of alleles. Mutation formation, importantly those induced by microsatellite regions with extended core-repeats, contributes to genetic variation among populations, showing how mutations and meiosis act in concert to ensure adaptability within evolving environments (Amos et al.,

2012). The interaction of mutations, genetic variation, and meiosis is very important to evolution and diversity. With organisms experiencing multiple environmental stresses, the resulting genetic diversity as a result of these processes enables populations to evolve with time. This variation provides a canvas on which natural selection works to eventually set the course of evolution for species. Both in the case of *Trypanosoma brucei* and *Pseudonitzschia multistriata*, pathology and environmental responses can illustrate the facilitative role of genetic exchange and mutation in supporting adaptability. Such outcomes not only enrich our conception of fundamental biological processes but also betray underlying complexity in inheritance, necessitating an increased exploration of genetic variation as a force behind the evolutionary agenda (Figure 10.6).

10.2.1. Role of Crossing Over in Genetic Diversity

The process of meiotic crossing over plays an important role in enhancing genetic diversity between offspring. Homologous chromosomes swap bits of genetic material during prophase I. The consequence of this is that alleles on the chromosomes get reshuffled to form new combinations of alleles that are not present in either parent. Thus. Crossing over creates genetic diversity among populations that is extremely crucial for evolution and adaptation, especially under conditions of environmental change. With greater genetic diversity generated through this process, organisms are better able to survive and reproduce, contributing to species being sustained and maintained. Besides, it is also accountable for the dynamic nature of genomes as organisms are capable of handling new challenges, thereby the function has an essential role to play in the overall picture of inheritance and population genetics.

Furthermore, the significance of crossing over can be observed in many species and situations. Genetic analysis of hybrid organisms , such as intergeneric hybrids of *Oreochromis niloticus* and *Sarotherodon melanotheron* marks the position of meiotic processes , such as crossing over in enabling genetic diversity. Mendelian predictability in the segregation of the hybrids also suggests crossover events as the key to maintaining genetic stability coupled with diversity. Such studies are important to determine the evolutionary importance of hybridization, particularly in environments where prezygotic barriers are nonfunctional. The implications are that cross-over not only enables genetic diversity but also assists in the development of new species and, as a result, aids the evolutionary tapestry of life on planet Earth. Thus, cross-over is a bridge between genetic diversity and evolutionary adaptive processes. Apart from the encouragement of genetic diversity, crossover also aids in population stability by allowing for favorable characters to be preserved and transmitted across generations.

The case of *Pseudonitzschia multistriata* is a perfect example of how both mutation in sexual reproduction as well as clonal growth are each able to generate more allelic diversity through crossover occurrences. The build-up of the mutations subsequently promotes genetic diversity, and there is a pool of potential adaptation to new environments. This dynamic interplay between crossing over and mutation demonstrates the intricacy of the major evolutionary processes. By creating genetically diverse offspring, populations can respond better to selective pressures, which is necessary for them to survive in the long term. To wrap it up, the role of crossing over in genetic diversity is crucial not only for single organisms but for the survival and resistance of whole populations within a dynamic environment (Figure 10.7).

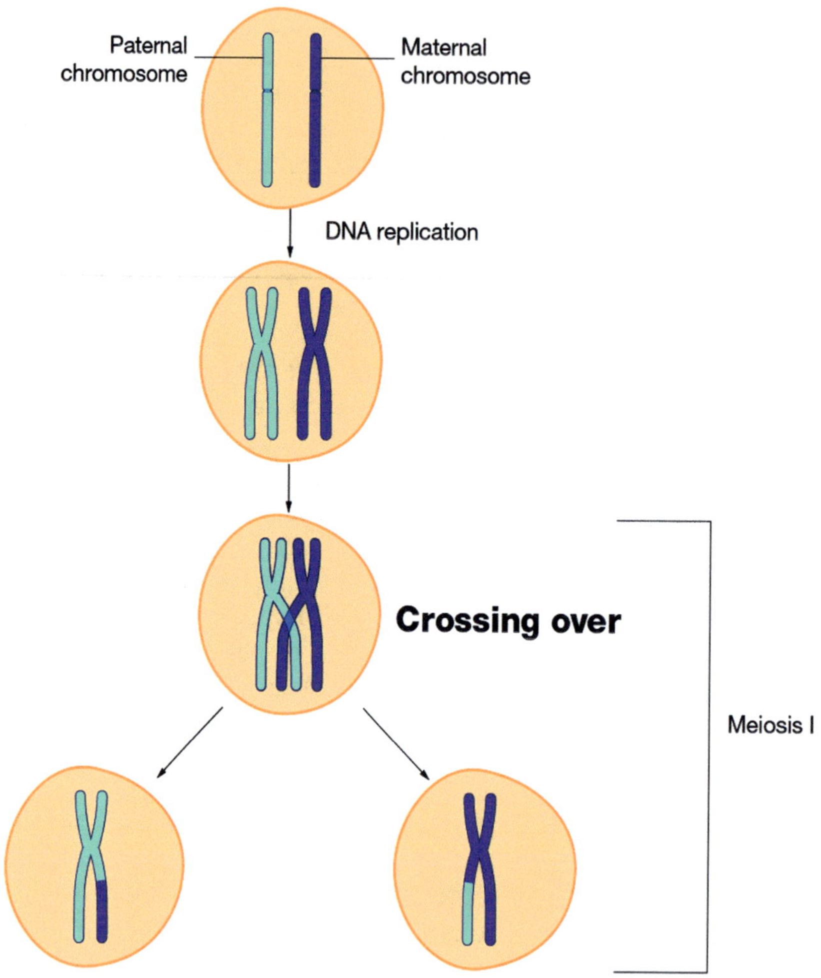

Figure 10.7. Role of crossing over in genetic diversity.

Source: https://www.genome.gov/sites/default/files/media/images/tg/Crossing-over.jpg.

10.2.2. Independent Assortment and Its Effects on Gametes

Independent assortment is a fundamental principle in the process of meiosis and is crucial in creating the genetic variation observed in gametes. During meiosis, homologous chromosomes align and separate from one another independently, and there is random separation of maternal and paternal chromosomes to gametes. This is done in a way that many combinations of alleles can occur, with the potential for offspring to have many different genetic traits. When alleles on distinct loci sort independently, the outcome is governed by Mendelian principles, causing differences in gametes that enhance genetic diversity in populations. The process not only maximizes the potential for adaptive characters but also ensures that core Mendelian principles are in the mechanisms of inheritance for future generations, providing a good foundation for genetic evolution.

In addition, independent assortment influence extends beyond mere genetic variation. For large panmictic populations with free interbreeding of individuals, purifying selection influences the effectiveness of recombination processes that lead to independent assortment. It promotes the preservation of advantageous alleles and the diminution of the spread of deleterious alleles to maximize the population's evolutionary track.

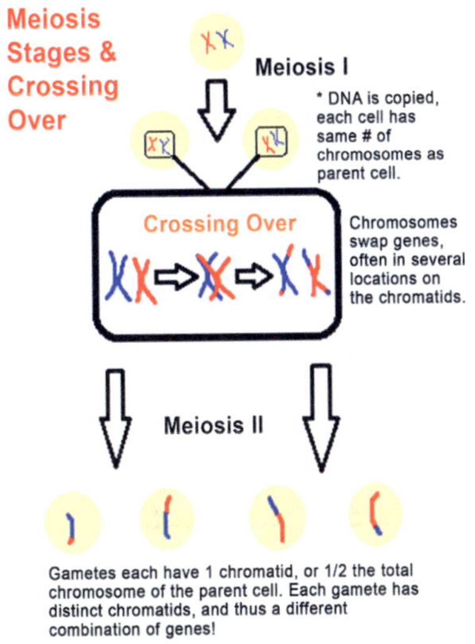

Figure 10.8. Independent assortment and its effects on gametes.

Source: https://study.com/cimages/multimages/16/meiosis_crossing_over.png.

The case is different in small populations where genetic drift may step in. Under such conditions, linked genes usually segregate together, and this results in a situation where clusters of genes act as genetic units as opposed to individual units. These cases demonstrate the intricacy of inheritance at the genetic level and show how effects of independent assortment can change enormously depending on population size and composition and would go on to shape the overall evolutionary genetics landscape of an organism (Cebrat et al., 2010). The implications of independent assortment also apply in the context of recent research that has assisted in explaining the mechanisms of genetic segregation in various species. For instance, studies in Trypanosoma brucei have demonstrated the independence of segregation of loci on different chromosomes, which provides empirical proof for Mendelian genetics in a greater biological context. In these findings, researchers indicated that segregation ratios observed in these experiments agreed with independent assortment expectation, further justifying the validity of classical genetics models. Unexpectedly, selection pressures deviations from expected proportions were also observed, suggesting that forces outside of the genetic process could actually influence gametic outcomes. These findings not only confirm the meiosis and inheritance laws but also offer avenues for further research on how independent assortment interacts with environmental factors and genetic selection processes and consequently dictates future population trends and evolutionary tendencies (Tweedie et al., 2005) (Figure 10.8).

10.2.3. Mutations and Their Contribution to Variation

Meiosis' complex processes form a critical component of population genetic variation, a process that is essential for evolution. At the core of the variance are mutations, which serve as the raw material for evolutionary forces. Mutations could result from errors in DNA replication or environmental factors and exist in numerous forms like single nucleotide polymorphisms (SNPs) and copy number variants (CNVs). Interestingly, CNVs cut across coding and noncoding DNA regions, offering a vast reservoir of genetic variation that can affect phenotypic impacts. While SNPs are single changes, it is more complex CNVs that most typically characterize functional variation in the human genome, thus explaining their importance in studies on inheritance and evolution, as seen in the methods that employ high-throughput genotyping platforms (Teslovich et al., 2010). Apart from irregular mutations occurring in meiosis, interaction between these genetic mutations and environmental stresses enhances population adaptability. For instance, some organisms display polyploidy, a condition brought about by the doubling of the entire genome, resulting in enhanced diversity and robust adaptations within evolving environments. The polyploid plant *Cerastium alpinum*, which thrives optimally in harsh serpentine soils, is a robust case illustrating how ecological specialization is promoted by mutations. Genetic studies of this species show a remarkable degree of variation among its populations, which have developed to deal with specific environmental pressures, conveying the point that mutations not only generate genetic variation but also facilitate localized evolution. This dynamic interaction of mutations and adaptive traits shows the complexity of inheritance mechanisms active in different environments (Berglund et al., 2005). Furthermore, the role of mutations is also extended to higher-order biological interactions, demonstrating their significance not just at the micro scale of gene modification but also at the macro scale of community ecology. Genetic diversification brought about by mutations enables populations to evolve in response to environmental pressures, , such as climate change and habitat disruption, to facilitate survival and continuity. The finding of de novo mutations adds yet another layer of this information, showing how new genetic changes accumulate variation in survival characteristics unique to environments. By looking at the influence of these mutations on inheritance, researchers can read the history of how populations have evolved, providing insight into how genetic variation governs species performance. Therefore, it is important to appreciate the versatile function carried out by mutations in variability because they are the foundation of ongoing processes of evolution during meiosis and inheritance.

Test Your Knowledge

1. What is the role of crossing over in genetic diversity?
2. What is independent assortment and its effects on gametes?

10.3. Gametogenesis

Learning Outcome

1. Formation of sperm and egg cells;
2. Differences between spermatogenesis and oogenesis;
3. Impact of environmental factors on gametogenesis.

Gametogenesis is a fundamental process in sexually reproducing animals, a critical phase of the entire life cycle and the hereditary system. This complex process may be separated into two broad classifications: spermatogenesis, male gamete (sperm) development, and oogenesis, female gamete (ova) production. Both are required in sexual reproduction, where genetic material is passed on by the combination of haploid gametes and concludes with the formation of a diploid zygote. Meiosis is activated during gametogenesis to provide genetic variation through recombination and random distribution of chromosomes and thereby facilitate inheritance of traits in ensuing generations. Control of these processes is tightly regulated by a series of molecular events and endocrine signals, a reflection of gamete development's complexity and significance in evolution and genetics. The linkage of gametogenesis to apomixis broadens our understanding of modes of alternative reproduction, particularly in the plant species. Gametogenesis in species like Hypericum perforatum can include both traditional sexual reproductive processes as well as the new process of apomixis. In this case, aposporic gametogenesis allows the creation of unreduced but functionally so embryo sacs, effectively bypassing the meiosis usually found in gamete production. The reproductive system is one that carries with it interesting questions regarding genetic diversity and heredity since it means that offspring can receive characteristics from the maternal lineage directly without the mixing of genes found in sexual reproduction. The broader implications of mechanisms like these are observed in agriculture, where genetic regulation of apomixis may eventually lead to improved methods in crop breeding and the creation of cultivars that have preferred traits without compromising genetic stability (Barcaccia et al., 2015). Molecular research based on the foundation of gametogenesis and apomixis reveals the complex gene interactions that occur in these processes. Investigation of the expression of some unigenes involved in the aposporous pathway in *H. perforatum* highlights the immediate function of genes involved in the control of sporogenesis and gametogenesis, demonstrating complex interaction between hormonal signals and genetic expression patterns. Notably, gene ontology term analysis identifies that processes at the cellular level are significantly involved in the control of key genes involved in sexual reproduction pathways. By opening up the regulatory mechanisms acting on gametogenesis, researchers aim to de-mystify the genealogical reason behind apomixis that could have revolutionary applications in inheritance and reproductive strategies in plants, paving the way for breakthroughs in plant biology and genetic engineering (Altschmied et al., 2017) (Figure 10.9).

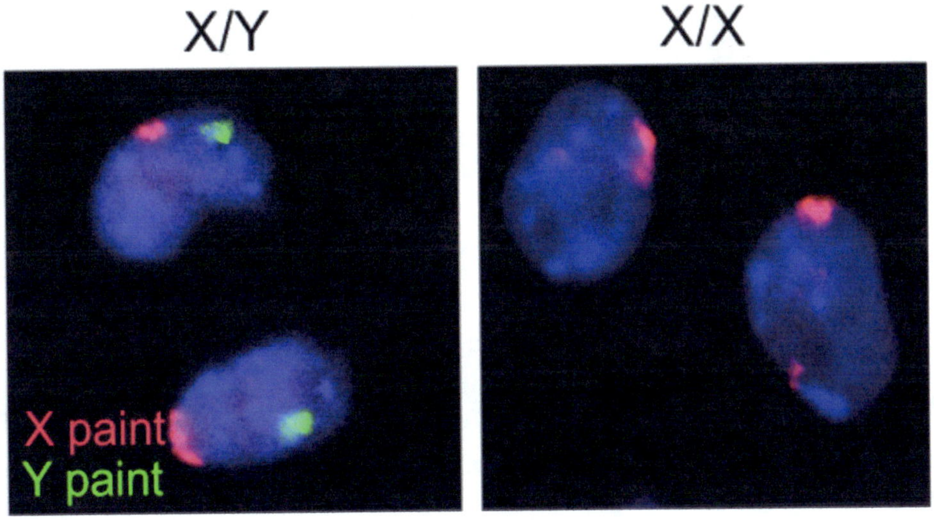

Figure 10.9. Gametogenesis illustrated by an ovum (left) and a sperm cell (right), representing the formation of male and female gametes.

Source: By Janice Y Ahn, Jeannie T Lee - Janice Y Ahn, Jeannie T Lee Retinoic acid accelerates downregulation of the Xist repressor, Oct4, and increases the likelihood of Xist activation when Tsix is deficient BMC Developmental Biology 2010, 10:90 doi:10.1186/1471-213X-10-90 http://www.biomedcentral.com/1471-213X/10/90, CC BY 2.0, https://commons.wikimedia.org/w/index.php?curid=43944729

10.3.1. Formation of Sperm and Egg Cells

The gametogenesis, or the formation of sperm and egg cells, is a vital part of sexual reproduction and inheritance of genes. The formation of sperm cells in male organisms is conducted by a strictly regulated process of meiosis known as spermatogenesis. Having begun within the seminiferous tubules of the testes, spermatogenesis involves spermatogonia being split into primary and secondary spermatocytes that both go through meiosis to produce ultimately four haploid sperm cells. Not only does it reduce the number of chromosomes, but it also produces genetic variation through crossing over and independent assortment, the foundation of meiosis that influences genetic variation in offspring. Such diversity is necessary for evolutionary growth and survival of the species, which in turn highlights the importance of gametogenesis in learning about hereditary characteristics (Bekish et al., 2017).

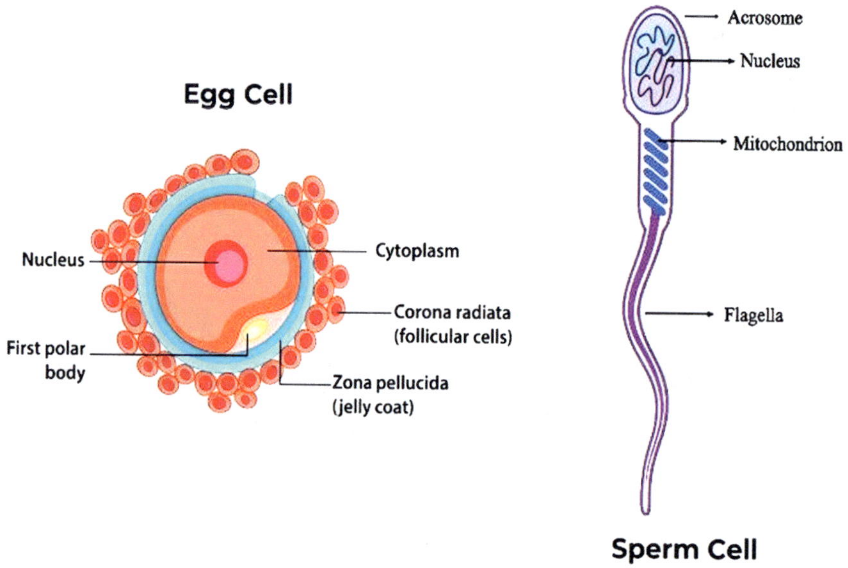

Figure 10.10. Ovum and sperm formation through gametogenesis.

Source: https://medlifemastery.com/wp-content/uploads/2024/11/Gametogenesis.jpg.

In contrast to spermatogenesis, oogenesis or development of the egg cell is a more complex process in the female ovary. Primary oocytes are at arrest in prophase I of meiosis during oogenesis until hormonal signals initiate further development. In contrast to sperm cell development, oogenesis yields a single viable ovum and polar bodies, which are typically non-viable. Asymmetrical division is not merely a function of cellular asymmetry but is thought to enhance the viability of the oocyte by minimizing cytoplasmic resources. Besides, successful fertilization involves sperm having to pass through several layers to get to the egg, and one sperm fuse with the ovum to preserve the haploid state of the zygote, hence preserving genetic integrity (Allen et al., 2017). Synchrony and regulation of meiotic activities that occur in the production of sperm and ovum emphasize the delicate balance of cell activities that regulate inheritance. Research on meiotic divisions, like those in marine animals , such as the oyster, has provided hints at centrosome and chromosomal dynamics behavior at these stages (Allen et al., 2017). Understanding these mechanisms is crucial since a failure in meiosis could lead to chromosomal diseases , such as aneuploidies. Thus, sperm and egg cell development are not merely cell division but the very essence of the establishment of genetic inheritance, a witness to life's continuity and diversity from one generation to another. This complex dance between meiosis and reproductive strategy finally formulates the genetic constitution of populations, highlighted by the importance of the knowledge of these processes in the whole context of evolution and heredity (Bekish et al., 2017) (Figure 10.10).

10.3.2. Differences Between Spermatogenesis and Oogenesis

The spermatogenesis and oogenesis are the foundation of sexual reproduction, which differ in their characteristics according to their adaptive evolution and functional significance. The spermatogenesis takes place in the male gonad, i.e., the testes, and is the continuous production of spermatozoa from the initial spermatogonia cells. Four haploid sperm cells result from each precursor cell via the process, ensuring greater fertilization potential and genetic diversity. Conversely, oogenesis is conducted in the female gonads, or the ovaries, where only a few oocytes are produced. One egg cell matures from each ovarian cycle while the remaining oocytes enter a dormant state of long-term suspended development.

This contrast in gamete production is a reflection of the distinct biological strategies employed by females and males, which emphasizes the role of sperm in quantity and eggs in quality in reproduction. In addition, the complexity of meiosis also separates spermatogenesis from oogenesis, most significantly their timing and efficiency. During spermatogenesis, meiosis is a continuous, fairly rapid process over a man's reproductive lifetime to accommodate an ongoing rate of sperm production. In contrast, oogenesis takes a more cyclical approach in which meiosis is initiated during fetal development but held in abeyance until ovulation occurs, and this could be years or decades. Besides, unequal cytokinesis during oogenesis yields one mature ovum and the smaller polar bodies that degenerate afterwards. Such unequal division benefits the delivery of nutrients to the egg by enhancing the viability of the egg to be fertilized and grow into an embryo. Such dissimilarities in time and efficiency in these processes influence reproductive strategy and capacity for genetic diversity in offspring. The interaction between environmental factors and genetics also highlights the difference between these two gametogenic processes. As per studies, epigenetics during gametogenesis plays a central role in determining the developmental fate of sperm and egg cells, where environmental stimulus plays a crucial role in gene expression as well as phenotypic traits in the offspring. For instance, environmental forces acting on oogenesis have implications not merely for the quality of the resultant oocyte but also for whether the oocyte can transmit epigenetic information to the next generation. Across spermatogenesis and oogenesis, these forces again evidence the complexity of transmission beyond gene transmission and point to the need to appreciate both processes within the broader realms of developmental biology and evolution. The more that is known about the connection between meiosis and inheritance, the more clear it becomes that such variations have an intense effect on reproductive success as well as genetic diversity among species, as illustrated through studies considering disparate meiotic patterns in various organisms (Maamar et al., 2021; Blaxter et al., 2025) (Figure 10.11).

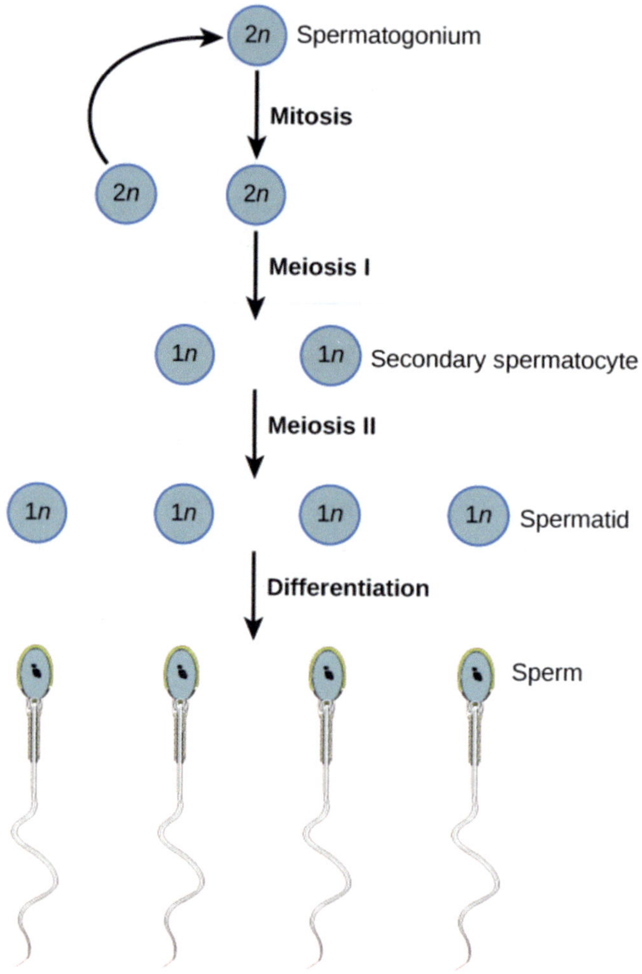

Figure 10.11. Difference between spermatogenesis and oogenesis.

Source: https://bio.libretexts.org/Bookshelves/Introductory_and_General_Biology/General_Biology_(Boundless)/43%3A_Animal_Reproduction_and_Development/43.03%3A_Human_Reproductive_Anatomy_and_Gametogenesis/43.3C%3A__Gametogenesis_(Spermatogenesis_and_Oogenesis)

10.3.3. Impact of Environmental Factors on Gametogenesis

The influence of the environment on gametogenesis is a significant area of study in genetics and developmental biology. There are various extrinsic factors, including temperature, toxins, and nutrient inputs, which have strong influences on gamete production. These environmental inputs act at various stages in the sequence of meiosis, with tendencies to affect epigenetic processes that control gene expression. For example, certain exposure to endocrine disruptors at key developmental stages can trigger irreversible changes in gametogenesis, with subsequent potential impacts on fertility and viability of offspring. Because gametes transmit genetic and phenotypic traits to future generations, an understanding of such influences is critical to unraveling complex inheritance patterns. Thus, environmental conditions not only specify the individual gametes but also profoundly influence the evolutionary trajectory of species as they can induce epigenetic modifications through generations (Maamar et al., 2021). In addition, variations in environmental conditions can lead to drastic alterations in the meiotic processes themselves. These influences are illustrated in the manner by which organisms adapt their reproductive strategies in compensatory mechanisms in response to contrasting environments. For example, in certain nematodes, temperature and resource availability variations can modify meiotic modes, resulting in heterogenous offspring types, male and hermaphroditic gametes (Blaxter et al., 2025). These diversities represent a meiotic plasticity of the organisms that allows them to enhance reproductive success under varying environmental conditions. Varying types of gametes produced under the effect of external conditions reflect the integrative nature of gametogenesis in evolutionary biology. These adaptive processes raise deep questions about the ecological and evolutionary impact of environmental variation on reproductive health and offspring development over generations. In the end, the impact of environmental influences on gametogenesis goes well beyond immediate reproductive success and extends to genetic diversity and population resilience. As environment dictates the epigenetic profile, they also complicate the inheritance patterns defined by classical Mendelian genetics. The interaction between the environment and epigenetic alteration could have transgenerational implications wherein phenotypic changes are retained in offspring, perhaps without altering the DNA sequence per se. Epigenetic transgenerational transmission of this type emphasizes the importance of critically examining the impact of exposure to the environment upon gametogenesis. Such information not only assists the world of science in comprehending meiosis and inheritance but also plays a crucial role in public health intervention towards reducing the dangers that come with environmental toxins, thus protecting reproductive health in generations (Maamar et al., 2021).

Test Your Knowledge

1. Illustrate the formation of sperm and egg cells.
2. What is the impact of environmental factors on gametogenesis?

10.4. Mendelian Genetics

Learning Outcome

1. Gregor Mendel's principles of inheritance;
2. Dominant and recessive traits;
3. Punnett squares and probability in inheritance.

The Mendelian genetics laws form the foundation for the description of sexually reproducing organisms patterns of inheritance. Gregor Mendel developed them in the 19th century, using pea plants as his test subjects and, thereby, was able to identify the rules of inheritance, including the concept of dominant and recessive characters. Mendel performed strict experiments illustrating how characters separate during gamete production and recombine at fertilization. This division is accompanied by the process of meiosis, when homologous chromosomes are separated into gametes, resulting in genetic variation. Implications of Mendelian genetics extend not only to pea plants but are a paradigm through which inheritance can be studied in various species. Mendel's research is also fundamental to the evolution of genetic counseling and new breeding technologies, illustrating the applicability of his findings to modern genetics (Ferguson-Smith et al., 2022). During meiosis, genetic recombination plays a crucial role in generating diversity within populations. Mendel's results were further refined by the idea of genetic linkage, where genes that are close to each other on a chromosome have a tendency to be inherited as a unit. Independent assortment, yet another of Mendel's laws, suggests that the inheritance of alleles for one gene will not affect the inheritance of alleles for a different gene located on another chromosome. Such interaction between independent assortment and segregation during meiosis generates novel sets of combinations of alleles in progeny. Thus, when studying inheritance, it is important to consider the genetic makeup of parents as well as the mechanisms of meiosis that cause variation.

These principles are the foundation not only in plant and animal breeding but also in human genetics inheritance pattern prediction (Alberts et al., 2006). Genetic inheritance has evolved from Mendel's early experimentation to the development of high-level genomic theory. Modern genetics recognizes the gene's complexity in defining it as instrumental, nominal, or post-genomic. The instrumental gene plays a central role in experimental designs that relate genotype and phenotype via hybridization. Conversely, the nominal gene forms an effective communication tool among researchers, founded on well-demarcated nucleotide sequences but not with reference to profound underlying theoretical structures. Finally, the post-genomic perspective encourages a refined appreciation of the influence of genome structure upon functions and attributes. Such an understanding outgrows conventional assumptions regarding genotype-phenotype correlation, validating the efficacy of Mendelian laws in the context of newly investigated genetic works. This changing knowledge of genetics adds depth to the discussion of inheritance, essential for applications from agriculture to medicine (Ferguson-Smith et al., 2022) (Figure 10.12).

Figure 10.12. Mendelian genetics.

Source: https://i0.wp.com/freescience.info/wp-content/uploads/2023/06/Mendelian-Genetics.jpg?fit=1000%2C666&ssl=1.

10.4.1. Gregor Mendel's Principles of Inheritance

Gregor Mendel's laws of inheritance more or less revolutionized the whole notion of genetic inheritance in living organisms. His ground-breaking experimentations with pea plants unveiled the concept of dominant and recessive traits, which are the backbone of Mendelian genetics. By controlled experiments of crossbreeding, Mendel observed that there were definite patterns of passing on of the traits, and from these observations, he derived three significant laws: the law of segregation, the law of independent assortment, and the law of dominance. All the above laws explain certain specific manners in which the alleles are passed on while the gametes are formed during meiosis, thereby elucidating how children inherit their characteristics from their parents. Mendel's precise approach and quantitative manner, examining traits in successive generations, provided the basis for modern genetics, establishing principles that remain valid in present-day genetic research and application.

The law of segregation informs us that alleles for a certain trait segregate when producing gametes so that every gamete will be furnished with one allele for every gene. This is a direct correlation to meiosis, in which homologous chromosomes separate in meiosis I, providing genetic variation in gametes. Mendel's experiments identified that when he cross-bred purebred plants with contrasting traits, the F1 generation produced only one of the parental traits consistently, signifying that the other trait was suppressed. It was in the following F2 generation that the hidden trait re-emerged to produce a ratio of close to 3:1 for dominant to recessive expressions. This careful observation not only provided evidence for the regularity of inheritance patterns but also provided an early description of genetic variation, so critical in fields , such as agriculture and pharmacogenetics, where Mendelian principles continue to guide breeding and selection of traits. Apart from this, Mendel's research on gene knowledge can additionally be explained by the theory of the law of independent assortment, which states that the alleles for different traits segregate independently of one another while the gametes are being produced. It is the combinatorial aspect of genetic heritage that emerges due to the immediate effect of random assortment of chromosomes in the process of meiosis. As Mendels work was applied to dihybrid crosses, he discovered that independently inherited characters were presented in a predictable proportion. Not only did this finding confirm the laws of segregation but also revealed the complexity of genetic interaction in heredity. The lessons learned from Mendel's application of quantitative analysis and systematic experimentation left serious questions that continued to shape genetic theory many years after his initial discoveries. Indeed, Mendels work remains robust to date as it provides a holistic paradigm for studying genetic mechanisms, as evident in contemporary studies of the complex relations between genes and phenotypes in various organisms (Ek et al., 2011; Shan et al., 2016) (Figure 10.13).

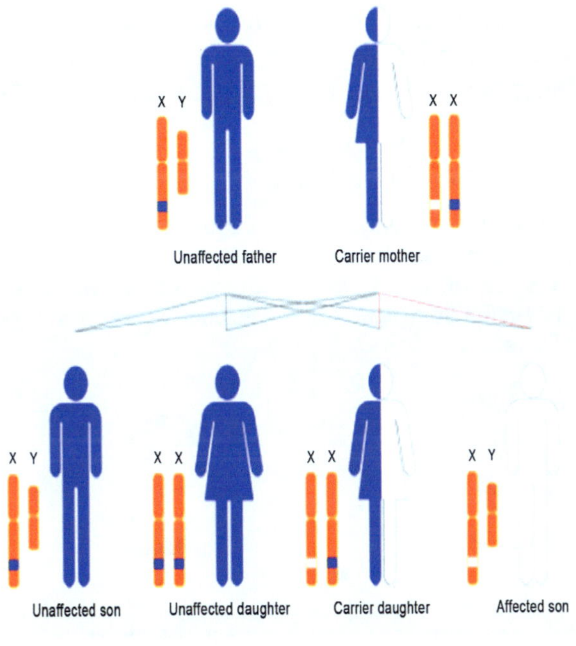

Figure 10.13. Mendelian inheritance.

Source: https://www.medicalexamprep.co.uk/wp-content/uploads/2018/12/Untitled-design-3.png.

10.4.2. Dominant and Recessive Traits

The hereditary foundation of dominant and recessive traits lies within the rules of inheritance set forth by Mendelian genetics. Dominant traits manifest in the phenotype irrespective of the presence of one allele copy, overpowering the expression of the recessive alleles that require two copies to be expressed. This process illustrates the complex interaction between alleles and their functions in trait expression. In meiotic organisms, the separation of homologous chromosomes assures that alleles segregate independently to produce a genetically variable combination in offspring. This fundamental process not only highlights predictability in inheritance but also the potential variability among populations, an important role in evolutionary biology. Moreover, the prevalence of dominant or recessive characteristics may involve selective pressure among populations. For instance, within large, panmictic populations, purifying selection will take hold, removing deleterious recessive alleles from the gene pool. Within such cases, Mendelian forces are more rigid, exercising tighter independent assortment of genes and fixation of dominant traits within a population. But within small populations, the effectiveness of recombination is diminished, leading to forces like genetic drift and increased linkage disequilibrium. Here, sets of genes may inherit collectively, and it is hard to use independent assortment directly. It may allow dominant traits to be inherited even in the re-emergence of recessive alleles (Cebrat et al., 2010). The interaction between dominant and recessive traits only becomes more complex when considering their roles as part of adaptation in evolution. Not only does allele dominance influence the phenotypic characteristics of a population, but complementation and purification processes also play major roles in determining genetic diversity. For example, adaptation traits may arise as a result of recombination events, which can split the patterns of expression of dominant and recessive alleles in gene clusters. These genetic architecture modifications have significant implications for population viability and speciation, especially in response to environmental changes. As a demonstration by computational models, interactions between purifying selection and complementation may exhibit phase transitions in response to environmental cues, thus highlighting the significance of an understanding of such processes in meiosis and inheritance (Bekish et al., 2006) (Figure 10.14).

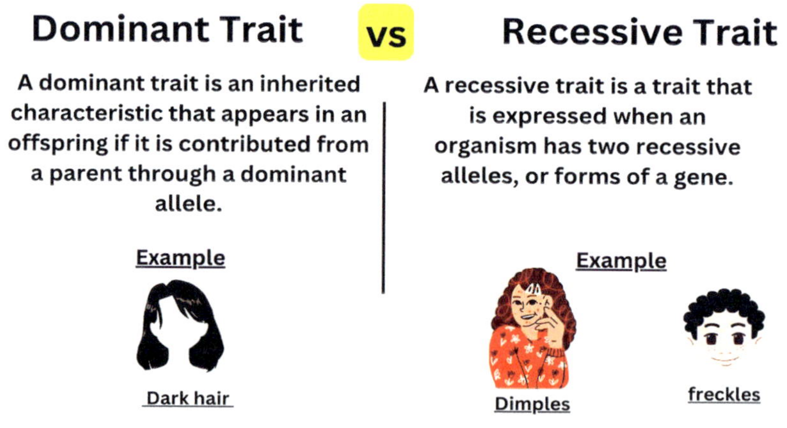

Figure 10.14. Differences between dominant and recessive traits.

Source: https://eduinput.com/storage/2023/05/Differences-Between-Dominant-And-Recessive-Traits-image.jpg.

10.4.3. Punnett Squares and Probability in Inheritance

Punnett square usage is a fundamental device to understand inheritance patterns, particularly in meiosis. They enable scientists and educators to predict the ratio of genotype and phenotype of offspring from a gene cross. Through absolutely representing the union of alleles in both the parents, Punnett squares clarify how a trait gets passed on from generation to generation. The application of probability within these squares represents the inherent randomness of genetic sorting during meiosis, in which alleles sort out independently. Therefore, when one studies the traits regulated by simple Mendelian inheritance, one can predict the likelihood of particular traits occurring within the offspring, revealing the complex dependence between genotype and phenotype as they cascade from one generation to the next. The incorporation of probability in inheritance studies transcends basic characters to complex genetic interactions, revealing the nuances of polygenic traits and environmental influences. While Punnett squares are traditionally employed to describe Mendelian inheritance, they are applied further to cover many types of inheritance, including incomplete dominance and epistasis. For example, in polygenic characteristics like skin color, in which many genes contribute to one phenotype, Punnett square diagrams facilitate the ability to make probabilistic sense of phenotypic variation. The knowledgeable comprehension of the inheritance of genes is aligned with the requirements outlined in the next generation science standards (NGSS), which emphasize statistical thinking and data interpretation as fundamental aspects in biology education. Employing these constructs in critical educational frameworks, educators can aid in the promotion of deeper understanding of genetic principles among student groups with diverse backgrounds, as underscored by the necessity of a culturally responsive pedagogy for the teaching of science (Quattrini et al., 2021). Moreover, the development of expert approaches to teaching genetics, and how they utilize Punnett squares, is essential to involve increasingly diverse learners in the sciences. As they transition to align their curricula with contemporary standards , such as the NGSS, the need for inquiry learning emerges. Incorporating real-world applications of genetic concepts as areas of focus for study not only heightens student engagement but also educates students with essential skills in analyzing and interpreting genetic variation data. Hands-on learning, which can involve technology and collaborative projects, has the capacity to make scientific knowledge more widely accessible, a process highlighted by efforts to develop an equitable science curriculum (Hogan et al., 2018). Thus, by widening the range of genetic instruction to include probabilistic thinking and ordinary contexts, educators can help build a future generation of scientifically literate and socially engaged citizens who are prepared to address the challenges of genetics in contemporary society.

Test Your Knowledge

1. Explain Gregor Mendel's principles of inheritance.
2. What are the Punnett squares and probability in inheritance?

10.5. Non-Mendelian Inheritance

Learning Outcome

1. Incomplete dominance and codominance;
2. Polygenic inheritance and its implications;
3. Epigenetics and environmental influences on gene expression.

Non-Mendelian inheritance is a topic of great significance in genetic variation beyond the classical Mendelian model. Though Gregor Mendel's laws were the foundation of inheritance, they do not define all genetic events. Genetic exchange in *Trypanosoma brucei*, a parasitic organism, is one such instance. Research indicates that this organism has maintained genetic diversity through new mechanisms of genetic recombination and segregation that do not involve classical Mendelian processes. The multigenic genomes of such parasites illustrate how independent assortment and crossing over can result in novel combinations of alleles, vindicating the complexities of eukaryotic inheritance patterns. By integrating the data of genetic mapping and polymorphic markers, researchers have constructed highly informative genetic maps of *T. brucei*, outlining how such non-Mendelian traits may inform more general studies of genetics and disease susceptibility (MacLeod et al., 2007). Further exploring the concept of non-Mendelian inheritance, the genus *Pseudonitzschia multistriata* demonstrates an intriguing relationship between sexual reproduction and vegetative cell division in its diplontic life cycle. Similar to Mendelian inheritance, significant aspects of inheritance, , such as segregation, can often appear fixed across generations.

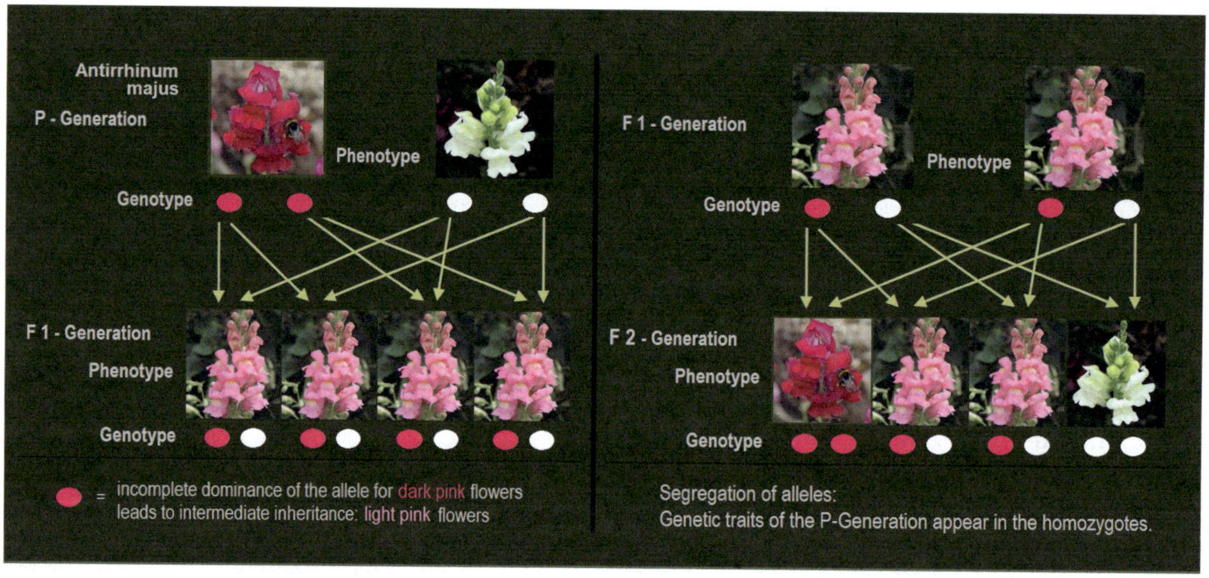

Figure 10.15. Non-Mendelian inheritance.

Source: By Jacek Halicki, I am Jim, Crochet.david and Sciencia58 - I made this inheritance pattern using the following images: File:2014 Kwiat lwiej paszczy.jpg, File:Snapdragon -- Antirrhinum majus.jpg, File:Amboise, fleur d'un parterre 4.jpg, File:Intermediate inheritance - incomplete dominance.png, CC BY-SA 4.0, https://commons.wikimedia.org/w/index.php?curid=82956143

Nevertheless, departure from predicted Mendelian ratios is common, resulting mainly from mutations caused by sexual reproduction. In research on this diatom, scientists noticed high rates of mutation during the early stages of maintenance of the culture, indicating that genetic diversity can build up very quickly. This observation reinforces that non-Mendelian processes, , such as mutation rates and allelic heterogeneity, can play an important role in structuring a population's genetic makeup, underpinning the need to take these phenomena into account in evolutionary and ecological frameworks (Amos et al., 2012). The significance of non-Mendelian inheritance transcends individual organisms to shape populations and ecosystems in important ways. Non-Mendelian mechanisms , such as cytoplasmic inheritance, genomic imprinting, and epigenetics, add dimensions to the predictability of genetic traits. Maternal effects, for instance, can affect expression in the offspring's phenotypic trait, which is different from the expected Mendelian outcome. Additionally, as documented in a number of case studies, environmental pressures interacting with genetic variability aid in adaptive evolution. Their study provides further insight into the overall problems of genetic variation, disease resistance, and adaptation in changing environments. By understanding the interaction of Mendelian and non-Mendelian mechanisms, scientists can build a more comprehensive picture of inheritance, which is crucial in domains , such as conservation biology, agriculture, and medicine (Figure 10.15).

10.5.1. Incomplete Dominance and Codominance

The genetic processes of incomplete dominance and codominance are the key non-Mendelian patterns of inheritance that add to our knowledge of phenotype expression. In incomplete dominance, the phenotypic expression of heterozygous individuals is an intermediate between the two parental forms. For example, in some flower varieties, when a red flower and a white flower are crossed, the result is pink offspring, showing how neither allele is completely dominant. This smearing of expression of qualities not only highlights the complexity of genetic interaction but also indicates that phenotypes emerge due to the complex mixture of allelic inputs, which is essential in grasping the diversity one observes in nature. Examining these patterns will enable us to uncover the laws of inheritance that work in simple and complex qualities. Conversely, codominance refers to a certain genetic process in which both alleles develop fully in the heterozygous state. It occurs in human blood groups, with someone with an A allele and a B allele having both A and B antigens on their red blood cells, resulting in them having type AB blood. This development substantiates that allelic interactions create distinct phenotypic effects without any kind of dominance. Codominance demonstrates the mechanism of genetic variation underlying important biological processes in populations and is pertinent to an understanding of organ transplant compatibility and heritable genetic health conditions. The study of these models of inheritance provides valuable insight into how characteristics are passed down and expressed through generations, relevant to both evolutionary biology and medicine. The value of learning incomplete dominance and codominance extends beyond the mere classification of traits; it has practical applications in areas , such as agriculture and medicine. For example, plant breeders would be able to apply such principles to develop new strains with desired phenotypic traits, enhancing crop yields and resistance. Even in human medicine, knowledge of inheritance can be applied in genetic counseling and risk prediction for hereditary diseases. As seen in the integration of school material with contemporary teaching standards, , such as those found in the NGSS, the inclusion of incomplete dominance and codominance within curricula for genetics deepens understanding of these processes among students. By emphasizing these principles, educators aim to improve the degree of student engagement, as illustrated in courses that utilized innovative teaching strategies for genetics (Warren et al., 1999; Hogan et al., 2018). Utilizing this interdisciplinary understanding of patterns of inheritance, the next generation of scientists will be able to solve more complex biological problems.

10.5.2. Polygenic Inheritance and Its Implications

Polygenic inheritance is a fundamental concept in genetics that addresses more than one gene that contributes to a single phenotypic trait. Unlike the concentration on single-gene traits and distinct phenotypes in Mendelian inheritance, polygenic inheritance captures the trait complexity in such traits as height, skin color, and intelligence, where numerous alleles play together to determine phenotypic outcomes. This multifactorial foundation not only highlights the importance of additive allelic effects but also suggests interaction between environmental and genetic components. The vast scope of phenotypic variation occurring in populations is a consequence of this rich genetic make-up, demonstrating the ways in which more than one gene can cooperate to produce numerous diverse results. Additionally, grasping polygenic inheritance is important for medicine and agriculture, where the selection of traits affects breeding techniques and interventions for health. The significance of polygenic inheritance is far-reaching in the domains of behavioral genetics and the social sciences, influencing our comprehension of human variation and ability. Genetic inclinations that result from polygenic traits, although manipulated by them, are not controlling. This sparks ongoing controversies over the interaction of environment and genetics, putting to rest simplistic conceptions of heredity and into the nature-nurture controversy. The implications for education are particularly relevant. With studies increasingly unveiling the pervasive intervention of genetic factors, teachers are confronted with having to provide an equitable education environment, recognizing and conforming to these differences. These kinds of educational reforms can encourage greater inclusivity and engagement in STEM fields, particularly among previously excluded groups, like with the request for curriculum reforms that right the discrepancy in interest among BIPOC students (Quattrini et al., 2021). Developments in genetic studies and the understanding of polygenic inheritance have far-reaching ethical implications, particularly in genetic testing and misuse. With its potential measurement and prediction, polygenic traits raise questions of privacy, discrimination, and ethics of dissemination of genetic data. The ethical dilemma is to balance pursuit of knowledge against the danger of misinterpreting or misusing genetic data. In addition, increased availability of genetic testing products could fuel the dissemination of genetic determinism, as individuals might perceive their potential and capabilities as having a significant portion determined by their genes.

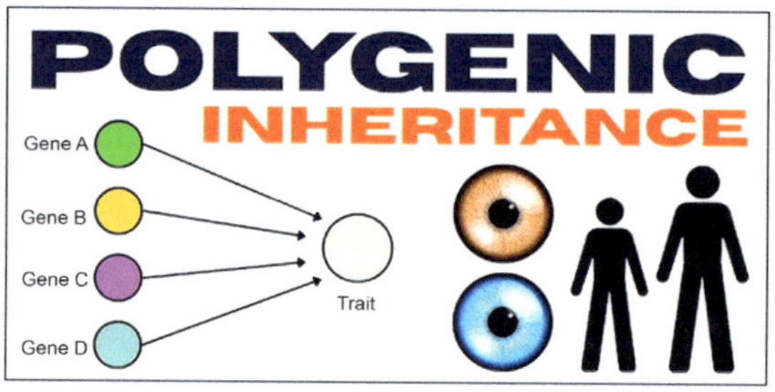

Figure 10.16. Polygenic inheritance.

Source: https://i.ytimg.com/vi/s9IBIDUzeG4/hq720.jpg?sqp=-oaymwEhCK4FEIIDSFryq4qpAxMIARUAAAAAGAElAADIQj0AgKJD&rs=AOn4CLBg6-KNTdYd57SaI5D6n0sUDyD9wg.

This type of reasoning could also reinforce social inequalities if genetic information is used to justify inequalities in education or access to health care. Hence, it is crucial that ongoing debate addresses these challenges while society grapples with the implications of polygenic inheritance, undermining a balanced appreciation which synthesizes both scientific knowledge and ethical thought (Figure 10.16).

10.5.3. Epigenetics and Environmental Influences on Gene Expression

The interaction of epigenetics and environmental stimuli is significant in the regulation of gene expression, particularly under conditions of meiosis and heredity. Epigenetic modifications , such as DNA methylation and histone alteration are adjustable to react to various environmental stimuli, and thus change the phenotypic traits without the modification of the associated genetic code. This process is typically accountable for so-called phenotypic plasticity, whereby the phenotype of an organism can adjust as a response to changing environmental conditions. For instance, there exist stressors that initiate epigenetic events that then influence patterns in gene expression, and thus traits that are very important for survival and reproduction. Understanding these mechanisms is crucial in separating the way offspring can inherit the adaptive traits shaped by the environmental pressures their parents went through, thereby linking epigenetics with inheritance laws as observed in the process of meiosis. The research has recently shown the importance of transgenerational epigenetic inheritance; whereby epigenetic modifications can be passed on over generations and influence phenotypic effects. This principle presumes that environmental interactions are not merely transitory; instead, they can leave profound imprints on the gene expression of subsequent generations. In a case, plants are seen to exhibit transgenerational priming of defense, which means that pest stress is likely to trigger epigenetic modification which prepares offspring for the same stress (Roberts et al., 2015). Furthermore, this capacity for epigenetic inheritance complicates our ability to use straightforward models of inheritance that are dependent only on Mendelian mechanisms, as the offspring's phenotype can now be defined not only by their genotype but also by the parental environmental past. This process of epigenetics thus better accounts for how stress-adaptive traits evolve in populations. In addition, epigenetic study shifts from an exclusive focus on biology to the domain of evolutionary theory, with the hypothesis that the interplay between gene expression and environment is at the forefront of how evolutionary direction is settled. Interestingly, reference to a zygotic unicellular stage underscores the fact that the development of the cell early in life is of critical importance and environmentally sensitive (Miller et al., 2016). This view proposes that mechanisms of inheritance are linked with complex environmental processes that can shape evolutionary change. With the analysis of how epigenetic signals integrate environmental experience into the hereditary framework, researchers can extract information on the adaptive range of organisms. Lastly, a careful exploration of epigenetics in relation to inheritance and meiosis provides insight into the dynamic relationship between an organism and its environment, revealing the complex mechanism by which traits are passed on.

Test Your Knowledge

1. What are incomplete dominance and codominance?
2. What is polygenic inheritance and its implications?

10.6. Applications of Meiosis and Inheritance

Learning Outcome

1. Implications in agriculture and selective breeding;
2. Role in genetic counseling and disease prevention;
3. Advances in biotechnology and genetic engineering.

Meiosis is key to sexual reproduction, producing genetic variation by independent assortment and crossing over. Through the production of gametes that are genetically distinct from one another, meiosis guarantees offspring a unique combination of characters from each parent. Genetic variation is essential in evolution because it allows populations to adapt to new environments. In a species like Trypanosoma brucei, genetic exchange processes expand the scope of meiotic processes beyond paradigms. Research has presented evidence that meiosis occurs in *T. brucei* and, as such, it is a valuable model for investigating genetic inheritance systems in protozoa. Sequencing of the genome has given researchers the capacity to construct a genetic map of *T. brucei*, again supporting the existence of meiosis and its application towards genetic study in disease etiology and potential therapy (MacLeod et al., 2007). The application of meiosis is also prestigious in agriculture, particularly in the improvement of plants through selective breeding. Traditional approaches would involve several generations of inbreeding to achieve true-breeding lines; however, advancements in haploid induction techniques have revolutionized the process. By causing haploids and doubling their chromosomes subsequently, breeders can produce homozygous lines in a single generation, greatly compressing the timescale of breeding. For instance, manipulation of centromere-specific histone CENH3 in *Arabidopsis* has been well-suited for haploid induction and has already been successfully altered in crops , such as maize. These improvements are not only making breeding routines more efficient but also increasing the genetic diversity from which crops can be bred. This makes way for more robust agricultural strains and better harvests (Britt et al., 2016). The knowledge gained from meiosis also translates to understanding and controlling genetic diseases in human beings. Meiosis ensures that chromosomal integrity is maintained, and any variation in this is responsible for genetic disorders like aneuploidy. Studies of such meiotic mistakes shed light on a variety of genetic disorders like Down syndrome and Turner syndrome. In addition, advances in genetic mapping and inheritance analysis make it possible for researchers to identify chromosomal defects related to these diseases and, eventually, direct genetic counseling and therapy. As the understanding of meiosis and its role in inheritance grows, so does its potential to benefit not only agricultural practices but human health as well, demonstrating the breadth of application across a variety of disciplines (MacLeod et al., 2007).

10.6.1. Implications in Agriculture and Selective Breeding

Use of selective breeding in agriculture has far-reaching implications for the inheritance pattern of cultivated organisms. By making careful choices of specimens with desired traits, farmers have influenced genetic outcomes over generations, thereby raising yields and resistance to disease and insects. The case of maize presents such revolutionary changes, with early Mesoamerican farmers having domesticated the wild grass teosinte into the domesticated maize species that are staples today. Such a shift did not come by chance; comparison of the genomes of maize and its wild relatives has shed light on the genetic basis of those traits that have been selected for during selective breeding. The knowledge gained helps inform breeders, enhancing efficiency in producing new cultivars that are better suited to meet agricultural needs (Hake et al., 2015). Also, the entry of epigenetics has emerged as a revolutionary element in selective breeding that challenges traditional ideas of genetic heredity. Epigenetic modification , such as DNA methylation, is heritably stable and can influence phenotypic characteristics without modifying the original DNA sequence.

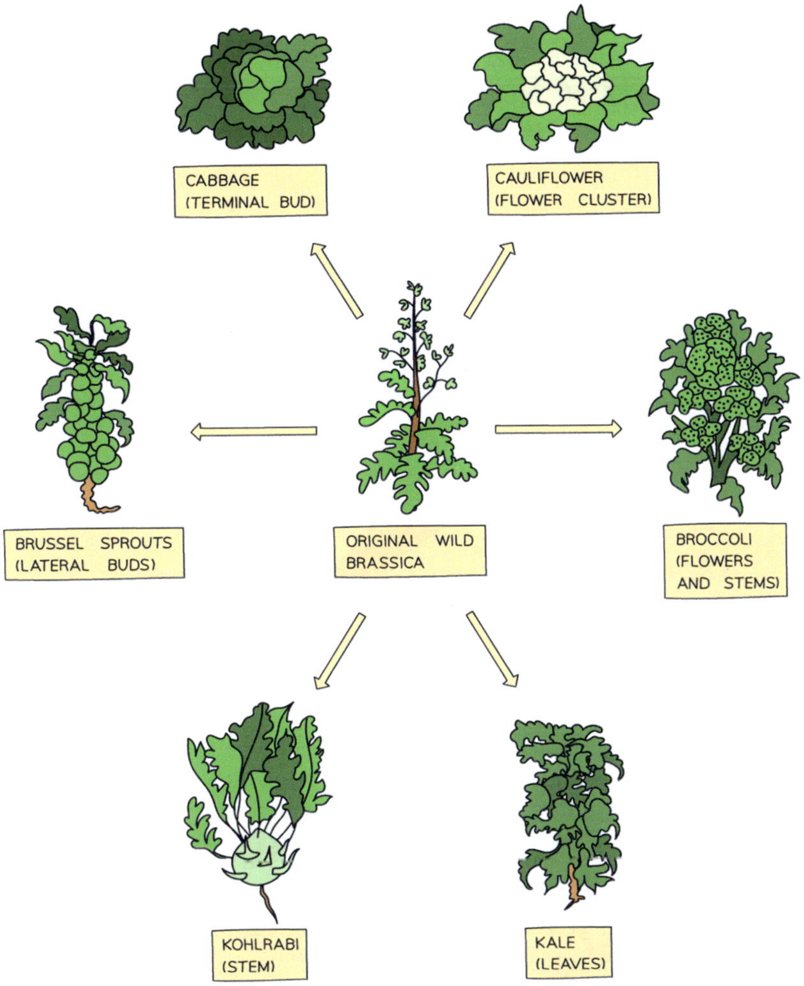

Figure 10.17. Selective breeding in plants.

Source: https://cdn.savemyexams.com/uploads/2020/01/Selective-breeding-plants.png.

Experimental evidence from studies on the woodland strawberry has shown how demethylating chemicals can generate more phenotypic variation, giving a new direction to selective breeding. The ability to create a population that is more epigenetically diverse but has a shared genetic origin expands the genetic pool available to breeders. Essentially, this method not only enables the preservation of favorable traits over time but also enables the generation of variability that can be harnessed in response to changing environmental conditions (Xu et al., 2016). Overall, the integration of meiosis and selective breeding culture has profound implications for agricultural sustainability. By utilizing the leverage of both genetic and epigenetic factors, breeders can produce crops that not only yield highly but also resist other stresses and climate change. Employing this two-pronged method makes the agricultural system more robust to withstand the strains of a growing global population and shifting climate patterns. As new farming methods continue to alter, understanding the mechanisms of inheritance from genetic to epigenetic will play a major role in designing effective breeding schemes. Combined with advances in contemporary technology, it can transform agriculture, ensuring food security without compromising the diversity of biodiversity (Hake et al., 2015; Xu et al., 2016) (Figure 10.17).

10.6.2. Role in Genetic Counseling and Disease Prevention

The genetic counseling service is becoming ever more central in the meiotic and inheritance context since it directly influences family decision-making and public health. Genetic counselors provide crucial information that pertains to the probable risk of inherited disease as a result of meiotic errors , such as nondisjunction, which leads to Down syndrome. Through the description of how genetic traits are passed down to generations, genetic counselors educate families on their genetic backgrounds and the risks of transmitting conditions to future generations. By integrating information from both genetic screening and novel techniques like amniocentesis and chorionic villus sampling, counselors can identify high-risk pregnancies and counsel pregnant parents on preventive measures and well-educated reproductive choices that honor their circumstances and values (Burada et al., 2018). In contemporary practice, the effectiveness of genetic counseling hinges on the complete elucidation of disease mechanisms and available testing protocols. Increasing genetic disease sophistication necessitates continuing education for health care professionals, as highlighted by a study of physicians' educational needs in the field of genetics (Alwan et al., 2007).

Many clinicians indicate feeling inadequately prepared to manage genetic risk and advise families effectively, potentially limiting disease prevention. Through investment in genetic training and education among healthcare workers, the health system can optimize the use of genetic counseling for early detection of genetic predispositions. Such early detection not only results in better health outcomes but also instills a culture of awareness of genetic determinants of health, which is instrumental to reducing genetic disease prevalence. Moreover, genetic counseling is also an important gateway to addressing ethical considerations of genetic testing and reproductive choices. As families struggle with the implications of their genetic heritage, counselors assist them in navigating the exploration of options , such as preimplantation genetic diagnosis and prenatal testing. These decisions can decrease the uncertainty that pregnant parents who have genetic risk go through, allowing them to make informed decisions regarding how they should handle their pregnancies (Burada et al., 2018). Simultaneously, genetic counseling makes ethics discussions possible, particularly on the choices that parents make in relation to what is learned. Where ethics, genetics, and counseling converge, it is evident that genetic counseling plays a complex role in disease prevention. By way of explanation and advice, genetic counseling not only aids in risk assessment but also helps to prepare families for the potential outcomes of their genetic make-up, an essential precursor to informed family planning in modern times (Figure 10.18).

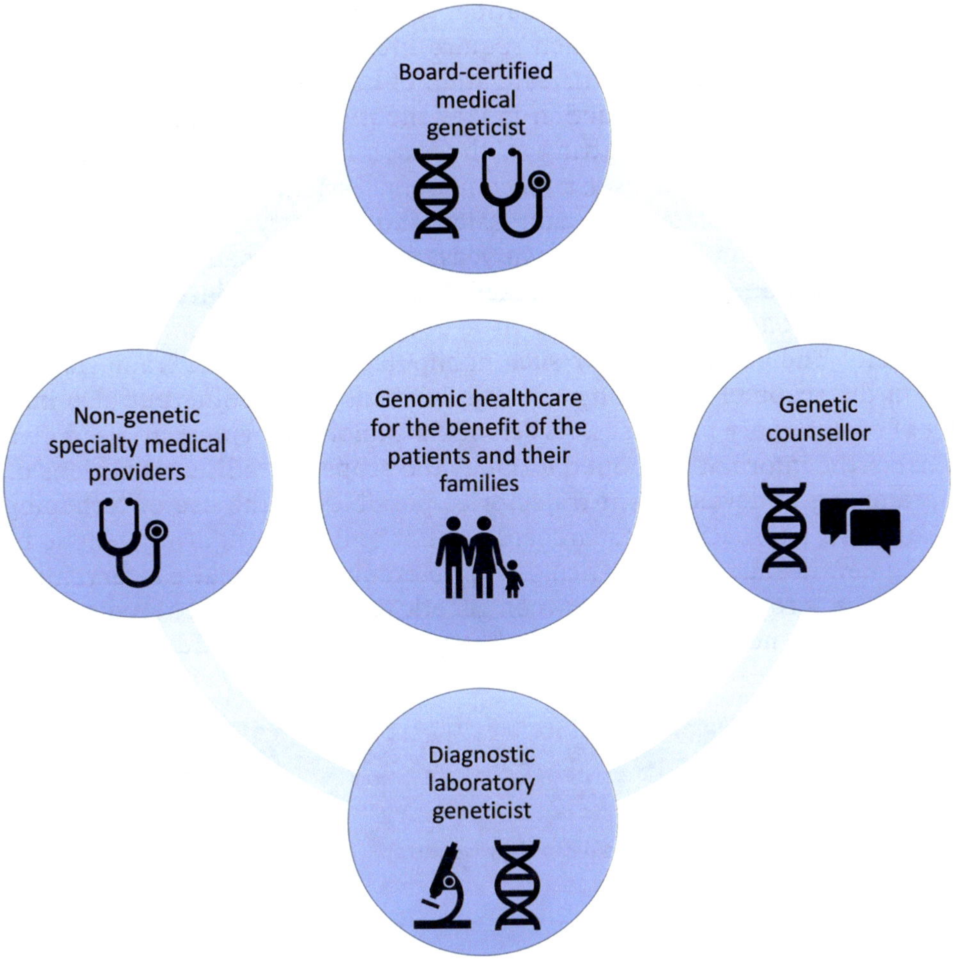

Figure 10.18. Role in genetic counselling.

Source: By DataBase Center for Life Science (DBCLS) - https://doi.org/10.7875/togopic.2022.198, CC BY 4.0, https://commons.wikimedia.org/w/index.php?curid=119380142

10.6.3. Advances in Biotechnology and Genetic Engineering

The recent advances in genetic engineering and biotechnology are revolutionizing the understanding of meiosis and heredity. Research into processes , such as apomixis has been of specific interest, demonstrating how certain organisms can bypass typical meiotic mechanisms. Nucellar cells in angiosperms, for instance, have the phenomenal capability to utilize or avoid meiosis depending on environmental cues and developmental patterns. The ongoing quest to introduce apomixis gene into sexual crop species is a bid to enhance agricultural output by providing a way to produce clones from seeds without the need for fertilization. This biotechnology project not just holds promise in food security but also disrupts existing paradigms of plant reproduction and genetic diversity, compelling researchers to rethink the role of epigenetic elements in inheritance and developmental programming, as has been acknowledged by latest genetic studies (Ranganath et al., 2004). Additionally, yeasts, particularly some of the species like *Saccharomyces cerevisiae*, are a valuable model system for genetic manipulation, with understanding of meiotic processes and modes of inheritance. These unicells, which require little in the way of diet and will grow very quickly, provide a robust system for the investigation of biochemistry and genetics. Their capacity to undergo sexual and asexual reproduction makes them perfect organisms to monitor meiosis-associated events in an experimental setting. With yeast strains genetically engineered to produce a variety of biopharmaceuticals and industrial compounds, scientists are learning valuable insights into the mechanisms that regulate gene expression and heredity. This capability to identify the foundation of genetic variation and heredity in yeast also has a broad impact from basic research to biotechnology application (Lachance & Walker, 2018). Lastly, the intersection of biotechnology innovation and genetic engineering results in a deeper understanding of the complexities of meiosis and heredity. The possibilities for such manipulation of genetic frameworks in organisms not only usher in the scope of examining new mechanisms of breeding but also impart a deeper understanding of evolutionary processes. As scientists explore the epigenetic and genetic domains of many organisms, the information gained expands the scope of traditional notions of inheritance, making reprogramming of development trajectories possible. By the use of technologies , such as CRISPR and transgenic approaches, researchers are not only reconfiguring genetic blueprints but are also unveiling new avenues for enhancing crop tolerance, therapeutic interventions, and green bioproduction. This reciprocal interplay between genetic engineering and heredity research explains why there is such a vital need for cross-disciplinary approaches in addressing world health and agriculture problems.

Test Your Knowledge

1. What are the implications in agriculture and selective breeding?
2. What is the role in genetic counseling and disease prevention?

Summary

The complex processes of meiosis and inheritance form the basis of the investigation of genetic variability and evolution in species. Meiosis not just enables sexual reproduction but is also responsible for genetic variation through mechanisms , such as independent assortment and crossing-over. The consequences of these processes for natural selection and adaptation are massive, as testified by a number of studies demonstrating the adaptive significance of polyploid organisms under diverse environments. A good example is the polyploid *Cerastium alpinum* that shows how meiotic-generated genetic diversity leads to survival in a specific niche, and particularly under stressed environments , such as serpentine soils. This further highlights the importance of meiosis not just in allowing genetic diversity but also in resilience of populations when subjected to environmental stresses and thus repeating the importance of understanding these processes in ecological environments (Berglund et al., 2005). The findings concerning genetic inheritance illuminate further the interaction of meiosis with evolutionary process. Experiments involving organisms , such as *Pseudonitzschia multistriata* have detailed how vegetative propagation and sexual reproduction cooperate in the contribution of genetic diversity through Mendelian inheritance and mutation frequency. Detections of deviation from the anticipated mode of inheritance point to the complexity within genetic transmission, revealing anomalies that can contribute to adaptation. These mutations, with rates varying based on genetic and environmental conditions, can diversify the genotypic landscape of the population. This mutual interaction between mutation and inheritance serves to underscore the role of meiosis in creating genetic variation integral to long-term species adaptability (Amos et al., 2012). In the larger context of evolution and genetics, what has been learned from the examination of inheritance and meiosis reinforces the complex interplay between genetic processes and ecological results. The processes of meiosis, , such as the proper segregation and reunion of alleles, operate not just to transmit genetic information but also to increase genetic diversity among offspring according to environmental stresses. With the organism adjusting to changing environments and novel threats to survival, the adaptability resulting from these mechanisms at the genetic level permits speciation and evolutionary survival. Continued research into the mechanisms of meiosis and heredity is thus essential, not merely for improving our concept of genetics, but also for supporting conservation efforts and the preservation of diversity in a rapidly changing world.

Meiosis is the sexual cell division process that is the cause of genetic diversity by the reduction in gametes of chromosome numbers. It is preceded by DNA replication and involves two successive divisions—meiosis I and meiosis II—with the production of four haploid daughter cells. Unlike mitosis, which gives rise to daughter cells that are the same as the parent cell, meiosis gives rise to genetic variation by processes , such as crossing over and independent assortment. At prophase I, the homologous chromosomes undergo genetic recombination, where pieces of DNA are exchanged with one another, yielding new allele combinations. Genetic shuffling is crucial to evolution because it determines the genotype and phenotype in subsequent generations, eventually leading to species survival in environments that are constantly changing (Matthews et al., 2016). Inheritance patterns, highly influenced by meiotic processes, determine how traits are passed from parents to offspring. The rules established by Gregor Mendel, that is, the conditions of segregation and independent assortment, provide the foundation for the understanding of inheritance. Mendels first law stipulates that alleles in a trait segregate during the formation of gametes such that offspring inherit one allele from each parent. This is followed up by the second law stipulating that segregation of alleles in unlike traits occurs independently of one another. Collectively, these statutes illustrate how genetic variation is generated in populations, with meiosis as the mechanism for Mendelian inheritance patterns. The random arrangement of chromosomes in meiosis once more illustrates the use of chance in generating genetic outcomes. Such mechanisms are essential in evolutionary forces because they create the variation that enables natural selection to work best (Beckers et al., 2009). In the context of modern genetics, understanding of meiosis and heredity goes beyond Mendelian principles to include the complex interplay of genetic and environmental forces. Advances in molecular biology have revealed how genetic variation contributes to disease susceptibility and reaction to environmental stress. For instance, recent research shows how host and pathogen interactions determine response to infection, with loci outlined influencing disease pathogenesis. This provides us with a new level to grasp inheritance, as traits are not simply explained by Mendels laws but are also affected by genetic interactions and environmental conditions. The interplay of environment and genetics works to reinforce the significance of genetic studies, including quantitative trait loci (QTL) mapping, as integral tools in explaining the intricacies of inheritance and disease process, thus shaping the focus of genetic research and implications on health and disease treatment (Beckers et al., 2009).

KEY TERMS FROM THIS CHAPTER

Allele	Crossing over	Diploid (2n)
Dominant trait	Gametes	Gene
Genetic variation	Genotype	Haploid (n)
Heterozygous	Homologous chromosomes	Homozygous
Independent assortment	Meiosis	Mendelian inheritance
Phenotype	Punnett square	Recessive trait

Critical Thinking Questions

1. What is the primary purpose of meiosis?

a. To produce identical cells

b. To repair damaged tissues

c. To produce gametes (sex cells) with half the number of chromosomes

d. To increase cell size

2. How many daughter cells are produced at the end of meiosis?

a. 1

b. 2

c. 3

d. 4

3. During which phase of meiosis does crossing over occur?

a. Prophase I

b. Metaphase I

c. Anaphase II

d. Telophase II

4. Which of the following best describes Mendel's Law of Segregation?

a. Traits are blended from both parents

b. Two alleles for a trait separate during gamete formation

c. One allele always dominates over another

d. Chromosomes are inherited in linked groups

5. If an organism has two identical alleles for a trait, it is said to be:

a. Heterozygous

b. Dominant

c. Recessive

d. Homozygous

Short Answer Questions

1. What is the purpose of meiosis, and how is it different from mitosis?
2. How many cells are produced at the end of meiosis, and how many chromosomes do they contain (in humans)?
3. What are homologous chromosomes?
4. During which phase of meiosis does crossing over occur, and why is it important?
5. What is independent assortment, and how does it contribute to genetic variation?
6. What is the difference between genotype and phenotype?

Answers to Critical Thinking Questions

1. (c); 2. (d); 3. (a); 4. (b); 5. (d)

References

1. Ales Pecinka, Christian Chevalier, Isabelle Colas, Kriton Kalantidis, Serena Varotto, Tamar Krugman, Christos Michailidis, María-Pilar Vallés, Aitor Muñoz, Mónica Pradillo, Chromatin dynamics during interphase and cell division: similarities and differences between model and crop plants, *Journal of Experimental Botany*, Volume 71, Issue 17, 17 August 2020, Pages 5205–5222, https://doi.org/10.1093/jxb/erz457 (accessed on 13th June 2025).
2. Alam, S., Almeida-Santos, T., Anderson, R., Ankel-Simons, F., Barritt, J., Bavister, B., Blok, A., et al. (2006). Transmission of mitochondrial DNA following assisted reproduction and nuclear transfer. https://core.ac.uk/download/29820080.pdf (accessed on 13th June 2025).
3. Alberts, B., Finta, C., Waters, C. K., Communi, D., et al. (2006). Genes in the postgenomic era. https://core.ac.uk/download/11921232.pdf (accessed on 13th June 2025).
4. Allen, S. K., Jr., De Sousa, J. T., Komaru, A., et al. (2017). Meiotic and early zygotic development in *Crassostrea virginica* observed through confocal microscopy. https://core.ac.uk/download/235399132.pdf (accessed on 13th June 2025).
5. Altschmied, L., Avesani, L., Barcaccia, G., Galla, G., et al. (2017). Pistil transcriptome analysis to disclose genes and gene products related to apos porous apomixis in *Hypericum perforatum* L. https://core.ac.uk/download/84922892.pdf (accessed on 13th June 2025).
6. Alwan, A., Tomatir, A. G., Baars, M. J. H., Akdağ, B., Burke, W., Caulfield, T., Christianson, A., et al. (2007). Genetics and genetic counseling: Practices and opinions of primary care physicians in Turkey. https://core.ac.uk/download/227060650.pdf (accessed on 13th June 2025).
7. Amos, W., Bengtsson, B. O., Brinkmann, B., Callen, D., Casteleyn, G., Chepurnov, V., et al. (2012). Mendelian inheritance pattern and high mutation rates of microsatellite alleles in the diatom *Pseudo-nitzschia multistriata*. https://core.ac.uk/download/9838856.pdf (accessed on 13th June 2025).
8. Barcaccia, G., Sharbel, T. F., Galla, G., Heiko, T., et al. (2015). *De novo* sequencing of the *Hypericum perforatum* L. flower transcriptome to identify potential genes that are related to plant reproduction *sensu lato*. https://core.ac.uk/download/53524393.pdf (accessed on 13th June 2025).
9. Baroiller, J. F., Bezault, E., Chevassus, B., Clota, F., et al. (2012). Analysis of the meiotic segregation in intergeneric hybrids of tilapias. https://core.ac.uk/download/28970602.pdf (accessed on 13th June 2025).
10. Beckers, A., Cooper, A., Khan, A., MacLeod, A., Martinelli, A., et al. (2009). A major genetic locus in *Trypanosoma brucei* is a determinant of host pathology. https://core.ac.uk/download/110298.pdf (accessed on 13th June 2025).
11. Bekish, V. J., & Nikulin, Y. T. (2006). *Practical Book on Medical Biology.* https://core.ac.uk/download/80151753.pdf (accessed on 13th June 2025).
12. Bekish, V. J., Logishinets, I. A., & Zorina, V. V. (2017). *Practical Book on Biology.* https://core.ac.uk/download/80152210.pdf (accessed on 13th June 2025).
13. Ben Maamar, M., Nilsson, E. E., & Skinner, M. K. (2021). Epigenetic transgenerational inheritance, gametogenesis and germline development. https://core.ac.uk/download/603337054.pdf (accessed on 13th June 2025).
14. Berchowitz, L., Jorgensen, V., Joshi, P., Sawyer, M., et al. (2018). Developmental regulation

of an organelle tether coordinates mitochondrial remodeling in meiosis. https://core.ac.uk/download/323081944.pdf (accessed on 13th June 2025).

15. Blaxter, M., Farrell, M. C., Koutsovoulos, G., Parihar, M., et al. (2025). Gender- and gamete-specific patterns of X chromosome segregation in a three-gendered nematode. https://core.ac.uk/download/110168630.pdf (accessed on 13th June 2025).

16. Britt, A. B., & Kuppu, S. (2016). Cenh3: An emerging player in haploid induction technology. https://core.ac.uk/download/323067865.pdf (accessed on 13th June 2025).

17. Burada, F., Chelu, G., Cimpoeru, A. L., Cucu, C., et al. (2018). Prenatal genetic counseling in congenital anomalies. https://core.ac.uk/download/322435467.pdf (accessed on 13th June 2025).

18. Cebrat, S., Stauffer, D., & Waga, W. (2010). The role of haplotype complementation and purifying selection in the genome evolution. http://arxiv.org/abs/1007.3357 (accessed on 13th June 2025).

19. Ek, W. (2011). Genetic analysis of autoimmune and metabolic traits in chickens. https://core.ac.uk/download/11697807.pdf (accessed on 13th June 2025).

20. Ferguson-Smith, A. C., Lorenz, A., & Wolf, J. B. (2022). Mendel's laws of heredity on his 200th birthday: What have we learned by considering exceptions? https://core.ac.uk/download/559320617.pdf (accessed on 13th June 2025).

21. Flynn, J. R., & McNally, F. J. (2017). A casein kinase 1 prevents expulsion of the oocyte meiotic spindle into a polar body by regulating cortical contractility. https://core.ac.uk/download/323069928.pdf (accessed on 13th June 2025).

22. Ghafari, F., Gutierrez, C. G., & Hartshorne, G. M. (2007). Apoptosis in mouse fetal and neonatal oocytes during meiotic prophase one. https://core.ac.uk/download/49355.pdf (accessed on 13th June 2025).

23. Hake, S., & Ross-Ibarra, J. (2015). Genetic, evolutionary and plant breeding insights from the domestication of maize. https://core.ac.uk/download/323069148.pdf (accessed on 13th June 2025).

24. Hogan, E. (2018). Genetics unit lesson plan for the Next Generation Science Standards. https://core.ac.uk/download/222998260.pdf (accessed on 13th June 2025).

25. Lachance, Marc-André and Walker, Graeme M (October 2018) Yeasts. In: eLS. John Wiley & Sons, Ltd: Chichester, (accessed on 13th June 2025).

26. MacLeod, A. and Turner, C.M.R. and Tait, A. The system of genetic exchange in Trypanosoma brucei and other trypanosomatids. In Barry, D. and McCulloch, R. and Mottram, J. and Acosta-Serrano, A. (Eds) Trypanosomes: after the genome, Chap 3, pages pp. 71-90. Horizon Press (2007) (accessed on 13th June 2025).

27. Matthews, L. J., Molter, D., Plutynski, A., Vernon, J., et al. (2016). Chance in the Modern Synthesis. https://core.ac.uk/download/186329988.pdf (accessed on 13th June 2025).

28. Miller, W. B., & Torday, J. S. (2016). The unicellular state as a point source in a quantum biological system. https://core.ac.uk/download/323079765.pdf (accessed on 13th June 2025).

29. Nyberg Berglund, A. B. (2005). Postglacial colonization and parallel evolution of metal tolerance in the polyploid *Cirsium alpinum*. https://core.ac.uk/download/11694882.pdf (accessed on 13th June 2025).

30. Quattrini, K. (2021). Creating an equity-based genetics unit using culturally relevant pedagogy and the Next Generation Science Standards. https://core.ac.uk/download/479075164.pdf (accessed on 13th June 2025).
31. Ranganath, R. M. (2004). Harnessing the developmental potential of nucellar cells: Barriers and opportunities. https://core.ac.uk/download/72797918.pdf (accessed on 13th June 2025).
32. Roberts, M. R., & Singh, P. (2015). Keeping it in the family: Transgenerational memories of plant defense. https://core.ac.uk/download/42415027.pdf (accessed on 13th June 2025).
33. Salem, M. S. (2014). Educational corner of the issue: Basic concepts of medical genetics. Formal genetics, part 4. https://core.ac.uk/download/478418894.pdf (accessed on 13th June 2025).
34. Shan, Y. (2016). Exemplarising the origin of a science: A path to genetics (from Mendel to Bateson). https://core.ac.uk/download/79499826.pdf (accessed on 13th June 2025).
35. Teslovich, T. M., & Zöllner, S. (2010). Using GWAS data to identify copy number variants contributing to common complex diseases. http://arxiv.org/abs/1010.5040 (accessed on 13th June 2025).
36. Tweedie, A., Tait, A., Cooper, A., MacLeod, A., Benson, C., Turner, C. M. R., Culleton, R., et al. (2005). Allelic segregation and independent assortment in *T. brucei* crosses: Proof that the genetic system is Mendelian and involves meiosis. https://core.ac.uk/download/1394865.pdf (accessed on 13th June 2025).
37. Warren, K. (1999). A study on behavioral objectives: Do they affect learning? https://core.ac.uk/download/638982074.pdf (accessed on 13th June 2025).
38. Xu, J. (2016). Developing a novel hypomethylated population and characterizing a stable early flowering epimutant in strawberry (*Fragaria vesca*). https://core.ac.uk/download/226147626.pdf (accessed on 13th June 2025).

DNA STRUCTURE AND REPLICATION

11

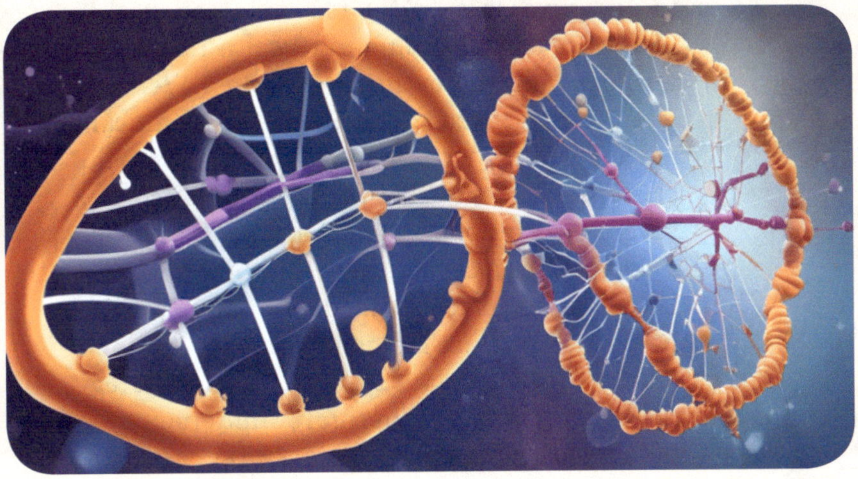

LEARNING OBJECTIVES

After studying this chapter, you will be able to:

1. Explore the discovery of DNA structure;
2. Describe the chemical composition of DNA;
3. Discuss about double helix model;
4. Explain the DNA replication process;
5. Illustrate the mechanisms of DNA repair;
6. List the applications of DNA research.

Introduction

DNA sequence and replication research is fundamental to our comprehension of genetic inheritance mechanisms and cellular functions. Since as early as the 1950s when it was found, the DNA double helical form has been at the forefront of molecular biology, showing the way genetic information is preserved, replicated, and transmitted from one generation to another. The complex building of DNA by its nucleotides and the complementary base pairing enables accurate replication during cell division. Also, the understanding of DNA structure has led the way to advancements in genetic engineering, biotechnology, and medicine. This preamble is merely the introduction to a thorough exploration of the DNA architecture and DNA replication process, in the light of both historical context and contemporary relevance in various scientific disciplines. To further situate DNA replication, it is also necessary to highlight the roles played by the main enzymes and proteins whose involvement is paramount to this complex process. Replication is the unwinding of the helical double helix with enzymes , such as helicase allowing the two strands of DNA to be separated, and then there is the action of DNA polymerases, which build new strands by adding nucleotides complementary to the template.

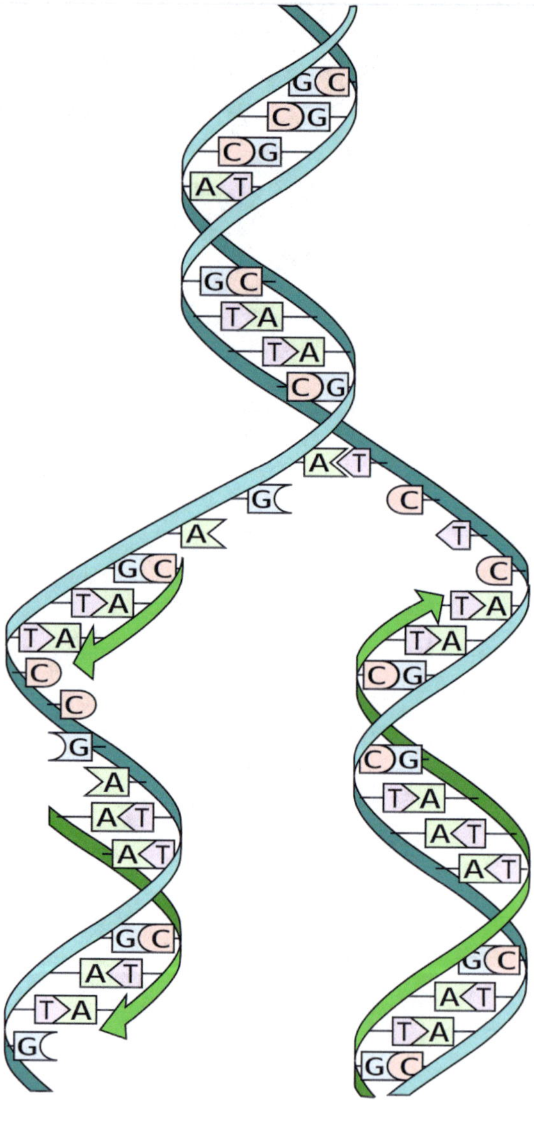

Figure 11.1. DNA structure and replication.

Source: By I, Madprime, CC BY-SA 3.0, https://commons.wikimedia.org/w/index.php?curid=2497221

Every step in the replication process is strictly controlled to prevent errors and ensure fidelity, illustrating the commitment of cells to genetic integrity. Furthermore, the coordinated effort of ligases, primases, and other cofactors illustrates the process's complexity of DNA replication. Studying these elements and their functions enhances the appreciation not only of how DNA is replicated but also of the effects of replication mistakes and how they contribute to genetic disease. DNA research also applies to its use in other different health fields, like virology and gene therapy, as in HIV-1 and latent reservoir research. Even with significant advancements in antiretroviral therapy, the existence of latent HIV-1 within host cells creates severe difficulties in achieving a cure, implying a complex association between viral and host DNA. Studies, , such as the development of PCR-based methods, are intended to characterize this latent reservoir to further inform the strategy for eradication. Furthermore, understanding the molecular machinery of DNA replication in viruses again highlights the interplay between host and viral genetic material, highlighted by the need for novel concepts in the treatment of viral infection. This expanded view of DNA not only brings richness to the discussion of genetic duplication but also interweaves itself into fundamental current health concerns, establishing its pervasiveness in both classic and modern scientific research (Figure 11.1).

Definition of DNA

Deoxyribonucleic acid (DNA) is a very complex macromolecule that serves as the hereditary material for all living things and for many viruses. DNA is a double helix structurally consisting of two strands in opposite directions that are coiled around each other. Each of the two strands consists of a backbone composed of alternating deoxyribose sugar and phosphate groups to which the nucleotide bases are bound.

These bases, adenine (A), thymine (T), cytosine (C), and guanine (G), pair with each other in a very specific manner, with A pairing with T, and C with G. This type of specific pairing is necessary not only for the structural integrity of the molecule but also for the accuracy of replication of genetic information during cell division. By serving as the blueprint for protein assembly, DNA becomes involved in having a pivotal role in determining the characteristics and functions of various organisms. Furthermore, the role of DNA extends beyond fundamental genetics to encompass vital biochemical functions, primarily replication. DNA replication is an essential process that leads to genetic continuity across generations of cells. When replication takes place, the double helix is unwound, each of the original strands serving as a template for the synthesis of a new complementary strand. It relies on a series of enzymes, one of which is DNA polymerase, which adds nucleotides in an incredibly ordered sequence. The process has to be precise; errors in replication can lead to mutations, which could have significant consequences for the organism, , such as disease. Furthermore, the precise regulation of DNA replication is also linked to most of the cell's processes that stress the harmony between genetic stability and responsiveness to environmental stimuli. Besides its structural and functional complexity, DNA is subject to numerous kinds of modifications, , such as DNA methylation, which are of significant relevance to the expression of genes. Methylation is the addition of a methyl group to cytosine bases, altering the expression of genes without changing the underlying DNA sequence. This epigenetic modification usually occurs at CpG islands, regions that consist of densely arranged nucleotides of cytosine and guanine that play crucial roles in gene transcription regulation. Interestingly, external stimuli like environmental signals and developmental cues can modify the patterns of DNA methylation, thus modulating gene accessibility and expression. Such complex regulatory processes bear witness to the dynamism of DNA, siting it not only as a static reservoir of genetic information, but as an active participant in cellular function and adaptation (Caiafa et al., 2005). Moreover, the interconnection of DNA structure, replication, and gene control lies at the core of significant advances in fields , such as genetics and molecular biology, which pave the way to biotechnology and therapy innovations (Kauffman et al., 2002).

Importance of DNA in Genetics

In genetics, DNA is the blueprint of life par excellence, containing the instructions for the growth, development, and functioning of all living organisms. The double helix structure of DNA, made up of nucleotides, allows it to store very large amounts of information in a compact way. This remarkable architecture not only enables the stable heredity of genetic information from one generation to the next but also plays an important role in other biological mechanisms like protein synthesis and cell division. With research uncovering the complexities of genetic diseases, insight into how the structure of DNA specifies genetic expression and replication comes to the forefront. For instance, defects in DNA replication and repair lead to severe disorders, which underscore the significance of the function and integrity of DNA to guarantee health in populations. Interconnectedness among different cell processes involving DNA is also elucidated by continued research in genetic disorders, , such as congenital microcephaly. It has been observed in recent studies that abnormalities in DNA repair and replication are significant factors in the pathomechanism of these conditions, which promotes the significance of DNA in genetics. Here, significant elements like the normality of microtubule organization and centrosomes become pivotal factors in cellular division and differentiation processes, expressing the way molecular complexities can be influential on phenotypic effects. By comparison of these interactions, scientists underscore the necessity of detailed investigations of the structure of DNAs and methods of replication. This endeavor not only climbs the ladder of genetic information but also paves the way for subsequent interventions and therapies of genetic disease, substantiating the claim that DNA is at the center of genetic health.

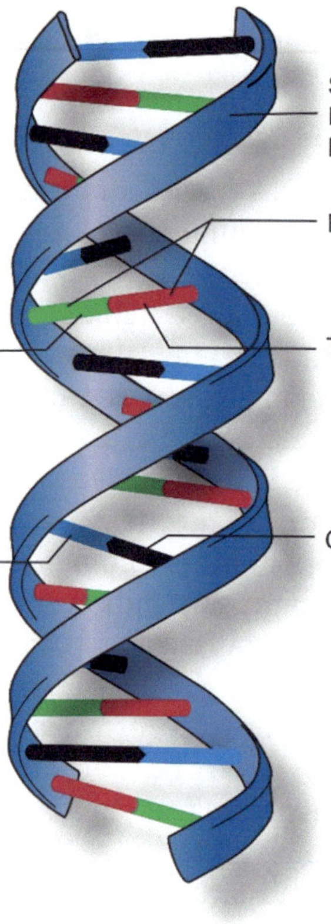

Figure 11.2. DNA in genetics.

Source: https://bio.libretexts.org/@api/deki/files/55374/Figure-4_DNA_Genetic_Material.png?revision=1.

Additionally, the mechanisms that govern DNA replication reveal still greater levels of sophistication that are pertinent to genetic heredity. New studies have uncovered the importance of asymmetrical segregation of chromosomal imprints during DNA replication that has a vital role to play in cellular differentiation. Asymmetrical segregation of imprints is also necessary to develop diverse cell types in multicellular organisms. For instance, the incorporation of ribonucleotides during lagging-strand synthesis highlights an innovative approach to understand how genetic imprinting takes place. The interaction between cis-acting sequences and replication arrest demonstrates that DNA is not an inert storehouse of genetic information. Instead, it is an active participant in cell decision-making processes in the determination of physical attributes and biological functions, which underscores the enormity of the role that DNA plays in the broader context of genetics and heredity (Figure 11.2).

11.1. Discovery of DNA Structure

Learning Outcome

1. Historical context of DNA research;
2. Contributions of Watson and Crick;
3. Role of X-ray crystallography in DNA discovery.

The determination of the structure of DNA is a milestone in the history of biological sciences, and it irreversibly transformed our understanding of genetics and heredity. This historic accomplishment was the outcome of collective work by a set of major scientists, James Watson and Francis Crick among them, whose identification of the double helix structure in 1953 was supported by conclusive X-ray diffraction photographs taken by Rosalind Franklin. Their model defined DNA in a twisted ladder structure, and this perfectly explained how genetic information is stored, replicated, and transmitted from generation to generation. The double helix form is not a static aggregation; it is indicative of active interplay of chemical interactions, which is the foundation of cellular processes. So, this discovery not only explained the structure of DNA but also cleared the way for future research into molecular biology and biochemistry. In the following years following the discovery of DNA's structure, an outstanding breakthrough in gene studies illuminated into DNA replication process. Being able to understand that the double helix could unwind into two strands during replication was important in envisioning how genetic information was copied inside cellular processes. Replication is an extremely synchronized interaction of proteins and enzymes like DNA polymerases and helicases, ensuring accuracy in the replication of the genetic information. The process of replication conveys a semiconservative model in which each new DNA molecule is made up of one original and one newly synthesized strand, preserving the integrity of the genetic information. It is a very coordinated process required for cell division and organism development.

It also opens up fields of research in genetic fidelity, mutation, and their health and disease consequences. Moreover, the discovery of mechanisms of DNA repair again emphasizes the complexity of DNA structure and its biological significance. Since cells are subjected to DNA damage by external or metabolic mechanisms at all times, cells have evolved sophisticated repair mechanisms for maintaining genomic integrity. A salient case is the SOS response in Escherichia coli, a bacterium that possesses a coordinated group of DNA repair genes for coping with stress. While the SOS system is predominantly found in prokaryotes, homologous systems exist in various kingdoms of life, which attests to the global relevance of DNA repair. A perception of these processes not only underpins the crucial significance of DNA architecture in maintaining genetic fidelity but also lays the groundwork for an examination of therapeutic strategies for the treatment of genetic disease and implications of mutagenesis for cancer biology (Alagramam et al., 2008; Janion et al., 2008) (Figure 11.3).

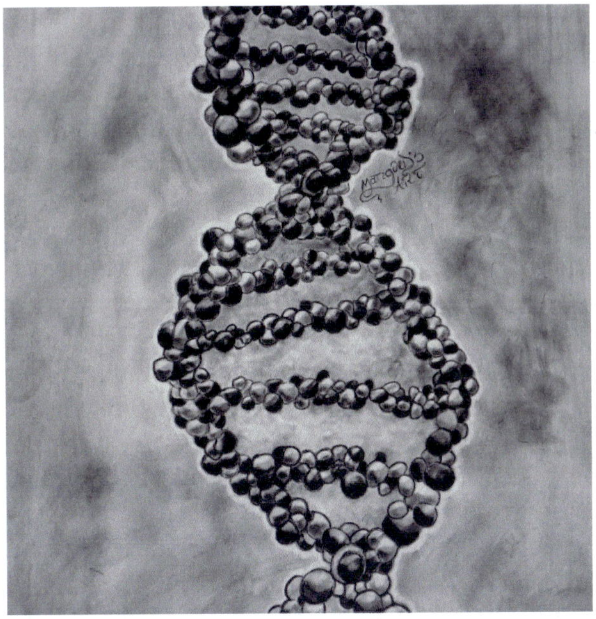

Figure 11.3. Artistic representation of the DNA double helix structure, emphasizing its molecular symmetry and complexity.

Source: https://images.squarespace-cdn.com/content/5e2a22a2ca128a771ade8d50/1596473986622-29SB1TC11MHOUYUSNFSP/DNA%2BArt.jpg?format=1500w&content-type=image%2Fjpeg.

11.1.1. Historical Context of DNA Research

The history of DNA research is marked by important milestones that have influenced our knowledge about genetics. It started at the beginning of the 20th century, with the rediscovery of Gregor Mendel's work on patterns of inheritance, which formed the foundation for contemporary genetics. Nevertheless, the discovery of the structure of DNA by Watson and Crick in 1953 transformed the field only towards the mid-century. Their model predicted the double helix structure of the DNA molecule and provided the replication and mutation processes. This revolutionary finding brought a research explosion on the attempted explanation of genetic material, resulting in efforts like the Human Genome Project, which hoped to map the entire human genetic blueprint. Such projects have pushed DNA study into the limelight of science, enabling us to achieve biological breakthroughs in biotechnology and targeted medicine. As research on DNA went on, it was understood that the understanding of genes was evolving. Traditional views that looked at genes as isolated components that could be worked with as molecular building units were disrupted. Findings from extensive research on genome editing showed that these sequences had interactions in complex ways, often governed by environmental factors and forces of evolution. Research highlighted by Witzany et al. (2017) contends that natural mechanisms of genome editing in organisms like viruses discredit the simple theory of genetic mutations as the sole source of diversity.

This paradigm shift highlights the need to understand the delicate balance between genetic sequences, their functions, and their adaptive roles. Consequently, the science has been compelled to reconsider the basic laws of genetics in light of DNA research. The consequences of having DNA knowledge are not just for GM but extend to broader societal dimensions. The recent study of simian foamy virus (SFV) transmission in Bangladesh, as reported by Pacheco et al. (2013), describe how viral evolution and genetic diversity are affected by geographical and anthropogenic factors. This complex interaction highlights the necessity of interdisciplinary study combining genetics, virology, and ecology to fully realize the complexity of DNA and its application across different environments. The historical development of DNA research advances along the trail of increasing complexity, culminating in an overall realization that genetic material is not static but rather changes within an ever-changing environment. These findings are not only relevant for the advancement of science but also for ethical questions in the realms of gene therapy and genetic engineering, in which the consequences of tampering with DNA are immense.

11.1.2. Contributions of Watson and Crick

James Watson and Francis Crick's trail-blazing research detailing the structure of DNA did not just usher in a milestone in molecular biology but also led to increased exploration of the processes of genetic replication. Their description of the double helix model back in 1953 was significant since it described how the shape of the molecules is intricately tailored for replication and transcription. The base pairing complementarity, in which adenine recognizes thymine and guanine recognizes cytosine, was of particular significance. It is this system of base pairing that is accountable for accurate DNA replication, which is the hub of the transmission of genetic information from generation to generation. As a result, the Watson-Crick model not only enhanced the biological understanding of molecular genetics but also provided a framework that researchers could work from, hence revolutionizing the field of genetics and cellular biology. Following on with the impact of their discovery, Watson and Crick's findings regarding the structure of DNA have had far-reaching effects on genetic engineering and biotechnology. By understanding DNA as a sequence of nucleotide bases, scientists have been able to manipulate genetic material and have made advances in gene therapy, cloning, and synthetic biology. Their work spurred interest in the possible applications of DNA information other than mere replication, like the creation of genetically engineered organisms and the invention of new therapeutic methods for disease. This increased interest has pointed out DNA's application not only as a storage of genetic information but also as an innovation agent in medicine and agriculture. Watson and Crick's fundamental work, therefore, revolutionized our perception of how genetic material could be harnessed to address difficult biological problems, stressing their pioneering efforts to modern science. In addition, the consequences of Watson and Crick's research are carried into the RNA study, where their early understanding laid the foundation for investigation into the role of RNA as a genetic intermediary. The insight into DNA structure facilitated further reflection upon the evolutionary consequences, namely the genetic takeover from RNA to DNA as the primary storehouses of genetic information. Their discovery not only contributed to an understanding of the mechanism of DNA replication but also posed fundamental questions regarding the stability and fidelity of genetic information. As revealed in subsequent studies, the transition from RNA to DNA as the predominant genetic material disclosed significant advantages, including greater chemical stability and reduced rates of error during replication, as seen in studies on RNA and DNA polymerization reactions. Thus, the contributions of Watson and Crick extend beyond their first report, illuminating a broader concept of molecular genetics and the evolutionary history of life itself, indicating a profound impact on scientific studies as well as practical applications of genetic data (Figure 11.4).

Figure 11.4. Contributions of Watson and Crick.

Source: https://www.sciencehistory.org/wp-content/uploads/2023/04/watson-crick-dna-model_crop3.jpg.

11.1.3. Role of X-Ray Crystallography in DNA Discovery

The invention of X-ray crystallography was a milestone in the discovery and exploration of the structure of DNA, significantly enhancing the scientific knowledge of this life-critical molecule. Through producing atomic-resolution images of crystallized biological specimens, X-ray crystallography unveiled the complex double-helix nature of DNA, an iconic feat first achieved by Watson and Crick in 1953. Scientists were able to ascertain the spatial geometry of nucleotides and complementary pairing, key to DNA reproduction and genetic stability, using this technique. The results from these studies lay the foundation for molecular biology, so that subsequent work can focus on how DNA is duplicated, repaired, and translated. In this manner, X-ray crystallography not only helped to demystify DNAs structure but also initiated a whole field of investigation focused on unveiling heredity and cellular function. With the improvement in the X-ray crystallographic technique, it became possible to study DNA interactions with other biological molecules in favor of the complex biological processes that DNA is involved in within living organisms. For instance, crystallography studies of DNA polymerases have indicated how the enzymes catalyze the synthesis of new strands of DNA during replication. Careful structural analyses have revealed the character of the interactions between the polymerase and the DNA, detailing the role of the enzymes in the efficient and accurate replication of the DNA. The study of the DNA polymerases, along with knowledge derived from structures of interacting proteins like helicases and DNA-binding proteins, has shown the cooperative character of DNA replication and repair processes. Collectively, these crystallographic studies have refined our understanding of enzymic processes which maintain genomic stability, shedding light on the crucial role of nucleic acids in cell physiology and pathology. Additionally, X-ray crystallography has become more widely used to analyze the nuances of DNA modification and its implications for epigenetics and molecular pathology. The ability to characterize structures at the level of atoms allows researchers to observe how chemical modifications of DNA, , such as methylation, influence gene expression, stability, and create diversity in genomic form. Highlighted here are the discoveries of how different structural forms of DNA exist under different sets of environmental conditions, showing a correlation between structure and function. Techniques using X-ray crystallography have also enabled studies of mechanisms through which alterations in DNA structure can result in oncogenic transformation, notably in the elucidation of genomic instability diseases. As continued research enhances, integration of X-ray crystallography with other imaging techniques will yield unmatched insight into cooperative functions of DNA in health and disease, further establishing the technique's value in molecular genetics (Figure 11.5).

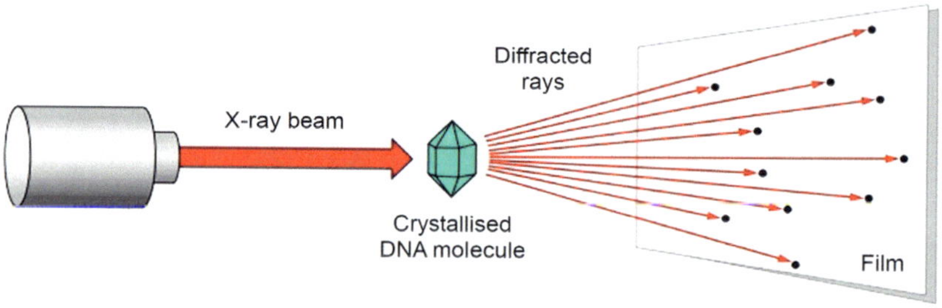

Figure 11.5. Role of X-ray crystallography in DNA discovery.

Source: https://old-ib.bioninja.com.au/_Media/x-ray-crystallography_med.jpeg.

Test Your Knowledge

1. Illustrate the historical context of DNA research.
2. What is the role of X-ray crystallography in DNA discovery?

11.2. Chemical Composition of DNA

Learning Outcome

1. Nucleotides and their components;
- The sugar-phosphate backbone;
- Base pairing rules (A-T and C-G).

The chemical composition of DNA is the foundation upon which its role in carrying and transferring genetic information is based. DNA, at its most fundamental, consists of nucleotides, each of which is a phosphate group, a sugar molecule known as deoxyribose, and a nitrogenous base. Four nitrogenous bases occur in DNA: adenine (A), thymine (T), cytosine (C), and guanine (G). These bases are paired in specific pairs by hydrogen bonds, where adenine pairs with thymine and cytosine pairs with guanine. The complementary pairing of the bases is required for proper replication of genetic information upon cell division. The stability in structure is contributed by the sugar-phosphate backbone, which allows the double helix to maintain its structure. Thus, the chemical structure of DNA is essential to its role in the storage and relay of genetic information in stable form, opening the way to the complexity of life. Replication of DNA is directly linked to the chemistry of the DNA molecule. The double helix structure of DNA allows separation of strands upon replication, a necessary procedure that avoids errors in genetic replication. Enzymes , such as DNA polymerase utilize the existing strands as a template on which to build new strands, essentially creating complementary copies. It has been hypothesized that the transition from an RNA-based world to a DNA-based genetic system evolved due to the greater fidelity in replication that is built into DNA. Lower DNA replication error rates, as highlighted by the study, also corroborate the fact that DNA is intrinsically more stable than RNA, suggesting that lower mutation rates might lead to greater genomic complexity and preservation of information during evolution (Chen et al., 2011). This biochemical adaptation bears witness to the evolutionary advantages of employing DNA for genetic information storage. Moreover, the chemical organization of DNA not only allows for replication but also facilitates influencing gene expression and environmental adaptability. DNA stability is of greatest significance when considering the barrier imposed by environmental conditions; the molecular arrangement allows it to withstand a number of biochemical stresses. Processes of heredity and gene expression, as governed by DNAs chemical constitution, are central to evolution and existence potential in diverse milieus. Apart from that, a study on the relationship between the structure of DNAs and environmental conditions extremes provides possible adaptations of living things (Jukes et al., 2025).

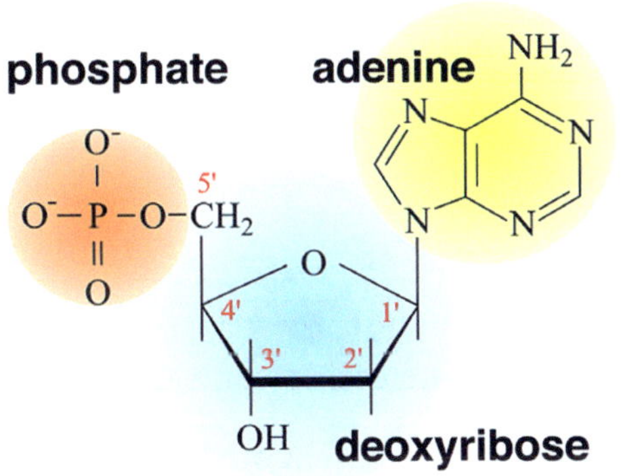

Figure 11.6. Chemical composition of DNA.

Source: https://lh4.googleusercontent.com/proxy/lHTy2ownmOi5LSmSW6-bynYBwb13eZKUh3xmJ_wMZravp4HmKQdw9434hTiwE93iml_5VP-tAmbKlRQQYv1TNVHLquI874GY7-8VD8ffcopCU46n_Yrio3o6ImMU.

Understanding the relationship between the chemical structure of DNA and its functional capabilities can clarify the evolutionary mechanism that led to life on Earth, casting spotlight on DNAs as a key factor in the history of biological organisms (Figure 11.6).

11.2.1. Nucleotides and Their Components

The elementary components of DNA, the nucleotides, are made up of three different constituents: a nitrogenous base, a five-carbon sugar, and one or more phosphate groups. The nitrogenous bases come in two forms: purines, which include adenine (A) and guanine (G), and pyrimidines, which include cytosine (C) and thymine (T). Each form of the nucleotides not only defines its role in the DNA molecule but also contributes to the overall functionality and stability of the molecule. The five-carbon sugar deoxyribose in DNA is different from ribose in RNA, influencing the overall stability and reactivity of the nucleic acid. Meanwhile, the phosphate groups link adjacent nucleotides to form the backbone of the DNA strand. This spatial structure enables nucleotides to polymerize into extended, complex strands required for genetic coding, replication, and cellular function. Understanding the dynamics of nucleotide incorporation and their function in DNA replication is crucial for cell processes.

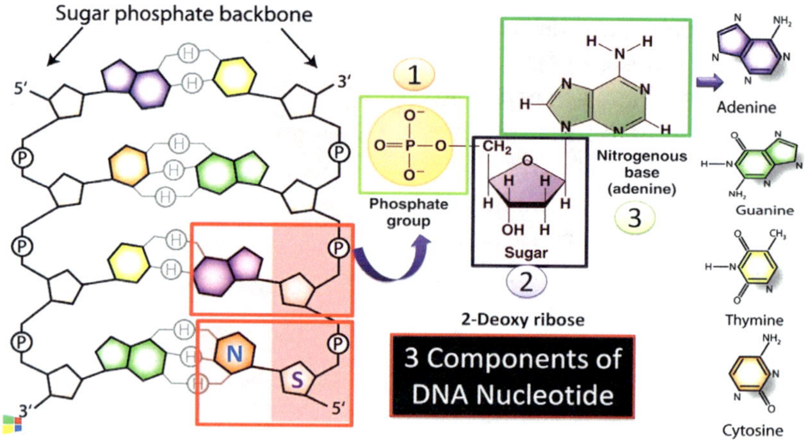

Figure 11.7. Nucleotides and their components of DNA.

Source: https://blogger.googleusercontent.com/img/b/R29vZ2xl/AVvXsEgIi7FLPJ_0do1qBwvHvePbzx1kENQO_h1Wo1UIUTKv8_tYgbAPA9Gz_g8J_EnCbT9kSvhie3-iN8pqPnOTOITTqZRdXhrxirp40P5obe4879rEMmESyCp1wXmrhlwkicmf7ZCDZqXVHMA/w1600/3-components-of-DNA.webp.

DNA polymerase, the catalytic enzyme that forms new DNA strands, relies primarily on nucleotides as a building block. It adds nucleotides in the complementary sequence of the template strand, following the innate rules of base pairing that demand adenine to pair with thymine and guanine with cytosine. The enzyme not only aids in the extension of the DNA strand but is also used to verify errors during replication. As highlighted in biophysical studies, these molecular machines operate with great accuracy, virtually assuring the replication of genetic information in a very accurate way. The regulation of nucleotides and their polymerization by DNA polymerase bears witness to the sophisticated biochemical processes sustaining life, playing crucial functions in genetic continuity and cell growth (Chowdhury et al., 2012). The role of nucleotide components is not limited to mere structural functions; they also carry out elementary roles in cellular differentiation and signaling processes. Current research has established fascinating associations between nucleotides and their involvement in cellular imprinting during replication. In species , such as fission yeast, ribonucleotides incorporated into the DNA during the lagging strand synthesis have been identified as playing significant roles in the asymmetrical segregation of sister chromatids, a situation critical for cellular differentiation (Kaykov et al., 2010). It not only showcases the flexibility of nucleotides as DNA's central building block but also dynamic participation in cell processes beyond genetic coding. The consequences of such findings go far beyond, revealing something about the mechanism of cellular differentiation, development, and the molecular structure of life in general. The complex roles of nucleotides as both part of DNA and controllers of cellular processes, thus, bear testimony to their pivotal position in the perpetuation of life (Figure 11.7).

11.2.2. The Sugar-Phosphate Backbone

The backbone of the DNA structure is the sugar-phosphate, crucial to its stability and integrity. Constructed by the alternating sugar (deoxyribose) and phosphate groups, the backbone takes the form of a long chain on which nitrogenous bases are attached. A nucleotide consists of a molecule of sugar bonded to a phosphate group, contributing to the global polarity of the DNA strand. The 5-phosphate end of one is complemented by the 3-hydroxyl end of the other, resulting in a directional antiparallel structure. This confirmation is not only responsible for the structural stability but also plays a crucial role in DNA replication and transcription processes. Through sequence-specific positioning of the nucleotides radiating outward from the backbone, genetic information is preserved, thus making the sugar-phosphate backbone an essential component not only for structure but also for function in biological systems. The function of the sugar-phosphate backbone extends beyond the furnishing of structural support; it also plays a part in a variety of fundamental biochemical processes. For instance, during DNA replication, enzymes like DNA polymerase travel along the backbone, synthesizing new strands by adding nucleotides to the growing chain. Backbone stability takes center stage, with a break or modification potentially leading to a replicational or transcriptional mistake. Recent studies bring into focus advanced electrochemical techniques to track the activity of DNA ligases, the enzymes responsible for closing breaks in the backbone and thus maintaining genomic integrity. By analyzing the electrochemical activity associated with the backbone structure, researchers can distinguish between ligatable and unligatable breaks, which highlights the importance of the backbone in maintaining DNA integrity in physiological conditions (Ahel et al., 2008). Such innovation not only enhances our understanding of DNA metabolism but also illuminates potential therapeutic implications in gene repair and editing. Besides, the sugar-phosphate backbone also forms the core of DNA interaction with other biomolecules like enzymes and drugs. For example, binding affinities are controlled based on the structural characteristics of the backbone, as exemplified by studies on ligand-dominated binding modes to DNA. Such interactions affect recognition of mismatched bases in processes like DNA repair, thus demonstrating the diversified functionality of the sugar-phosphate backbone (Barton et al., 2006). With the potential for multiple binding modes, scientists are moving towards developing molecules that selectively bind DNA regions for drug use, e.g., in the combat against genetic ailments or cancer. Thus, the sugar-phosphate backbone not only contributes to the underlying structure of DNA but also to the dynamic forces that control cell function and stability, thus emphasizing its importance in the general theory of molecular biology (Figure 11.8).

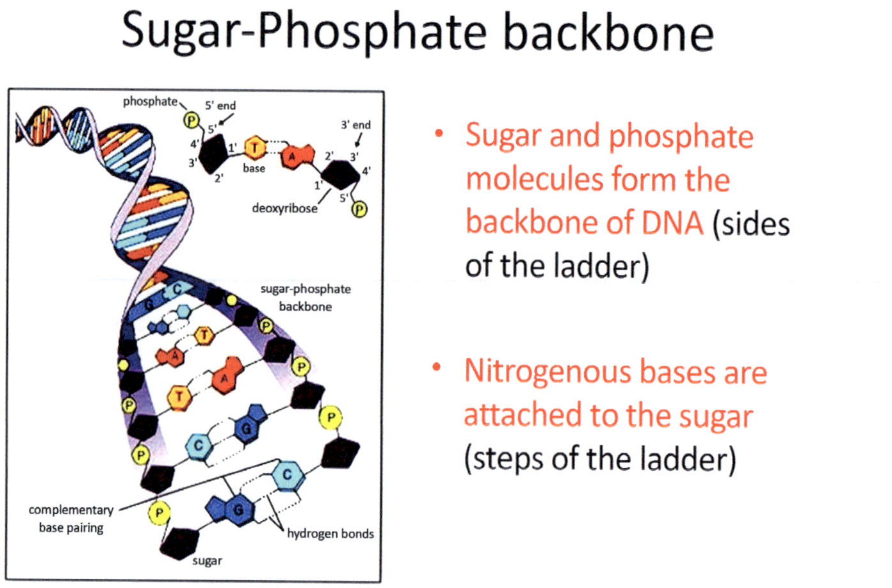

Figure 11.8. DNA sugar-phosphate backbone.

Source: https://i.pinimg.com/736x/d2/f2/77/d2f2772c6d878451def1a3b78716683d.jpg.

11.2.3. Base Pairing Rules (A–T and C–G)

The base pairing rules, which consist of the highly specific interactions between adenine (A) and thymine (T) and cytosine (C) and guanine (G), are fundamental to DNA structure and replication. These rules demonstrate the highly specific complementary nature of nucleotide interaction, where A is consistently with T by two hydrogen bonds and C with G by three hydrogen bonds. This specificity is significant in maintaining the integrity of genetic information by DNA replication, whereby the genetic code is replicated from one generation to another precisely. The uniform pairing stabilizes the double helix structure of DNA and provides room for the repair and replication mechanisms responsible for cellular function and heredity. Knowing these rules of base pairing is important to theoretical models and to applications in genetics and biotechnology. The implications of the rules of base pairing extend beyond mere nucleotide interactions to affect the broader landscape of genetic engineering and manipulation. Old ideas about genetic sequences as linear strands have evolved as new information regarding how viruses and RNA structures perform complex editing of genetic material has come to light. These naturally occurring genome-editing processes contradict the theory of random mutations as the sole driving force for genetic diversity. Instead, they suggest a more sophisticated interplay between sequences and structural stability, with an emphasis on the significance of the A-T and C-G pairings in connection with adaptability and evolutionary processes. To the extent that scientists explore further into such dynamics, they can redefine existing paradigms on genetic manipulation and innovate new approaches to artificial genome editing (Witzany, 2017). Besides, the study of base pairing can merge with mathematical and logical systems, demonstrating the symbiotic confluence of biology and formal systems. By way of illustration, in this setting, containers and extainers define the ideas that allow for a profound understanding of genetic substances interacting and functioning in cellular settings. Such boundary algebra agrees with the laws governing DNA replication, as it not only stipulates the spatial orders of nucleotides but also their dynamic purposes in biological settings. With transdisciplinary synapses between the realms of biology and logic, scientists can construct advanced models that mirror the workings of DNA complexity and replication processes. A cross-disciplinary model may yield greater insight not just into the form of DNA but also into its future applications in emerging fields , such as synthetic biology and computational genomics (Kauffman et al., 2002).

Test Your Knowledge

1. What are the nucleotides and their components?
2. What is the sugar-phosphate backbone?

11.3. Double Helix Model

Learning Outcome

1. Description of the double helix structure;
2. Antiparallel strands and their significance;
3. Major and minor grooves in the DNA structure.

The Double Helix Model of DNA, which was first proposed by James Watson and Francis Crick in 1953, provides a classic model of understanding and explaining genetic material. The double helix is composed of two twisted strands made up of a nucleotide sequence of genetic information. The process of complementary base pairing, by which adenine binds with thymine and cytosine binds with guanine, creates a stable structure required for DNA replication. This configuration not only accounts for the three-dimensional arrangement of nucleotides but also complies with the biochemistry of heredity. The helical configuration allows effective packing in the nucleus without sacrificing access for replication machinery and other proteins required for gene expression and fixation, hence producing a founding principle of molecular biology. With cellular processes ongoing, double helix stability is tested, particularly during DNA replication. Uncoiling of such complexity necessitates proper untangling of twisted strands, or chromosomal intertwines, that can play a pivotal role in chromosome segregation during mitosis. Although most of these intertwines are loosened before cell division, there remain leftovers that can contribute to genomic instability. For instance, the presence of mitotic sister chromatid intertwines (SCIs) can lead to problems , such as aneuploidy if not properly controlled. Various mechanisms, including the action of topoisomerases, function to counteract such problems so that the DNA double helix structure does not disrupt the fidelity of chromosome segregation on cell division (Bartkova et al., 2014). Such events function to emphasize the dynamism of DNA and functional significance of its structures. In addition to its structural relevance, the double helix is fundamental to the mechanism of DNA replication. Observing the entropic character of information transfer during the process highlights the energetics of proper DNA synthesis. This process operates off equilibrium, demonstrated through the work of researchers applying statistical theory to analyze the dynamics operating in biomolecular systems. By creating a rigid structure on Shannon Information, scholars maintain that the process of replication produces entropy dependent on the interactions between nucleotides as they are incorporated into the growing strand. This nuance not only emphasizes the subtlety inherent in DNA replication but also serves as a model for estimating the accuracy of the reproduction in terms of the structural controls imposed by the template strand (Gonzalo et al., 2012). The double helix is thus not merely a structural aspect; it is an active participant in replication, blending information theory with molecular biology (Figure 11.9).

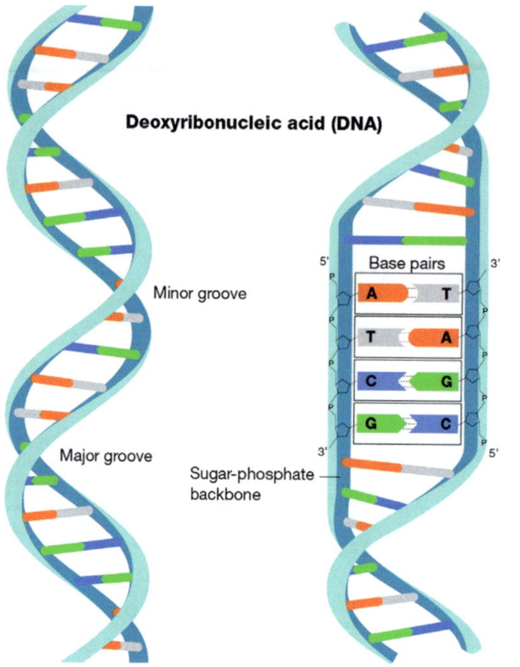

Figure 11.9. Double helix model.

Source: https://www.genome.gov/sites/default/files/media/images/tg/Double-helix.jpg.

11.3.1. Description of the Double Helix Structure

The familiar double helix structure of DNA, first characterized by James Watson and Francis Crick in the early 1950s, forms the basis of an understanding of its role in gene replication and information storage. It consists of two antiparallel strands of nucleotides twisted together, held in place by base pairing complementarity, with adenine binding to thymine and cytosine to guanine. The order of these bases along the strands contains genetic information, while the sugar-phosphate backbone helps in structural stability and complex protection of the base pairs. Not only does the helical twist of the DNA molecule aid in its compactness, facilitating efficient packing within the cellular nucleus, but it also to mechanisms of DNA replication. Through such a structure, the double helix allows reliable transmission of genetic information between generations. Double helix stability is vital in DNA duplication and cell division processes. Proteins interact with the DNA to unwind or preserve its unique structure to replicate or for replication. Topoisomerases, for instance, relax the helical tension generated by the helix unwinding, allowing replication forks to move without problems. The entwining of the sister chromatids also creates huge challenges in cell division, as detailed in DNA catenation mechanisms and SCIs (Bartkova et al., 2014). These intertwines, although essentially resolved in replication, become problematic during mitosis if not well addressed. Redressing these structural complexities is vital to the assurance of accurate segregation of chromosomes to evade potential genomic instability as well as disorders , such as aneuploidy, which are greatly rampant in various cancers. Furthermore, the relationship between DNA's structural qualities and its functional dynamics is vast when one takes replication accuracy and efficiency into perspective. When DNA polymerase builds a new strand, the fidelity of nucleotide addition directly determines the entropic consequences of replication, a testament to the central role of the DNA double helix structure in information processing (Gonzalo et al., 2012). Where replication is occurring more quickly, this kind of accuracy can decline, potentially leading to mutations. Understanding the double helix stability mechanisms and replication nuances, therefore, is crucial in an effort to appreciate how DNA maintains genetic integrity across millions of cell divisions. Such structural stability mirrors the success of DNA as the molecular foundation of life to allow diverse organisms to exist while preserving their unique informational blueprints (Figure 11.10).

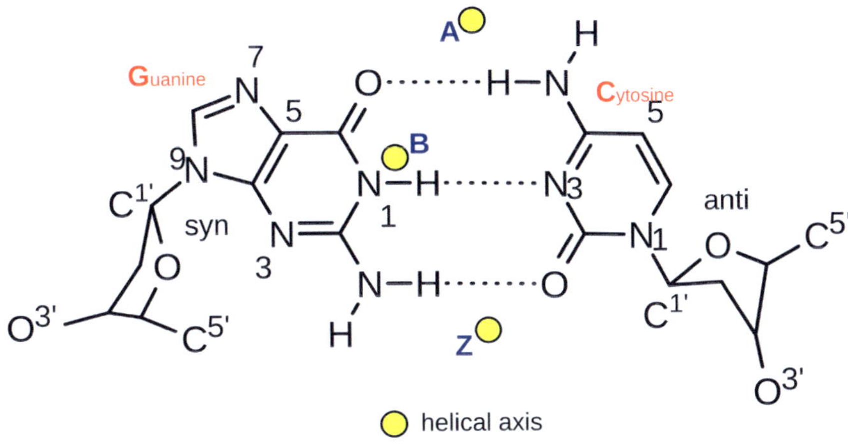

Figure 11.10. The double helix structure of DNA.

SourceBy Lankenau - File:B&Z&A_DNA_formula.jpg, CC0, https://commons.wikimedia.org/w/index.php?curid=94913697

11.3.2. Antiparallel Strands and Their Significance

The antiparallel nature of the double helix structure of DNA into strands has intrinsic significance to double helix stability and operation. Each of the two strands runs in opposite directions, which are referred to as 5 to 3 and 3 to 5 directions. Orientation is significant in processes like DNA replication, where DNA polymerases require a template strand for the synthesis of a new complementary strand. Due to the antiparallel structure, enzymes can catalyze reactions using both strands as a template in a well-controlled manner. Furthermore, the structure preserves the hydrogen bonding among complementary nucleotide bases and therefore genetic accuracy is guaranteed while stability for replication and transcription sessions is promoted. It is thus necessary to study how these antiparallel strands interact to appreciate the entirety of DNA's chemical nature and roles in cellular processes. In addition, antiparallel organization of strands of DNA is key to various biological processes , such as transcription and repair. As RNA polymerase initiates synthesis of RNA, it uses one of the strands of DNA as a template, hence relying on the strands orientation in synthesizing complementary RNA sequences. Antiparallel direction also results in special arrangements, for example, secondary structures which may be a major factor in gene regulation and expression. For example, certain structures like G quadruplexes, which form when guanine-rich sequences assemble in a parallel mode, require proteins and helicases for dissolution in transcription and replication. Secondary structures are likely to impede regular enzymatic processes, hence the requirement of understanding the orientation of the strand in the right way and its effect on molecular mechanisms. The intricacy additionally justifies the need to investigate the antiparallel mode against the broader functionality of DNA. The importance of antiparallel strands also applies to their interactions with other biomolecules participating in DNA metabolism. For example, differences in strand orientation contribute significantly to the function of helicases in DNA replication. Helicases are designed to unwind DNA strands for replication and tend to show differential preferences for strand orientation as a result of antiparallel DNA structure. In processes , such as telomere maintenance, proteins like Dna2 are of particular preference for G4 DNA structures, which are stabilized by antiparallel stacks of guanine tracts. These features depict DNA's mechanism of dealing with structural stresses in processes of repair and replication. Thus, research on the activity of antiparallel strands enhances their understanding of biochemical interactions and sheds light on the complex dance of enzymes and DNA in cellular functions that involve replication and repair, thereby contributing to the broader account of the unavoidability of DNAs structural properties (Ambjoernsson et al., 2006; Campbell et al., 2008) (Figure 11.11).

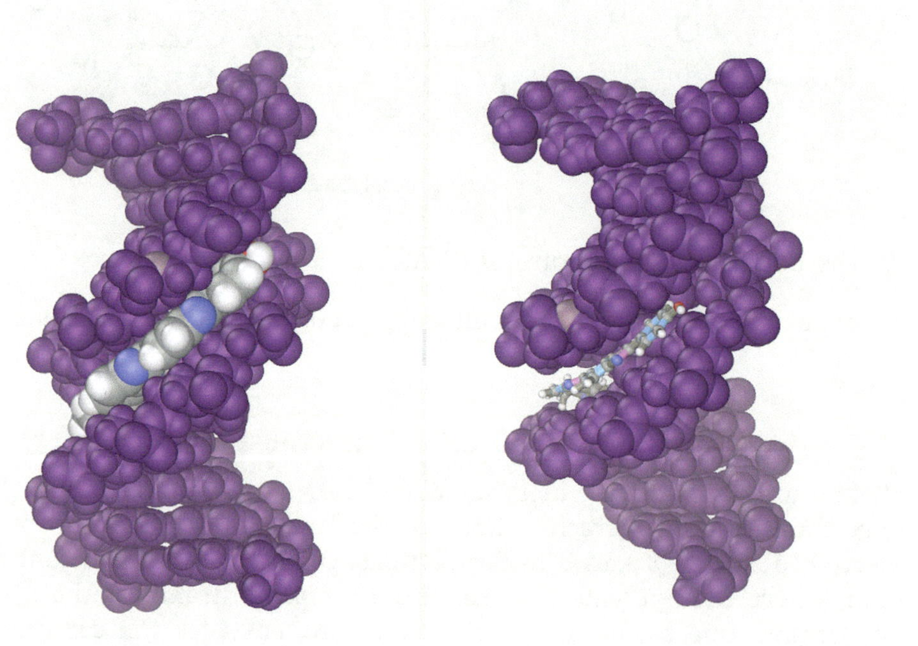

Figure 11.11. Antiparallel strands.

Source: By P99am - Own work, CC BY-SA 3.0, https://commons.wikimedia.org/w/index.php?curid=12404079

11.3.3. Major and Minor Grooves in the DNA Structure

Inside the DNA structure, the major and minor grooves are of primary significance to molecular interactions necessary for genetic processes , such as replication and transcription. The grooves are a result of the arrangement of the nucleotide bases, with the helical shape not being an even width. The larger major groove can fit more extensive protein-binding interactions, enabling TFs and other proteins to bind to DNA's genetic information. The smaller minor groove has fewer access points but can still regulate protein binding by way of specific interactions. This varying accessibility results in selective protein binding, capable of reading the DNA sequence by perceiving the distinctive conformational properties of the grooves, impacting cellular functions and control mechanisms essential for life (Abi et al., 2015). Perhaps one of the most fascinating aspects of DNA's grooves is their effect on protein-DNA interaction. Proteins are inclined to contact DNA in the major groove, where structural features like hydrogen bonds and electrostatic interactions facilitate binding. Proteins like Fis, a characterized nucleoid protein, provide examples of the significance of such grooves as they tend to bind DNA primarily via the minor groove, depending on the DNA sequence inherent structural features. Experiments have shown that the conformation and width of such grooves can control binding affinities, and that conformational adjustment is an integral part of high-affinity site recognition (Cascio et al., 2016). This is an expression of the flexibility of DNA as it responds to changing biological conditions, showing how its structure has a direct impact on protein function. Recognition of the significance of major and minor grooves extends from structural biology into fields of genetic control and adaptation at an evolutionary level. Sequence differences that influence the size and shape of these grooves can potentially regulate the binding rates of other proteins, influencing in turn DNA replication mechanisms. For example, the availability of particular base pairs can suppress or promote the binding of a protein to DNA and thereby tailor the reaction of genetic pathways to diverse cell contexts. This flexibility in protein-DNA interaction highlights the evolutionary pressure on nucleotide sequences to be structurally stable yet functionally diverse, illustrating DNA's capacity to be both a stable repository of genetic data as well as a dynamic participant in cellular control (Abi et al., 2015).

Test Your Knowledge

1. What are antiparallel strands and their significance?
2. Explain major and minor grooves in the DNA structure.

11.4. DNA Replication Process

Learning Outcome

1. Overview of the replication mechanism;
2. Role of enzymes in DNA replication;
3. Leading vs. lagging strand synthesis.

DNA replication is a highly controlled and complex process, critical for the preservation of genetic fidelity during cell division. At the center of this process lies the function of enzymes referred to as polymerases, which catalyze new DNA strand synthesis through the addition of complementary nucleotides onto the template strand. These polymerases not only perform the vital task of elongation but also maintain high fidelity in replication through mechanochemical coupling mechanisms that reduce error in nucleotide incorporation. Computational studies in recent times have highlighted the importance of mechanochemical coupling in the cycle of polymerase elongation, illustrating how fluctuations and specific control mechanisms contribute to transcription and replication fidelity. The outcomes of the above studies are a great step towards comprehending the dynamic action of polymerases, whereby their functions as molecular motors essential for proper replication of genetic material are emphasized (Yu et al., 2015). Structural complexity of DNA also plays a critical role in the replication process, particularly in the retention of the double helix structure integrity. During DNA replication, double helix unwinding presents possible tension and entwining of the strands, which must be resolved lest tangling and catenation occur.

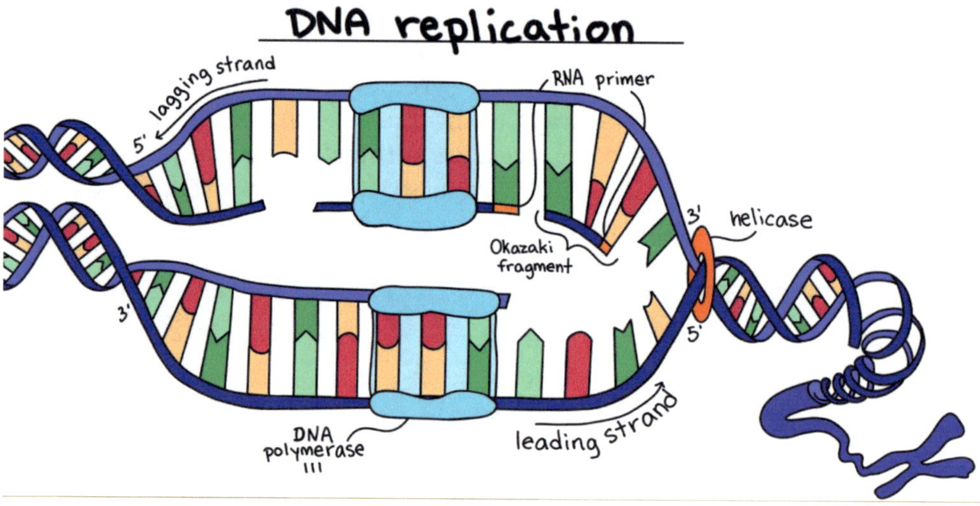

Figure 11.12. DNA replication process.

Source: https://d20khd7ddkh5ls.cloudfront.net/dna_replication_edit1.jpg.

Such is here that the services of some enzymes become necessary. For instance, topoisomerases introduce transient cuts into the DNA strands, relieving torsional tension that accumulates ahead of the replication fork. Moreover, specific mechanisms like Holliday junction resolvases are crucial in melting DNA intermediates that are formed as a byproduct of HR events. Melting entwined sister chromatids before the onset of mitosis is crucial, as failure to melt such issues may lead to drastic genetic consequences, including aneuploidy, which has normally been associated with numerous forms of cancer (Bartkova et al., 2014). As the replication fork proceeds, coordination of multiple proteins and enzymes becomes critical for facilitating effective and continuous synthesis of DNA. The leading strand is continuously synthesized, while the lagging strand is synthesized in the form of Okazaki fragments, and this creates extra trouble for the overall process of replication. Fragmented synthesis , such as this, requires the coordinated action of a set of proteins, which involves primases that add RNA primers to fragments and ligases that join fragments together. Additionally, accessory proteins present aid in unwound DNA stabilization and shielding it from degradation. The tight association between these components, apart from illustrating the complexity of the replication machinery, further indicates how such processes evolved to perfection, thus allowing cells to duplicate their genomic information accurately under varying biological conditions. Together, these processes carry out the miraculous feat of making certain that there is efficient and accurate DNA replication, safeguarding the preservation of hereditary information (Yu et al., 2015) (Figure 11.12).

11.4.1. Overview of the Replication Mechanism

The mechanism of DNA replication is a vital process that ensures genetic fidelity in the reproduction of a cell. It involves the unwinding of the double helix of DNA, which is a process that is mainly performed by the enzyme helicase. It disrupts hydrogen bonds that connect pairs of nucleotides, resulting in two single-stranded templates of DNA. On unwinding of the strands, single-strand binding proteins cover such regions to avoid re-annealing and thus creating a replication fork. Synthesis of the leading strand is continuous in the direction of the replication fork, while synthesis of the lagging strand is discontinuous in short segments known as Okazaki fragments. These fragments are later connected to create a single continuous strand by DNA ligase. This elaborate orchestration by enzymes and binding proteins reiterates the complexity of the process since it highlights its importance to maintain genomic stability during cell growth. Central to DNA replication is the activity of DNA polymerase to incorporate complementary nucleotides onto the growing strand of DNA. The enzyme requires a primer, which is typically synthesized by primase, to initiate synthesis. DNA polymerase catalyzes phosphodiester bond formation between nucleotides so that the new DNA strand formed is complementary to the template strand. Precision in DNA replication is crucial; thus, the polymerase contains proofreading activities that enable it to detect and correct mismatched bases to lower mutation rates to extremely low levels.

Such high accuracy is necessary because errors during DNA replication can lead to genetic illness or cancers. Moreover, the various mechanisms like microhomology-mediated repair have been associated with the genetic etiology of disease, i.e., even the replication process influences organism health and genetic stability (Beysen et al., 2013). Regulation of the replication process is necessary because any type of aberration leads to genomic instability. Initiation of replication is regulated by a collection of regulatory proteins that cause the process to occur most probably once in a cell cycle. The proteins bind to particular origins of replication where the formation of the pre-replicative complex occurs. After the activation of the complex, assembly of helicase and the replication machinery at the fork occurs. In addition, recent results demonstrate that genomic structural features, , such as non-B DNA conformations, may participate in the replication fork dynamics and, thus, determine the process efficiency (Bennett, 2005). Disruption of such regulatory mechanisms may render cells more susceptible to replication stress, again underlining the importance of proper control mechanisms. Knowledge of these complex regulatory networks illuminates not only normal cellular physiology but also the etiology of most of the genetic diseases.

11.4.2. Role of Enzymes in DNA Replication

DNA replication is reliant in most fundamental terms upon the action of several enzymes that facilitate accurate and efficient duplication of genetic material. Central to the process is DNA polymerase, an enzyme that synthesizes new DNA strands by adding nucleotides to a pre-existing template. In some viral systems, , such as *Bacillus subtilis* phage Φ29, new features of DNA polymerase facilitate more efficient function. This polymerase has good processivity since it extends the DNA strand both by displacing the strand and by polymerizing, processes that are required for the maintenance of replication process integrity.

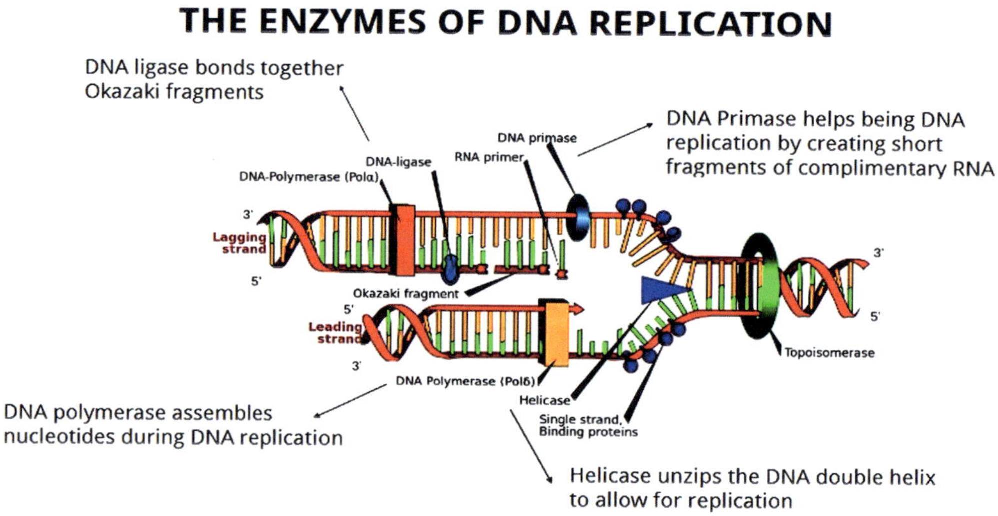

Figure 11.13. Enzymes involved in DNA replication.

Source: https://microbconlinc.com/wp-content/uploads/2022/09/enzymes-feature-image.png.

Therefore, understanding the complex processes of various DNA-binding proteins like terminal protein (TP) and viral proteins p5 and p6 helps scientists to make sense of how these components interact with DNA polymerase for effective and accurate DNA replication (Holguera et al., 2016). Additionally, the role of helicases in DNA replication underscores the enzymatic complexity for surmounting the challenge of unwinding the double helix. While helicases are broadly tasked with the unzipping of both strands of DNA to create access for replication, there are essential roles for them in replication fork stability as well. The Lhr helicase family, for instance, targets specifically structures of the replication fork to allow DNA strand unwinding. This action is precious, particularly when the replication fork is facing obstacles , such as damaged DNA or structural distortions. Apart from that, Lhr helicases have also been found to be involved in mutation repair and oxidative stress response processes, again testifying to their dual role during replication and maintenance of genomic stability (Buckley, 2023). Such helicase functions highlight their role within the overall context of DNA replication and repair, emphasizing cooperative enzymatic processes. Finally, the interaction between several enzymes offers not only redundancy of the replication apparatus but also the capacity of the apparatus to respond to barriers presented by DNA synthesis. Coordination between DNA polymerases, helicases, and repair enzymes ensures faultless continuation of replication even when there is damage to DNA or structural hindrances. For example, replication-coupled DNA repair enzymes are required to remove replication blockages, thereby enabling uninterrupted synthesis to be maintained. When more such enzymatic activities are known, investigators can further delineate pathways that ensure fidelity and stability of DNA replication. The subtle reactions and functions of each enzyme in this complex situation reflect the evolutionary strategies employed to safeguard genetic continuity and stability from generation to generation, finally underscoring the elegant complexity of the DNA replication mechanism (Figure 11.13).

11.4.3. Leading versus Lagging Strand Synthesis

DNA replication refers to the method by which the two complementary strands, the lagging strand and the leading strand, are developed. The leading strand is constructed in a continuous manner in the direction of 5 to 3 while the replication fork is opening, allowing the DNA polymerase to continue adding nucleotides. This is because of the direction of the elongated DNA template that makes it simple for the replication machinery to move without any complications. On the other hand, the lagging strand, since it is synthesized discontinuously, produces short fragments referred to as Okazaki fragments. The fragments are produced as the replication fork proceeds, having to undergo repeated priming and synthesis, thus complicating the process. Understanding these mechanisms of divergence is significant since it determines DNA replication fidelity and speed, and shows the complex choreography required for proper genomic duplication and maintenance. Replication errors can lead to widespread genomic instability, and many factors involved in leading and lagging strand synthesis have been documented in recent studies. For instance, replication blocks can cause helicase-polymerase uncoupling that directly affects the lagging strand synthesis, and even that continues for hundreds of bases away from the block region. Such perturbations can lead to the formation of secondary structures like G-quadruplexes, thereby contributing to replication fidelity issues. Research indicates that p53-deficient cells are more susceptible to such structures, and this results in deletions as well as mutations that compromise genomic integrity. Such correlation between lagging strand synthesis obstacles and damage associated with replication underscores the role of cellular mechanisms in preventing mutations and emphasizing the importance of having both strands for overall genome stability (Abdelhamid et al., 2020).

The distinction between leading and lagging strand synthesis also becomes evident in their interactions with DNA replication and repair proteins. For instance, flap endonuclease 1 (FEN1) has been discovered to play a significant role in the integrity of leading strand replication, notably in mitigating telomere fragility linked to RNA:DNA hybrid structures. This development shows a new function for a usually strand-linked protein, one that underlines the subtle interaction between replication and transcription functions. Furthermore, ribonuclease H1 (RNH1) has been discovered to be vital in resolving RNA:DNA hybrids, thereby allowing unobstructed passage of replication forks. Interference with RNH1 levels increases the level of such hybrids, which leads to replication stress and increased DNA damage. Generally, the complex roles of these proteins highlight the intricacies of DNA replication and repair mechanisms, delineating how they are uniquely designed to cater to individual requirements of leading and lagging strand synthesis (Parajuli et al., 2017).

Test Your Knowledge

1. What is the replication mechanism?
2. What is the role of enzymes in DNA replication?

11.5. Mechanisms of DNA Repair

Learning Outcome

1. Importance of DNA repair mechanisms;
2. Types of DNA damage and repair pathways;
3. Consequences of unrepaired DNA damage.

It is essential to the integrity of an organism's genetic information that there are robust DNA repair mechanisms. These mechanisms are employed to fix various types of DNA damage that could arise due to environmental exposure, metabolism, or errors during DNA replication. One such mechanism that is essential for the repair of double-strand DNA breaks is HR. HR facilitates the exchange of DNA strands between homologous sequences, providing accurate repair by employing the sister chromatid as a template. Among the premier models to describe HR is bacteriophage T4, which offers great insights into protein-protein interactions and dynamics involved in DNA repair and replication. The preservation of recombination proteins' activity across species reinforces the evolutionary significance of these events in genomic stability, emphasizing that a deficiency in HR can lead to severe cellular consequences and genomic instability (Conway et al., 2010). In addition to HR, cells employ several other mechanisms to repair DNA, including non-homologous end joining (NHEJ) and base excision repair (BER). NHEJ, which is commonly utilized in the instance of DSBs, is independent of a homologous template and is thus critical to rapidly repairing DNA, particularly in post-mitotic cells. While efficient, NHEJ is capable of introducing mutations, thus indicating its support role and not that it is a completely faultless process. BER, however, focuses on the nonspecific damage of individual bases in the DNA structure. It consists of various enzymes that remove the damaged bases and replace them with the correct nucleotides. It is important to understand how these repair processes intermingle, as they all collaborate to maintain genetic integrity and prevent resulting diseases, including cancer, most of the time caused by defective repair mechanisms (Esashi et al., 2024). Cellular quiescence is a unique state where cells do not proliferate, and it needs adaptive DNA repair mechanisms to prevent genomic instability. Although quiescent cells do not divide, they remain susceptible to endogenous DNA damage produced by metabolic reactions and normal cellular processes. The most important aspect is maintaining genomic integrity in the state to ensure tissue function and longevity. New evidence suggests that, in addition to general downregulation of DNA repair mechanisms in quiescent cycles, cells may take on alternative or specialized mechanisms distinct from those of active proliferation. Further, disruption in these mechanisms leads to pathological consequences, which underscores the necessity of targeted repair response without compromising cell features and genomic stability in quiescent cells

(Esashi et al., 2024). This versatility reflects the intricacy of DNA repair mechanisms and their applicability to normal cell function as well as disease prevention (Figure 11.14).

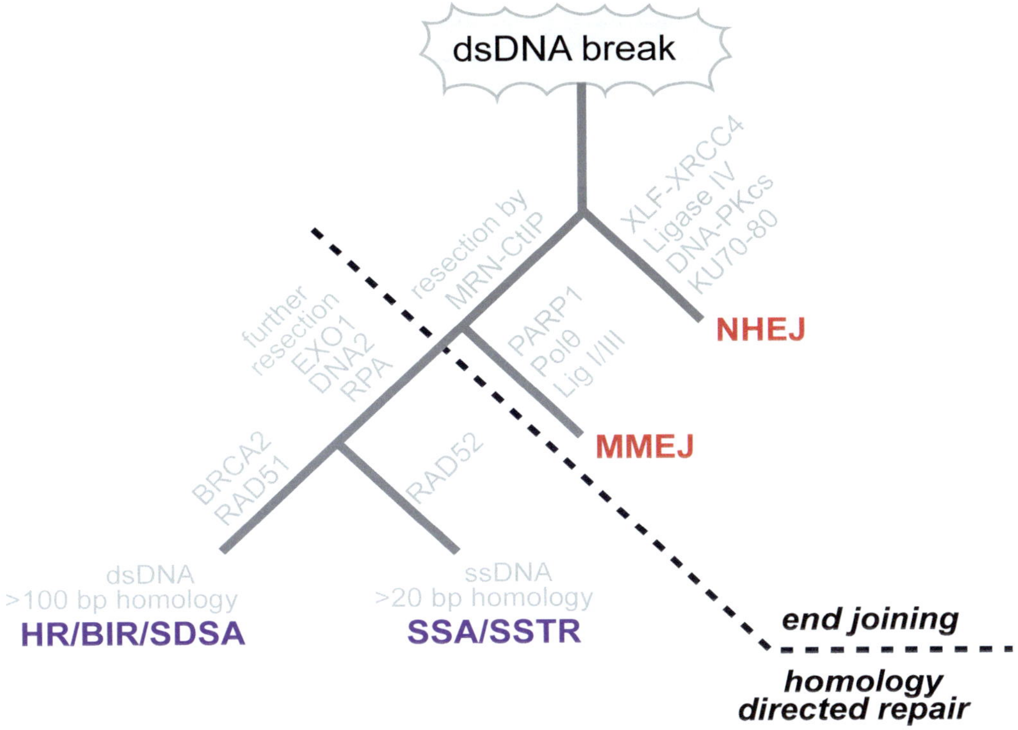

Figure 11.14. Mechanisms of DNA repair.

Source: By J.J.Froehlich, CC BY-SA 4.0, https://commons.wikimedia.org/w/index.php?curid=122458929

11.5.1. Importance of DNA Repair Mechanisms

Genomic integrity is in large measure preserved by effective DNA repair processes that are critical in preventing the accumulation of mutations that could lead to cellular dysfunction or malignancy. Dynamic chromatin structure, enabling access to DNA for repair processes, plays a pivotal role in this aspect. In response to damage in DNA, chromatin structure relaxes, thereby enhancing the access of repair proteins to the damaged site. The process emphasizes close coordination between DNA repair and chromatin dynamics. Research indicates that the best genomic maintenance involves some regulating mechanisms working synergistically to ensure that necessary DNA transactions, , such as replication and transcription, are done with unobstructed ease, hence ensuring cellular homeostasis and organismal overall health (Nair et al., 2017). Interstrand crosslinks (ICLs) are among the most toxic forms of DNA damage because they disrupt vital cellular functions like replication and transcription. Both endogenous metabolic reactions and exogenous assaults can create ICLs, leading to severe dangers if left unrepaired. Failure to repair ICLs may result in hypergenomic instability, typically causing cell death. This genomic stability-ICL relationship justifies the significance of learning various repair mechanisms for cell survival and therapeutic use. The promise of DNA repair pathway targeting in using ICLs in anti-cancer treatment is indicated, for example, by the employment of ICL-inducing compounds as clinical trials in breast cancer therapy

(Kullberg et al., 2018). The entanglement of treatment responses by resistance processes, however, reflects the importance of dissecting the complex web in ICL repair (Kato et al., 2018). Moreover, the diversity of strategies employed in DNA repair mechanisms indicates their flexibility and exigibility under various cellular conditions, most notably at various phases of the cell cycle. For instance, in dividing cells, ICL repair occurs during the S-phase in replication-coupled repair, while in resting or slowly growing cells, variant pathways , such as replication-independent repair, are utilized. This difference in repair machinery ensures that cells can respond appropriately to DNA damage, thereby maintaining genomic integrity in different cellular situations. As investigations proceed, a greater understanding of these subtle repair processes will not just sharpen our knowledge of cellular physiology but also hasten the development of precision therapies for DNA repair disorders that underlie many diseases, including cancer and genetic diseases (Kato et al., 2018; Nair et al., 2017).

11.5.2. Types of DNA Damage and Repair Pathways

DNA, as the repository of genetic information, is susceptible to various forms of damage that disrupt its structural integrity. Among the types of DNA damage are single-strand breaks (SSBs), DSBs, and more evolved lesions like DNA–protein cross-links (DPCs). SSBs, resulting from oxidative damage or replication errors, are reparable through the BER pathway, which recognizes and removes damaged or non-canonical bases. On the other hand, DSBs, as more catastrophic in their capacity to cause chromosomal rearrangements, need more advanced repair processes , such as HR and NHEJ. Cellular capacity to identify and correct such damage is crucial to genomic integrity and cell function, which underscores the significance of understanding such damage and the associated pathways of repair in the context of DNA structure and DNA replication. Efficient repair of DNA damage is regulated by various molecular pathways, and each pathway is geared towards repairing a specific type of lesion. The DNA damage response (DDR) represents a very complex network involved in cellular recognition, signaling, and repair of DNA lesions. For example, DPCs are a very poisonous form of DNA damage where proteins are covalently attached to DNA, physically obstructing vital processes like replication and transcription. Three distinct repair pathways for DPCs were discovered in recent research, illustrating the intricacy of cellular response to damage. Endonucleolytic cleavage by the enzyme MUS81 is one of the excision pathways for damaged DNA. A second pathway is proteolytic cross-link protein degradation by metalloproteases, and this is indicative of the multiple pathways available to cells. This multi-faceted repair is a reflection of the evolutionary developments in DNA repair pathways protecting organisms from disease and genetic instability (Dorn et al., 2023). The complex interplay among various DNA repair pathways is also extended to their regulation, which is critical for the response to specific types of DNA damage. For instance, upon the detection of DNA lesions, cells activate checkpoint pathways that can arrest the cell cycle to allow for the completion of repair and prevent inheritance of mutations. The SMC5/6 complex, which is now known to be a significant player in DPC repair, interacts with different repair mechanisms, which confirms an elaborate layering mechanism in regulating repair efficiency. The SUMOylation of damage sites, which is associated with the SMC5/6 complex, also testifies to a high level of regulation in repair mechanisms. While researchers continue to learn about the role of multiple players along these pathways, more knowledge on how they regulate will provide crucial information regarding how repair mechanism defects lead to disease , such as cancer and neurodegenerative disorders, emphasizing the importance of robust DDR systems (Singh et al., 2022) (Figure 11.15).

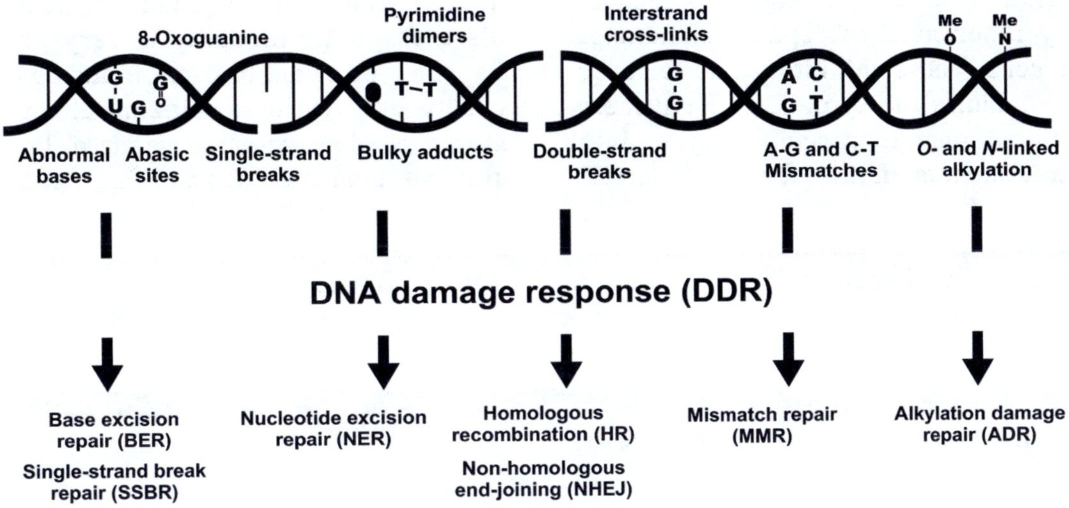

Figure 11.15. Types of DNA damage and the associated repair pathways.

Source: https://www.researchgate.net/profile/Eisuke-Tasaki/publication/328247889/figure/fig1/AS:681970333925376@1539605806303/Types-of-DNA-damage-and-the-associated-repair-pathways-Examples-of-DNA-lesions-top.png.

11.5.3. Consequences of Unrepaired DNA Damage

The consequences of unchecked DNA damage are broad-ranging and multifaceted, impacting cellular integrity and organismal health. DNA, as its complex double helix, is susceptible to endogenous sources of damage , such as ROS and to exogenous compounds like UV light and mutagenic chemicals. When not repaired, this damage can lead to the generation of mutations that disrupt normal cellular function, often initiating a cascade of genomic instability. This instability is linked to numerous pathological processes, including cancer, where defective DNA cells are free to proliferate uncontrolled without functional checkpoints for repair and apoptosis. The failure of DNA repair processes becomes particularly important in cancer treatments, where accumulation of DNA damage will ultimately dictate the efficacy of treatment modalities, and therefore highlight the critical need for robust repair processes in cellular homeostasis. Other than cancer, there has been evidence of unrepaired DNA damage leading to a range of neurological disorders, again demonstrating the relevance of DNA repair to brain function. Those diseases caused by deficits in DNA repair are usually disabling and can arise at any point in development, ranging from congenital failure through to late-onset progressive disease. These conditions could be caused by defects in repair processes, i.e., double-strand break repair, in the proliferation of neuroprogenitor cells. Abnormalities in these mechanisms lead to aberrant neuronal development and functioning and result in symptoms that could have significant impacts on cognitive and motor activities. For example, certain DNA repair diseases have been linked to conditions like ataxia, demonstrating the critical connection between DNA health and neurological well-being (Abraham et al., 2013). Understanding these implications indicates an urgent need to investigate how DNA damage impacts neural networks. Moreover, the study of mechanisms of DNA repair signifies therapeutic strategies with hopeful potential, particularly within oncology. Particular repair protein inhibitors, like PARP, aim to exploit the weaknesses formed by inherent DNA repair deficiencies in cancer cells. These inhibitors may also be applied to enhance the cytotoxic effectiveness of conventional chemotherapeutic drugs so as to enhance treatment modalities for a wide variety of tumor types. For instance, preclinical studies demonstrate that the combination of PARP inhibitors and established chemotherapeutics significantly enhances cell death in BRCA1- and BRCA2-mutant cancers (Chalmers, 2009). While this therapeutic modality is very promising, care must be taken to avoid inducing systemic toxicity due to the sensitizing effect of such inhibitors. More research needs to be done to optimize these methods and find out about the specific settings in which such inhibitors will have maximum therapeutic effect and minimum adverse effects.

Test Your Knowledge

1. What is the importance of DNA repair mechanisms?
2. What are the types of DNA damage and repair pathways?

11.6. Applications of DNA Research

Learning Outcome

1. Genetic engineering and biotechnology;
2. Forensic science and DNA profiling;
3. Medical advancements through DNA studies.

The applications of DNA research extend far from the realm of classical genetics into numerous fields , such as medicine, agriculture, and environmental science. In medicine, DNA analysis provides critical information to diagnose genetic disease and individualize treatment protocols. For example, advances in genomic tools have accelerated the development of targeted treatments based on one's unique genetic profile, improving the efficacy of treatments. Besides, DNA research has shown the way to technological breakthroughs like gene therapy, which attempt to correct genetic errors by introducing functional genes. Scientists can tailor these treatment techniques to individual patients based on what they have learned about DNA structure and replication. This ultimately means changing healthcare into an even more individualized system with much improved outcomes for patients. Agricultural applications of DNA research are a case of the potential of genetic engineering to enhance crop resistance and food security. Since the population across the globe is projected to reach close to 10 billion by 2050, food demand will rise, and new technology has to be created for agriculture. Scientists employ methods , such as CRISPR and marker-assisted selection (MAS) to engineer GMOs that can resist extreme climatic conditions, pests, and diseases. These are founded on the comprehension of DNA replication and its regulation of trait inheritance, enabling the ability to perform precise alterations in crop genomes. Moreover, these applications allow for the generation of biofortified foods, which will help fight malnutrition by improving nutritional content and guaranteeing not just agricultural sustainability, but also public health interventions aimed at eliminating hunger and nutritional inadequacies. Environmental science also benefits from progress in DNA research, particularly regarding biodiversity conservation and ecosystem management. Techniques like environmental DNA (eDNA) analysis enable tracking of the presence and diversity of species within ecosystems without observation or capture. Scientists can gain valuable information about habitat integrity and species interactions by studying DNA fragments shed by organisms into their habitats using eDNA analysis. Furthermore, DNA sequencing and barcoding technologies can be used to aid in the determination and recording of genetic diversity, a vital aspect to take into account for conservation strategies employed to safeguard endangered species. The complexity of DNA structure and replication renders these monitoring systems more effective by ensuring conservation measures rely on sound genetic data, ultimately being a factor in preserving global biodiversity (Figure 11.16).

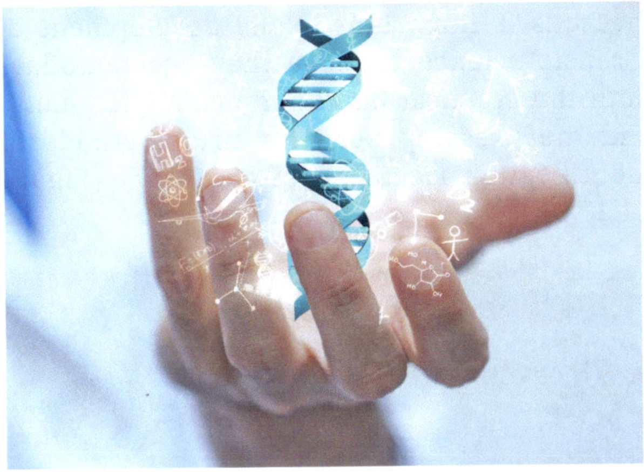

Figure 11.16. Conceptual representation of genetic science, showing a digital DNA double helix surrounded by scientific symbols, illustrating the integration of biology with modern technology.

Source: https://endeavordna.com/wp-content/uploads/2023/04/AdobeStock_103537025-scaled.jpeg.

11.6.1. Genetic Engineering and Biotechnology

The complex dependence of genetic engineering on biotechnology and vice versa is partly based on the existence of DNA structure and replication. Yet, comprehension of the double-helix nature of DNA allows scientists to play with genetic code and make exact alterations in genetic material, producing innovations in industries from agriculture to medicine. The ability to replicate DNA segments has existence-wide consequences, particularly in synthetic biology, where researchers engineer organisms to exhibit desired traits. This manipulation typically employs enzymes, i.e., those of serine integrases, which facilitate site-specific recombination. By exploiting the natural mechanisms of DNA recombination, scientists can generate targeted solutions in gene therapy and novel biotechnological uses, thus bridging the gap between pure genetic investigation and actual, practical applications (Stark, 2017). Additionally, the advent of CRISPR technology has rendered genetic engineering more accurate to an unprecedented extent in genome editing. The CRISPR-Cas system, a prokaryotic native adaptive immune system, has been repurified as a form of flexible gene modification agent in other organisms. Its capacity to target genome sequences selectively makes not only insertion and deletion of genes possible but also gene defect correction. This reframing of cellular functions underscores the relevance of understanding how DNA is replicated and repaired. As research continues to advance on CRISPR, technologies that leverage this technology have the capability to produce new strategies in addressing complex biological problems, , such as disease combat and food security (Bonomo et al., 2018). As genetic engineering and biotechnology merge, this ultimately paves the way for groundbreaking developments in some sectors. As a result of a clearer insight into DNA dynamics, researchers can develop valid methods for gene editing and synbio applications. The implications are gigantic, ranging from producing GMOs to additional gene therapy breakthroughs in genetic disorders. Further, the ethical implications surrounding these technologies augment the need for controlled utilization and regulation.

As the science matures, ongoing research and innovation will not only enhance our procedures but also lead to debate regarding the ethical considerations linked with genetic manipulation, and biotechnology will remain in accordance with societal values and health practices. In these developments, the interwebs of DNA structure, replication, and genetic engineering will continue to shape the future of science and society.

11.6.2. Forensic Science and DNA Profiling

In the field of forensic science, DNA profiling has been an indispensable tool in criminal conviction and identification. The unique composition of DNA, in the form of nucleotide sequences that make each individual distinct from all others, allows law enforcers to create a genetic fingerprint. It not only helps identify suspects but also acquits wrongfully accused individuals. As crucial as it is, the application of DNA evidence in court proceedings is typically confronted with adversity. Legal experts must stay up to date with the sophistication of DNA analysis to make judges and jurors fully aware of its scientific basis and significance to justice, so that DNA can become more amenable to the judicial process is essential to maintain its utility as forensic information (Griffith et al., 2014). But the success of DNA profiling can also be compromised by the quality and quantity of the available DNA sample. When degradation has occurred, for instance, in the case of human bones, standard genotyping techniques might prove inadequate. The issues raised by low-template DNA can lead to partial profiles that hinder certain identification. For instance, efforts to improve genomic DNA recovery from bones highlight the limitations faced in forensic examination. Procedures , such as employing high-performance DNA kits are being researched to optimize the recovery of recoverable DNA, which is vital for successful downstream application in DNA profiling (Griffith et al., 2014). This once again highlights the need for ongoing research and innovation in forensic methods. The marriage of various technologies in DNA profiling, , such as capillary electrophoresis and massively parallel sequencing, is the new face of forensic science. These methods not only promise to enhance the discriminatory power of DNA profiles but also attempt to reveal other genetic data, including biogeographical ancestry and Y-haplogroup. Although massively parallel sequencing offers revealing data that conventional methods may be unable to offer, it often at the cost of time and intractability. Hence, the best strategy for interrogation of recalcitrant samples, including those obtained from bone, must be discerned carefully. Lastly, the creation of DNA profiling methods is important to the development of forensic science because they accommodate both the practical limitation of sample quality and the pressing need for firm methods of identification (McLamb et al., 2017) (Figure 11.17).

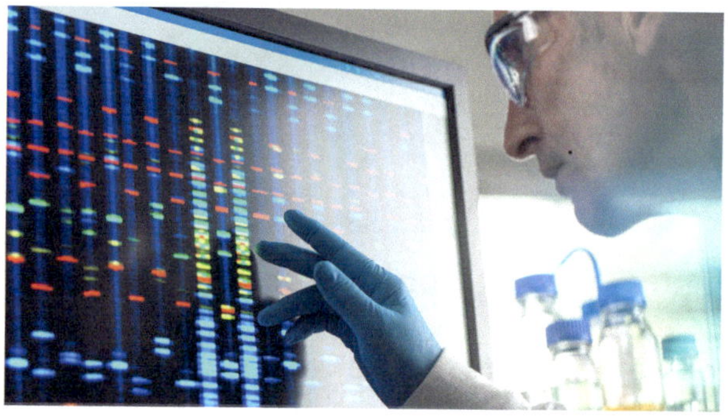

Figure 11.17. Forensic science and DNA profiling.

Source: https://cdn.hswstatic.com/gif/gettyimages-1428615121.jpg.

11.6.3. Medical Advancements Through DNA Studies

Medical practice has significantly changed due to advancements in DNA research, including genetic disease causes and cancer therapy.

Description of the structure of DNA has provided avenues for scientists to explore genetic mutations causing various diseases. For instance, advancements in genomic sequencing allow for identification of some genetic markers that are causes of inherited conditions so that preventive health care measures can be taken. It is, therefore, genetic testing that has become an integral aspect of early diagnosis and individualized treatment protocols, allowing clinicians to devise plans that are targeted towards the unique genetic profile of patients. Through such testing, clinicians are presently in a position to administer personalized treatments whose efficacy is amplified while reducing potential side effects, a monumental shift towards precision medicine in modern medicine. Oncolytic viruses, , such as genetically modified forms of the measles virus, are a cancer breakthrough that is DNA research-based. The specific modification of the measles virus genome is utilized to enhance its ability to selectively infect and kill and destroy cancer cells without compromising the normal tissue. The method is based on the general understanding of the interaction between the viral DNA and the cellular processes of the hosts, which has shown promising results in clinical trials. Most notably, the Edmonston strain's safety profile, a viral derivative of the measles vaccine, is an exemplary case of the relevance of DNA research in the development of therapeutic advances. Applying such viral technologies, researchers are now more and more approaching the cancer treatment complexities that would otherwise have relied on invasive methods. The developments in this area show how the study of DNA is not only significant with respect to disease knowledge but also significant to formulate targeted therapy that pin's hopes for cancer patients. Furthermore, the recent emergence of synthetic antibodies as a possible alternative to monoclonal antibodies shows the effect of DNA research on the evolution of medicine. Traditional monoclonal antibodies have serious ethical and economic concerns based on their complex production processes. But research with synthetic antibodies, particularly those that have unnatural amino acids , such as L-DOPA, is poised to increase binding affinity for specific biomarkers , such as prostate-specific antigen (PSA). The new approach could generate purer and more consistent antibodies, avoiding the limitations of existing monoclonal antibodies. Significantly, advances in DNA technology allow for rapid design of such synthetic antibodies, something that would transform diagnostic and therapeutic applications in oncology and other disciplines. So, further investigation into DNA structure and replication mechanisms is one of the principal forces for key breakthroughs in medical science, ultimately contributing to expanded patient care and treatment.

Test Your Knowledge

1. Explain genetic engineering and biotechnology.
2. What is forensic science and DNA profiling?

Summary

In molecular biology, the understanding of DNA structure and replication not only demonstrates the machinery by which genetic information is preserved, but also the fundamental rules that ensure genomic stability from generation to generation. Double helix DNA is a robust architecture for rapid replication. This regulation, through regulatory pathways , such as posttranslational modifications, signifies the delicacy of cellular machinery operating to maintain genetic integrity. The study indicates coordination of DNA replication factors, particularly through ubiquitin-mediated control, as a sign of sophistication of such biochemical processes, as shown by the central role of CDC-48 in coordinating licensing and elongation processes during replication in *C. elegans*, as highlighted by Franz et al. (2012). Moreover, the existence of mitochondrial DNA in other organisms, including the linear mt genomes of *Medusozoa*, presents an interesting revelation of DNA replication beyond the common circular models of most eukaryotes. The study of *Hydra magnipapillatas* damaged mt chromosomes presents novel means of adaptation beyond the original expectations of mitochondrial replication processes. These accommodations raise a necessity to reanalyze the mechanisms regulating mt DNA replication, with the possibility of relevance to evolutionary functions in other taxa with similarly organized genomes. The organization of inverted terminal repeats and following identical oriented sequences within mitochondrial chromosomes is a mechanism that plays a pivotal role in replicating and transcribing, as described by Erpenbeck et al. (2008). These discoveries contribute to the more precise understanding of how diverse genomic architectures influence replication strategies. In conclusion, DNA structure and replication studies uncover profound secrets about the fidelity of the transmission of the genes and the evolutional processes that take place in different organisms. The complex interaction of various replication factors and their precise regulatory roles, like in the case of CDC-48, indicate the dynamic nature of cellular processes. With continuous growth in genomic studies, the significance of the findings will continue to make our comprehension of genetic stability even more precise and the chance for errors during replication. Hence, further studies of the processes of nuclear and mitochondrial DNA replication will not merely elucidate current paradigms of genetic integrity but will also direct forthcoming studies in evolutionary biology and the ability of organisms to adapt to various environments.

The revelation of DNA structure reveals a stunning but complex organization that is at the core of its role as a storage and transmission system for genetic information. At the center of the structure is the double helix composed of two antiparallel strands of nucleotides. Each nucleotide contains a sugar, phosphate group, and a nitrogenous base, with four possible bases: adenine, thymine, cytosine, and guanine. The particular pairing of these bases, adenine with thymine and cytosine with guanine, is responsible for fidelity of replication and is crucial to the correct transmission of genetic information from generation to generation. Comprehension of this complex molecular blueprint not only sheds light on the very mechanisms of heredity itself but also provides the basis for investigating how structural variations in DNA can lead to a broad spectrum of genetic diseases or disorders (Prokopenko et al., 2008). As DNA replication is considered, one comes to understand that such activity is needed in order for cells to replicate and for organisms to grow. When replication is taking place, the double helix is undone and separated so that both strands serve as templates to form new complementary strands. It is a highly regulated process performed by several key enzymes, i.e., helicase, DNA polymerase, and ligase, each of which plays a unique role in synthesizing and proofreading the new DNA strands. The semi-conservative mechanism of replication—each new molecule of DNA one initial and one made—this is what ensures genetic continuity. Not only is the process central to cellular reproduction, but it is also what sustains genomic stability; error in replication leads to mutations with extensive biological impacts, from susceptibility to cancer to hereditary diseases (Johnston et al., 2015). This is why replication fidelity is ensured through multiple proofreading mechanisms. Also, the interconnectedness of DNA replication and its integrity points to regulatory mechanisms that pervade the cell cycle. Coordination of events—spanning from DNA synthesis to checkpoint control—remains integral to removing mistakes and hence guaranteeing fidelity of genetic information. Furthermore, recent research has identified other aspects of complexity, , such as the involvement of epigenetic events and mitochondrial DNA dynamics, which must be considered in the overall context of replication fidelity. As an illustration, the integration into the model of the phenomenon of the mtDNA bottleneck establishes a conceptual framework for investigating how mitochondrial DNA populations are regulated during development to influence the onset of inherited disease. This new face of knowledge requires the need to integrate results from all areas of molecular biology and thereby making our understanding of DNA replication and structure as a cornerstone of life more complete (Prokopenko et al., 2008; Johnstonet al., 2015).

KEY TERMS FROM THIS CHAPTER

Adenine (A)	Antiparallel strands	Complementary base pairing
Cytosine (C)	DNA (deoxyribonucleic acid)	DNA polymerase
Double helix	Guanine (G)	Helicase
Hydrogen bonds	Lagging strand	Leading strand
Ligase	Nitrogenous base	Nucleotide
Okazaki fragments	Replication fork	Semi-conservative replication
Sugar-phosphate backbone	Thymine (T)	

Critical Thinking Questions

1. **What is the shape of a DNA molecule?**
   a. Single strand
   b. Straight ladder
   c. Double helix
   d. Triple helix
2. **Which of the following are the correct base-pairing rules in DNA?**
   a. A–G, C–T
   b. A–C, G–T
   c. A–T, G–C
   d. A–U, G–C
3. **The sugar found in DNA is:**
   a. Glucose
   b. Ribose
   c. Deoxyribose
   d. Sucrose
4. **DNA replication is described as:**
   a. Conservative
   b. Dispersive
   c. Semi-conservative
   d. Destructive
5. **Which enzyme is responsible for adding new nucleotides during DNA replication?**
   a. DNA helicase
   b. RNA polymerase
   c. DNA ligase
   d. DNA polymerase

Short Answer Questions

1. What does DNA stand for, and what is its primary function in cells?
2. Describe the shape of a DNA molecule. What is this shape called?
3. What are the four nitrogenous bases found in DNA, and how do they pair?
4. What components make up a DNA nucleotide?
5. Who are the scientists credited with discovering the double helix structure of DNA?

Answers to Critical Thinking Questions

1. (c); 2. (c); 3. (c); 4. (c); 5. (d)

References

1. Abdelhamid, M. A. S., Amparo, C., Bedell, V., Clark, K., et al. (2020). Duplex DNA from sites of helicase-polymerase uncoupling links non-B DNA structure formation to replicative stress. https://core.ac.uk/download/293754853.pdf (accessed on 13th June 2025).
2. Abi, A., Aravin, A., Danhel, A., Re, A., Simonova, A., Szabo, A., et al. (2015). Biophysical and electrochemical studies of protein-nucleic acid interactions. https://core.ac.uk/download/29112106.pdf (accessed on 13th June 2025).
3. Abraham, A., Ahel, I., Alano, P., Alderton, W., Alter, B., Barbi, D., Barlow, C., et al. (2013). DNA strand break repair and neurodegeneration. https://core.ac.uk/download/30608004.pdf (accessed on 13th June 2025).
4. Agha, A., Ahmad, B., Arai, Y., Banin, S., Bettencourt-Dias, M., Bicknell, L. S., et al. (2014). Congenital microcephaly. https://core.ac.uk/download/20327855.pdf (accessed on 13th June 2025).
5. Ahel, I., Bellamy, S. R., Borodina, T. A., Bullard, D. R., Cahova-Kucharikova, K., Cao, W. G., Doherty, A. J., et al. (2008). Label-free electrochemical monitoring of DNA ligase activity. https://core.ac.uk/download/2755414.pdf (accessed on 13th June 2025).
6. Iain G Johnston, Joerg P Burgstaller, Vitezslav Havlicek, Thomas Kolbe, Thomas Rülicke, Gottfried Brem, Jo Poulton, Nick S Jones (2015) Stochastic modelling, Bayesian inference, and new in vivo measurements elucidate the debated mtDNA bottleneck mechanism eLife 4:e07464
7. Alagramam, K., Varshavsky, A., Alfageme, C., An, H., Aulner, N., Bachmair, A., et al. (2008). Discovery of cellular regulation by protein degradation. https://core.ac.uk/download/4879072.pdf (accessed on 13th June 2025).
8. Ambjoernsson, T., Hanke, A., Levene, S., Metzler, R., et al. (2006). Single DNA conformations and biological function. https://core.ac.uk/download/44139099.pdf (accessed on 13th June 2025).
9. Arana, M. E., Bernstein, K. A., Braithwaite, D., Buttner, M. J., Cancer Genome Atlas Research Network, Cannavo, E., Ceccaldi, R., et al. (2015). Structure of the helicase domain of DNA polymerase theta reveals a possible role in the microhomology-mediated end-joining pathway. https://core.ac.uk/download/30734177.pdf (accessed on 13th June 2025).
10. Bartkova, J., Baumann, C., Baxter, R., Bermejo, R., Bester, A. C., et al. (2014). "Breaking up is hard to do": The formation and resolution of sister chromatid intertwines. https://core.ac.uk/download/30611328.pdf (accessed on 13th June 2025).
11. Barton, J. K., Kaiser, J. T., & Pierre, V. C. (2006). Insights into finding a mismatch through the structure of a mis paired DNA bound by a rhodium intercalator. https://core.ac.uk/download/4876061.pdf (accessed on 13th June 2025).
12. Batiuk, E., Kiyohara, C., & Nguyen, T. (2017). Engineering synthetic antibody by expanded genetic code. https://core.ac.uk/download/84118620.pdf (accessed on 13th June 2025).
13. Bennett, N. J. (2005). *Unlocking the M13 (F1) Virion: Investigation into The Role of pIII C Domain of F Specific Filamentous Bacteriophage in Infection: A Thesis Presented in Partial Fulfillment of The Requirements for the Degree of Master of Science in Biochemistry at Massey*

University, Palmerston North, New Zealand. https://core.ac.uk/download/162617266.pdf (accessed on 13th June 2025).

14. Beysen, D., Carvalho, C. M. B., D'haene, B., De Baere, E., et al. (2013). Microhomology-mediated mechanisms underlie non-recurrent disease-causing microdeletions of the *FOXL2* gene or its regulatory domain. https://core.ac.uk/download/55755250.pdf (accessed on 13th June 2025).
15. Bonomo, M. E., & Deem, M. W. (2018). The physicist's guide to one of biotechnology's hottest new topics: CRISPR-Cas. http://arxiv.org/abs/1712.09865 (accessed on 13th June 2025).
16. Buckley, R. J. (2023). Biochemical and genetic analysis of *Lhr*, a mysterious helicase/glycosylase conserved across species. https://core.ac.uk/download/573838686.pdf (accessed on 13th June 2025).
17. Caiafa, P., & Zampieri, M. (2005). DNA methylation and chromatin structure: The puzzling CpG islands. https://core.ac.uk/download/54305860.pdf (accessed on 13th June 2025).
18. Campbell, J. L., Chen, L., Masuda-Sasa, T., Peng, G., et al. (2008). Processing of G4 DNA by Dna2 helicase/nuclease and RPA provides insights into the mechanism of Dna2/RPA substrate recognition. https://core.ac.uk/download/4878686.pdf (accessed on 13th June 2025).
19. Cascio, D., Hancock, S. P., Johnson, R. C., Stella, S., et al. (2016). DNA sequence determinants controlling affinity, stability and shape of DNA complexes bound by the nucleoid protein Fis. https://core.ac.uk/download/323080256.pdf (accessed on 13th June 2025).
20. Chalmers, A. J. (2009). The potential role and application of PARP inhibitors in cancer treatment. https://core.ac.uk/download/2709754.pdf (accessed on 13th June 2025).
21. Chen, I. A., Gerland, U., Leu, K., Obermayer, B., et al. (2011). The prebiotic evolutionary advantage of transferring genetic information from RNA to DNA. https://core.ac.uk/download/323080491.pdf (accessed on 13th June 2025).
22. Chowdhury, D., & Sharma, A. K. (2012). Template-directed bio polymerization: Tape-copying Turing machines. http://arxiv.org/abs/1212.6827 (accessed on 13th June 2025).
23. Cole, B., Lambrechts, L., Rutsaert, S., Trypsteen, J., et al. (2020). Emerging PCR-based techniques to study HIV-1 reservoir persistence. https://core.ac.uk/download/287941778.pdf (accessed on 13th June 2025).
24. Conway, A. B., Pierce, A. J., Nanduri, B., Mickelson, C., Jongeneel, C. V., Kong, D., Giedroc, D. P., et al. (2010). Assembly and dynamics of the bacteriophage T4 homologous recombination machinery. https://core.ac.uk/download/pdf/8510155.pdf (accessed on 13th June 2025).
25. De Polavieja, G. G., & Arias-Gonzalez, J. R. (2012). Entropy involved in fidelity of DNA replication. https://core.ac.uk/download/18727162.pdf (accessed on 13th June 2025).
26. Doherty, A. J., Barry, P., Bebenek, K., Connolly, B. A., Biles, B., Boiteux, S., Braithwaite, D., et al. (2014). An archaeal family-B DNA polymerase variant able to replicate past DNA damage: Occurrence of replicative and translesion synthesis polymerases within the B family. https://core.ac.uk/download/30608668.pdf (accessed on 13th June 2025).
27. Dorn, A., Dvořák Tomaštíková, E., Finke, A., Jemelkova, L., et al. (2023). SMC5/6 complex-mediated SUMOylation stimulates DNA–protein cross-link repair in *Arabidopsis*. https://core.ac.uk/download/581125439.pdf (accessed on 13th June 2025).
28. Erpenbeck, D., Voigt, O., & Wörheide, G. (2008). A fragmented metazoan organellar genome. https://core.ac.uk/download/12174129.pdf (accessed on 13th June 2025).

29. Esashi, F., Leung, C. W. B., & Wall, J. (2024). From rest to repair: Safeguarding genomic integrity in quiescent cells. https://core.ac.uk/download/649472545.pdf (accessed on 13th June 2025).

30. Forsburg, S., Kearsey, S. E., Lydeard, J., Nishitani, H., Riedel, C., Cotterill, S., et al. (2008). DNA Replication: A database of information and resources for the eukaryotic DNA replication community. https://core.ac.uk/download/9536084.pdf (accessed on 13th June 2025).

31. Franz, A. (2012). Role of the ubiquitin-selective CDC-48/UFD-1/NPL-4 chaperone in DNA replication. https://core.ac.uk/download/12011230.pdf (accessed on 13th June 2025).

32. Griffith, J. T., & Leclair, S. L. (2014). DNA in the courtroom: The 21st century begins. https://core.ac.uk/download/233916649.pdf (accessed on 13th June 2025).

33. Holguera, I., Redrejo-Rodríguez, M., Salas, M., Vega, M. A., et al. (2016). DNA-binding proteins essential for protein-primed bacteriophage Φ29 DNA replication. https://core.ac.uk/download/84780072.pdf (accessed on 13th June 2025).

34. Hutzen, B., Raffel, C., & Studebaker, A. W. (2015). Advances in the design and development of oncolytic measles viruses. https://core.ac.uk/download/323063446.pdf (accessed on 13th June 2025).

35. Janion, C. (2008). Inducible SOS response system of DNA repair and mutagenesis in *Escherichia coli*. https://core.ac.uk/download/pdf/8195665.pdf (accessed on 13th June 2025).

36. Jih, J., Jin, L., Shih, C., Yu, L., et al. (2013). 3.5 Å cryoEM structure of hepatitis B virus core assembled from full-length core protein. https://core.ac.uk/download/323079656.pdf (accessed on 13th June 2025).

37. Jukes, T. H. (2025). *Chemistry of Living Systems: Semiannual Report, May 1–Sep. 30, 1966*. https://core.ac.uk/download/pdf/85249377.pdf (accessed on 13th June 2025).

38. Kato, N. (2018). The molecular mechanism of replication-independent repair of DNA interstrand crosslinks. https://core.ac.uk/download/161459282.pdf (accessed on 13th June 2025).

39. Kauffman, L. H. (2002). Biologic. https://core.ac.uk/download/25335191.pdf (accessed on 13th June 2025).

40. Kaykov, A., Klar, A. J., Holmes, A. M., Arcangioli, B., et al. (2010). Identification of a novel type of spacer element required for imprinting in fission yeast. https://core.ac.uk/download/12428.pdf (accessed on 13th June 2025).

41. McLamb, J. M. (2017). Application of enhancement strategies for the improvement of discriminating forensic DNA profiles from human bones. https://core.ac.uk/download/345084962.pdf (accessed on 13th June 2025).

42. Nair, N., Shoaib, M., & Sørensen, C. S. (2017). Chromatin dynamics in genome stability: Roles in suppressing endogenous DNA damage and facilitating DNA repair. https://core.ac.uk/download/269296074.pdf (accessed on 13th June 2025).

43. Pacheco, A., Barry, P. A., Craig, K. L., Engel, K., et al. (2013). Population dynamics of rhesus macaques and associated foamy virus in Bangladesh. https://core.ac.uk/download/323068585.pdf (accessed on 13th June 2025).

44. Parajuli, S. (2017). DNA replication challenges: Telomeres and R loops. https://core.ac.uk/download/233214603.pdf (accessed on 13th June 2025).

45. Phan, A. T., Bakalar, B., Heddi, B., Butovskaya, E., Hud, N. V., Nadai, M., & Richter, S. N. (2018). Major G-quadruplex form of HIV-1 LTR reveals a (3 + 1) folding topology containing a stem-loop. https://core.ac.uk/download/168406216.pdf (accessed on 13th June 2025).

46. Prokopenko, M., Polani, D. and Wang, P. (2008) Optimizing Associative Information Transfer within Content-addressable Memory. International Journal of Unconventional Computing, 4 (3). pp. 273-296. ISSN 1548-7199 (accessed on 13th June 2025).

47. Singh, J. K. (2022). Identification and characterization of novel factors in the DNA damage response. https://core.ac.uk/download/543775366.pdf (accessed on 13th June 2025).

48. Stark, W. M. (2017). Making serine integrases work for us. https://core.ac.uk/download/84148269.pdf (accessed on 13th June 2025).

49. Witzany, G. (2017). Artificial and natural genetic information processing. https://core.ac.uk/download/186332662.pdf (accessed on 13th June 2025).

50. Yu, J. (2015). Computational investigations on polymerase actions in gene transcription and replication combining physical modeling and atomistic simulations. http://arxiv.org/abs/1511.08392 (accessed on 13th June 2025).

GENE EXPRESSION AND BIOTECHNOLOGY

12

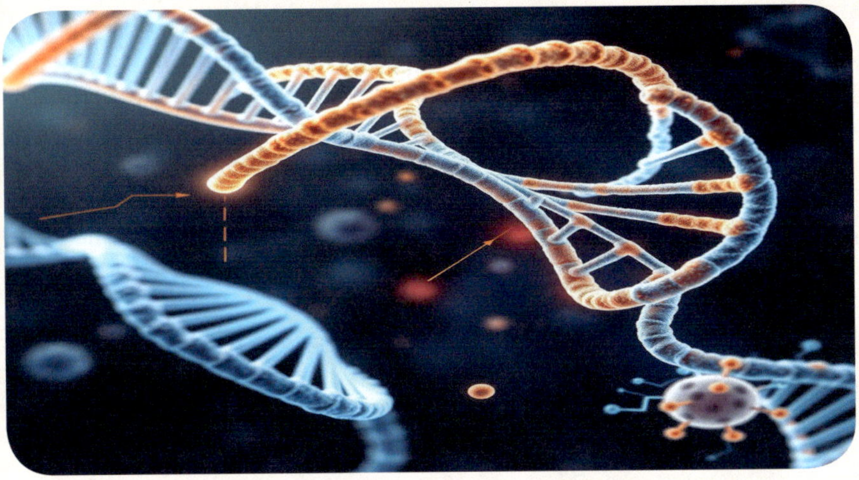

LEARNING OBJECTIVES

After studying this chapter, you will be able to:

1. Define the central dogma of molecular biology;
2. Explain the techniques in biotechnology;
3. List the applications of gene expression in medicine;
4. Discuss about agricultural biotechnology;
5. Focus on the ethical considerations in biotechnology;
6. Illustrate the future directions in gene expression research.

Introduction

The complex process of gene expression is a central element in the field of molecular biology and affects the operation of cellular systems. Gene expression entails the translation of genetic code in DNA into functional cellular units, primarily proteins, that indirectly determine the phenotype of an organism. This is regulated with great care to produce the genes at the appropriate time and amount necessary for the maintenance of homeostasis in biological systems. Gene expression deciphering is not only required for the understanding of simple biological processes but also to drive biotechnological applications towards improving agricultural production, the identification of therapeutics, and mitigating environmental problems. As the importance of this field grows, the integration of genetic engineering techniques enhances the possibility of regulating gene expression, making it possible to find new answers to biotechnology. While looking for gene expression, many experimental techniques have been established, contributing to the vast body of information regarding genes being expressed and regulated. Techniques , such as quantitative PCR (qPCR), microarray analysis, and RNA sequencing have revolutionized the way scientists quantify and contrast gene expression levels in numerous contexts. By employing these techniques, researchers can identify which genes respond to particular stimuli or stresses, and this illuminates pathways that are implicated in development, stress responses, and disease mechanisms. Furthermore, advances in bioinformatics technology have enabled sophisticated analysis of gene expression data, enabling genotype-phenotype correlations to be established with ever increasing specificity. Their integration is the foundation of continuing research efforts and drives continuing biotechnology advances that enable transgenic organisms and new agricultural techniques. As recent research illustrates, gene expression findings have applications beyond the laboratory to everyday biotechnology use. For instance, genetic improvement in forest trees illustrates how an understanding of gene expression can lead to increased growth rates, disease resistance, and improved biomass yield from forest products (Diouf et al., 2004). These applications reflect the utility of biotechnology to address global problems, including food safety and sustainable resource management. Furthermore, interdisciplinary research, for example, marine biotechnology, highlights cooperative endeavor by academia and industry to identify new biotechnological applications resulting from exploring gene expression in diverse organisms (Seaman & Whitehouse, 2001). This dialogue demonstrates the versatility of gene expression research, as it contributes not only to fundamental biology knowledge but also enables one to create solutions specifically geared toward meeting immediate societal needs (Figure 12.1).

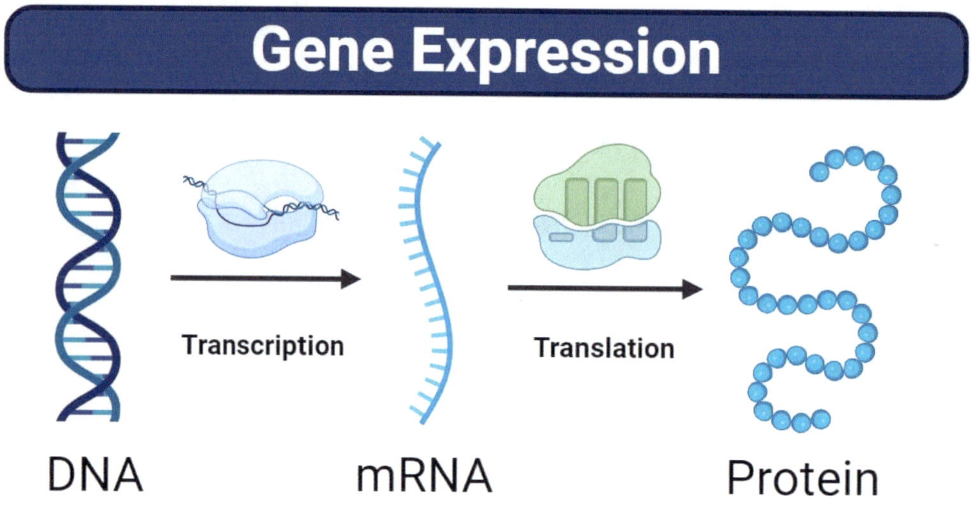

Figure 12.1. Gene expression.

Source: https://microbenotes.com/wp-content/uploads/2023/10/Gene-Expression.jpeg.

Definition of Gene Expression

Gene expression is a fundamental biological process through which the information in DNA is decoded into proteins to make phenotypic characters expressible. It is a multi-step process that typically encompasses transcription, where mRNA is synthesized from DNA, and translation, where messenger RNA (mRNA) is employed to synthesize polypeptides. These regulatory mechanisms that manage gene expression are complex and multifunctional with several factors, , such as TFs, enhancers, and silencers, to guarantee that the genes are activated at the appropriate time, in the appropriate cell type, and in the right quantities. Gene expression is essential because it enables us to learn about developmental biology, cell differentiation, and the molecular basis of disease, , such as cancer and genetic disease. From the knowledge of these processes, researchers can pave the way for novel therapeutic interventions in medicine and biotechnology. In biotechnology, gene expression manipulation is highly promising in advancing agriculture and health. CRISPR-Cas9 is one of the genetic engineering techniques where specific gene expression modification is possible, thereby enhancing favorable traits in plants or producing therapeutic proteins in microorganisms. For instance, GMOs are engineered to contain genes for pest resistance or drought tolerance, thereby boosting agricultural productivity. Alternatively, biotechnology is also utilized to produce biopharmaceuticals, whereby improved gene expression leads to the mass production of significant proteins and enzymes for the pharmaceutical industry. The study of gene expression not only illuminates our understanding of biological systems but also fosters innovations to address worldwide issues , such as food security and access to medicine. Moreover, as emerging biotechnological approaches develop, ethical issues of modulating gene expression still have spaces for discussion. The domain of gene expression reaches beyond its biological foundations, influencing practice in the life sciences. Advancements in computational power and data analysis enabled researchers to sort through great collections of literature to identify and associate gene function, expression levels, and associated biological phenomena. For example, text mining for life sciences has made database curation and knowledge summarization possible by enabling the automation of biomedical text analysis and association of findings with well-curated biomolecular databases (Van Landeghem et al., 2013). These unifications allow the expansion of genomic data, linking gene expression measurements to quantitative phenotypic change. Further, integration of automated techniques with rigorous experimental methods can enhance gene expression analysis precision, finally delivering a comprehensive understanding of complex biological interactions. With these techniques continuing to evolve, they will enable more breakthroughs that merge basic science and useful application, and further cement gene expression as a part of the broader picture of biotechnology science.

Importance in Biological Processes

Gene expression is a core process for many biological processes, serving as a fundamental means through which genetic information is translated into action proteins. Proteins form the scaffolding of virtually all cell functions, from enabling metabolic reactions to imparting structure in cells. This complex process of gene expression involves numerous steps , such as transcription, mRNA processing, translation, and post-translational modification. The controlling factors of gene expression dictate that genes are only expressed when needed so that adaptive responses may be produced to changes in the environment. By disassembling these control loops, researchers uncover how genes respond to stimuli, ultimately laying out potential targets for biotechnology interventions. Gene expression dynamics are crucial to comprehending to design new solutions in synthetic biology, where microbes can be made to produce desired products or respond to specific stimuli. Apart from being the foundation of cellular life, gene expression is also crucial for evolutionary biology. Gene expression variation generates phenotypic variation, providing raw material for natural selection. These kinds of flexibility can shape populations through generations and influence traits , such as development, disease resistance, and reproductive success. Gene expression can also be influenced by epigenetic mechanisms, which allow conditions leading to temporary changes without the delay of gene mutation to the underlying DNA sequence. Such epigenetic changes enable organisms to respond rapidly to environmental conditions without the delay of genetic mutations. The fact that gene expression patterns can be inherited gives greater significance to the role of regulatory networks in evolution, showcasing the interweaving of gene expression with large-scale biology. Insights from these evolutionary considerations can assist in furthering biotechnology, particularly its application in conservation as well as agriculture. Biotechnology in investigating gene expression has consequences that extend beyond basic biology. For instance, in marine biotechnology, researchers are researching gene expression in aquatic life to identify biochemical pathways that might lead to new drugs or enzymes. Research is particularly critical in habitats supporting extreme environments, where researchers identify the adaptation mechanisms that can be utilized to guide biotechnological innovation. By documenting and analyzing gene expression profiles for such organisms, researchers can identify potential academic collaborations and partners who employ marine biodiversity for industrial applications. Such research not only contributes to the understanding of biological processes but also drives economic advancement through the production of new bioproducts. The intersection of gene expression studies in biotechnology thus constitutes a critical facilitator for scientific advancement as well as use, bridging the divide between academic investigation and practical dividends in many different fields.

Overview of Gene Regulation Mechanisms

Gene regulation is an active, complex process governing gene expression, determining when and how a given gene is expressed or silenced in cells. TFs, proteins that bind to specific DNA sequences close to the promoter region of genes, are one such central mechanism of gene regulation. These can support or hinder the recruitment of RNA polymerase, the enzyme responsible for synthesizing the messenger RNA (mRNA) from DNA. TFs aside, epigenetic changes, , such as histone acetylation and DNA methylation, play significant roles in gene regulation by altering the chromatin structure. These alterations can enable a permissive or restrictive state for transcription, thus affecting levels of gene expression. The interactions of these different regulatory factors highlight the complexity of the control mechanisms of gene expression, which are critical for cellular differentiation and the organism's ability to respond to environmental stimuli. Another critical dimension of gene regulation is post-transcriptional regulation, through which gene expression is controlled after mRNA is produced. Alternative splicing is an essential one of these mechanisms by which diverse mRNA transcripts are produced from a single gene to enable the synthesis of multiple isoforms of proteins with different functions. This is regulated by several various splicing factors which have specific binding to specific splice sites within pre-mRNA. RNA stability and degradation play a central role in regulating gene expression levels because the half-life of an mRNA molecule can dictate the amount of protein being synthesized. MicroRNAs (miRNAs) play a major role in post-transcriptional regulation by binding to complementary sequences in mRNAs, leading to their degradation or inhibition of translation. Together, these mechanisms reflect the complexity of gene regulation and emphasize its significance in biotechnological applications, where gene expression manipulation is employed to achieve enhanced crop yields and resistance to disease. In biotechnology, knowledge of gene regulation mechanisms is a key factor in developing successful genetic engineering strategies. Agrobacterium-mediated transformation applies the principles of gene regulation to transfer novel traits to plants. This process was revolutionized in the 1980s, with researchers like Prof. Domenico Mariotti laying the foundation for the manipulation of plant genomes to maximize crop yield. Through the use of specific gene promoters and regulatory components, researchers can regulate the expression of inserted genes so that they can function optimally in their host plant. Also, developments in plant biotechnology have enabled the identification of networks of genes and their effects on metabolic pathways to develop crops with improved nutritional values and resistance to biotic and abiotic stress factors. This complex connection between gene regulation and biotechnology holds promise for change in controlling the expression of genes in achieving sustainable agriculture (Seaman & Whitehouse, 2001; Caretto et al., 2019) (Figure 12.2).

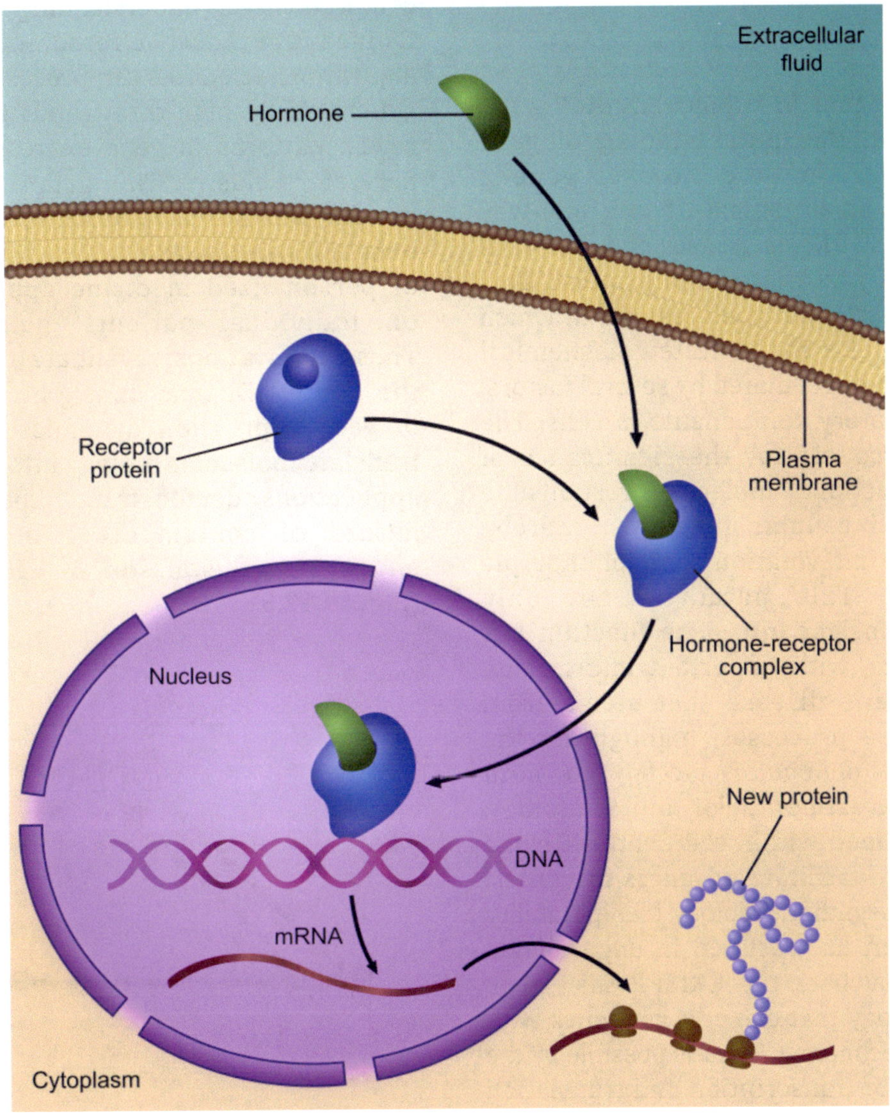

Figure 12.2. Gene regulation mechanisms.

Source: By Ali Zifan 03:07, 10 July 2016 (UTC) - Own work; Used information from:Campbell Biology (10th Edition) by: Jane B. Reece & Steven A. Wasserman., CC BY-SA 4.0, https://commons.wikimedia.org/w/index.php?curid=50049531

12.1. The Central Dogma of Molecular Biology

Learning Outcome

1. DNA replication;
2. Transcription process;
3. Translation and protein synthesis.

The Central Dogma of Molecular Biology is a foundation in the understanding of gene expression, directing the flow of genetic information in an organism. It is a one-way process in which DNA is transcribed into RNA and subsequently translated into proteins. This set is at the heart of gene expression, in which each step involves a sophisticated biochemical process that can be regulated by several factors, including regulatory components in cells. The dogma also accounts for the significance of proteins as functional molecules responsible for the myriad cellular functions, thereby linking genetic information with phenotypic manifestation. This paradigm not only increases our insight into gene function but also provides us with important information regarding genetic disease that stems from defects in these processes, highlighting the importance of molecular biology in both research and treatment. For biotechnology, the ideas contained within the Central Dogma are employed to facilitate advances in genetic engineering and synthetic biology. Manipulability of RNA and DNA has resulted in unparalleled innovations, , such as the CRISPR-Cas9 gene editing technology that can edit genomics with precision. The technology is made possible by the understanding of transcription and translation to create targeted changes, which means scientists can engineer organisms with specified features. Reprogramming cellular routes also provides revolutionary therapies for genetic disease, highlighting the applied implications of the Central Dogma. As biotechnology continues to develop, interdisciplinarity, including the systems biology integration type, is becoming more and more necessary to further utilize complex interactions between biomolecules and drive efficiency in the field's applications (Fernandes et al., 2014). Additionally, integrating computer methods with biological data analysis reflects the strength of modern biotechnology in comprehensively understanding gene expression. By means of inductive reasoning and advanced algorithms, scientists can process large datasets generated by high-throughput technologies to reveal patterns in gene expression that were heretofore elusive. This data-driven paradigm not only enables us to better forecast biological behaviors but also enables the construction of personalized medicine approaches based on individual patients' genetic profiles. These innovations revalidate the strength of the Central Dogma, as a growing knowledge of gene expression dynamics allows one to translate molecular biology into useful medical applications, demonstrating the multi-faceted nature of contemporary biotechnological endeavors (Gustafsson & Vallverdú, 2016) (Figure 12.3).

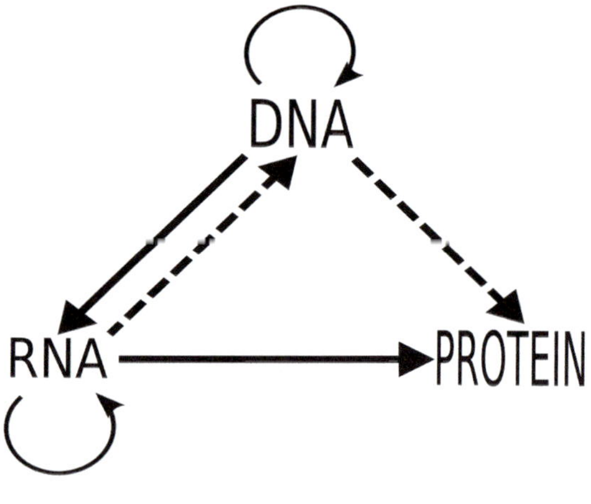

Figure 12.3. Central dogma of molecular biology.

Source: Public Domain, https://commons.wikimedia.org/w/index.php?curid=1851043

12.1.1. DNA Replication

Replication of DNA is the focal point of the examination of gene expression and biotechnology overall. The DNA molecule unfolds as it is replicated, and each strand is utilized as a template in creating new complementary strands. This semi-conservative replication guarantees that each daughter cell acquires an exact copy of the genetic material, crucial to maintaining the integrity of the genome from one generation to the next. Replication mistakes can give rise to mutations, which influence the expression of genes and potentially the development of a variety of diseases. A thorough understanding of molecular mechanisms that control DNA replication is therefore essential to develop biotechnological interventions to control gene expression with therapeutic purposes, , such as gene therapy and synthetic biology. Furthermore, mechanisms of DNA replication are crucial for various biotechnological advances, including the utilization of CRISPR-Cas systems for genome editing. The CRISPR-Cas toolbox revolutionized gene engineering by providing the means to edit genomic sequences precisely, enhancing our ability to analyze gene function and interaction. Lying at the foundation of CRISPR's effectiveness is actually the capacity of organisms to guarantee that the process of DNA replication is faithful so that the crRNAs replicate faithfully along with the rest of the genome. Understanding mechanisms of DNA replication not only contributes to the improvement of the CRISPR-Cas technology, but also to the development of robust cellular systems, which can survive and thrive under manipulative genetic engineering intervention, further unlocking the fields of applications for genetic biotechnology in medicine and agriculture (Bonomo et al., 2018). In addition, advancements in synthetic biology are significantly dependent on accurate DNA replication mechanisms, particularly when synthetic organisms are designed with success relying on optimal gene expression. A scientific understanding of how DNA replication impacts transcriptional and translational processes is important in facilitating the optimization of such synthetic units. For instance, cell cycle regulatory and DDR genes were of the most critical nature when exposed to stressful conditions because some of the disturbances alter replication rate and fidelity. Understanding how these genes are upregulated or downregulated upon environmental cues, including ionizing radiation (Aerts et al., 2014), allows researchers to enhance the stress tolerance of engineered organisms. Consequently, integrating DNA replication information into biotechnological models not only improves our skill in genetic engineering but also advances our understanding of cellular reactions to varied external stimuli.

12.1.2. Transcription Process

Transcription is a crucial component of the overall process of gene expression, whereby genetic information encoded within DNA is copied into messenger RNA (mRNA). Transcription begins with the assembly of RNA polymerase onto a promoter of a gene by many TFs. These elements are the major players in determining sequences of DNA and enabling the double helix to unwind, which ultimately allows RNA polymerase to synthesize a complementary RNA strand. As the transcription process goes on, the elongation phase comes into play where RNA polymerase continues to add ribonucleotides to the growing mRNA chain, moving along the DNA template. An understanding of this complex process brings to light how genes are regulated and expressed in various biological contexts, opening the door for further developments in biotechnology to regulate gene expression for various applications. Transcriptional regulation is sophisticated and there are many factors that determine how and when genes become expressed. There are silencers and enhancers, which are located within the DNA sequence, which interact with TFs to promote or inhibit transcription. One of the most intriguing aspects of this regulation is apparent in hematopoietic differentiation, where specific TFs dynamically modulate chromatin access and gene expression patterns. Such types of regulatory networks are essential for cell fate determination during development and have immense significance for biotechnology, particularly cell reprogramming and therapeutic use. These findings emphasize the importance of transcription factor binding kinetics because they uncover the regulatory mechanisms governing differential gene expression, allowing researchers to apply this knowledge towards future biomedical applications (Assi et al., 2016). In addition, better understanding about transcription mechanisms can lead to new applications in biotechnology, particularly in synthetic biology. For instance, lactic acid bacterial quorum sensing shows how transcription can be finely regulated by intercellular communication. Here, peptides are sensed by membrane-bound histidine kinases, which transmit the signal to response regulators that activate the transcription of target genes. Such two-component signaling pathway plays a pivotal role in the regulation of production of crucial metabolites, e.g., bacteriocins, and optimizing fermentation processes. The creation of systems , such as the nisin-controlled expression (NICE) illustrates the ways in which knowledge of transcriptional regulation can be used to drive microbial function in food science, revealing the intersection of genetic and applied biotechnological innovation (Brurberg et al., 1998) (Figure 12.4).

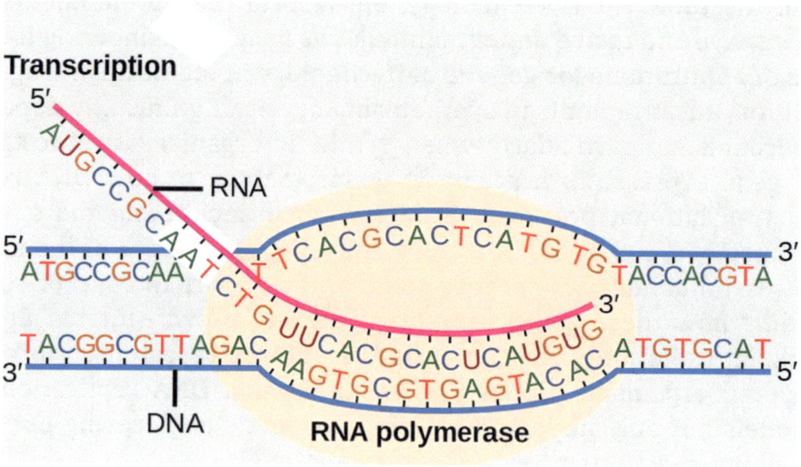

Figure 12.4. Transcription process.

Source: https://library.fiveable.me/_next/image?url=https%3A%2F%2Fstorage.googleapis.com%2Fstatic.prod.fiveable.me%2Fsearch-images%252F%2522DNA_to_RNA_transcription_process_RNA_polymerase_gene_expression_base-pairing_rules_molecular_biology_diagram%2522-Figure_15_02_02a.png&w=1080&q=75.

12.1.3. Translation and Protein Synthesis

Translation is an important process in gene expression since it provides functional proteins from the genetic code carried by messenger RNA (mRNA). The multi-step translation process starts with the initiation of the ribosome, the cellular machine responsible for translation. The subunits of the ribosome and mRNA and transfer RNA (tRNA) work in concert to decipher the sequence of the mRNA. Each tRNA molecule carries a specific amino acid corresponding to an individual codon in the mRNA. As the ribosome travels along the mRNA strand, tRNA molecules attach successively, carrying amino acids that are linked by peptide bonds, forming polypeptide chains. This protein synthesis is involved in various cell processes and finally contributes to determining the phenotype of the organism, indicating the integration of genetics and cells' bioactivity. Moreover, complicated regulation of translation is needed to guarantee that proteins are produced according to cellular needs. Several conditions , such as the availability of nutrients and the presence of signaling molecules, can influence the initiation of translation.

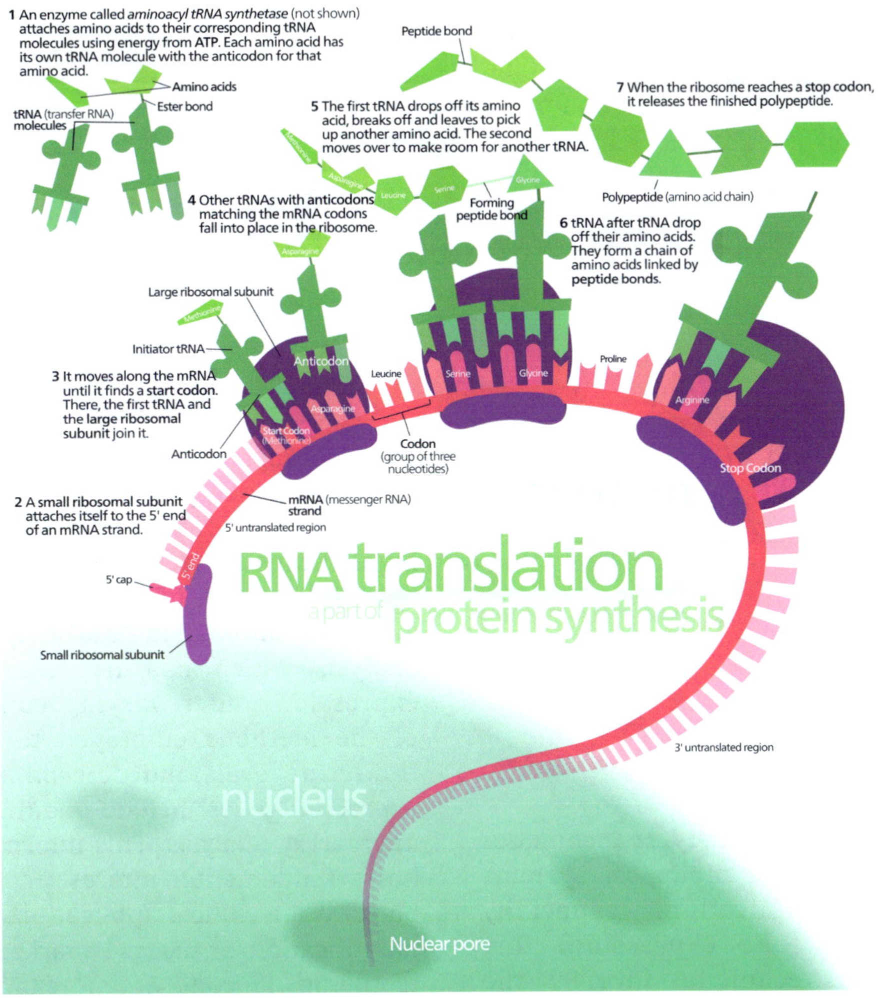

Figure 12.5. The translation process in protein synthesis.

Source: By Kelvinsong - Own work, CC BY 3.0, https://commons.wikimedia.org/w/index.php?curid=23198939

Regulatory proteins and small non-coding RNAs are involved in controlling the translation efficiency so that cells may respond rapidly to their environment. Some translation initiation factors, for instance, can promote or inhibit ribosome assembly. This control not only reflects the cell status but also integrates external signals with cellular processes, which is particularly evident in stress response mechanisms. Biotechnology advances tap into these regulatory processes to optimize protein production in industry, from drugs to enzymes. Biotechnology advances have also pushed the frontiers of translation and its multifaceted connection to genetic manipulation. CRISPR and

RNA interference (RNAi) have revolutionized the ability to manipulate gene expression at the level of translation entirely, enabling finely tuned modifications in protein synthesis. These methods may enable researchers to explore the role of single proteins in cellular processes and disease, indicating targets for therapy. The integration of genomics, proteomics, and metabolomics enables an interdisciplinary research strategy towards the investigation of protein synthesis to offer a tool for the design of experiments to improve our understanding of complex biological systems. For instance, our understanding of fruit ripening mechanisms has been refined through these discoveries, revealing the intricacy of gene regulation and its agronomic importance (Bouzayen et al., 2006). The continued exploration of translation processes, therefore, holds not only implications for basic biology but also for practical biotechnological applications (Ciandrini et al., 2013) (Figure 12.5).

Test Your Knowledge

1. What is DNA replication?
2. What is transcription?

12.2. Techniques in Biotechnology

Learning Outcome

1. Recombinant DNA technology;
2. CRISPR-Cas9 gene editing;
3. Polymerase chain reaction (PCR).

Biotechnology entails a variety of techniques employed to manipulate biological systems towards various applications, especially in understanding gene expression. One of these essential techniques involves the characterization of gene function, which has a fantastic implication in genetics, genomics, and medical diagnosis. As DNA sequencing technologies have improved, researchers can now determine the nucleotide sequence of genes, which provides the foundation for gene function. Computer programs further enhance the process by allowing for predictive gene behavior and protein interaction analysis that are essential for establishing links between genetic sequences and their phenotypic expressions. The identification of gene function guides pharmaceutical development and biotechnological innovations, hence highlighting biotechnology's pivotal position in today's science (Ongus et al., 2017). Another key biotechnology method is gene expression analysis, which measures the levels of gene expression under varying conditions. Some of the methods employed to quantify RNA transcript levels and determine differentially expressed genes include quantitative PCR and microarray analysis. This information not only holds fundamental interest in basic research but also for clinical applications, including the identification of disease markers or response to therapy. Secondly, gene expression studies contribute to an understanding of post-translational modifications and promoter activity, revealing the sophisticated regulatory networks that govern cellular behavior. By correlating with gene expression profiles and biological response, researchers can design targeted treatments that harness the potential of biotechnology to treat disease (Ongus et al., 2017). Genome-editing tools have revolutionized the landscape of biotechnology by allowing for precise modification of genetic material.

Techniques like CRISPR-Cas9 enable ways of delivering targeted changes to gene sequences, and scientists can investigate the role of genes using gene knockout or other types of editing. Such advances not only create new fundamental science knowledge but also pave the way for new agricultural biotechnologies and therapy of genetic disorders. With the development of biotechnology, the integration of advanced editing technologies into research work encourages collaboration between scientists and industry, as evidenced by research work in many fields of biotechnology, including marine biotechnology. These collaborations propel our understanding of gene expression and its application, further cementing the significant role played by biotechnological approaches to addressing prevailing scientific challenges.

12.2.1. Recombinant DNA Technology

Recombinant DNA technology has transformed the industry of biotechnology into a key tool for gene expression manipulation and protein production. With the capability to locate isolated DNA segments and combine them with other genetic material, the technology provides for introducing a new gene that contains a desired characteristic. Recombinant DNA uses the manufacture of important therapeutic proteins , such as insulin and growth hormones, which are crucial in the treatment of various medical conditions.

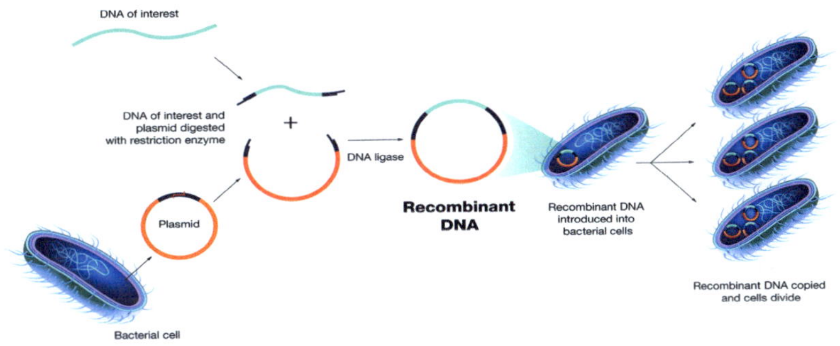

Figure 12.6. Recombinant DNA technology.

Source: https://www.genome.gov/sites/default/files/media/images/2022-04/Recombinant-DNA-technology.jpg.

In agriculture, recombinant DNA has led to the manufacture of GMOs with enhanced traits, , such as pest resistance and nutrient enrichment. The integration of recombinant DNA technology with molecular biology has resulted in huge strides in our ability to react to pressing global challenges, including food security and disease management, thus accentuating its multi-dimensional importance in contemporary society (Ghosh et al., 2023). The scope of recombinant DNA technology is not confined to medicine and agriculture but extends to industrial applications supporting environmental sustainability. For instance, the production of biofuels through genetically altered microorganisms shows how recombinant DNA technology can be applied to produce renewable sources of energy. Such altered microorganisms can efficiently convert biomass to fuel, which later replaces fossil fuel and reduces greenhouse gas emissions. Furthermore, bioplastics produced through recombinant methods show an innovative solution towards plastic waste reduction and environmental issues. As more industries embrace recombinant DNA technology, the future of innovation in biomanufacturing processes becomes increasingly evident, with this reflecting the pivotal role that biotechnology plays in seeking sustainable production methods and addressing environmental challenges (Ghosh et al., 2023). In the case of vaccine discovery, recombinant DNA technology offers encouraging avenues of research for infectious disease control, including novel challenges like pandemic influenza. New methods, , such as the expression of recombinant viral antigens in host cells, facilitate the formulation of efficient and safe vaccines. Employing reverse genetics in the production of live attenuated vaccines or DNA vaccine platforms, scientists can boost immune responses and provide wider protection against viral strains. The surge in recent patent applications for vaccine technologies is a testament to the growing demand for employing recombinant DNA against crisis public health challenges. As can be gleaned from the patent family examination, a significant proportion of innovations in vaccine technology are from prominent drug companies, which brings out the connection between recombinant DNA technology and global health efforts towards crisis preparedness (Figure 12.6).

12.2.2. CRISPR-Cas9 Gene Editing

The invention of CRISPR-Cas9 gene editing technology is a significant breakthrough in biotechnology, allowing one to make unprecedented precision in genomic alterations. Unlike other traditional gene modification methods that are not so efficient and specific, CRISPR-Cas9 uses a guide RNA to direct the Cas9 nuclease to the desired DNA sequence for purposeful changes like insertions, deletions, or replacements. This technology has afforded new avenues of genomic research and applications in various organisms, from plants and animals to microbes. This level of precision in working with genes has long-term implications not only for medicine and agriculture but also for understanding complex biological processes at the molecular level. The ease and efficiency of CRISPR-Cas9 have drawn much attention for its application in addressing significant problems in human healthcare and agriculture. For instance, researchers have thought about applying it to develop cures for genetic diseases, improve crop resistance, and manage diseases. This technology has the potential to correct mutations causing inherited disease, therefore paving the way for advances in customized or personalized medicine. In addition, the farm sector can benefit a great deal from this, as gene alteration can generate disease-resistant crops, environmental tolerance, and enhanced nutritional content. Yet, in becoming stronger, these uses have important ethical and regulatory concerns involving safety, environmental aspects, and implications of altering the genetic makeup of organisms. Despite the promise of CRISPR-Cas9, its application remains controversial to ensure ethical regulation and responsible control. The prospect of human germline editing has ignited controversy regarding the ethics of designer children as well as the unforeseen consequences of heritable alterations. Additional controversy exists regarding the classification of gene-edited organisms, are they to be regarded as GMOs? This differentiation holds immense importance in regulatory paradigms and may hold sway public acceptance and opinion. As countries have struggled to find a balance with these regulations, global consensus is paramount, especially with further innovation making it increasingly difficult for us to keep up with gene editing limits and guardianship. Discussion around CRISPR-Cas9 therefore, represents the dynamic between technological possibility and societal values in building the future of biotechnology (Beccari et al., 2017; Bonomo et al., 2018) (Figure 12.7).

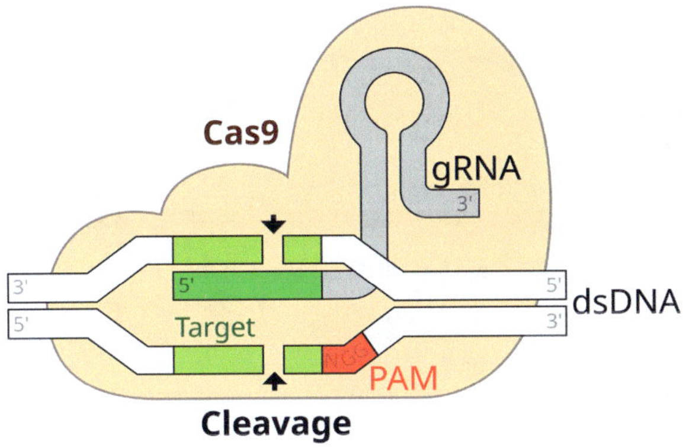

Figure 12.7. CRISPR gene editing.

Source: https://upload.wikimedia.org/wikipedia/commons/thumb/5/57/GRNA-Cas9.svg/1200px-GRNA-Cas9.svg.png.

12.2.3. Polymerase Chain Reaction (PCR)

The polymerase chain reaction (PCR) is a cutting-edge technique of molecular biology that makes it possible to amplify DNA sequences with high specificity and efficiency. It was invented by Kary Mullis in the late 1980s. PCR is a temperature cycle that enables double-stranded DNA to be denatured, primers to be annealed, and extension by DNA polymerase. It is the method used for the exponential amplification of target DNA to a point where it becomes feasible to produce millions of copies from a very small template. The high specificity and sensitivity of PCR have seen it used extensively in gene sequencing, cloning, and, importantly, the study of gene expression in a number of biotechnological applications. The technique is employed as a core technology of genetic engineering to allow researchers to manipulate and examine genetic material with unimagined accuracy and ease. PCR versatility is greater than mere simple DNA replication; it is a key participant in many diagnostic and study applications, including plant genetic transformation. For instance, efficient conversion of azuki bean plants in studies has greatly relied on PCR to confirm the presence of transgenes following Agrobacterium tumefaciens-mediated transformation. By the application of PCR, researchers have been able to authenticate the incorporation of desirable agronomic characteristics into the plant genome and thus enhance GMO breeding process. Furthermore, reverse transcription PCR (RT-PCR) has enabled the quantification of levels of gene expression, and this provides data on the functionality of transgenes, e.g., the bar and hygromycin phosphotransferase transgenes that increase herbicide resistance and overall viability in transformed plants (El-Shemy et al., 2005).

This capacity to fine control the regulation of gene expression further enhances agricultural biotechnology as well as emphasizes the importance of PCR in genetic research. Apart from its use in plant biotechnology, PCR also has crucial functions in distinguishing between GMOs and non-GM organisms. Multiplexed quantitative PCR (qPCR) assays, for example, have enabled effective discrimination between GM plants and virus-infected plants naturally, e.g., the Cauliflower mosaic virus (CaMV). This added specificity resolves the issue of false positives that have been encountered with traditional detection methods, which are prone to using general components like the P35S promoter in the majority of GM constructs (Bak et al., 2019).

By utilizing various plant-specific and viral DNA targets within a single qPCR reaction, researchers have been able to enhance the accuracy of detection of GM traits. Thus, PCR not only facilitates genetic engineering practice but also facilitates the regulatory processes required for the testing of safety and conformity in biotechnology application (Figure 12.8).

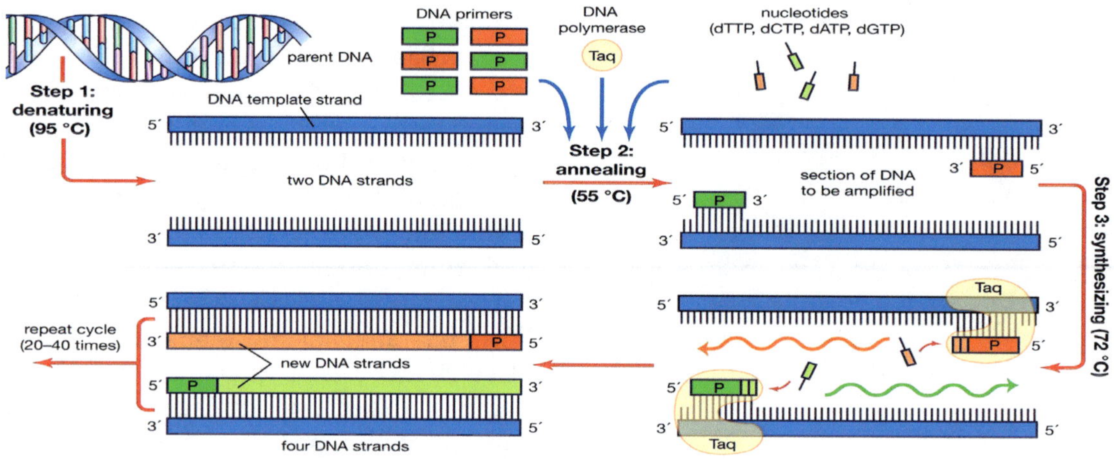

Figure 12.8. Polymerase chain reaction (PCR).

Source: By Enzoklop - Own work, CC BY-SA 4.0, https://commons.wikimedia.org/w/index.php?curid=96042657

Test Your Knowledge

1. What is recombinant DNA technology?
2. What is CRISPR-Cas9 gene editing?

12.3. Applications of Gene Expression in Medicine

Learning Outcome

1. Gene therapy for genetic disorders;
2. Development of vaccines;
3. Personalized medicine and pharmacogenomics.

Gene expression in medicine has transformed therapeutic strategies and ushered in a new age of precision medicine. Recombinant gene-based gene technology has been at the forefront of therapeutic protein synthesis, which forms the basis for essential agents used to treat disease. A good example is the production of recombinant insulin through gene expression systems, which has revolutionized diabetes treatment and the bulk production of this vital hormone. Further advances in monoclonal antibody therapies have also been released as targeted drugs for various types of cancers, significantly enhancing patient survival. These developments underscore the enormously important contribution of gene expression technology to disease therapy, previously a major challenge. As researchers continue to develop such systems, new medical applications expand, with potential treatments for a wide variety of health ailments emerging through new applications while also introducing concerns of safety and regulatory oversight into the deployment of such technologies (Ghosh et al., 2023). Gene expression is not just significant in therapeutic proteins but also in vaccine design since a response to new emerging public health issues can be addressed. The ability to tailor vaccines using recombinant DNA technology enhances immunity with less disease hazard associated with traditional vaccine production. For example, viral vectors can be genetically engineered to efficiently deliver antigens without causing disease, allowing for rapid reaction to outbreaks. Besides, there is increased interest in making use of mRNA technology, which has been prompted by the recent success of COVID-19 vaccines that demonstrated the capability of gene expression to be applied for inducing a robust immune response. This new strategy not only makes vaccine development easier but also holds future potential against numerous infectious diseases. Overall, the union of gene expression and vaccine technology is the dynamic capability of biotechnology in enhancing public health programs (Ghosh et al., 2023). The recent advances in synthetic biology have expanded the scope of applications of gene expression in medicine even more. Through the utilization of cell-free systems of gene expression, researchers can explore and engineer biological circuits without the limitations provided by living organisms. These techniques allow for more flexible experimentation, with the potential to create complex molecular programs for the treatment or diagnosis of disease. Current cell-free systems can rapidly produce proteins, potentially accelerating drug discovery and personalized medicine efforts. Additionally, the integration of riboregulators and DNA nanostructures also offers new opportunities for the highly accurate modulation of gene expression that impinges on therapeutic outcomes. With the growth of synthetic biology, its collaborative nature will merge synthetic and biological systems, opening doors for new therapeutic candidates that promise to change the face of modern medicine.

12.3.1. Gene Therapy for Genetic Disorders

Gene therapy has emerged as a revolutionary method for the treatment of genetic diseases, taking advantage of biotechnology advances to repair the fundamental causes of the diseases. The vast majority of genetic disorders have previously been caused by mutations that disrupt appropriate gene function, most often leading to debilitating health complications. Since gene therapy was developed, doctors are now able to consider treatments that introduce precise copies of genes or correct malfunctioning ones, and this normalizes function at a molecular level. These treatments can be transformative because they are capable not only of alleviating symptoms but also of curing the genetic defects behind most disorders. As the region advances, implications for patients are extensive, and it calls for ethically aware practices in implementing these novel treatments. Gene therapy has a wide array of approaches, which primarily belong to the general categories of *in vivo* and *ex vivo* modifications. *In vivo* gene therapy is the delivery of therapeutic genes directly to the body of a patient, typically by using viral vectors that have been engineered with high efficiency for target cell transduction. This strategy allows for immediate therapeutic effect, but with fear of immunogenicity and efficiency of delivery. *Ex vivo* gene therapy is the removal of cells from a patient, modification in the laboratory to incorporate the target gene, and readministration of these cells to the patient. While *ex vivo* approaches are more highly regulated, they are usually resource-intensive. Each approach has special strengths and limitations, as reported through the current research on refining these delivery systems in the best way possible and assessing their long-term efficacy for gene therapy for genetic diseases (Akki et al., 2019). The ethical challenges posed by gene therapy for genetic diseases are no easier than the science itself. As the practice becomes increasingly popular, patient consent has become an issue, particularly when germline modifications are considered because these modifications would have effects on future generations. Furthermore, when prenatal genomic screening becomes an option, a rich ethical controversy occurs; parents now may be able to obtain data regarding their child's potential future genetic weaknesses at birth and therefore make difficult decisions regarding treatment or intervention (Felice et al.,

2002). The intersection of medical science, biotechnology, and ethical practice demands ongoing discussion among physicians, scientists, and ethicists. There must be systems that maintain the patient's rights while promoting innovation in gene therapy so that such radical treatment is responsible and just in its application (Figure 12.9).

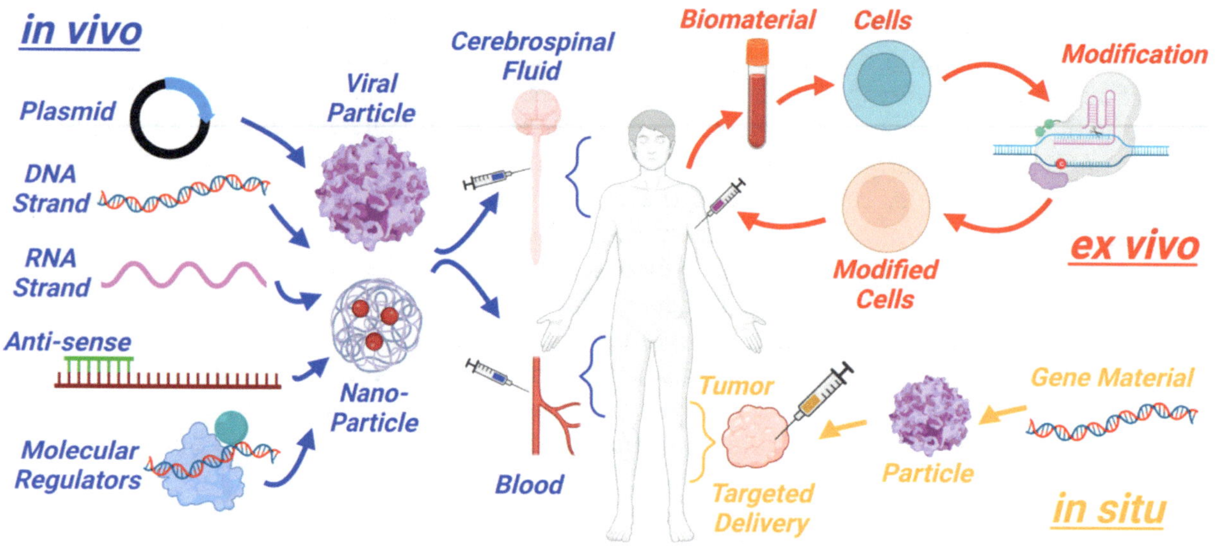

Figure 12.9. Gene therapy for genetic syndromes.

Source: Henderson ML, Zieba JK, Li X, Campbell DB, Williams MR, Vogt DL, Bupp CP, Edgerly YM, Rajasekaran S, Hartog NL, et al. Gene Therapy for Genetic Syndromes: Understanding the Current State to Guide Future Care. *BioTech*. 2024; 13(1):1. https://doi.org/10.3390/biotech13010001

12.3.2. Development of Vaccines

Vaccine development is a profound overlap of gene expression and biotechnology, where discoveries in molecular biology are being converted into effective public health interventions. Traditional vaccine production traditionally uses attenuated or inactivated pathogen forms. Breakthroughs in genetic engineering have led to recombinant and DNA vaccines, which utilize man-made or manipulated genetic material to initiate an immune response. This novel approach not only enhances the efficiency of vaccine development but also allows vaccines to be developed for targeting more than one pathogen simultaneously. For instance, using agroinfiltration techniques, scientists have been able to express multiple antigens within plant leaves, making the process much easier to produce and with increased immunogenicity, which has future use in urgent infectious disease response (Gallo et al., 2018). Technological advances in vaccine technology have evolved significantly in response to global health crises, especially the global threat from pandemic influenza viruses. Patent landscape analyses reveal an evolving new family of technologies dedicated to the development of prophylactic vaccines that utilize complex genetic manipulation and expression systems. The acceleration of the trend of patent applications, particularly after significant pandemics , such as H1N1 and H5N1, is a demonstration of a keen interest among the biotech fraternity to create new vaccine preparations and production methods. This is done by employing reverse genetics for the production of reassortant viruses and the utilization of genetically manipulated plants as a production vessel for vaccine antigens (Barton et al., 2012). These methods, in addition to improving vaccine efficacy, provide a good platform for mass production to be rapidly undertaken, which is required in the case of epidemic outbreaks. Incorporation of gene expression and biotechnology in vaccine research also offers the potential of personalized medicine and disease targeting therapies. With information on the genetic makeup of pathogens, scientists can formulate vaccines that are in harmony with the immune system of a given individual.

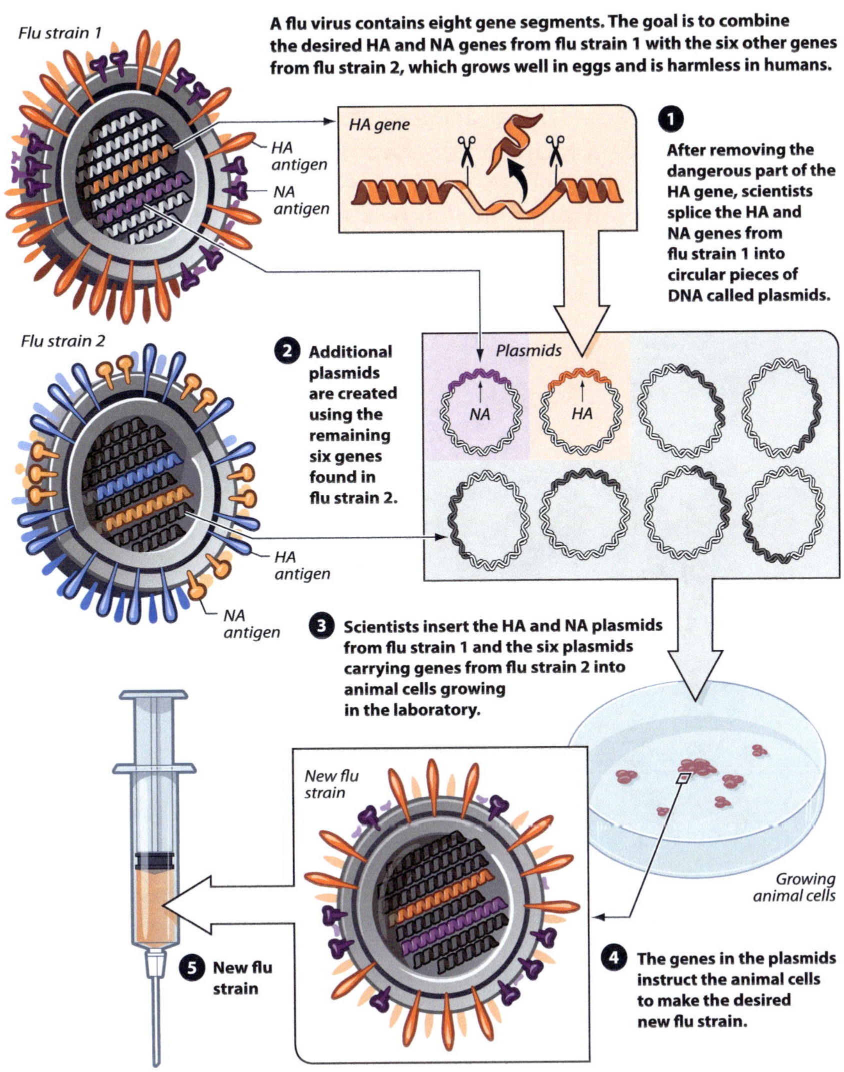

Figure 12.10. Development of vaccines.

Source: By Mouagip - Own work(Link Studio for NIAID), Public Domain, https://commons.wikimedia.org/w/index.php?curid=13869477

This cutting-edge strategy for discovering vaccines has the potential to improve overall efficacy and decrease undesirable side effects. Ultimately, the application of next-generation gene expression technologies in vaccine development is a new paradigm in the battle against infectious disease. As the international health environment continues its evolution, embracing these new biotechnological modalities is essential to safeguarding public health and addressing tomorrow's diseases. These initiatives underscore the compelling need for continued investment and collaboration in vaccine development and research (Figure 12.10).

12.3.3. Personalized Medicine and Pharmacogenomics

The introduction of customized medicine, which is established on the pillars of pharmacogenomics, marks an important advancement in therapeutic medicine. Personalized medicine tailors the medical treatment to the individual characteristics of each patient, targeting specifically genetic aspects. Pharmacogenomics examines how drug metabolism, efficacy, and safety are influenced by differences in genes, leading to a more precise application of drugs. With information about a patient's genetic constitution, clinicians can maximize drug option and dosage, minimizing harm from adverse drug reactions and maximizing therapeutic effects. This precision medicine not only improves patient outcomes but can also lower the cost of healthcare by avoiding trial-and-error prescribing that leads to ineffective treatment.

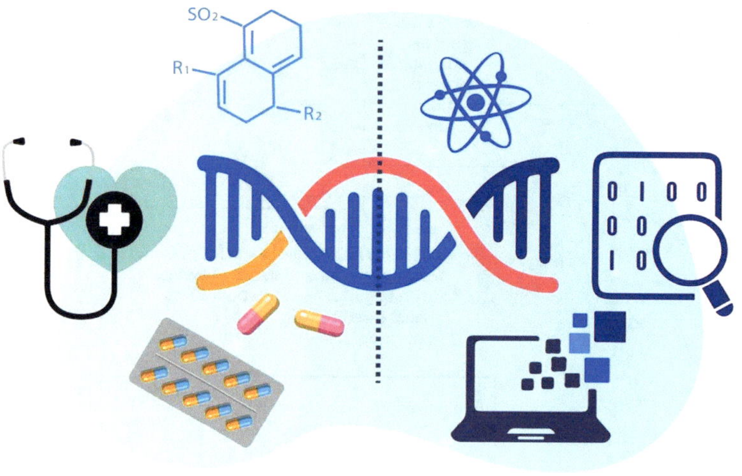

Figure 12.11. Personalized medicine and pharmacogenomics.

Source: https://coreprescribingsolutions.co.uk/wp-content/uploads/2024/03/PHARMACOGENETICS-VS-PHARMACOGENOMICS-1024x679.jpg.

As the science evolves, the integration of pharmacogenomic data into the everyday clinical practice is being more and more considered as essential in treatment strategies in most medical specialties, again pointing out the pivotal role of genetic information in the therapeutic model. Gene expression and biotechnology are the pillars of personalized medicine, allowing the assessment of individual genomes and transcriptomes. Techniques , such as next-generation sequencing (NGS) facilitate rapid and accurate analysis of genetic variation, which is central to pharmacogenomics. Such methods allow identification of biomarkers that may be used to predict patient response to specific drugs. Furthermore, the identification of gene expression profiles can elucidate disease mechanisms, informing targeted therapeutic interventions that account for the unique genetic make-up of patients. Through the application of biotechnological advancements, clinician researchers can utilize the promise of gene expression information to inform drug development choices and patient management. Its incorporation into practice remains a challenge, and there will have to be concerted collaboration among geneticists, clinicians, and bioinformaticians to realize the optimal potential of personalized medicine. Along with the progress of personalized medicine, ethical considerations regarding the use of pharmacogenomics also have to be determined. Access to genetic testing also raises problems with privacy, data security, and discrimination. Patients may be placed in a dilemma as to whether or not to seek genetic testing that can inform treatment, but at the expense of disclosing information about their well-being. Additionally, disparities in access to personalized medicine based on economic and social factors can exacerbate healthcare disparities. Guidelines and policies must evolve to ensure the safeguarding of the patient while promoting the appropriate use of pharmacogenomic data. It is needed that the stakeholders, including the healthcare professionals, the regulatory bodies, and patient advocacy organizations, discuss these ethical concerns. Harmonization of innovation with ethical responsibility is inevitable to ensure that the benefits of pharmacogenomics and personalized medicine are reaped equally across various populations, so the health benefits touch everyone who is part of this new era of medicine (Figure 12.11).

Test Your Knowledge

1. What is gene therapy for genetic disorders?
2. What are personalized medicine and pharmacogenomics?

12.4. Agricultural Biotechnology

Learning Outcome

1. Genetically modified organisms (GMOs);
2. Enhancing crop resistance;
3. Impact on food security.

The use of agricultural biotechnology in food systems shows a cutting-edge approach to enhancing the problems of food security with the rising population of the world. The emerging industry utilizes gene modification and biotechnological methods to enhance food production, nutritional content, and tolerance against environmental stress. As discussed in continuing studies, biotech crops have the potential to reduce pesticide use by a large extent and therefore reduce the environment impact associated with traditional agriculture. When scientists can decipher the gene expression mechanisms of these crops, they can create traits to improve both yield and resistance, thereby making agricultural systems strong to adapt to climate fluctuations and resource availabilities. Their application holds not only increased agricultural productivity but also environmental sustainability and hence is imperative in today's discourse on global food security (Gartland et al., 2018). Emerging technologies of crop biotechnology , such as genome editing and RNA spraying hold new prospects for crop performance and responsiveness improvement. These new methods allow scientists to easily make precise changes in the plant's genetic material, enabling them to target specific traits with great precision. For example, genome editing tools can create plants that are drought- or salt-resistant, a valuable development since climate change threatens conventional agriculture. Other novel applications like biofortification enrich the micronutrient content of staple foods to prevent deficiency among various populations. As this occurs, consideration of the meaning for gene expression will be required for their best use and for guaranteeing that the benefits of agricultural biotechnology will become accessible to multiple stakeholders at a broader level (Baianu et al., 2004). Debate between biotechnology and policy is central to charting the direction of future agricultural strategy across the globe. While technological advances and scientific understanding are the major push factors, there exist complementing regulatory frameworks and public acceptability to adopt these technologies. Policymakers must confront the issues in biotechnology applications to facilitate innovation in solving public fear, as well as environmental considerations. Proper guidelines can ensure safe application of GMOs into trade, hence preventing likely losses or service disruption in agricultural value chains. Moreover, ongoing discussions among scientists, policymakers, and the public are required to align agricultural biotechnology progress with the needs of society. By promoting an educated reaction to agricultural biotechnology, stakeholders can cooperate in a way that will realize its potential for improving food security and sustainability in a changing world (Gartland et al., 2018).

12.4.1. Genetically Modified Organisms (GMOs)

The diffusion of GMOs in farming is a critical point of intersection between gene expression and biotechnology that has triggered scientific breakthroughs and social controversy. GMOs are organisms with their genetic material, for example, DNA, altered through biotechnological procedures to include desired characteristics , such as resistance to insects and drought tolerance. Such alteration can lead to higher yields and reduced agricultural loss, finally responding to food security challenges for an emerging global population. However, the application of GMOs in farming has raised ecological and nutritional questions about their utilization. Regulation, particularly by the EU, has been rigorous towards applying GMOs, in line with consumer sentiments depending on ethical concerns associated with biotechnology and the integrity of organic farming practices (Clarke et al., 2001). While these controversies continue, the need for robust critiques of genetic traits in GMOs is to understand their broader ramifications. Consumer opinion governs the GMO regulatory regime and seeks to govern the direction of debate towards ethics and rights. Several state ballot initiatives and court cases regarding mandatory labeling of GMOs underscore the greater consumer concern with being aware of what is being made (Peck, 2017). The sentiment underscores the fundamental requirement of knowing what they are eating, which demonstrates the tension between scientific advancement and public confidence. Even as most consumers are concerned about the long-term health impacts and safety of the environment from GMOs, research has usually established that GMO foods are safe for ingestion. Nevertheless, the subtleties of gene expression and alteration require a dynamic system of regulation that will be in a position to effectively counter the public concerns without disregarding the potential advantages of biotechnology. This discussion emphasizes the need for equilibrium between technological development in agriculture and the protection of consumer rights and interests. The scientific controversy around GMOs is also reflective of the overall impact of gene expression in biotechnology, particularly the environmental sustainability aspect. Research into the fate of genetically modified DNA in the soil and potential absorption into non-GMO crops raises genuine issues about biodiversity and the integrity of ecosystems. For instance, the impact of herbicide-resistant crops may have unforeseen impacts on non-target organisms and lead to increased herbicide use, harming adjacent ecosystems (Clarke et al., 2001). Moreover, HGT issues represent a significant challenge in unravelling the longer-term impacts of GMOs on natural populations. As biotechnology continues to advance, it remains crucial to perform a diligent assessment of both scientific information and public perception to implement good policies. It will be essential to balance agricultural technological advancement with environmental protection in establishing standards that harmonize innovation and sustainability in the biotechnology age (Figure 12.12).

Figure 12.12. Genetically modified organisms (GMOs).

Source: https://drearth.com/wp-content/uploads/GMOs-iStock-1018188230.jpg.

12.4.2. Enhancing Crop Resistance

Crop resistance enhancement has become a core area of interest in agri-biotechnology, by necessity, to counter the risks of environmental stresses and pathogens. Gene expression technologies should assist scientists in identifying and controlling genetic characters that confer resilience. Such characters may be critical in maintaining crop yields under adverse conditions , such as drought, salinity, and high temperatures. For instance, a full review and analysis of genetic prospects for staple food crops initiated during the early 1980s provides central data about the promise of such techniques, setting a precedent for present-day innovations in crop hardness through targeted GM (Gary et al., 2001). Utilization of such genetics allows the application of biotechnology in the development of crops capable of withstanding ever more volatile climates. Perhaps the most significant area of research in developing crop resistance is the integration of QTL and MAS approaches. Such technology allows breeders to identify genomic locations associated with desirable traits, , such as disease resistance or drought tolerance. Using such advanced methods, breeders can reduce the time needed to develop improved crop varieties significantly, significantly reducing the time it would take with traditional breeding methods. Furthermore, transgenic technology development unlocks the doors of directly adopting the stress-resistance characteristics, thus evading the shortcomings normally encountered with conventional breeding processes. The contemporary field now operates at the interface of conventional approaches and advanced biotechnological paradigms, advocating an integrated system for breeding stress-tolerant crops to address future food security requirements. The significance of enhancing crop resistance cannot be overemphasized, particularly as the populations in the world continue to grow, potentially up to 9.6 billion by 2050. For such staple foods as rice, which is staple for half the world's population, the production of varieties that are resistant to stresses , such as salinity, drought, and flooding becomes ever more pressing. Research has come forward with mechanisms of salt tolerance in rice and suggested the necessity of large-scale screening of rice germplasm for the possible discovery of traits fundamental to successful crop improvement. This multifaceted approach incorporating conventional techniques with innovative biotechnology advancements has the promise to produce salinity-resistant rice varieties that can increase food output while establishing agriculture as a sustainable enterprise (Jahan et al., 2016). Finally, they provide an important conduit for achieving food security under uncertainty as a result of climate change and resource limitations (Figure 12.13).

Figure 12.13. Agricultural scientist examining tomato plant leaves for signs of disease or stress, illustrating plant health monitoring in crop research.

Source: https://d1g9yur4m4naub.cloudfront.net/images/Article_Images/ImageForArticle_844_16527085096098556.jpg.

12.4.3. Impact on Food Security

Gene expression and biotechnology intersect with profoundly significant implications for food security, especially in the Third World developing nations that are confronting climactic and environmental pressures. Genetic enhancement of crops using biotechnological processes facilitates the production of robust plant species that can tolerate extreme conditions like drought and disease, which otherwise can wipe out yields. In Africa, for instance, biotechnology in modern agriculture can address the most significant issues, including food and nutrition security, environmental stewardship, and soil conservation. Employing plant biotechnology efficiently could transform African agriculture by balancing high productivity with sustainability principles. Through the integration of advanced scientific data with native farm methods, nations are in a position to increase their food systems and offer a safer food supply to their respective populations, thus contributing to more extensive-scale socioeconomic development (Bauw et al., 2019). The possibility to augment farm crops through genetic engineering is not exclusive to any single nation; it is an international opportunity for improving food security as the expansion of the population continues to grow. Projecting, populations like the Philippine population will increase exponentially come 2050, hence it is a necessity to have significant increases in levels of food production to meet basic demands. On this note, the role of intellectual property rights (IPR) is most significant because they can trigger research, development, and utilization of agricultural innovations aimed at ensuring food security. By establishing an environment conducive to biotechnological innovation in agriculture, nations can improve productivity and become self-sufficient, particularly in staple food grains like rice. IP system management as a strategy can enable universal access to advanced agricultural practices and thereby empower nations to respond better to food insecurity (Alfonso et al., 2013). Furthermore, the application of gene expression techniques facilitates a dynamic interaction between useful scientific discovery and efficient agricultural utilization and has a direct impact on food security. With advancing biotechnology, the production of crops with enhanced nutritional quality, reduced allergenicity, or disease resistance is both feasible and necessary. These advances not only enhance food supply but enhance nutritional success, most notably among vulnerable groups where undernutrition exists. Additional frameworks must be established so that such biotechnological progress trickles down to the neediest while ensuring appropriate ethical consideration with regards to food manufacture and delivery. Prioritizing a coordinated effort among stakeholders, , such as farmers, consumers, policymakers, and researchers, will be critical in forging a sustainable future for agriculture that integrates food security for everybody (Bauw et al., 2019).

Test Your Knowledge

1. What are genetically modified organisms (GMOs)?
2. What is the impact on food security?

12.5. Ethical Considerations in Biotechnology

Learning Outcome

1. Concerns about genetic modification;
2. Implications for biodiversity;
3. Regulatory frameworks and public perception.

The rapid advancements in biotechnology, particularly in the realm of gene expression, have raised serious ethical issues deserving proper attention. As biotechnological applications increase, ethics in gene manipulation and its impacts on human health and environmental balance are causes of concern. Imbalanced access to biotechnological innovations raises issues of justice, particularly when developing countries are taken into consideration which may lack sufficient financial resources to benefit from such developments. While the proponents of biotechnology advance the capability of enhancing food security and sustainability of agriculture, its ethical implications regarding IPR are also to be addressed. Specifically, the development and application of GURT as a vehicle of appropriation may enhance investment in genetic breeding, where there are possible positive effects in terms of increased crop and consumer product yield, and possible adverse effects that require precise screening by regulatory bodies (Beekwilder et al., 2025). Another key area of ethical concerns in biotechnology is the development of regulations whose aim is to ensure public safety and scientific responsibility. The European Union (EU), for instance, has made notable strides in building regulatory frameworks for the management of gene therapy and other biotechnologies. Such initiatives are an expression of the commitment to equip human beings with parameters that are not only scientifically acceptable but also ethically and morally acceptable to society.

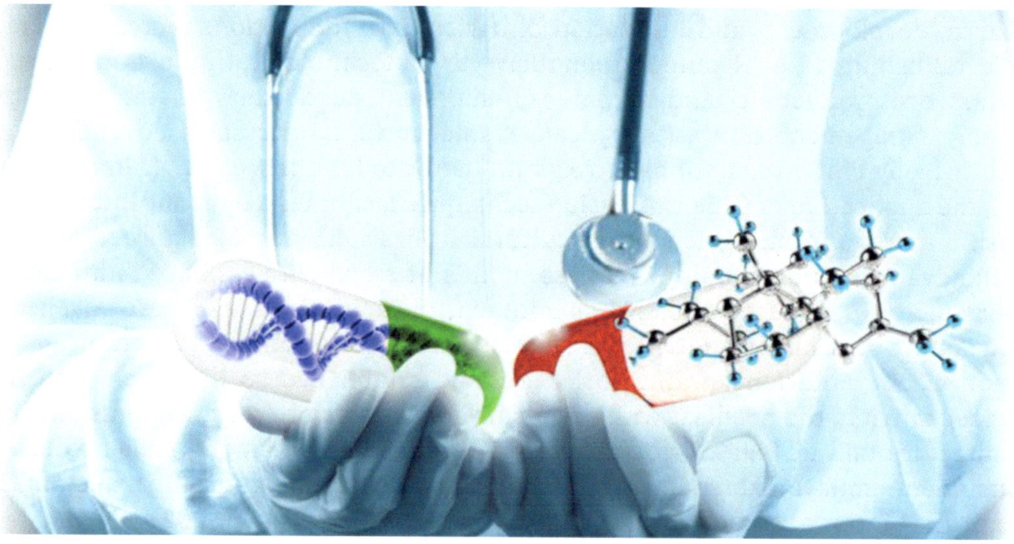

Figure 12.14. Ethical considerations in biotechnology.

Source: https://media.licdn.com/dms/image/v2/D5612AQHgJS_l0b7XBg/article-cover_image-shrink_600_2000/article-cover_image-shrink_600_2000/0/1670963506165?e=2147483647&v=beta&t=U8Syq82Pi0zq1LbW5deGYQiiIw5Jf5QkGmR_7XS6Xjc.

The complexity of the task to regulate biotechnology is further enhanced by the necessity to incorporate scientific advancement into social values and interests that may broadly vary from one socio-economic and cultural environment to another. It is thus essential that there is a constant dialogue involving stakeholders like scientists, policymakers, and the public to effectively navigate the ethics, to the extent that goals of biotechnology are incorporated with ethical responsibilities as much as with practical needs (Jager et al., 1994). Second, the ethics of gene expression and biotechnology include broader concerns of biodiversity and ecological integrity. With new biotechnology technologies revolutionizing farms and food plants, it is time to consider potential ecological disruptions. Deployment of GMOs compromises genetic diversity and ecological health. Pressure to yield more must be weighed against the necessity to support healthy ecosystems, as the long-term effects of biotechnological intervention are unpredictable. As ethical issues emerge in concert with scientific potential, so regulatory frameworks must adjust, showing a commitment to balanced regulation that prioritizes not only human welfare but also environmental health. These interconnected ethical, scientific, and environmental facets highlight the need for open discussion of the consequences of biotechnology in gene expression and agriculture (Figure 12.14).

12.5.1. Concerns About Genetic Modification

Concerns over genetic modification (GM) are multifaceted, including scientific, ethical, and consumer perspectives. Behind these concerns is the specter of possible unknown consequences that arise from GM, which has been a public debate topic since the start of the technology. The majority of consumers express unease with the idea of playing with the fundamental units of life, concerned with potential long-term health effects and effects on the environment that are unknown. Notably, the present controversy about mandatory labeling of GMOs testifies to a strong consumer protection and labeling demand.

Figure 12.15. Genetic modification (GM).

Source: https://media.greenmatters.com/brand-img/ZZmyp_WDv/0x0/gmos-1610386402681.jpg.

Existing controversies are pointing towards the reality that the existing regulatory scheme, instituted under the Coordinated Framework for Regulation of Biotechnology, has been unable to adequately address these concerns, and therefore, a lot of consumers still feel that their GMO-related concerns remain unaddressed by federal authorities (Peck, 2017). Furthermore, the morality of genetic engineering cannot be overlooked, especially thinking about its effect on biodiversity and food security. Genetically modified plants have generated apprehensions over monopolization of agri-resources and effects on small farmers, especially in the developing world. The critics are of the view that GM can result in a skewness, favoring large industries at the expense of conventional methods of cultivation. The potential of genetic use restriction technologies (GURTs) brings with it promises of possibilities as well as pitfalls in terms of providing additional intellectual property protection that can drive innovation in crop cultivation. However, the same technologies can also be used to inhibit rural farmers' access to seeds, thus ultimately aggravating food insecurity and unequal resource allocation (Beekwilder et al., 2025). Thus, the ethical argument of GM needs to consider seriously its socio-economic impact on vulnerable communities. In light of these ongoing problems, it is clear that re-consideration of the regulatory policy governing GM is inescapable. The limitations of the Coordinated Framework, based on ancient statutes, highlight the importance of more direct response to the evolving nature of biotechnology. With advances in science further pushing the boundaries of genetic alteration, regulatory authorities must adapt to being able to respond appropriately to both valid concerns of consumers and the accelerating technology. Closing the divide between scientific innovation and public trust is key to the future of biotechnology. Public participation in discussions involving consumer inputs and ethical reflection can create an inclusive regulatory regime, enabling societies to harness the potential of GM while keeping all interested parties' safety and rights secure (Peck, 2017) (Figure 12.15).

12.5.2. Implications for Biodiversity

The interface of gene expression with biotechnology has a significant influence on biodiversity, particularly in agricultural production and conservation practice. Gene editing and modification, among other sophisticated biotechnological tools, facilitate maximization of crop yields as well as the generation of climate-tolerant and pest-resistant plant varieties. Such technologies can be used to improve food security and reduce the environmental impact of agriculture. But reliance upon a limited array of GMOs can be highly risky to biodiversity. As noted, the use of gene restriction technologies (GURT) can trigger more investment in breeding with intellectual property protection in mind. While such investments can raise farm production and sustainability, they can, on the other hand, lead to the loss of genetic diversity needed for ecosystem resilience, thus necessitating strict regulatory management to balance these opposing results (Beekwilder et al., 2025). For animal conservation, genetic diversity within local and landrace species raises fundamental issues for conserving biodiversity. Native equine populations within East and Southeast Asia are the genetic wealth that can be utilized for the advantage of conservation strategies while simultaneously being utilizable for commercial use. With the advancement of biotechnology, the prospect of such gene resources being harvested for veterinary purposes, diagnostics, and therapy continues to grow. The only obstacle is establishing a system capable of generating fair returns for genetic resources, for ensuring sustainable conservation in addition to responding to commercial demands. A market-oriented approach that recognizes the inherent worth of such genetic records can promote conservation efforts and protection of biodiversity, suggesting a change of paradigm in terms of how we perceive and utilize genetic resources in biotechnology (Kowalski et al., 2014). Applications of gene expression and biotechnology have consequences that extend beyond crop productivity and animal conservation to the overall ecological landscape. Genetic control holds potential unanticipated ecological consequences, particularly when foreign or modified organisms are introduced into indigenous ecosystems. Such interruptions have the potential to result in mutated species interactions, competition for resources, and loss of native species diversity. Interdisciplinary research and rigorous regulatory frameworks are needed immediately to identify the environmental impact of biotechnology advancements. By incorporating ecologic principles in biotechnological advancement, policymakers and researchers can foster sustainable activities that enhance biodiversity rather than diminish it. Public debate is fundamental in guaranteeing that the benefits of gene expression and biotechnology are balanced with the common need to maintain and safeguard the complex web of life that occupies our planet.

12.5.3. Regulatory Frameworks and Public Perception

The regulatory systems governing biotechnology, particularly gene expression and utilization, play a critical role in influencing public opinion. In Europe, for instance, the complexities surrounding plant molecular farming (PMF) and new plant breeding techniques (NPBTs) are topics of immense concern for questions of feasibility and social acceptance. Stakeholders involved in endeavors like the Pharma-Factory and NEWCOTIANA projects indicate that current EU regulations are massive barriers to the advancement of such technologies. They argue that public opinion on biotechnology is largely influenced by regulatory discourse, and within regulatory discourse, GM is often depicted as unsafe or morally suspect. Consequently, the gap between public opinion and scientific knowledge remains large, and clear communication to drive home the advantages and goals of PMF and NPBTs is necessary. Its success in scaling up biotechnological innovations in the European context depends on breaking down such barriers in perception (Drake et al., 2020). In the African context, the future of agricultural biotechnology in meeting pressing issues is bright, but regulatory frameworks have a vital role to play in influencing public perception and acceptance. As African nations increasingly aspire to exploit the application of biotechnology for food security, environmental protection, and economic growth, establishing robust regulations is the utmost priority. These frameworks not only assure efficacy and safety but also assist in fostering trust among the stakeholders, including policymakers, farmers, and consumers. The reputation of biotechnology can greatly be improved with different actors contributing to research and development as it facilitates higher participation of the community and appreciation of the benefits of the technology. By addressing issues of safety and regulatory complexity, biotechnology can be reframed as a key technology for transforming farming approaches on the continent. Facilitating public acceptance through transparent regulatory processes will be critical as Africa navigates its own challenges (Bauw et al., 2019). Biotechnology is reinforced through media presentation, which adds to the complexity of interaction between public attitudes and regulatory systems. Media accounts typically educate and shape people's views and can facilitate or hinder adoption. All stakeholders argue that there is communication required to bridge the evidence uptake gap. Highlighting the positive impacts, , such as the potential of PMF to produce high-quality proteins or increase food security, will assist in breaking the myths around biotechnology. Also, the role played by the media in passing on these messages cannot be ignored; good storylines can lead to an enlightened citizenry that is more receptive to biotechnological innovation. As in the instance of the biotechnology sector in Europe, the view can be turned for the better where there is a unified approach to highlight the intention and value of GM technologies, thereby bringing more people-centered policy-making and discussion to biotechnology (Drake et al., 2020) (Figure 12.16).

Figure 12.16. Scientist analyzing a plant sample in a laboratory setting, symbolizing the intersection of biotechnology and agriculture.

Source: https://www.webnx.in/wp-content/uploads/2023/07/Regulatory-Framework-and-Public-Perception.jpg.

Test Your Knowledge

1. What is genetic modification?
2. What are the implications for biodiversity?

12.6. Future Directions in Gene Expression Research

Learning Outcome

1. Advances in synthetic biology;
2. Potential for regenerative medicine;
3. Role of bioinformatics in gene expression studies.

With the developments in gene expression research, the application of synthetic biology offers the key direction for future development. Synthetic biology, a multidisciplinary field, enables researchers to design and engineer complex biological systems, which can lead to significant therapeutics and biotechnology advances. Synthetic biology enables researchers to de novo design genes and proteins and permits control over genetic pathways in a precise manner. This capability can make way for the creation of tailored solutions to solve pressing problems from across industries like medicine and agriculture. Moreover, the sophisticated interaction of synthetic biology and computational analysis can precision-tune the engineering of functional biological circuits to speed up the identification of optimal genetic designs for certain traits. As gene expression research intertwines with synthetic biology, it promises transformative innovations and novel methods for enhancing organismal capabilities and productivity in diverse applications (Abbas et al., 2024). Another promising avenue in gene expression research is the emerging role of CRISPR-Cas technology. This groundbreaking tool allows for precise genome editing, enabling researchers to activate or deactivate specific genes within various organisms. The uses of CRISPR-based technologies are vast, particularly in gene therapy, where fixing genetic diseases by correcting faulty genes could hold revolutionary treatments. Furthermore, the enhancement of CRISPR gene drives could regulate pest organisms and therefore, reduce reliance on poisonous pesticides and enhance agricultural sustainability. As scientists develop safer and more efficient CRISPR methods, they stand at the vanguard of unveiling new frontiers in gene expression research to be able to better manipulate genetic traits in different species. Not only do such developments transform the landscape of therapeutic interventions, but they also create windows to innovations towards environmental conservation and food security (Kowalski et al., 2010). In the future, gene expression studies integrating with big data and AI technologies present an exciting prospect for biotechnological breakthroughs. With massive sets of genomic and transcriptomic data available, advanced computational approaches can analyze diverse genetic information to uncover complex networks of gene interactions. ML algorithms may also be applied to forecast what changes in protein expression would be able to do to phenotypic effects, aiding scientists in identifying biomarkers and disease therapeutic targets. Through the integration of bioinformatics and gene expression studies, scientists can better manage the complexity of gene regulation and apply their findings to practical uses. This technology-based strategy is likely to make the process of discovery in gene expression research faster, ultimately leading to more focused diagnostics, tailored medicine, and biotechnological processes optimized. As the discipline moves forward, harnessing the power of technologic advancements will certainly redefine the gene expression research and biotechnology landscape (Kowalski et al., 2010).

12.6.1. Advances in Synthetic Biology

The technology of synthetic biology has undergone astounding advancements that expand the boundaries of genetic engineering and biotechnology applications. With a profound understanding of the mechanism of gene expression, scientists are now increasingly capable of engineering and designing new biological components, devices, and systems. Among the successes is spider silk gene engineering for enhancing its tensile strength. As identified in recent studies, spider silk is especially strong and elastic, hence a perfect material for use in a broad range of applications, ranging from medical devices to sports textiles. With the help of advanced genetic engineering approaches, scientists can now circumvent production problems by producing organisms that can produce spider silk industrially, which will accordingly revolutionize materials science as well as biotechnology. This union of biological understanding and new engineering is an exemplary instance of how synthetic biology is transforming genetic applications in research and industry (Gupta et al., 2023).

Figure 12.17. Advancements in synthetic biology.

Source: https://www.biocom.org/wp-content/uploads/2022/10/gene-editing-LL.jpg.

Besides its application in materials science, synthetic biology plays a crucial role in agricultural biotechnology with the new conversion of plants. Scientists have targeted chloroplasts, the cell organelles that facilitate photosynthesis and regulate many cell processes, as the optimal sites for transgene expression. Progress in molecular biology during the past few years has allowed scientists to build favorable characteristics in plant chloroplasts, leading to improved performance and enhanced tolerance towards environmental stress. The plastid transformation system, by which industrial enzymes and drugs can be expressed, demonstrates the immense potential of chloroplast biotechnology. Based on past studies, dsRNA synthesis in chloroplasts increases the efficiency of agronomic changes, paving the way for the production of biosafe engineered crops. These advancements not only ensure food security but also represent the emerging interface between synthetic biology, agriculture, and biotechnology (Joyia et al., 2019). Moreover, the interface of synthetic biology with law and ethics is indicative of a complex framework around genetic engineering. When scientists develop new technology to induce genetic changes, patentability, ownership, and IPR are questions that come to mind. Particularly when it involves spider silk enhancement, understanding the legal implications of genetic advancements can guide further research and business pursuits. Where law and science intersect, it demands open regulatory policies that ensure moral concerns are reconciled alongside fostering innovation. Through the critical scrutiny of the current legal paradigm, researchers can more effectively tackle concerns regarding patenting genetic alterations and intellectual property, such that developments in synthetic biology will be of benefit to society and the economy. This three-way interaction between science, law, and ethics demonstrates the sophistication of developments in synthetic biology and their undertows in gene expression and biotechnology (Gupta et al., 2023) (Figure 12.17).

12.6.2. Potential for Regenerative Medicine

The potential of regenerative medicine is largely enhanced by the development of gene expression technologies and biotechnology, which results in new therapeutic applications. At the heart of regenerative medicine is the ability to modulate stem cells to repair or substitute for damaged tissue. This has enormous potential for the treatment of degenerative disease and injury. Techniques allowing for effective culture and preservation of embryonic stem cells (ESCs) are crucial. For instance, recent studies have tackled the development of biomaterials to facilitate ESC expansion without the need for traditional, animal-based substrates. The use of synthetic nanofibers, designed for ESC culture, has shown better proliferation and pluripotency, a time-saving factor removing the need for traditional measures while improving the effectiveness of stem cell use in therapeutic contexts (Anderson et al., 2016). This progress not only rationalizes ESC culture but also aligns it with ethical principles by minimizing the use of animal byproducts. Further, elastin-like recombinamers (ELRs) are a new engineering innovation in the field of regenerative medicine, particularly applicable to tissue engineering and biomaterial design. ELRs are recombinant molecules designed to mimic the intrinsic properties of elastin, offering a general-purpose platform for biomedical applications. These polymers can be created with intrinsic disorder, which is beneficial in the design of hydrogels that can serve as scaffolds for cell adhesion and growth. Their capability to stimulate cell proliferation and guide differentiation is especially important for regenerative therapy, particularly in cases where complicated tissue structures are required. ELRs can also be applied to drug delivery systems to enhance therapeutic agents and reduce potential side effects. Their multifold properties emphasize the biotechnology potential to make cellular environments optimal, ultimately helping in tissue injury regeneration (Rodr et al., 2019). Regenerative medicine's potential is also complemented through the utilization of gene editing technologies , such as CRISPR-Cas9, which allows for precise editing within gene expression in differentiated and stem cells. By gene editing some of the genes, researchers can enhance the regenerative capacity of cells or correct genetic defects leading to a number of disorders. Integrating gene expression manipulation with the stem cell technology offers a strong platform for designing patient-specific medicine strategies. The integration of biotechnology underscores the importance of tailored therapy, tailored to meet the requirements of individual patients, and providing optimal therapeutic efficacy. With every new revelation about gene activity and cellular milieu, the revolutionizing potential of regenerative medicine continues to be a driving force for treating heretofore incurable conditions, eventually yielding revolutionary gains in health outcomes and quality of life for patients from a broad spectrum of medical specialties.

12.6.3. Role of Bioinformatics in Gene Expression Studies

The role of bioinformatics in gene expression studies has grown increasingly vital as it enables researchers to handle large amounts of biological data collected through high-throughput sequencing technologies. These technologies yield vast amounts of data representing the complexity of gene expression in various conditions and treatments. Bioinformatics programs provide the computational platform essential to process, handle, and analyze this data to transform raw sequence information into useful insight. Bioinformatics can help identify highly upregulated or downregulated genes under specific experimental conditions through the application of differential expression analysis algorithms and thereby pave way towards understanding associated underlying biological processes. It facilitates multi-dimensional integration of data, enhancing the degree of analysis and supporting innovation in personalized therapy and treatment strategies to specific patterns of gene expression. One of the primary advantages of bioinformatics in gene expression studies is that it can model interaction in biological systems. Traditionally, approaches have been trained on single gene expression profiles without accounting for the complex web of interactions that govern cellular processes. Bioinformatics, however, integrates diverse forms of data like genomic, epigenomic, transcriptomic, and proteomic data to provide a holistic view of biological phenomena. With the current research even more focused on a systems biology perspective, the application of bioinformatics tools is even more appropriate. Not only can they be applied for hypothesis-free identification of biological pathways but also be used to support data mining and re-stratification of existing datasets. As pointed out, "this latter strategy eliminates assumptions that can predict how a biological system ought to react, thereby stimulating the development of fresh hypotheses that can guide subsequent research on gene expression and biotechnology (Fernandes et al., 2014). Secondly, the continuous advancement in bioinformatics strategies also handles the increasing difficulty in integrating biological data. With emerging technology, it is essential to maintain integrative analytical strategies." Optimal bioinformatics techniques enhance the ability to combine heterogeneous datasets, allowing an unambiguous representation of gene expression variations in reaction to disease conditions or various stimuli. The main challenge is developing good models that can effectively combine data insights and facilitate decision-making in clinical applications and biological research. Combining multidimensional datasets allows researchers to tackle complex queries, , such as disease mechanisms and patient-specific responses to treatments. Through the power of bioinformatics in these unifications, scientists not only find major aspects of gene expression but also make predictive conclusions about what may be, thereby engineering the future of gene expression studies in biotechnology (Goldenberg et al., 2018).

Test Your Knowledge

1. What are the advances in synthetic biology?
2. What is the role of bioinformatics in gene expression studies?

Summary

In an examination of the intersection of gene expression and biotechnology, it is evident that the rate of genetic alteration has profound implications for various industries, particularly agriculture and medicine. The potential to crack the code and manipulate gene expression profiles has begun to allow for the design of crops with better qualities, , such as increased pest and environmental stress resistance. This is particularly the case in the wake of the ongoing challenges of food security and climate change. A case in point is the identification of single genes, as was the case with investigating β-glucosidase in thermophilic bacteria, and how it illuminates how microbial genetics can be utilized to enhance agricultural productivity, thus bringing out the revolutionary capability of biotechnological innovations in agriculture and sustainability (Ahmed et al., 2018). Furthermore, the character of ongoing crop biotechnology research is one of intensive interaction with abiotic and biotic stresses that crops normally experience. Not only is the approach applied one of enhancing intrinsic plant performance, including efficiency in photosynthesis, but it is also focused on making crops more resilient to environmental stresses. This dual focus serves to demonstrate the comprehensive character of genetic studies for crop resilience and productivity. Furthermore, as highlighted in recent literature, the transgenic crop development pipeline holds a promise of an abundance of innovative solutions committed to food security improvement. These innovations are ever so relevant since they assist in addressing the urgent need for sustainable agriculture, ending up with a safer food supply chain for an ever-growing global population (Dunwell et al., 2010). To summarize, the complex balance of biotechnology and gene expression holds immense potential for assisting with some of the world's most significant issues of today. Utilizing advances made in biotechnological tools, researchers are now capable of creating crops that not only withstand environmental stresses but also enhance adaptation under varied conditions. This agricultural biotechnology advance illustrates the larger movement toward the confluence of scientific research and applied uses. Since research keeps on developing the field, it is clear that gene expression remains at the core of focus, spearheading the future of biotechnology. The understanding from current research will definitely set the stage for the development of new uses, and biotechnology will remain at the forefront of enhancing world health as well as food security (Ahmed et al., 2018; Dunwell et al., 2010).

In the study of the complex interactions of gene expression in biotechnology, it becomes evident that traditional molecular biology techniques are fast becoming woefully substandard. Systems biology has emerged as a revolutionary paradigm, realizing that research on isolated molecules is not adequate to fully describe the intricacy of biological systems. The system not only demands a holistic approach to molecular interactions but also shifts from hypothesis-driven strategies to hypothesis-free strategies. The latter allows scientists to create new, testable hypotheses on the basis of objective analysis of data, extending the range of investigation into gene expression function in various cellular environments. However, the progress in systems biology is faced with many challenges, mostly the lack of potent analytic tools and large databases, which hinder the utilization of these approaches to address many biological questions from fundamental research to individualized medicine (Fernandes et al., 2014). Moreover, biotechnology has also made tremendous progress in the generation of recombinant proteins through new agricultural technologies, namely in PMF. PMF is known for its cost-effectiveness and efficiency in the production of high-value pharmaceutical and other health product proteins. New plant breeding technologies (NPBTs) have now facilitated the progress of PMF systems, which enable scientists to apply more speed and accuracy in what they do. However, the widespread use of PMF and NPBTs is discouraged by regulatory issues and public perception concerns. Stakeholders' views show that prevailing legislation within the EU and bio-cultural values related to biotechnology create real challenges to scalability for these technologies. Ultimately, there is dependence upon public acceptance based on clear communication about the benefits and safety of PMF and NPBTs to bring such biotechnological advancements to accepted usage (Drake et al., 2020). The marriage of gene expression science to state-of-the-art biotechnological techniques has a profound impact on future medicine and agriculture. By leveraging systems biology and PMF, scientists are poised to address some of the most critical world concerns , such as access to health care and food security. With advancing methods in gene expression research, the possibility of revealing new therapeutics and crop resistance increases. However, the transition from lab findings to application in the real world is subject to overcoming the technical as well as sociopolitical barriers that influence public policy and regulatory frameworks. Stakeholder outreach, transparent communication strategies, and public education initiatives need to be undertaken to legitimize such advances. With the development of the discipline further, it is important that the scientific community reflects on these issues to achieve the full potential of biotechnology in addressing global as well as local problems.

KEY TERMS FROM THIS CHAPTER

Anticodon	Cloning	Codon
CRISPR-Cas9	Gel electrophoresis	Gene expression
Genetic engineering	Genetically modified organisms (GMOs)	
mRNA (Messenger RNA)	Mutation	Operon
Polymerase chain reaction (PCR)	Promoter	Recombinant DNA
Regulatory genes	RNA polymerase	rRNA (ribosomal RNA)
Transcription	Translation	tRNA (transfer RNA)

Critical Thinking Questions

1. **What is the process by which DNA is copied into mRNA called?**
   a. Translation
   b. Transcription
   c. Replication
   d. Mutation
2. **During translation, what molecule carries amino acids to the ribosome?**
   a. mRNA
   b. tRNA
   c. rRNA
   d. DNA
3. **Which technique is used to make many copies of a specific DNA segment?**
   a. Gel electrophoresis
   b. PCR (Polymerase Chain Reaction)
   c. DNA sequencing
   d. Cloning
4. **In biotechnology, what is a genetically modified organism (GMO)?**
   a. An organism that has been selectively bred
   b. An organism with a changed DNA sequence introduced artificially
   c. An organism cloned from a parent
   d. An organism infected by a virus
5. **Which molecule forms the backbone of the DNA strand?**
   a. Phosphate and sugar
   b. Nitrogenous bases
   c. Amino acids
   d. Lipids

Short Answer Questions

1. What is gene expression? Describe the two main processes involved.
2. What is the role of mRNA in gene expression?
3. Explain the process of transcription and where it occurs in a eukaryotic cell.
4. Describe translation and its significance in protein synthesis.
5. What are codons, and how do they relate to amino acids?
6. What is the genetic code, and why is it considered universal?

Answers to Critical Thinking Questions

1. (b); 2. (b); 3. (b); 4. (b); 5. (a)

References

1. Abbas, Z., Abbas, Z., Afzal, A., Bin Muhammad Ishaq, M., et al. (2024). Commands of synthetic biology to modernize and re-design the biological systems. https://core.ac.uk/download/616936549.pdf (accessed on 13th June 2025).
2. Acosta Rodríguez, S., Flora, T., Ibáñez Fonseca, A., Rodríguez Cabello, J. C., et al. (2019). Trends in the design and use of elastin-like recombiners as biomaterials. https://core.ac.uk/download/286693424.pdf (accessed on 13th June 2025).
3. Aerts, A., Baatout, S., Beck, M., De Vos, D., et al. (2014). Modulation of gene expression in endothelial cells in response to high LET nickel ion irradiation. https://core.ac.uk/download/55763293.pdf (accessed on 13th June 2025).
4. Agarwal, S., Choi, S., Dohyun, J., Jeong, F., et al. (2019). Cell-free synthetic biology platform for engineering synthetic biological circuits and systems. https://core.ac.uk/download/210536168.pdf (accessed on 13th June 2025).
5. Ahmed, A., Batool, K., Bibi, A., Rafi, M., et al. (2018). Characterization of thermophilic β-glucosidase of rhizosphere bacterial strain (LSKB15) isolated from Cholistan Desert, Pakistan. https://core.ac.uk/download/188828041.pdf (accessed on 13th June 2025).
6. Akki, R. (2019). A review article on gene therapy. https://core.ac.uk/download/230749290.pdf (accessed on 13th June 2025).
7. Albert, R., Gupta, A., Barabási, A. L., Christiansen, F. B., Christensen, C., Christensen, C., Maranas, C. D., et al. (2006). Large-scale inference and graph theoretical analysis of gene-regulatory networks in *Bacillus subtilis*. http://arxiv.org/abs/q-bio/0607024 (accessed on 13th June 2025).
8. Alfonso, A., Estigoy, R., Grimes, H., Jones, R., et al. (2013). Intellectual property and opportunities for food security in the Philippines. https://core.ac.uk/download/72052053.pdf (accessed on 13th June 2025).
9. Van Landeghem, S., Björne, J., Wei, C. H., Hakala, K., Pyysalo, S., Ananiadou, S., Kao, H. Y., Lu, Z., Salakoski, T., Van de Peer, Y., & Ginter, F. (2013). Large-scale event extraction from literature with multi-level gene normalization. *PloS one*, *8*(4), e55814. https://doi.org/10.1371/journal.pone.0055814(accessed on 13th June 2025).
10. Anderson, D. G., Bellis, L. J., Chiefari, D., Duncan, R., Evans, C. W., Fussell, R., Geckil, H., et al. (2016). Novel triblock co-polymer nanofiber system as an alternative support for embryonic stem cells growth and pluripotency. https://core.ac.uk/download/61524530.pdf (accessed on 13th June 2025).
11. Assi, S. A., Batta, K., Bonifer, C., Cauchy, P., et al. (2016). Dynamic gene regulatory networks drive hematopoietic specification and differentiation. https://core.ac.uk/download/77411828.pdf (accessed on 13th June 2025).
12. Baianu, I. C., Lin, H. C., Lozano, P., Prisecaru, V., et al. (2004). Novel techniques and their applications to health foods, agricultural and medical biotechnology: Functional genomics and basic epigenetic controls in plant and animal cells. https://core.ac.uk/download/pdf/85962.pdf (accessed on 13th June 2025).
13. Bak, A., Bak, A., Bak, A., Wong, A. Y. T., Maghari, B. M., Wolf, C., Kamau, E., et al. (2019). Multiplex quantitative PCR for single-reaction genetically modified (GM) plant detection and

identification of false-positive GM plants linked to Cauliflower mosaic virus (CaMV) infection. https://core.ac.uk/download/323068505.pdf (accessed on 13th June 2025).

14. Barton, J., Cavicchi, J. R., Janovetz, J., Kowalski, R., et al. (2012). Patent landscape of influenza A virus prophylactic vaccines and related technologies. https://core.ac.uk/download/72052387.pdf (accessed on 13th June 2025).
15. Beccari, T., Dundar, M., Gartland, J. S., Gartland, K. M. A., et al. (2017). Advances in biotechnology: Genomics and genome editing. https://core.ac.uk/download/293859551.pdf (accessed on 13th June 2025).
16. Beekwilder, J., Eaton, D., Louwaars, N., Meer, I. M. van der, et al. (2025). Potential impacts of genetic use restriction technologies (GURTs) on agrobiodiversity and agricultural production systems. https://core.ac.uk/download/pdf/29302568.pdf (accessed on 13th June 2025).
17. Bonomo, M. E., & Deem, M. W. (2018). The physicist's guide to one of biotechnology's hottest new topics: CRISPR-Cas. http://arxiv.org/abs/1712.09865 (accessed on 13th June 2025).
18. Bouzayen, M., Flores, F. B., Latché, A., Pech, J. C., et al. (2006). Mechanisms of fruit ripening: Retrospect and prospects. https://core.ac.uk/download/12040271.pdf (accessed on 13th June 2025).
19. Brurberg, M. B., de Ruyter, P. G. G. A., de Vos, W. M., Diep, D. B., Eijsink, V. G. H., et al. (1998). Quorum sensing-controlled gene expression in lactic acid bacteria. https://core.ac.uk/download/pdf/12904453.pdf (accessed on 13th June 2025).
20. Caretto, S., Frugis, G., Giannino, D., Iannelli, M. A., et al. (2019). Plant cellular and molecular biotechnology: Following Mariotti's steps. https://core.ac.uk/download/188831409.pdf (accessed on 13th June 2025).
21. Castle, D. (2007). Genomic nutritional profiling: Innovation and regulation in nutrigenomics. https://core.ac.uk/download/28966982.pdf (accessed on 13th June 2025).
22. Ciandrini, L., Romano, M. C., & Stansfield, I. (2013). Ribosome traffic on mRNAs maps to gene ontology: Genome-wide quantification of translation initiation rates and polysome size regulation. https://core.ac.uk/download/25497447.pdf (accessed on 13th June 2025).
23. Clarke, B., Dale, P., Fontes, E., Martin, P., et al. (2001). A review of knowledge of the potential impacts of GMOs on organic agriculture. https://core.ac.uk/download/10923231.pdf (accessed on 13th June 2025).
24. De Bauw, V., De Buck, S., De Oliveira, D., Diasolua Ngudi, D., et al. (2019). Scientific innovation for the sustainable development of African agriculture. https://core.ac.uk/download/287941303.pdf (accessed on 13th June 2025).
25. De Jager, C. F. (1994). The development of regulatory standards for gene therapy in the European Union. https://core.ac.uk/download/144225997.pdf (accessed on 13th June 2025).
26. Diouf, D. (2004). Genetic transformation of forest trees. https://core.ac.uk/download/478321604.pdf (accessed on 13th June 2025).
27. Drake, P. M. W., Fuller, S. S., Hobbs, M., Ma, J. K. C., et al. (2020). Shotguns vs lasers: Identifying barriers and facilitators to scaling-up plant molecular farming for high-value health products. https://core.ac.uk/download/288436395.pdf (accessed on 13th June 2025).
28. Dunwell, J. M. (2010). Crop biotechnology: Prospects and opportunities. https://core.ac.uk/download/9050.pdf (accessed on 13th June 2025).

29. El-Shemy, H. A., Ishimoto, M., Khalafalla, M. M., Mizanur, R., et al. (2005). Recovery of herbicide-resistant *Azuki bean* [*Vigna angularis* (Wild.), Ohwi & Ohashi] plants via *Agrobacterium*-mediated transformation. https://core.ac.uk/download/478321807.pdf (accessed on 13th June 2025).

30. Felice, A. (2002). Molecular biology at the bedside: The impact of genomics on the practice of medicine. https://core.ac.uk/download/46601830.pdf (accessed on 13th June 2025).

31. Fernandes, M., Husi, H., & Robinson, S. W. (2014). Current advances in systems and integrative biology. https://core.ac.uk/download/42354777.pdf (accessed on 13th June 2025).

32. Gallo, R. L., Huang, C. M., Liu, P. F., Simmons, K., et al. (2018). Leaf-encapsulated vaccines: Agroinfiltration and transient expression of the antigen staphylococcal enterotoxin B in radish leaves. https://core.ac.uk/download/323064086.pdf (accessed on 13th June 2025).

33. Gartland, J. S., & Gartland, K. (2018). Contributions of biotechnology to meeting future food and environmental security needs. https://core.ac.uk/download/293884179.pdf (accessed on 13th June 2025).

34. Ghosh, B., Das, D., Ghosh, S., Sarkar, S., & Sarkar, S. (2023). Recombinant protein production: Advancements and applications. https://core.ac.uk/download/603897859.pdf (accessed on 13th June 2025).

35. Goldenberg, A., Hoffman, M. M., Leskovec, J., Nguyen, T., et al. (2018). Machine learning for integrating data in biology and medicine: Principles, practice, and opportunities. http://arxiv.org/abs/1807.00123 (accessed on 13th June 2025).

36. Gupta, V., & Rastogi, S. (2023). Engineering spider genes for high-tensile silk and patenting challenges. https://core.ac.uk/download/588566912.pdf (accessed on 13th June 2025).

37. Gustafsson, C., & Vallverdú, J. (2016). The best model of a cat is several cats. https://core.ac.uk/download/pdf/82478912.pdf (accessed on 13th June 2025).

38. Jahan, N., Javed, M. A., Khan, S., Manan, F., et al. (2016). A short review on the development of salt tolerant cultivars in rice. https://media.neliti.com/media/publications/300688-a-short-review-on-the-development-of-sal-f7507002.pdf (accessed on 13th June 2025).

39. Joyia, F. A., Khan, M. S., & Mustafa, G. (2019). Technical advances in chloroplast biotechnology. https://core.ac.uk/download/322440960.pdf (accessed on 13th June 2025).

40. Kowalski, S. P., & Priess, M. R. (2010). Algae and biodiesel: Patenting energized as green goes commercial. https://core.ac.uk/download/72052027.pdf (accessed on 13th June 2025).

41. Kowalski, S., & McClory, H. (2014). Horses as sources of proprietary information: Commercialization, conservation, and compensation pursuant to the Convention on Biological Diversity. https://core.ac.uk/download/72056339.pdf (accessed on 13th June 2025).

42. McEvilly, M. P. (2013). Lack of transparency in the premarket approval process for AquAdvantage salmon. https://core.ac.uk/download/62559747.pdf (accessed on 13th June 2025).

43. Ongus, J. R. (2017). Biotechnological advances in methods for functional analysis of genes. https://core.ac.uk/download/234657369.pdf (accessed on 13th June 2025).

44. Ozdemir, V., & Williams-Jones, B. (2008). Pharmacogenomic promises: Reflections on semantics, genotype and global justice. https://core.ac.uk/download/151544094.pdf (accessed on 13th June 2025).

45. Alison Peck, *The Failure of Federal Biotechnology Regulation*, 51 Val. U. L. Rev. 483 (2017). Available at: https://scholar.valpo.edu/vulr/vol51/iss2/6 (accessed on 13th June 2025).

46. Seaman, W., & Whitehouse, J. H. (2001). Florida marine biotechnology: Research, development and training capabilities to advance science and commerce. https://core.ac.uk/download/11016354.pdf (accessed on 13th June 2025).

47. Toenniessen, G. H., O'Toole, J. C., Harris, R. R., Herdt, R. W., & Murashige, T. (2001). The Rockefeller Foundation's international program on rice biotechnology. https://core.ac.uk/download/71354545.pdf (accessed on 13th June 2025).

INDEX

C

D

E

H

I

K

L

M

N

O

P

Q

R

S